Innovative Nanocomposites for the Remediation and Decontamination of Wastewater

Azad Kumar
M.L.K. P.G. College, India

IGI Global
PUBLISHER of TIMELY KNOWLEDGE

A volume in the Advances in Industrial Ecology (AIE) Book Series

Published in the United States of America by
IGI Global
Engineering Science Reference (an imprint of IGI Global)
701 E. Chocolate Avenue
Hershey PA, USA 17033
Tel: 717-533-8845
Fax: 717-533-8661
E-mail: cust@igi-global.com
Web site: http://www.igi-global.com

Library of Congress Cataloging-in-Publication Data

Names: Kumar, Azad, 1988- editor.
Title: Innovative nanocomposites for the remediation and decontamination of wastewater / Azad Kumar, editor.
Description: Hershey, PA : Engineering Science Reference, 2022. | Includes bibliographical references and index. | Summary: "This book presents contributed chapters that describe the various nanocomposites used for the degradation of toxic and hazardous chemicals"-- Provided by publisher.
Identifiers: LCCN 2022004119 (print) | LCCN 2022004120 (ebook) | ISBN 9781668445532 (h/c) | ISBN 9781668445549 (s/c) | ISBN 9781668445556 (eISBN)
Subjects: LCSH: Sewage--Purification--Materials. | Nanocomposites (Materials)
Classification: LCC TD745 .I5638 2022 (print) | LCC TD745 (ebook) | DDC 628.3--dc23/eng/20220404
LC record available at https://lccn.loc.gov/2022004119
LC ebook record available at https://lccn.loc.gov/2022004120

This book is published in the IGI Global book series Advances in Industrial Ecology (AIE) (ISSN: pending; eISSN: pending)

British Cataloguing in Publication Data
A Cataloguing in Publication record for this book is available from the British Library.

All work contributed to this book is new, previously-unpublished material. The views expressed in this book are those of the authors, but not necessarily of the publisher.

For electronic access to this publication, please contact: eresources@igi-global.com.

Advances in Industrial Ecology (AIE) Book Series

Syed Abdul Rehman Khan
Xuzhou University of Technology, China

ISSN:pending
EISSN:pending

Mission

The **Advances in Industrial Ecology (AIE) Book Series** examines current, state-of-the-art studies in the areas of Industrial ecology and Circular economy. Industrial ecology is a rapidly growing field that systematically examines local, regional, and global materials along with energy uses and flows in products, processes, industrial sectors and economies. It focuses on the potential role of industry in reducing environmental burdens throughout the product life cycle from the extraction of raw materials, to the production of goods, to the use of those goods and to the management of the resulting wastes. Industrial ecology is ecological in that it (1) places human activity -- industry in the very broadest sense -- in the larger context of the biophysical environment from which we obtain resources and into which we place our wastes, and (2) looks to the natural world for models of highly efficient use of resources, energy, and byproducts. By selectively applying these models, the environmental performance of industry can be improved. Industrial ecology sees corporate entities as key players in the protection of the environment, particularly where technological innovation is an avenue for environmental improvement. As repositories of technological expertise in our society, corporations provide crucial leverage in attacking environmental problems through product and process design.

The **Advances in Industrial Ecology (AIE) Book Series** encompasses empirical, analytical, theoretical, reviews, communications, and conceptual research, and comprehensive reviews of relevant research case studies of effective applications in the areas of Industrial ecology and circular economy. It provides a practical and comprehensive forum for exchanging scientific knowledge and ideas which bridge the technological advancement and sustainable practices with industrial ecological management. The goal of this series is to bring together the expertise of professionals who are working in the field of industrial ecology and sustainability across the globe.

Coverage

- Environmental Science
- Technological Advancement in Industrial Ecology
- Industrial Ecology
- Resource Management
- Waste Management
- Industrial Symbiosis
- Green Practices
- Material Flow Analysis (MFA)
- Blockchain and Circular Economy Practices
- Material Efficiency

IGI Global is currently accepting manuscripts for publication within this series. To submit a proposal for a volume in this series, please contact our Acquisition Editors at Acquisitions@igi-global.com or visit: http://www.igi-global.com/publish/.

The Advances in Industrial Ecology (AIE) Book Series (ISSN pending) is published by IGI Global, 701 E. Chocolate Avenue, Hershey, PA 17033-1240, USA, www.igi-global.com. This series is composed of titles available for purchase individually; each title is edited to be contextually exclusive from any other title within the series. For pricing and ordering information please visit http://www.igi-global.com/book-series/advances-industrial-ecology/270446. Postmaster: Send all address changes to above address.

Titles in this Series

For a list of additional titles in this series, please visit: http://www.igi-global.com/book-series/advances-industrial-ecology/270446

Cases on Circular Economy in Practice
Pietro De Giovanni (Luiss University, Italy)
Business Science Reference • © 2022 • 300pp • H/C (ISBN: 9781668450017) • US $215.00

Green Chemistry for the Development of Eco-Friendly Products
Kavita Shakya Chahal (Government College, Bichhua, India) and Twinkle Solanki (Government College, Bichhua, India)
Engineering Science Reference • © 2022 • 276pp • H/C (ISBN: 9781799898511) • US $250.00

Handbook of Research on SDGs for Economic Development, Social Development, and Environmental Protection
Cristina Raluca Gh. Popescu (University of Bucharest, Romania & The Bucharest University of Economic Studies, Romania)
Engineering Science Reference • © 2022 • 531pp • H/C (ISBN: 9781668451137) • US $325.00

Integrating Blockchain Technology Into the Circular Economy
Syed Abdul Rehman Khan (Xuzhou University of Technology, China)
Business Science Reference • © 2022 • 307pp • H/C (ISBN: 9781799876427) • US $250.00

IGI Global
PUBLISHER of TIMELY KNOWLEDGE

701 East Chocolate Avenue, Hershey, PA 17033, USA
Tel: 717-533-8845 x100 • Fax: 717-533-8661
E-Mail: cust@igi-global.com • www.igi-global.com

Table of Contents

Detailed Table of Contents

Sangeeta Ghosh, Indian Institute of Engineering Science and Technology, Shibpur, India
Paramita Hajra, Indian Institute of Engineering Science and Technology, Shibpur, India
Debasis Sariket, Indian Institute of Engineering Science and Technology, Shibpur, India
Debasish Ray, Indian Institute of Engineering Science and Technology, Shibpur, India
Swarnendu Baduri, Indian Institute of Engineering Science and Technology, Shibpur, India
Chinmoy Bhattacharya, Indian Institute of Engineering Science and Technology, Shibpur, India

The photoelectrochemical water splitting process has a lucid and efficacious impact, which emulates the natural photosynthesis process by converting solar energy into chemical energy. The construction of a PEC system can convert H_2O to H_2 or CO_2 to C-based fuels. To achieve artificial photosynthesis, rate-determining kinetics of the OER is regarded as a highly efficient photo-anode. $BiVO_4$ grabbed strong attention as a photoanode in the communal of PEC. Owing to a moderate bandgap and the Earth-abundant nature of the constituents, it is considered an inexpensive n-type semiconductor for PEC H_2O splitting. This chapter discussed the recent progress of $BiVO_4$-based photoanodes fabrication, including control in the surface, effects of dopant, different synthesis techniques, co-catalyst, etc. Typical unbiased tandem devices of a photoanode system in the presence of $BiVO_4$ are also reflected. The report also demonstrated the photocatalysis principles regarding the degradation of organic pollutants.

Azad Kumar, M.L.K. P.G. College, India

In this research, a proficient method for the synthesis of TiO_2/PPy and TiO_2/PPy/GO nanocomposites is explored. These nanocomposites were prepared by one-step in situ deposition oxidative polymerization of pyrrole hydrochloride using Ammonium persulfate (APS) as an oxidant in the presence of TiO_2 nanoparticles cooled in an ice bath. The obtained nanocomposites were characterized by XRD, TEM, SEM, UV-Vis, and FTIR techniques. The obtained results showed that TiO_2 nanoparticles have been encapsulated by PPy with a strong effect on the morphology of TiO_2/PPy and TiO_2/PPy/GO nanocomposites. The photocatalytic degradation of rose Bengal, and Victoria blue dye was done under different conditions regarding concentration of dye, time of illumination, pH, and dose of the photocatalyst. The maximum photodegradation was found at 7 pH, 20 ppm concentration of VB and 25 ppm of RB dye solution, 800 mg/L for VB and 1600 mg/L for RB amount of photocatalyst, and 120 min irradiation of visible light. Kinetics of photodegradation were investigated for Victoria blue and rose Bengal dye and found first order kinetics.

Two-dimensional (2D) MXene has been considered as a hotspot toward environmental photocatalysis because of its outstanding structural stability, highly efficient conductivity, and versatile hydrophilicity. As an efficient photocatalytic candidate, MXenes offer rapid photogenerated charge carrier isolation, thereby providing plentiful availability for surface functional groups in respect of light-harvesting promising materials, and additionally executing a suitable foundation in favor of superior photoconversion proficiency. This chapter summarizes a comprehensive analysis of recent studies on fabrication method for MXene photocatalysts and photocatalytic performance for contaminant degradations. More significantly, MXenes are frequently employed as cocatalysts to boost the efficacy of photocatalytic activities when combined with other traditional photocatalysts such as metal oxide, metal sulfide, g-C3N4, and so on. Furthermore, in an effort to disclose the unique qualities of MXene-based nanocomposites, the stability of MXene-based nanocomposite photocatalysts is briefly examined.

Heavy metals have come up as a threatening pollutant in aqueous media, leading to life threatening consequences. Biomaterial has been a novel and innovative wing of Green Chemistry, eradicating the threats in a cost effective and clean manner. The present study has been focused on the successful removal of a life-threatening heavy metal Cd (II) from aqueous solution using biosorbent created using selected plants. The present study establishes the fact that carboxylic acid group plays an important role in the metal binding process using protection of COOH group by propylamination and esterification. We could also conclude that the enrichment of COOH group onto the biomaterial using synthetic modifications succination leads to the increase in the sorption efficiency.

Developing innovative technologies for the effective treatment of organic contaminants comprising agricultural wastes, industrial dyes, and chemicals is gaining extraordinary importance across the globe. In the last few years, photocatalytic degradation has become an effective and established route to eliminate these pollutants from aqueous solution relative to simple adsorption. 2D nanomaterials exhibit great potential as an effectual photocatalyst in degradation of contaminants, especially hybridization with other functional components due to wide-ranging band structures, sufficient active sites, and significant specific surface area. Herein, the unique hybridization of 2D nanomaterials with numerous functional species is reviewed comprehensively by highlighting their improved photocatalytic performances and remarkable environmentally friendly activity. The chapter outlines the mechanism of photocatalytic degradation to explore the advantages/disadvantages of regular 2D materials and discover the significance of developing hybrid 2D photocatalysts.

Chapter 6
Metal Oxide-Cellulosic Nanocomposite for the Removal of Dyes From Wastewater........................ 126
Suneeta Bhandari, G.B. Pant University of Agriculture and Technology, India

Water is a vital component of life, and its availability is critical for all living things. Due to rising water demand, traditional water/wastewater treatment methods are inefficient in supplying adequate safe water. The leaching of harmful compounds into the process water is a problem with most commercial and chemically manufactured materials for water treatment. As a result of research into developing better materials that could achieve high efficiency without posing a health concern, non-toxic composite materials made of cellulose and metal oxides were investigated. Due to its great physical, chemical, and mechanical qualities, cellulose is one of the materials gaining popularity. Nanocomposites have been approved as a solution for water purification that avoids the issues associated with using simply metal oxides. The purpose of this study is to review the potential applications of cellulose integrated with metal oxides for wastewater treatment and harmful metal removal from dyes via industrial waste.

Chapter 7
Biopolymer-Based Nanocomposite Materials for Detection and Removal of Pollutants in Wastewater.. 141
Ratnesh Das, Dr. Harisingh Gour Central University, India
Arunesh Kumar Mishra, Dr. Harisingh Gour Central University, India
Pratibha Mishra, Dr. Harisingh Gour Central University, India
Megha Das, Indira Gandhi National Tribal University, India

Biopolymer-based nanocomposites, particularly chitosan, cellulose, alginate, starch, and carrageenan, are increasingly being employed as reinforcements for composite materials because they are biodegradable, recyclable, renewable, abundant, conveniently available, cost-effective, and non-abrasive to processing equipment. These biopolymer nanocomposite materials are also lightweight, stiff, and have good mechanical properties. Biopolymer nanocomposites have interfacial limitations because all nanocomposite biopolymers are hydrophilic. Water recycling has been made possible by biopolymer-based nanocomposite materials, which have a variety of applications for cleaning wastewater, making it a viable and cost-effective solution to water scarcity. The growing concern about heavy metal contamination has necessitated the development of new and better-suited sorbent materials for effective detoxification.

The $TiO2/PAni$ and $TiO2/PAni/GO$ nanocomposites were prepared by one-step in situ oxidative polymerization of aniline hydrochloride using ammonium persulphate as an oxidant in the presence of powder of TiO2 nanoparticles cooled in an ice bath. The obtained nanocomposites were characterized by XRD, TEM, SEM, BET, FTIR, and DRS. The obtained results showed that TiO2 nanoparticles have been encapsulated by PAni. The FTIR characterization confirms that the TiO2/GO molecules are well combined with polyaniline structure. The maximum photodegradation of Thymol blue was found in TiO2/PAni/GO at 25 ppm concentration of dye, 1600 mg/L amount of photocatalyst, pH 7, and 120 min irradiation of visible light. Hence, the photocatalytic activity of Titania has been increased by the coating of PAni and Graphene oxide.

Access to safe drinking water is one of the most pressing challenges in the 21st century. New and better technologies for the treatment of wastewater are critically needed. Carbon nanotubes are emerging as effective and environmentally friendly alternative adsorbents for water purification due to their porous structure, relatively large specific surface areas, and strong hydrophobicity. Nevertheless, carbon nanotubes also suffer the inherent challenges of nanomaterials with potential health risks. This chapter presents a detailed review of the progress made in the utilization of carbon nanotubes and their composites in the sequestration of organic and inorganic pollutants from water. The factors affecting performance, the adsorption capacities, and mechanisms are concisely discussed. Additionally, the associated health risks of carbon nanotubes are highlighted, and risk assessment strategies are recommended. Overall, carbon nanotubes are shown to be suitable candidates for water treatment regimes.

Microencapsulation is a well-established process of enveloping or surrounding one substance into another substance that provides capsules having range from one micron to many hundred microns in size. One among the highly efficient methods is microencapsulation. The encapsulation efficiency of microcapsules, micro particles, or microspheres depend on various factors like solubility of polymer in solvent, concentration of polymer, solubility of organic solvent in water, rate of solvent removal, etc. Substances are often encapsulated in such a way that the core material is confined within capsule shells (coating material) for a particular interval of time. Different types of techniques are used for preparation of microcapsules. These techniques are utilized in different fields like pharmaceutical, agriculture, textile, food, printing, and defence. This text covers a review on microencapsulation and materials involved, microencapsulation techniques, and use of microencapsulation in textiles.

Carbon nanotubes (CNTs) are a unique carbon material with physical, chemical, mechanical, optical, structural, and electrical characteristics researched and tested for a wide range of uses. The safeguards of environmental health have been identified as one of the most critical sustainability goals in recent decades. When it concerns identifying atmospheric toxins, carbon nanotube-based detectors offer great sensibility and precision, along with carbon nanotubes displaying the ability for adsorption to remove impurities with great rates and excellent amelioration competency. Carbon nanotubes have made essential contributions to a responsible future in wastewater treatment, air pollution management, biotechnologies, nano sensors, and sorbents. Carbon nanotubes are also utilized as a reinforcing material in green nanocomposites, which are essential for achieving desired characteristics and are ecologically benign. The utilisation of carbon nanotubes as hybrid filters, nano sensors, sorbents, and other materials is covered in this chapter, as well as its advantages for the environment.

Nanotechnology is broadly used in the different fields of science such as biomedicine, pharmaceuticals, electronics, diagnostic instruments, and environmental detection. Nanoparticles have great potential to purify wastewater and decontaminate wastewater. Nanoparticles can eliminate inorganic/organic pollutants, heavy metals, and chemical dye from contaminated water. Nanoparticles are synthesized with various methods such as physical, chemical, and biosynthesized. Plant extract is used for the synthesis of metallic nanoparticles because plant extract contains different types of primary and secondary metabolites. These metabolites act as stabilizing and reducing agents in the synthesis of novel metallic nanoparticles. The size and shape of nanoparticles have unique properties; thus, they are widely used for removing pollutants from water. The chapter discussed green synthesized metallic nanoparticles and their application in the treatment of wastewater.

Nanocomposites are a class of nanomaterials where there is more than one phase in nano dimension. Simply, they are multiphase solid materials in nano dimension. Nanocomposites are found to be attractive because of their large aspect ratio. There are different types of nanocomposites like metal, organic, metal-organic polymer, carbon nano tubes, and nano fibres. Different methodologies were adopted to synthesize the nanocomposite which includes sol-gel, wet chemical method, thermal decomposition, Pechini method, insitu polymerisation, solution blending methods, etc. The nanocomposites are found to be used mainly in photodegradation, drug delivery, sensors, biomedical applications, artificial implants, and batteries. Significant research has been carried out to analyse the dye degradation property of the nanocomposites. This chapter particularly concentrates to explain the synthesis of nanocomposite and their catalytic activity towards dye degradation.

Arun Kant, Kirori Mal College, University of Delhi, India
Gyanendra Kumar, Swami Shraddhanand College, India
Mohd Ehtesham, Jamia Millia Islami University, India
Sudipta Ghosh, Kirori Mal College, University of Delhi, India
M. Ramananda Singh, Kirori Mal College, University of Delhi, India
Panmei Gaijon, Kirori Mal College, University of Delhi, India

Water is a vital component of life. It is naturally available as earth hydrosphere and plays an important role in the world economy, and it essential for balancing of the ecosystem. Numerous microbes and other toxins such as chemicals and heavy metals are integrated into rainwater and flowing water, resulting in water pollution. This chapter examines the numerous ways in which nanomaterials can be used to remove various kinds of contaminants from polluted water. In this chapter, carbon-based adsorbents material, that is, carbonaceous materials, has described. Carbonaceous materials such as stimulated carbon, carbon nanotubes, and graphene oxide have good performance and high adsorption value for medicinal active chemicals. In present-day investigations, researchers have found that carbon-based nanomaterials have been located progressively being applied in recycling of wastewater treatment research with overwhelmingly positive results.

Deepak Vasant Nagarale, VVM's S.G. Patil Arts, Science, and Commerce College, India

The word "nano" is from the Greek word "nanos" meaning "dwarf." It is a prefix used to describe "one billionth" of something. A nanometer (nm) is a billionth of a meter or a millionth of a millimeter. This chapter started with an introduction to nanoscience followed by what nanostructure is and its applications of nanotechnology (basic idea) and various size-dependent properties of nanomaterials. In this chapter, some unique properties like 1) semiconducting nanoparticles and 2) metallic nanoparticles are explained with examples. Synthesis aspects of nanomaterials also need to be understood using bottom-up and top-down approaches including mechanical alloying and shape and size control of nanomaterials. In the current scenario, the research and development of nanotechnology is very active globally, and nanotechnologies are already used in many products. Further, nanotechnologies are also being developed for use in environmental applications (e.g., clean-up of environmental pollutants).

Preface

Water pollution is defined as the contamination of water sources by contaminants that render the water unfit for drinking, cooking, cleaning, swimming, and other uses. Chemicals, waste, germs, and parasites are all examples of pollutants. All types of pollutants ultimately end up in water. Pollution from the atmosphere settles on lakes and seas. Pollution from the land can flow into an underground stream, then into a river, and eventually into the ocean. As a result, rubbish thrown on an empty lot might ultimately damage a water source.

Human activities generate wastewater, which may be harmful to the environment and result in water loss in areas where water is limited. When wastewater contaminates rivers and groundwater tables, the water resource is rendered useless. As a result, wastewater must be treated before it is discharged into the environment, and if feasible, treated to make it drinkable.

The goal of wastewater treatment is to decrease pollutants to fewer than the maximum allowable levels in order to protect the environment and human health. To do this, wastewater is collected and processed in massive plants before being discharged back into the environment. Water that goes into drains or the sewage system from dwellings is referred to as wastewater. Large amounts of wastewater are routinely contributed to sewage collecting systems by industries and enterprises.

The discovery of new and novel materials capable of improving the efficiency of industrial wastewater treatment processes, and even the manipulation of these materials' characteristics to increase pollutant recovery, has made steady progress. Secondary effluents including heavy metals and radionuclides are produced by anthropogenic activities such as mining, manufacturing, and energy generation. Given their potential influence on water quality, the development of innovative technologies aimed at recovering such pollutants is a top priority. Adsorption is regarded a useful approach in water pollution prevention because of its basic design, universal nature, high efficacy, and ease of operation and regeneration.

We seek articles in this book that highlight research findings in the creation of novel materials for the removal of soluble forms of hazardous chemicals, dyes, and heavy metals. Aside from the scientific originality of the recommended materials, the authors should underline the possibility of implementing their technique in full-scale facilities working under actual liquid effluent treatment circumstances. We accept contributions from several disciplines of research in this book, including material science, chemical engineering and processing, chemistry, adsorption, and photochemistry.

The focus of this book is the production of novel materials (bulk, composites, and hybrids) through the improvement/transformation of certain wastewater treatment procedures. Papers on the following themes will be given special consideration among the areas of interest:

- Development of novel procedures for the synthesis of nanomaterial, nanocomposites, carbon based nanomaterials, and other nanomaterials by fine-tuning synthesis conditions to get the optimum adsorption characteristics;
- Innovative composite materials are being developed for use in liquid-phase adsorption processes.
- Elaboration of bio-sourced materials (biochars, hydrochars, chars, activated carbons) from various biomasses or carbon materials for the particular removal of dyes, hazardous chemicals, and heavy metals;
- High selectivity polymers or polymer composites are being developed.

SOLUTION

The efforts of scientists are the key driving force behind nanotechnology innovation, via the assessment of the techniques used to generate Nano materials for water and wastewater treatment, and these technologies must be demonstrated in the future. Without a question, nanoparticles have played a significant role in the advancement of wastewater treatment technology. To minimize the possibility for these nanomaterials to become a source of environmental contamination, researchers must concentrate understanding of the related negative hazardous and environmental effects. Effective learning in the use of nanotechnology for sustainability may be a critical step in developing learning experiences that cultivate the knowledge and skills required to drive change in a sustainable manner.

BOOK ORGANIZATION

In this book, we have selected the 15 research and review articles for publication. The chapters in this book reflect a wide range of fundamental and applied research in water and wastewater treatment by the nanomaterials and interdisciplinary subjects. This book is a unique collection of full research papers as well as reviews.

In the 1st chapter, the photoelectrochemical water splitting process as well as the photocatalysis process for pollutant degradation of BVO photoanode has been discussed. Growth of visible light active efficient and stable photocatalyst can degrade pollutants with an environmental impact. Since the BVO-based semiconductors can be activated under visible light irradiation, these materials would gain much popularity, especially for in water splitting process as well as degradation of pollutants in air and surface water.

In the 2nd chapter, describes a proficient method for synthesis of TiO_2/PPy and TiO_2/PPy/GO nanocomposites. The Photocatalytic degradation of Rose Bengal and Victoria blue dye was done at different condition viz concentration of dye, time of illumination, pH and dose of photocatalyst. The maximum photodegradation were found at 7 pH, 20 ppm concentration of Victoria blue and 25 ppm of rose bengal dye solution, 800 mg/L for VB and 1600 mg/L for RB amount of photocatalyst and 120 min irradiation of visible light.

In the 3rd chapter, as an efficient photocatalytic materials, MXene offers rapid photogenerated charge carrier isolation, thereby providing plentiful availability for surface functional groups in respect of light-harvesting promising materials, and additionally executing a suitable foundation in favour of superior photoconversion proficiency. This chapter summarizes a comprehensive analysis of recent studies on fabrication method for MXene-based photocatalysts and photocatalytic performance for contaminant degradations. More significantly, MXenes are frequently employed as cocatalysts to boost the efficacy of photocatalytic activities when combined with other traditional photocatalysts such as metal oxide, metal sulfide, g-C_3N_4 and so on.

In the 4th chapter, heavy metals have come up as a threatening pollutant in aqueous media, leading to life threatening consequences. Biomaterial has been a novel and innovative wing of Green Chemistry, eradicating the threats in a cost effective and clean manner. The present study has been focused on the successful removal of a life threatening heavy metal Cd (II) from aqueous solution, using biosorbent created using selected plants. The present study establishes the fact that carboxylic acid group plays an important role in the metal binding process using protection of COOH group by propylamination and Esterification. We could also conclude that the enrichment of COOH group onto the biomaterial using synthetic modifications Succination leads to the increase in the sorption efficiency.

In the 5th chapter, describe 2D materials, intensive efforts have been devoted in recent years. The daunting quest for unique 2D materials remains ongoing and is primarily intended to discover novel 2D materials and their remarkable properties. In this chapter, we aimed to represent a thorough analysis of the latest innovations made in the field of photocatalytic degradation by 2D materials. In addition to current progress in photocatalysis, a throwback of basic knowledge is outlined. Various combination of fabrication methods employed for preparing novel 2D NMs is also illustrated. It is widely believed that 2D materials exhibit excellent photocatalytic performance. The potential for various 2D nanomaterials has been reported at length to remediate aqueous systems contaminated with dyes.

In the 6th chapter, cellulose is a biodegradable, non-toxic, low-cost material that can be found in a wide range of natural resources and agricultural waste. To remove a variety of impurities, including harmful metals and dyes, cellulose could be used in a variety of water treatment methods. As a result, due to their high surface area, light stability, and low toxicity, the application of cellulose-metal oxide composite as an efficient dye adsorption and photo degradation in water. Incorporation of metal oxide into cellulose improves the stability of the material, prevent desorption of the nanoparticles into the water system, reduce toxicity effects and also helps the material to be long-lasting. Moreover, multi-metal or polymetal oxides can be used with the cellulose instead of pure metal oxides in the composites. Overall, when all procedures are relatively safe, would exhibit less aggregation and offer advance dye removal from waste water purification process.

In the 7th chapter, biopolymer-based nanocomposites, particularly chitosan, cellulose, alginate, starch, and carrageenan, are increasingly being employed as reinforcements for composite materials because they are biodegradable, recyclable, renewable, abundant, conveniently available, cost-effective, and non-abrasive to processing equipment. These biopolymer nanocomposite materials are also lightweight, stiff, and have good mechanical properties. Biopolymer nanocomposites have interfacial limitations because all nanocomposite biopolymers are hydrophilic. Water recycling has been made possible by biopolymer-based nanocomposite materials, which have a variety of applications for cleaning wastewater, making it a viable and cost-effective solution to water scarcity. The growing concern about heavy metal contamination has necessitated the development of new and better-suited sorbent materials for effective detoxification.

In the 8th chapter, the TiO_2/PAni and TiO_2/PAni/GO nanocomposites were prepared by one-step in situ oxidative polymerization of aniline hydrochloride using ammonium persulphate as oxidant in the presence of powder of TiO_2 nanoparticles cooled in an ice bath. The obtained nanocomposites were characterized by XRD, TEM, SEM, BET, FTIR and DRS. The obtained results showed that TiO_2 nanoparticles have been encapsulated by PAni. The FTIR characterisation confirms that the TiO_2/GO molecules are well combined with polyaniline structure. The maximum photodegradation of Thymol blue was found in TiO_2/PAni/GO at 25 ppm concentration of dye, 1600 mg/L amount of photocatalyst, pH 7 and 120 min irradiation of visible light. Hence, the photocatalytic activity of Titania has been increased by the coating of PAni and Graphene oxide.

In the 9th chapter, carbon nanotubes (CNTs) have the potential to support point of use-based treatment approach for removal of water hardness, chemical, and biological contaminants from water. Generally, CNTs exhibit higher adsorption capacities in the removal of heavy metals, dyes and emerging contaminants relative to other adsorbents. This is attributable to their fibrous shape with high aspect ratio, large surface area and well developed mesopores. The relatively high cost of CNTs stands as a major constraint towards application of CNTs o industrial scale for water purification. Additionally, the release of contaminant-laden unrecovered CNTs into the environment and concomitant human exposure to CNTs remains contentious due to the associated health risks. Adsorbent recovery strategies need to be further explored including the use of life cycle assessment (LCA) tool in adsorption studies both at laboratory and pilot scale experiments.

In the 10th chapter, in today's world of developing technologies, the technique of microencapsulation is applied in most the fields. It's become a prominently effective technique which reinforces the property imparted to the material and assures its durability. It's fascinating that our clothing is now ready to actively moisturize, heal and even can release fragrances to scale back anxiety. The growing health awareness among consumers is further propelling researchers to undertake and test all possible ingredients to deliver expected performance. New materials are being explored and a serious shift is towards the utilization of organic compounds both in sheath and core. There's little question that this technology features a promising future, however, one aspect that seems critical is that the intended delivery of the encapsulated core on particular external stimulus. There's a requirement to optimize the methods of manufacturing microcapsules and extend the time period of treated materials to realize large scale industrial production for every specific application. A huge use of this system is often witnessed in functional finish fabrics, medical and healthcare textiles, aromatherapy, cosmetic textiles and lots of more.

In the 11th chapter, carbon nanotubes (CNTs) are a unique carbon material Because of their unique physical, chemical, and electrical characteristics. CNTs show tremendous promise as a viable material for usage in various environmental sectors when evaluated for specific uses. When it concerns identifying atmospheric toxins, carbon nanotube-based detectors offer great sensibility and precision, along with carbon nanotubes displaying the ability for adsorption to remove impurities with great rates and excellent amelioration competency. Carbon nanotubes have made essential contributions to a responsible future in wastewater treatment, air pollution management, biotechnologies, nanosensors, and sorbents. Carbon nanotubes are also utilized as a reinforcing material in green nanocomposites, which are essential for achieving desired characteristics and are ecologically benign. The utilisation of carbon nanotubes as hybrid filters, nano sensors, sorbents, and other materials is covered in this article, as well as its advantages for the environment.

In the 12th chapter, nanotechnology is broadly used in the different fields of science such as biomedicine, pharmaceuticals, electronics, diagnostic instruments, and environmental detection. Nanoparticles have great potential to purify wastewater and decontaminate wastewater. Nanoparticles can eliminate inorganic/organic pollutants, heavy metals, and chemical dye from contaminated water. Nanoparticles are synthesized with various methods such as physical, chemical, and biosynthesized. Plant extract is used for the synthesis of metallic nanoparticles because plant extract contains different types of primary and secondary metabolites. These metabolites act as stabilizing and reducing agents in the synthesis of novel metallic nanoparticles. The size and shape of nanoparticles have unique properties thus they are widely used for removing pollutants from water. The chapter discussed green synthesized metallic nanoparticles and their application in the treatment of wastewater treatment.

In the 13th chapter, we have discussed briefly about the various types of nanocomposites and its role in pollution prevention. As stated in the introduction, increasing population and rapid industrialization causes a serious threat to the environment by water pollution. The effluents from dye industries are a serious problem, which can be solved to some extent by science and technology developments. Various types of nanomaterials and nanocomposites play a very vital role in dye degradation. Nanocomposite preparation characterisation, degradation methodology and its efficiency towards degradation of various types of dyes are discussed in this chapter.

In the 14th chapter, water is a vital component of life, it naturally available as earth hydrosphere and play an important role in world economy and it essential for balancing of the ecosystem. Numerous microbes and other toxins such as chemicals and heavy metals are integrated into rainwater and flowing water, resulting in water pollution. This chapter examines the numerous ways in which nanomaterials can be used to remove various kinds of contaminants from polluted water. In this chapter, carbon-based adsorbents material that is carbonaceous materials has described. Carbonaceous materials such as stimulated carbon, carbon nanotubes and graphene oxide have good performance and high adsorption value for medicinal active chemicals. In present-day investigations, researchers have found that carbon-based nanomaterials have been located progressively being applied in recycling of wastewater treatment research, with overwhelmingly positive results.

In the 15th chapter, the word nano is from the Greek word 'Nanos' meaning Dwarf. It is a prefix used to describe "one billionth" of something. A nanometre (nm) is a billionth of a meter or a millionth of a millimeter. This chapter started with an Introduction to Nanoscience then what is nanostructure and its Applications of Nanotechnology (basic idea), various size-dependent properties of nanomaterials. In this chapter some unique properties like a) Semiconducting nanoparticles b) Metallic nano-particles are explained with examples. Synthesis aspects of nanomaterials also need to understand using Bottom-up and Top-down approaches include mechanical alloying, shape and size control of nanomaterials. In the current scenario, the research and development of nanotechnology is very active globally, and nanotechnologies are already used in many products, Further, nanotechnologies are also being developed for use in environmental applications, e.g., clean-up of environmental pollutants.

Azad Kumar
M.L.K. P.G. College, India

Acknowledgment

Water pollution has become a huge concern to living organisms in recent years as a considerable amount of hazardous industrial waste comprising dyes, pigments, pharmaceutical goods, industrial chemicals, and numerous organic compounds is thrown into bodies of water, polluting the water. Because these wastes are resistant to aerobic digestion and are stable to light and oxidising chemicals, they pose major ecological hazards. According to some studies, roughly 12 percent of synthetic textile dyes used each year, such as Caramine and Indigo Red, are lost during manufacturing and processing activities, and 20 percent of these toxic dyes enter water through effluents. While chemical, biological, and adsorption treatments have been used in the past to remove toxic dyes from industrial waste water, these methods are ineffective for eliminating dye, pigments, pharmaceutical goods, industrial chemicals, and numerous organic compounds pollutants.

Advanced materials are made of distinct materials that differ in their nature, properties, shape, and size. Because of its uses in the design and production of novel goods in a variety of disciplines, the development of advanced materials has become a key scientific and technical focus in recent years. One of these is the use of modern water technologies. Examples include water filters, pollutants adsorption materials, and photocatalysis, among others. These innovative technologies are required due to the increasing demand for high-quality water. These concerns, together with the growing consequences of climate change, necessitate the search for novel solutions to these issues.

The purpose of this book is to offer a wide range of studies on hot issues in advanced materials for application in water quality and wastewater treatment. However, innovative materials save energy, such as energy cells and the creation and optimization of materials derived from forest leftovers such as nanocellulose and other natural polymers. This advancement in highly innovative materials applications in both water and wastewater treatment in order to optimise the energy system will lead to increased sustainability in the future.

I would like to express my deep and sincere gratitude to Pro. Ratnesh Das, Department of Chemistry, Dr. Harisingh Gour Central University, Sagar- 470003 (M.P.), Victor Shikuku, Kaimosi Friends University College, Kenya, Wilfidah Nyairo, Maseno University, Kenya, Dr. Asma Rafiq, Government College University Lahore, Pakistan, Chinmoy Bhattacharya Department of Chemistry, Indian Institute of Engineering Science & Technology (IIEST), Shibpur, Howrah-711103, West Bengal, Gao Li, Dalian Institute of Chemical Physics, Chinese Academy of Sciences, for their support and guidance during the publication of this book.

I am also thankful to all authors who contribute the chapter for this book.

I am also thankful to publisher, IGI Global publisher and their entire team for their invaluable support and encouragement.

I find no way to express in word my deep gratitude and profound reverence to my Parents and all my relatives. This undertaking would not have even been contemplated, let alone be completed, without their prayers, their blessings, their encouragement and the confidence they have showed in me. I recollect all the moments, they have shared with me all these years and waited patiently to see my research this destination in life.

Editorial Advisory Board

Chapter 1

Modifications of BiVO4 Semiconductors for Oxidation of Water and Detoxification of Organic Waste:

Photoelectrochemical Applications of Semiconductors

Sangeeta Ghosh
Indian Institute of Engineering Science and Technology, Shibpur, India

Paramita Hajra
Indian Institute of Engineering Science and Technology, Shibpur, India

Debasis Sariket
Indian Institute of Engineering Science and Technology, Shibpur, India

Debasish Ray
Indian Institute of Engineering Science and Technology, Shibpur, India

Swarnendu Baduri
Indian Institute of Engineering Science and Technology, Shibpur, India

Chinmoy Bhattacharya
Indian Institute of Engineering Science and Technology, Shibpur, India

ABSTRACT

The photoelectrochemical water splitting process has a lucid and efficacious impact, which emulates the natural photosynthesis process by converting solar energy into chemical energy. The construction of a PEC system can convert H2O to H2 or CO2 to C-based fuels. To achieve artificial photosynthesis, rate-determining kinetics of the OER is regarded as a highly efficient photo-anode. BiVO4 grabbed strong attention as a photoanode in the communal of PEC. Owing to a moderate bandgap and the Earth-abundant nature of the constituents, it is considered an inexpensive n-type semiconductor for PEC H2O splitting. This chapter discussed the recent progress of BiVO4-based photoanodes fabrication, including control in the surface, effects of dopant, different synthesis techniques, co-catalyst, etc. Typical unbiased tandem devices of a photoanode system in the presence of BiVO4 are also reflected. The report also demonstrated the photocatalysis principles regarding the degradation of organic pollutants.

DOI: 10.4018/978-1-6684-4553-2.ch001

INTRODUCTION

With the advancement of technology and ever amplifying population creates a huge energy demand in the 21st century, whereas the primary source of energy remains coal, oil, and natural gas, which are non-renewable and also during the production and use of energy from these sources increases the amount of carbon dioxide (CO_2) which is a potential Greenhouse gas and brought adverse effect on the environment. (Crabtree et al., 2004; Ginley et al., 2008) Since these conventional energy sources are going to finish once and they are concentrated in certain regions of the world, so scientists have introduced some energy sources which are clean, renewable, and significant to meet the global energy demand and also address the environmental issues at the same time. Solar energy, wind energy and tidal energy are introduced as alternatives to conventional energy sources. Immense research should be undertaken to develop of clean, renewable, and sustainable energy sources that can replace fossil-fuel energy and are commercially accessible across the globe. Nevertheless, the most developed renewable energy source that can produce electricity, or fuel that is transportable and storable, remains a challenge still now.

Solar energy, the never-ending energy on the earth considered the most promising candidate for clean, sustainable energy sources. (Park & Holt, 2010) A photoelectrochemical solar energy conversion device primarily focuses on the conversion of sunlight into chemical energy in the form of hydrogen gas. The solar water splitting comes with several advantages like ample sunlight on the earth, an abundance of salty water, minimal reaction potential, and zero greenhouse gas emission. (Walter et al., 2010)

Origin of PEC Water Splitting

TiO_2 and $SrTiO_3$ were mainly used as photocatalysts to explore water splitting during the 1970s and the first half of the 1980s. The TiO_2 photocatalyst acts as an active material with the modification of the co-catalyst. The reaction is performed in a suitable aqueous medium or a gas phase. There is no activity after the dispersion of native TiO_2 powder in water. Beginning in the post-1980s, some new photocatalyst materials, e.g., $K_4Nb_6O_{17}$ (Kudo et al., 1989), $K_2La_2Ti_3O_{10}$ (Ikeda et al., 1997), $BaTi_4O_9$ (Inoue et al., 1998), ZrO_2, and Ta_2O_5 (Sayama & Arakawa, 1994), other than TiO_2 and $SrTiO_3$, were introduced for the water splitting reaction. Many tantalite photocatalysts have been developed since the post-1990s (Kato & Kudo, 2003). In addition, few metal oxide photocatalysts having d^{10} electronic configuration, e.g., Ga(III), In(III), Ge(IV), Sn(IV), Sb(V) etc. and assisted with RuO_2 co-catalyst, have been reported very recently. Non-oxide Ge_3N_4 also behaves as a promising photocatalyst after the addition of some co-catalyst like RuO_2. (Sato et al., 2005) Hence, the photocatalytic water-splitting process of photocatalysts has been grown of immense importance in the last two decades.

Generally, photocatalysts comprising only d^0 and d^{10} metal cations respond in the presence of UV light wide band gaps (i.e., BG $\geq$ 3.0 eV). Highly active photocatalysts for H_2 or O_2 evolution under visible light irradiation, even in the presence of sacrificial reagents, were only Pt/CdS (Reber & Rusek, 1986) and WO_3 (Darwent & Mills, 1982) during the earlier era of the 1990s. There had been quite limited materials to become visible-light-driven photocatalysts. However, many oxides, oxy-nitrides, and oxy-sulfides have recently been reported to be active materials for H_2 and O_2 evolution in the presence of visible light irradiation with the help of sacrificial reagents. Recently, several two-photon photocatalyst processes, Z-scheme, as seen in photosynthesis by green plants have been reported for water splitting. Pt/$SrTiO_3$: Cr, Ta or Pt/TaON, is an example of combined systems, for the hydrogen evolution photocatalyst and WO_3 for the oxygen evolution photocatalyst, are responsible for water splitting into hydrogen and oxygen respectively in stoichiometric amounts in the presence of an iodate-iodide redox couple. For the overall process of water splitting reaction, Cr-Rh oxide / GaN:ZnO is also an active photocatalyst, developed by combining the solid solution of GaN and ZnO.

Basic Principles of Photocatalytic Water Splitting

Natural photosynthesis is a process in which plants converts carbon dioxide and water into carbohydrates and oxygen by utilizing sunlight is termed. Whereas solar hydrogen production is done using a similar kind of technique in semiconductors, utilizing sunlight that splits water into hydrogen and oxygen is known as artificial photosynthesis. (Osterloh & Parkinson, 2011)

Solar-to-chemical energy conversion can be achieved by PEC device, the most popular approach combining the electro-catalytic and light-absorbing functions. The process occurs in sunlight by the photoactive materials, producing e^- - h^+ pairs. The electron in the conduction band is responsible for the water reduction ($2H^+ + 2e^- \rightarrow H_2$), and the hole in the VB will involve the four-electron water oxidation process ($2H_2O \rightarrow 4H^+ + O_2 + 4e^-$).

The overall process of water-splitting reactions (Fujishima & Honda, 1972) take place in three consecutive steps, as presented in Fig. 1-

- Absorption of sunlight by semiconductor; the energy of the photon must be higher than the band-gap of the semiconductor.
- Generation of electron and hole and charge separation on the interface.
- Surface chemical reaction where oxidation and reduction of water take place simultaneously.

The semiconductor absorbs photons having higher energy than the band gap of it. It is also noted that the position of VB must be more positive compared to the oxygen evolution potential.

An electrochemical workstation is implemented via a 3-electrode configuration system for PEC measurements to detect the photoactive current by applying a bias voltage.

The water splitting reaction follows the mechanism mentioned below:

$$\text{Semiconductor} \rightarrow e^- + h^+ \quad (1)$$

Generation of excitons

$$2H^+ + 2e^- \rightarrow H_2 \quad (2)$$

Hydrogen evolution reaction (HER)

$$2\,H_2O + 4h^+ \rightarrow 4H^+ + O_2 \quad (3)$$

Oxygen evolution reaction (OER)

Figure 1. Fundamental principle of photocatalytic water splitting on semiconductor

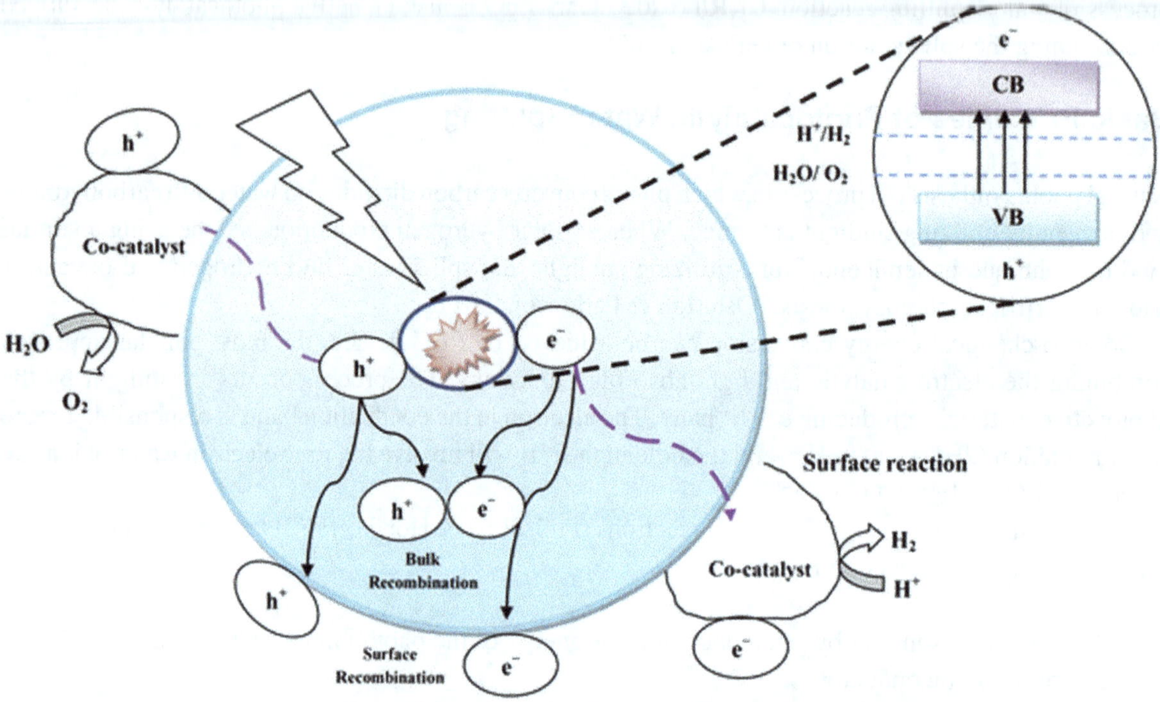

JUSTIFICATION FOR CHOOSING $BiVO_4$

n-type bismuth vanadate (BVO)semiconductor has been paid a lot of attention as one of the most encouraging photoanode based on the following requirements: its valance band edge situated at ~2.4 V vs. RHE, and the conduction band position is situated just little of the thermodynamic level for H_2 to provide sufficient over potential to photo-oxidize water with holes. Although BVO, having a band gap of 2.4 eV, it is to some extent larger than the desired photoanode. In 1998, Kudo et al. first identified the practical utilization of BVO electrodes in the solar oxidation process. (Kudo et al., 1998) Fundamental work of BVO essentially concentrated on a system consisting of small particles kept dispersed by agitation for water splitting reaction or photodegradation of organic compound. As the conduction band does not permit water reduction reaction, Ag (I) generally behaves as a sacrificial electron scavenger in these studies.

Naturally occurring crystal BVO is the pucherite mineral having an orthorhombic configuration. (Yao et al., 2008) Moreover, when it is synthesized in the laboratory, it crystallizes either in a zircon (z) or scheelite(s) like structure despite adopting the pucherite structure as presented in Figure 2 (Park et al., 2013). The scheelite (s) structure can exist in three different arrangement; tetragonal (t) crystal configuration with a = b = 5.1470 Å & c = 11.7216 Å or a monoclinic (m) crystal configuration (a = 5.1935 Å, b = 5.0898 Å, c = 11.6972 Å) whereas the zircon (z)-type structure has a tetragonal (t) crystal configuration with a = b = 7.303 Å & c = 6.584 Å. (Dreyer & Tillmanns, 1981)

In the scheelite structure, as presented in Figure 2 (a), each vanadium ion is coordinated with four oxygen atoms in a tetrahedral site, and each bismuth ion is coordinated by eight oxygen atoms from eight different VO4 tetrahedral units. The vanadium centre (coordination number 4) alternates towards

Figure 2. Crystal structures of (a) tetragonal scheelite and (b) zircon-type $BiVO_4$ (red: V, purple: Bi, and grey: O). The crystal structure of monoclinic scheelite is very similar to what is shown in (a) and local coordination of V and Bi ions in (c) tetragonal scheelite, (d) monoclinic scheelite, (e) and zircon-type $BiVO_4$ structure with bond lengths
Reprinted with permission from Ref. (Park et al., 2013) copyright from Royal Society of Chemistry.

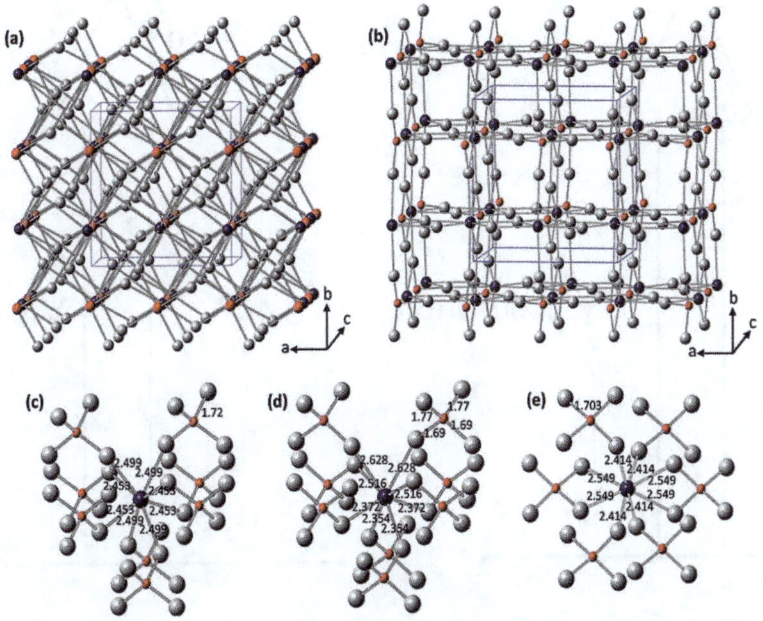

the (Crabtree et al., 2004) direction with the bismuth centre (coordination number 8); each oxygen atom is also connected with two bismuth centres and one vanadium centre in this arrangement. This bismuth and vanadium centres hold together composes a 3-D arrangement. In between the tetragonal scheelite (t-s) and monoclinic scheelite (m-s) arrangement, the vicinal environments of Bi and V ions are distorted more significantly in the monoclinic arrangement to remove the four-fold symmetry, which is very obvious for a tetragonal arrangement.

In the zircon-type arrangement, bismuth is coordinated by eight oxygen atoms; four oxygen atoms still stabilize vanadium. Moreover, only six VO_4 units are surrounded by each Bi because two VO_4 units provide two oxygen atoms to bismuth. All oxygen atoms also have connectivity between two bismuth centres and one vanadium centre to allot the vanadium and bismuth centres together to compose a three-dimensional arrangement, as presented in Figure 2.

The fabrication of BVO photoanode at low temperatures leads to a zircon-type arrangement. However, kinetics plays a vital role at low temperatures to determine products or structure type. Depending on the synthesis technique and reaction environments, it may vary. It has also been reported that an irreversible phase transition occurs to achieve monoclinic scheelite configuration from tetragonal zircon structure in between the temperature region 670-770 K. The high temperature tetragonal phase includes all scheelite structures. The reversible phase transition is stuck with monoclinic in addition to tetragonal scheelite BVO at 528 K. (Kudo et al., 1999)

The photocatalytic (PC) activity of BVO has strong connectivity with its crystal structure. Figure 3 represented the schematic band structures of $BiVO_4$ having tetragonal zircon and monoclinic scheelite-

Figure 3. Band structures of tetragonal $BiVO_4$ (zircon) and monoclinic $BiVO_4$ (scheelite)

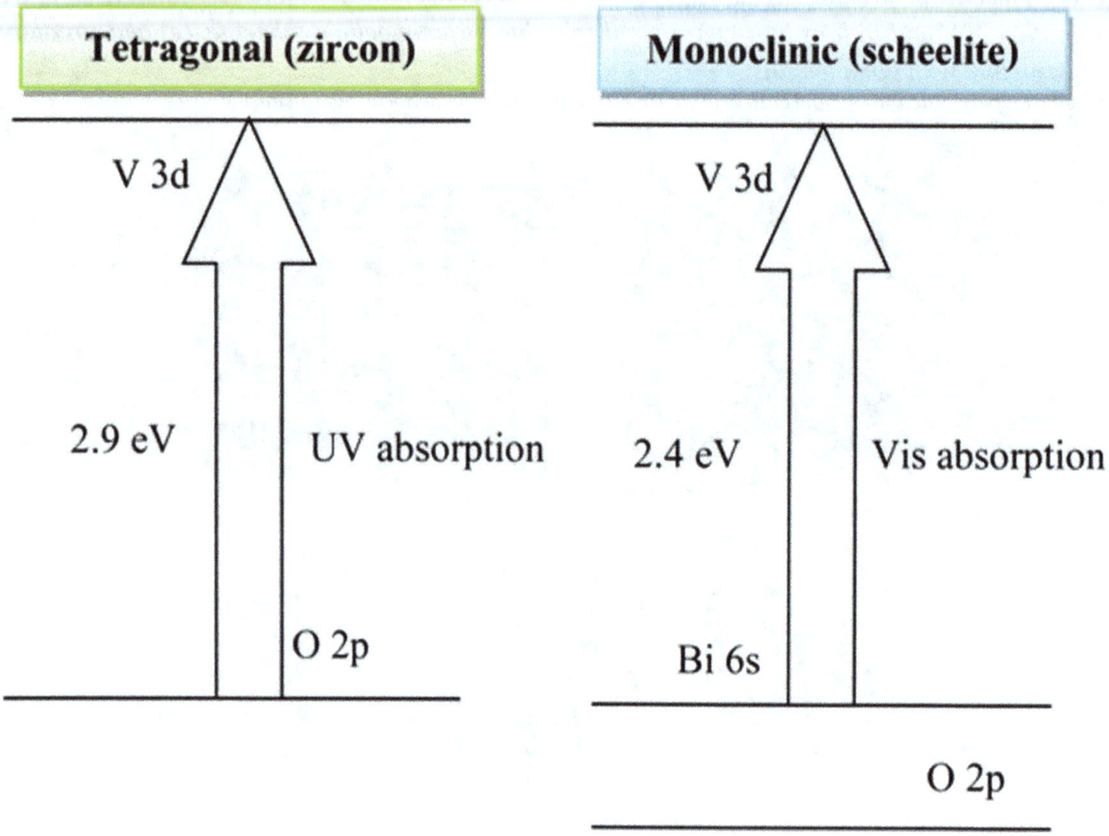

type arrangement. Kudo et al. described a monoclinic scheelite BVO achieved the better performance of photoactivity for water splitting reaction to convert O_2 in the presence of Ag(I) ions as an electron scavenger when prepared as powder type photocatalysts. Due to the enhancement of photon absorption in the case of scheelite(s) BVO, the improved photoactivity of this structure appears over zircon-type BVO. This is because the band gap energies for scheelite type BVO appear in 2.4 eV whereas zircon-type BVO is 2.9 eV.

The band gap transition occurs due to the charge transportation from 2p orbitals of an oxygen atom to vacant 3d orbital of vanadium atom in the zircon type of arrangement. The band gap decreased due to the appearance of Bi-6s state over the O-2p state. The transition from Bi-6s orbital or may be hybridization of Bi-6swith O-2p orbitals to the V-3d is also responsible for the reduction of band gap in the scheelite like system.

A monoclinic-scheelite (m-s) system (Figure 2 (c)) contributes much better performances towards the photocatalytic (PC) water oxidation process in comparison to a tetragonal-scheelite (t-s) structure (Figure 2 (d)) among all BVO having scheelite structures, as reported by Tokunaga et al. Although the band gap energies are comparable for the different arrangements of scheelite type BVO, in m-s BVO, superior performance was observed due to the more severe distortion of the metal polyhedron. In a scheelite like arrangement, the neighbour environment of the bismuth atom in the monoclinic phase is too distorted compared to the tetragonal phase. Tokunaga et al. further stated that the effects of the e^- - h^+ separation are favourable with distortion of the provincial polarization.

Unbiased Tandem Device Using $BiVO_4$ as Photoanodes

Unassisted tandem device configurations are classified into three different types presented in Fig. 4. There are two absorbers to complete the cell in Type A and B, which comprises a photoanode to occur water oxidation (modified BVO) and a photocathode with a small band gap semiconductor (Cu_2O) for water reduction. In Type A, any conductive material, e.g., graphite or metal is used as the conductive matrix, whereas an e^- - h^+ pair recombination layer is responsible for forming a transparent conductive membrane in Type B.

In both types, to fulfil a complimentary light absorption, the most important is the position of the CB of the photoanode (minimum E_{CB}) must situate at a more negative potential than the valance band of the photocathode (maximum E_{VB}). The oxidation and reduction potentials of water must overlap with its band gap so that photo-activated electrons and holes thermodynamically can effort the hydrogen evolution reaction (HER) and oxygen evolution reaction (OER), respectively, as discussed.

On the other hand, in Type C, a photoanode system is in combination with a photovoltaic device. The device with less energy than the band gap of BVO mainly affords an external bias to the photoanode after modification of BVO by harvesting the transmitted photons. (Andrei et al., 2018) To become a high transmittance photoanode, BVO is required, while a TCO layer, considered as the recombination layer, is required for photoactive holes (from Photovoltaic device) and electrons (from PEC system) of BVO.

Meanwhile, the PEC system, in combination with the photovoltaic system (PEC/PV tandem cell), is more flexible to choose the photoanode materials to compare to the PEC system in connection with the PEC system (PEC/PEC tandem cell), which is generated by the PV cell. To match the energy levels in the photovoltaic system is not an essential requirement, which is a basic criterion for the PEC system junction with PEC cell (PEC/PEC tandem cell).

In practical applications, the water-splitting process on the basis of the PEC system should behave as an isolated device. The utilization of solar light is to supply the external energy without any bias voltage. The PEC-based device can play dual roles of energy converter and light absorber and is a more straightforward approach compared to the photovoltaic electrolyzer by the interface of semiconductor and electrolyte of the one PEC system. The ultimate goal for modifications of BVO is to fabricate an efficient photo-electrolysis is to achieve the hydrogen evolution and oxygen evolution reaction in an unbiased tandem device.

The PEC system in combination with photovoltaic cell (PEC/PV tandem cell) of any tandem device is mainly determined by two factors: one is photoanode i.e., the front light absorber which can achieve a better photocurrent (I_{ph}) by applying lower bias or not and the other is the photocathode performance or rear photovoltaic system. To become a more efficient PEC tandem device junction with the photovoltaic system (PEC/PV tandem cell), there is a correlation with short circuit I_{ph} and fill factor, whereas in the case of the photovoltaic system, it correlates with only increased open circuit photovoltage. For a photocathode, to become a more efficient PEC tandem device in combination with PEC system (PEC/PEC tandem cell), a higher I_{ph} with more positive onset potential can contribute effectively. This is due to the determination of I_{ph}, which can be achieved by both photoanode and rear counterpart.

Figure 4. Different devices of the tandem cell for water splitting.

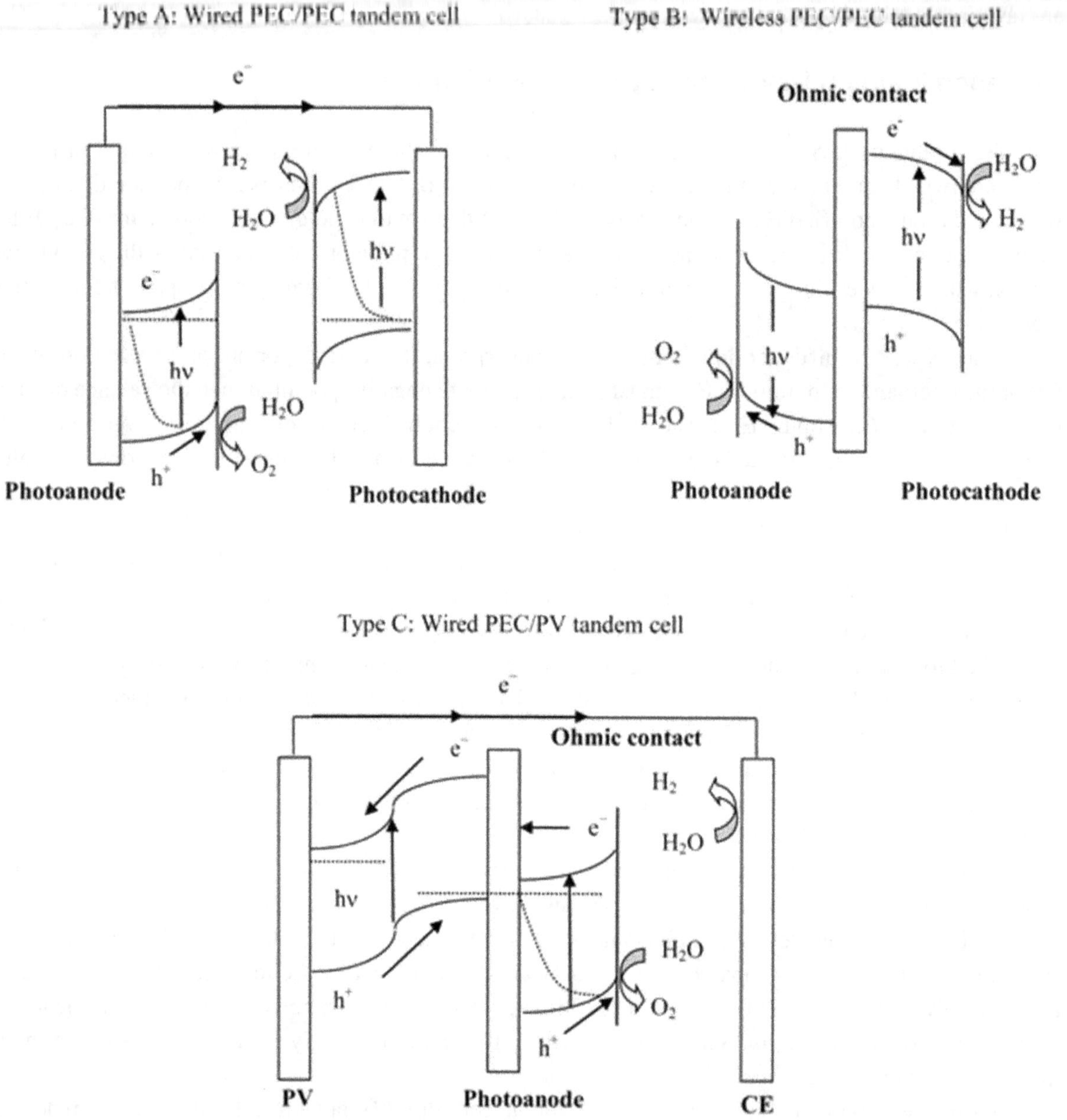

COMMON SYNTHESIS METHODS TO PREPARE $BiVO_4$ SEMICONDUCTOR

There are so many routes that have been proposed to prepare a BVO semiconductor. We classify these methods into the electrospinning method and depending upon phases; these are divided into three major categories; the solid phase technique, the liquid phase technique, and the vapour phase technique. Next, we will demonstrate the technique to fabricate a semiconductor in detail accordingly.

Electrospinning

This method is especially effective for preparing nanofiber materials. Cheng et al. synthesized BVO nanofiber via electrospinning technique using bismuth nitrate and ammonium metavanadate as starting material in the presence of citric acid as a chelating agent, (Cheng et al., 2015) the phase junction structure of scheelite, monoclinic and tetragonal phases of electrospun BVO were obtained by the controlled heating. FR. P. Antony et al. prepared BVO and Mo-doped BVO nanoparticles by the electrospinning process (Antony et al., 2016) to measure PEC performance which indicates that due to the doping of Mo, the water oxidation photocurrent of BVO increases by four times that of pure semiconductor.

Solid Phase Technique

This method has great advantages with simple technique and favourable process, uniformity in particle size, and adjustable force. Li et al. synthesized BVO powder by the solid phase technique. (Li et al., 2019) The BVO prepared by heating at 500°C showed the best photocatalytic reduction efficiency against Cr (VI).

Sol-Gel Technique

The sol-gel method is the most important wet chemical synthetic approach for the preparation of nanoparticles. Drisya et al. prepared TiO_2/BVO nanocomposite via sol-gel method, (Drisya et al., 2020) the composites exhibit significant PC performance under the visible light irradiation; it was also noted that the PC efficiency is increased when percentage of BVO decreased in the nanocomposite. Pookmanee et al. developed BVO powder by sol-gel method, (Pookmanee, Kojinok, Punthaeod et al, 2013) and reported that the particle size, crystallinity and purity of the sample depend upon the calcined temperature. Wang et al. prepared molybdenum doped BVO by the sol-gel method, which shows enhanced photocatalytic activity than the bare BVO. (Wang, Shan, Wu et al, 2017)

Precipitation or Co- Precipitation Technique

For synthesizing nanomaterials, the most popular method is the precipitation method, where different chemical components are mixed together. The composite prepared by the precipitation technique demonstrated significant light absorption potentiality and also promote the separation of electrons and holes to improve PC performance. Pérez et al. synthesized BVO powder by surfactant-assisted co-precipitation technique. (Pérez et al., 2012) The PC activity for the photodegradation of Rhodamine B (Rh B) of the as-prepared m-BVO in the presence of Pluronic non-ionic surfactants performs a greater activity than the samples synthesized via solid-state reaction. The PC efficiency of BVO which was prepared by Cruz et al. (Cruz & Pérez, 2010) through the co-precipitation process for the photodegradation of Rh B revealed the high capability to bleach the dye solution. Ganeshbabu et al. synthesized BVO nanoparticles via chemical precipitation technique. BVO nanoparticles calcined at 400°C exhibit the highest PC performance as it degrades Methylene blue (MB) dye about 92.25% within the duration of 120 min. (Ganeshbabu et al., 2020)

Hydrothermal Technique

It is the most useful technique for preparing nanoparticles which is basically an approach of solution-based reaction. Lei et al. employed BVO photocatalyst via the hydrothermal method; the sample prepared at the pH3.0 shows the best photocatalytic activity. (Lei et al., 2014) Jiang et al. synthesized BVO photocatalyst with various kinds of morphologies via hydrothermal method in the presence or absence of poly (vinyl pyrrolidine), in which the spherical shaped BVO demonstrated the highest photocatalytic activity. (Jiang et al., 2012) J. Yu fabricated BVO nanofiber by this method where the BET surface area of the nanofibrous was higher compared to bulk BVO. (Yu & Kudo, 2005) H. Jiang examined the PC performances for degradation of MO using BVO photocatalyst synthesized hydrothermally to achieve improved performance under visible light irradiation. BVO photocatalyst was also prepared by Ran et al. through this approach. (Ran et al., 2015) The BVO photocatalyst prepared through hydrothermal techniques in comparison with other techniques revealed improved surface morphological properties of the photocatalyst.

Microwave-Assisted Synthesis Technique

A major innovation and a dramatic change were observed through microwave synthesis for BVO nanomaterials. Pookmanee et al. synthesized BVO powder through microwave-assisted synthesis technique. (Pookmanee, Longchin, Kangwansupamonkon et al, 2013) Multi-phase monoclinic and tetragonal structures of BVO (surface area ~ 4.89-15.90 m^2 g^{-1}) were produced by the microwave irradiation of 600-800 W for duration 4-6 min. BVO photocatalyst was also developed by Intaphong et al. in microwave-assisted technique with irradiation of 500 W for 2 min, 4 min and 6 min. (Intaphong et al., 2016) The best photocatalytic performance is shown for the sample which is prepared for 4 min. Souza et al. obtained BVO nanoflowers modified with gold nanoparticles through microwave irradiation to increase the photocatalytic efficiency. (Souza et al., 2019) Another example for BVO heterojunction formation was revealed by Yan et al. through facial microwave-assisted technique. (Yan et al., 2015) They explored higher photo catalytic efficiency when tetracycline is degraded due to faster transfer of charges between the heterojunction of the various phases of BVO.

Electrodeposition

Electrodeposition is a flexible, low-cost method for the fabrication of a wide variety of two and three dimensional materials such as coatings and films. Ye et al. synthesized $Bi_2O_3/BiVO_4$ hetero-structure nanospheres via electrodeposition method with a photocurrent of 2.58 mA cm^{-2} at 1.2 V vs. Ag/AgCl which is almost 5 times higher than the pristine BVO. (Ye et al., 2015) Kong et al. prepared Ni-doped BVO photoanode by the electrodeposition method, where 5% Ni-doped BVO exhibits the highest photocurrent of 2.39 mA cm^{-2} at 1.23 V vs. RHE which is about 2.5 times higher than the pure BVO. (Kong et al., 2019)Cho et al. synthesized W doped BVO photoanodes by the electrodeposition technique; the photocurrent of W doped BVO is increased three times than that of pure BVO. (Cho et al., 2013) A. J. Bard et al. electrodeposited amorphous TiO_2 on W: BVO/F: SnO_2 resulting in almost 5.5 times higher water oxidation photocurrent than pure BVO. (Eisenberg et al., 2014)

Spin Coating

It is generally used for the development of nanocomposite film. Tayyebi et al. synthesized monoclinic BVO photocatalyst via spin coating followed by calcination method, the photocatalytic activity of monoclinic BVO is increased with an increase in the basicity of the medium. (Tayyebi et al., 2019) Sitaraman et al. synthesized $WO_3/BiVO_4$heterojunction by spin-coating technique; from the PEC studies it was evident that the $WO_3/BiVO_4$ shows higher photocurrent density (0.64 mA cm^{-2} at 1.23 V vs. RHE) than the bare WO_3 and $BiVO_4$. (Sitaaraman et al., 2021) Wang et al. prepared BVO thin film by spin coating followed by annealing, the thin film annealed at higher temperature (500°C-540°C) shows better PEC performance compared to the others prepared at a lower temperature. (Shi et al., 2018) Russo et al. synthesized BVO thin film via spin coating technique, from the PEC study it is evident that the charge transfer kinetics is three times faster than that of the porous film. (Hernándeza et al., 2016).

STRATEGIES FOR ENHANCING PEC PERFORMANCE

Various approaches such as control in surface morphology, composite construction, the effect of dopant, and heterojunction of photocatalysts have been employed to improve the overall PEC water splitting performance.

Heterojunction

Kalanur et al. synthesized WO_3/BVO via a combination of hydrothermal and spin-coating technique; different amount of ethyl cellulose is added during the spin coating, whereas the highest photocurrent (I_{ph}) is observed in the presence of 200 mg of ethyl cellulose which acts as an organic binder (Kalanura et al., 2017). Later they electrodeposited TiO_2 on the optimized WO_3/BVO heterojunction, fabricated WO_3/BVO/TiO_2 and studied surface modification by using Co-phosphate as a co-catalyst. Photoelectrochemical studies of WO_3/BVO electrode shows higher photocurrent (~3.9 mA cm–2 at 1.23 V vs. RHE) compared to individual WO_3 and BVO electrode whereas WO_3/BVO/TiO_2 composite shows maximum photocurrent of ~4.2 mA cm^{-2} at 1.23 V vs. RHE. Jang et al. designed a triple planar heterojunction SnO_2/ WO_3/BVO films via electron beam deposition followed by a wet chemical method and the thickness of the $SnO_2$and WO_3 layer are varied during the deposition. (Bhat et al., 2018) Among the different thicknesses of the SnO_2 and WO_3 layer, SnO_2 50 nm/ WO_3 50 nm /BVO shows the highest I_{ph} of ~2.01 mA cm^{-2} and 1.80 mA cm^{-2}at 1.23 V under front & back illumination, respectively.

Doping Approach

Doping is also one of the most intentional tactics to tune the properties of a bulk semiconductor. Due to the enhancement of e^-- h^+ recombination and reducing the depletion layer bandwidth, the effect of doping will not always be favourable. Park et al. (Park et al., 2011) demonstrated that BVO photoanode with co-dopants 2 at% Mo and 6 at% W performed more upgraded activity than the bare or only W modified BVO. The addition of Mo to the W-doped BVO sample indicates twice the carrier density compared to only the W-doped sample, as measured through Mott-Schottky analysis. A structural change occurs for Mo/W-doped as well as W-doped BVO from monoclinic scheelite to tetragonal scheelite phases. Berglund et al. (Berglund et al., 2012) reproduced the results of Park et al. where W, Mo and Mo/W-doped BVO photoanodes were synthesized via ballistic deposition technique. An important technique - inductively coupled plasma mass spectroscopy (ICPMS) was used to verify the doping composition. Compared to BVO photoanodes containing different percentages of W & Mo or the sample doped with either of W or Mo individually, the BVO electrode coupled with 2 at% Mo and 6 at% W achieved the best performance. The modification of Mo and W became quite popular owing to their unique effects whose addition is more facile and no further improvement in I_{ph} with the increase in the doping level of either Mo or W compared to co-doping of Mo and W for the identification of the synergistic effect of Mo and W.

Luo et al. investigated the effect of BVO electrodes doped in the presence of different metal ion W, Zr, La, Mo, Sr, Ag, Ta, Si, Zn, Ti as prepared through modified metal-organic decomposition method. The I_{ph} was found to enhance via doping with either of Mo^{6+} or W^{6+}- ion (Luo et al., 2011). A modified BVO sample prepared with 3 at% Mo exhibited significant improvement in I_{ph} as well as IPCE in natural seawater due to photo-oxidation of Cl^-. The Raman spectral study of the modified samples indicated that Mo^{6+} ions were located within the V^{5+} sites.

Phosphorous, non-metal constituent was also incorporated into BVO through the urea-precipitation method as reported by Jo et al. (Jo et al., 2012) where a small fraction of VO_4^{3-}-oxoanions was replaced by PO_4^{3-}-oxoanions. In the host structure, the incorporation of PO_4^{3-} oxoanions, which were added as a phosphorous precursor, caused shifts in the XRD peaks indicating an increase in cell parameters. The electrophoretic deposition of $Bi(VO_4)_{0.998}(PO_4)_{0.002}$ achieved a better performance to generate an appreciably higher photocurrent compared to the pure BVO electrode. The charge transfer resistance of BVO remarkably decreases upon the addition of PO_4^{3-}.

Co-Catalyst

Chiang et al. demonstrated the effect of photochemically modified nano porous BVO with Co^{2+} in acetate buffer on PEC water oxidation. The surface modification in the presence of Co-acetate gives a cathodic shift of onset potential (~0.45 V); the I_{ph} of the modified BVO is 2.05 mA cm^{-2} at 1.23 V vs. RHE which is almost 3.4 times higher than that of the pure one. Wang et al. synthesized modified BVO photoanode by using iron-cobalt oxide as a co-catalyst and the photoanode shows the I_{ph} of 4.82 mA cm^{-2} at 1.23 V vs. RHE, whereas the IPCE value reaches up to ~ 86% upon loading of $FeCoO_x$ layers on BVO. Wu et al. synthesized $Co_3(PO_4)_2$decorated BVO via solid-state reaction and observed I_{ph} of 0.30 mA cm^{-2} at 1.23 V vs. RHE which is higher than pure BVO. (Wu et al., 2020) Gamelin et al. (Zhong et al., 2011) demonstrates low onset potential of 0.31 V vs. RHE at pH8 for Co-Pi modified tungsten doped BVO photoanode. Liu et al. (Tang et al., 2018) synthesized a CoOOH-over layer coated coral-like BVO photoanode which also has a low onset potential of 0.2 V vs. RHE, a maximum I_{ph} of 4.0 mA cm^{-2} at 1.23 V vs. RHE. Pilli et al. synthesized Co-Pi catalyst modified Mo doped BVO, showing I_{ph} of 1.0 mA cm^{-2} at 1.0 V vs. Ag/AgCl. (Pilli et al., 2011) Nam et al. synthesized W-Mo doped BVO photoanode modified with iron oxy-hydroxide electrocatalyst via a two-step process drop-casting followed by photo-assisted electro-deposition, the co-catalyst modified doped BVO photoanode attained at least two-fold higher I_{ph} (at 0.3 V vs. Ag/AgCl) than that of the doped BVO photoanode. (Joo et al., 2016)

Figure 5. (a) Linear sweep voltammograms of undoped $BiVO_4$ (blue), W-doped $BiVO_4$ (red), and W/Mo-doped $BiVO_4$ (black) in the 0.1 M Na_2SO_4 aqueous solution (pH 7) with chopped light under UV-visible irradiation and XRD patterns of (i) 2 at % W & 6 at % Mo-doped $BiVO_4$, (ii) 5 at % W-doped $BiVO_4$, and (iii) undoped $BiVO_4$.
Reprinted with permission from Ref. (Park et al., 2011) copyright from American Chemical Society.

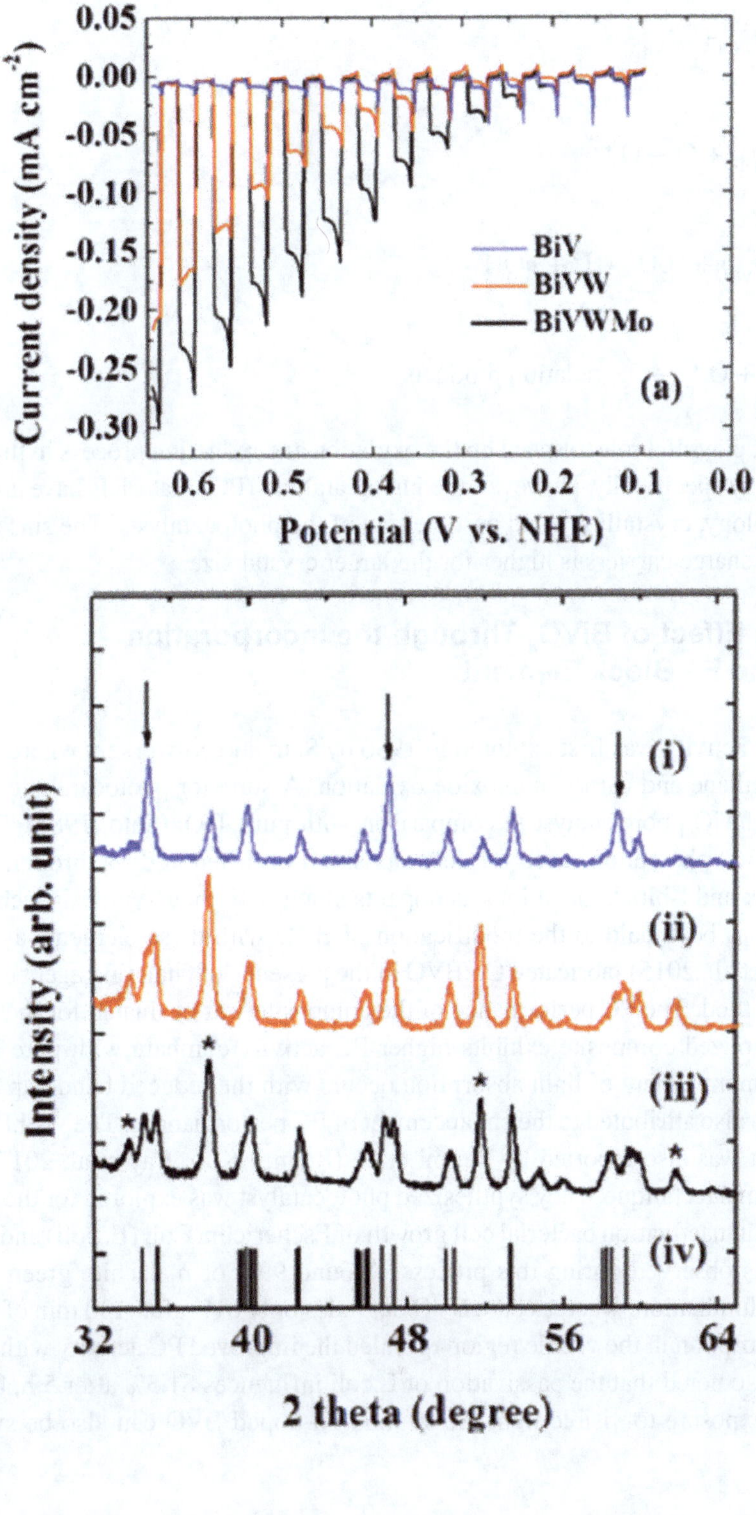

GENERAL PRINCIPLES AND MECHANISM OF PHOTOCATALYSIS FOR THE SEMICONDUCTOR

A photocatalytic reaction usually proceeds through four basic steps mentioned below:

$$\text{Photocatalysts} \xrightarrow{h\vartheta} e^{-}_{CB} + h^{+}_{VB} \quad (1)$$

$$\text{Photocatalysts } (e^{-}_{CB}) + O_2 \rightarrow O_2^{\bullet -} \quad (2)$$

$$\text{Photo catalysts } (h^{+}_{VB}) + H_2O \rightarrow H^{+} + \bullet OH \quad (3)$$

$$\text{Pollutants} + \bullet OH + O_2^{\bullet -} \rightarrow \text{Degradation products} \quad (4)$$

A photocatalyst plays the role to speed up the oxidation and reduction process in the presence of light as a source of energy specifically. However, the photocatalytic (PC) materials have an intimate association to the morphology, crystallinity and particle size of the photocatalysts. The surface recombination of photogenerated charge carriers is higher for the larger crystal size.

Photocatalytic Effect of $BiVO_4$ Through the Incorporation of D - Block and F - Block Element

The photocatalytic activity was first explored in 1986 by Sato and co-workers where N/TiO_2photocatalyst was used for ethane and carbon monoxide oxidation. A superior photocatalytic (PC) activity was demonstrated by N/TiO_2photocatalyst in comparison with pure TiO_2.(Sato, 1986) There are so many examples of the BVO photoanode used for decomposition of different dyes through the incorporation of different d-block and f-block metal ions as dopants along with their synthesis technique. Thereafter a lot of attention has been paid to the modification of BVO with these elements as suitable dopants. Geng et al. (Geng et al., 2015) fabricated Co/BVO in the presence of chelating agent e.g. EDTA through a hydrothermal method. The PC performance of the composite was evaluated for the decomposition of MB. Co/BVO optimized composite exhibits higher PC activity compare with pure BVO. It has been reported that the improvement of light absorption occurs with the reduced band gap and deliberate recombination rate is also attributed to the enhancement of PC performances. The visible-light-driven Co/BVO photocatalyst was also reported by Regmi et al. (Regmi, Kim, Ray et al, 2017) via microwave-assisted hydrothermal technique. They synthesized photocatalyst was explored for the decomposition of malachite green. An inactivation bacterial cell growth of Escherichia Coli (E. coli) and Chlamydomonas Pulsatilla (CP) was observed during this process. Around 99% of malachite green was decomposed within 90 min of illumination, whereas bare BVO degrades only 61% after 130 min of illumination. The improved light absorption in the visible region revealed the improved PC activity with lowering of band gap energy. It is also noted that the passivation of E.coli influences 81.3% after 5 h, but CP influences only 65.6% after exposure to visible light for 60 min. Ni-doped BVO can also be synthesized by the microwave-assisted hydrothermal process which was reported by Regmi et al. (Regmi, Kshetri, Kim, Pandey, Ray et al, 2017) The PCA of 1 wt% Ni/BVO composite was revealed 92% photocatalytic degradation proficiency when it exposed towards the ibuprofen. With the decreasing migratory time period of the excited carrier, the separation proficiency of the e^- - h^+ pair increases and the oxygen vacancies of the composite behave as an electron-trapping centre. Wang et al. (Wang, Guo, Chai et al, 2017) prepared Cu/BVO in the presence of chelating agent EDTA via solvothermal technique. The optimized Cu/BVO composite showed 5 folds higher PC performance than bare one in the presence of Rh B dye. It was assigned to lower band gap with the improvement of light absorption, surface morphology reduced rate of recombination reaction of the photogenerated charge transporters. The electrons (e^-) of the Cu/BVO photocatalyst from VB are activated under visible-light illumination, and a situation arises where the promotion of e^- from the valence band to the conduction band takes place with h^+ left in the valence band. On the surface reactive sites of Cu/BVO produce $O2^{\bullet-}$radicals and the electrons are confined by adsorbed oxygen molecules. To produce strong oxidizing agents like •OH radicals, the photoinduced h^+ of the VB can interact with H_2O and OH^-. Any active species (e.g. •OH radicals) holes can attack the Rh B molecules. However, Cu^{2+} and V^{4+} captured the excited electrons of the CB to enhance the charge separation effectiveness of photoactivated charge transporters. Enhanced photocatalytic degradation efficiency for Rh B was observed due to the promotion and involvement of sufficient amounts of electrons and holes in the photocatalytic reaction. Chen et al. (Chen et al., 2015) developed Cu-doped BVO to obtain the highly active optimized composite via the microwave-assisted hydrothermal process for not only UV light but also visible-light photodegradation of MB. Due to electrostatic attraction e^-s in the conduction band and h^+s in the valence band are adjacent to each other during the irradiation process. As Cu was inefficient compared to BVO, the charge transporters were identified only when the amount of Cu incorporation was appreciable. However, the Fermi's level played an important role. There is no shifting of photo-generated electrons from an upper Fermi level of BVO to a lower Fermi level of metal until the energy levels became the same. Thus, Cu (II) behaves as a resourceful dopant to significantly improve PC performance. Chen et al. (Chen et al., 2013) developed an Ag-modified BVO composite in the presence of L-lysine with the help of the sol-gel method to obtain a porous structure composite. They evaluated the catalyzed activity of the Ag modified BVO composite for the MB and Rh B decomposition. The 6.5 wt% modified composite was assessed as an optimized photocatalyst due to its enormous surface area in comparison with bare BVO. Xu et al. (Xu et al., 2016) also studied the role of Ag-doped BVO for the photodegradation of MB with high efficiency. A red shift occurred for Ag/BVO than that of pure BVO, as confirmed through DRS analysis and the band positions were evaluated from the band diagram. The VB and the CB of BVO loaded with Ag nanoparticles shifted towards the negative direction in comparison with pure BVO. These behaviours proposed a unique electronic structure which increases the capability of photo-oxidation of the excited charge transporters to encourage visible-light activity. Booshehri et al. (Booshehri et al., 2014) prepared an Ag-based BVO composite to investigate the disinfection activity of photocatalyst. The authors evaluated this activity for E.Coli after exposure to visible light where bacterial growth was around 100% after 3 h of illumination in the presence of Ag-BVO composite. Due to the presence of Ag nanoparticles, it acts as a trapping agent of electrons on the surface of BVO. Therefore, PC activity of Ag-based BVO enhanced with the separation of photoexcited charge transporters to generate reactive species like oxygen. The photocatalytic performance of Cu modified BVO and Ag modified BVO composite was examined by Bian et al. (Bian et al., 2014) as developed through the hydrothermal path. After exposure to visible light the Cu modified BVO photoanode exhibited 89% decay of ibuprofen, whereas the Ag modified sample assessed

Figure 6. (a) XRD patterns of $BiVO_4$ with various wt% of Fe and (b) enlarged view of (121) peak at $2\theta=28.5°$.

Reprinted with permission from Ref. (Regmi, Kshetri, Kim, Pandey, & Lee, 2017) copyright from Elsevier

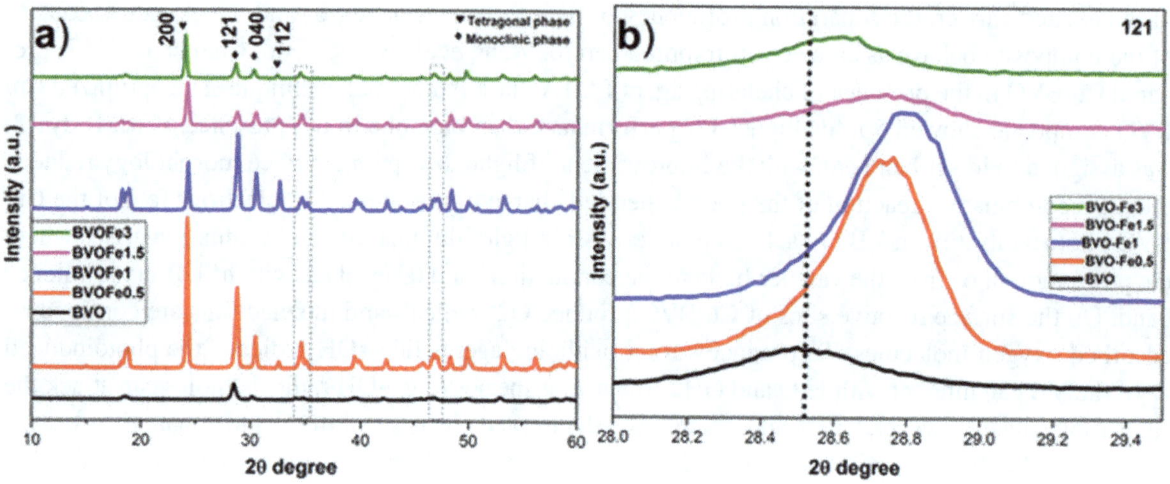

96% decomposition efficiency after 5 h irradiation. Lower band gap energies with greater surface areas indicate the slower recombination of charge transporters. Consequently, photocatalytic activity increased with the absorption of additional active sites and degraded constituents are also excluded in the presence of Ag metal as a dopant. Zhang et al. (Zhang & Zhang, 2010a) made hydrothermally synthesized Au/BVO composite to explore the photocatalytic behaviour under visible-light irradiation for the decolourization of methyl orange (MO). Among all samples, optimized 1.48wt% Au-BVO photoanode presented highest photocatalytic activity where Au metal ion exhibits a crucial role to suppress combination effect and increase the charge carrier mobility. Recently, Chala et al. (Chala et al., 2014) prepared Fe-doped BVO to evaluate the photodecomposition of MB via hydrothermal technique. The sample exhibited outstanding photocatalytic behaviour than that of bare BVO. In the presence of visible-light illumination, the maximum photodegradation proficiency is ~81% with an optimal (5 mol %) Fe-loading. After the addition of iron, the photocatalytic performance improved due to more proficient electron capturing by Fe(III) ions resulting in an enhancement of e^--h^+ separation. Regmi et al. (Regmi, Kshetri, Kim, Pandey, & Lee, 2017) also prepared Fe-modified BVO photoanode via microwave-assisted hydrothermal technique for use in the decomposition of ibuprofen in the presence of visible-light irradiation. The author assessed that 80% of ibuprofen is degraded with optimized 1 wt% Fe-modified BVO after 3 h exposure to visible-light. The development of monoclinic and tetragonal hetero structures endorses the interfacial charge transfer between these two phases and the photoinduced charge transporters are emphasized due to suppress the ion of the internal recombination to facilitate the PC performance. Additionally, it has been reported that there is an intimate connection between Fe^{3+} and BVO in the composite so that diffraction peaks in the XRD plots are modified to some extent, as mentioned in Figure 6.

By using microwave-assisted method Pt-modified BVO was produced by Shi et al. (Shi et al., 2013) The PC performance of 2 wt% Pt/BVO photocatalyst demonstrated showed 92% degradation efficiency of ciprofloxacin were recorded under visible-light exposure, which is 25% greater than that of pure BVO. The charge collection and transportation lead a more reactive surface site to enhance photodegradation

Figure 7. Photocatalytic mechanism of palladium oxide doped $BiVO_4$ composite photocatalysts under visible-light illumination.
Reprinted with permission from Ref. (Ge, 2008) copyright from Elsevier

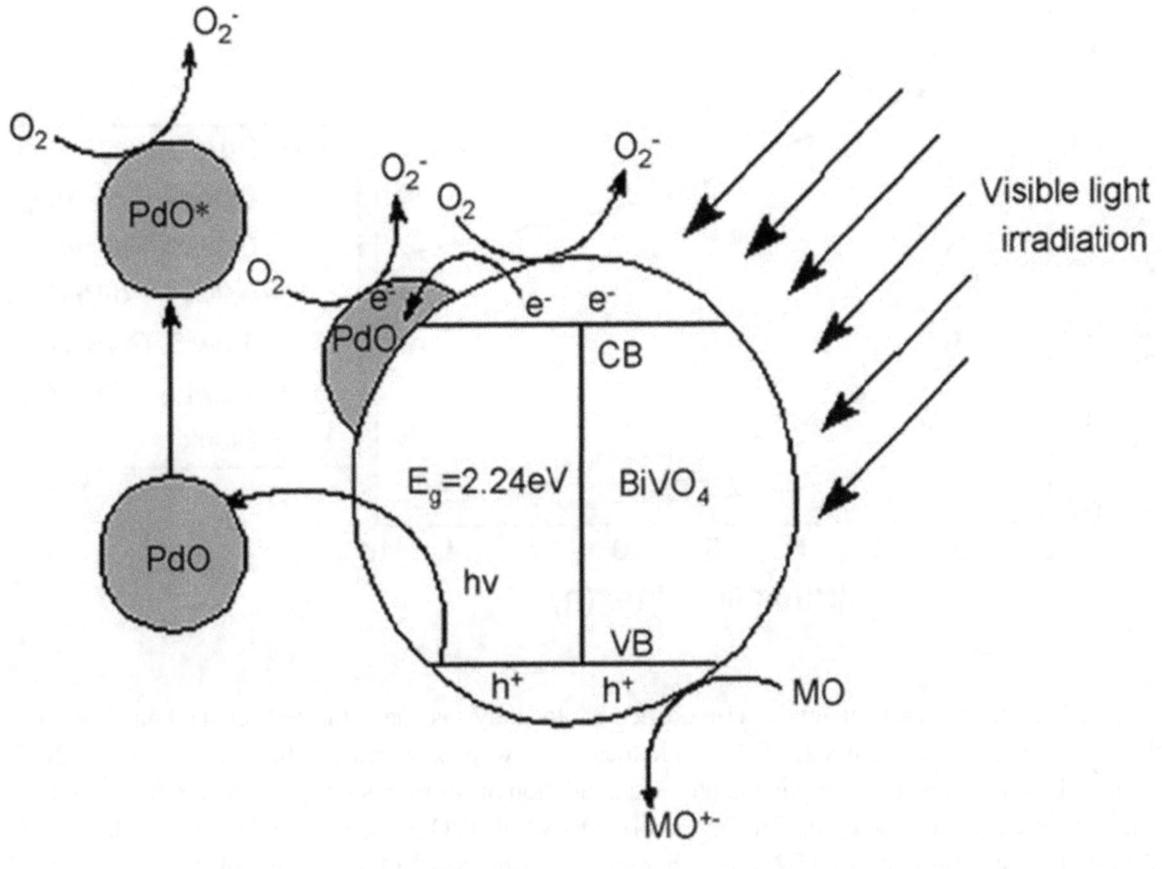

of the organic molecule through the addition of Pt. Ge et al. (Ge, 2008) synthesized visible-light-driven Pd-incorporated BVO photocatalyst via impregnation method. Figure 7 represents the photocatalytic mechanism of palladium oxide doped $BiVO_4$ composite in the presence of visible-light. The photo-decomposition of MO was examined with 1.0 wt% Pd/BVO which indicates higher efficacy (~100%) in comparison than that of pristine BVO (~42%) after 15 h of irradiation with visible-light as shown in Figure 8. This added element can act as an electron capturing agent for boosting the charge carrier separation and promoting the PC reaction rate using a Pd-based BVO sample.

The band gap energy of pure BVO and Pd/BVO can be calculated from DRS analysis, as shown in Figure 9. The light absorption ability of the composite increased and a red shift occurred after Pd species was incorporated, indicating improved PC performance for BVO samples after introducing the Pd species. The change in phase structure of the Pd/BVO photocatalyst was explored using XRD analysis. (Figure 10) The monoclinic phase of BVO exhibited improved PC activity under visible light irradiation and this phase must be invariant after Pd was added as a dopant.

Tl-modified BVO was developed by Karunakaran et al. (Karunakaran & Kalaivani, 2014) via the hydrothermal route to study the decay of MB in the presence of visible light. The authors reported that 19.3 at% Tl-modified BVO showed much better PC performance than others at % Tl-modified BVO and pure

Figure 8. The decrease of MO over different $BiVO_4$ and $Pd/BiVO_4$ composite photocatalysts (initial MO concentration 10 mg L^{-1}; pH = 3)
Reprinted with permission from Ref. (Ge, 2008) copyright from Elsevier

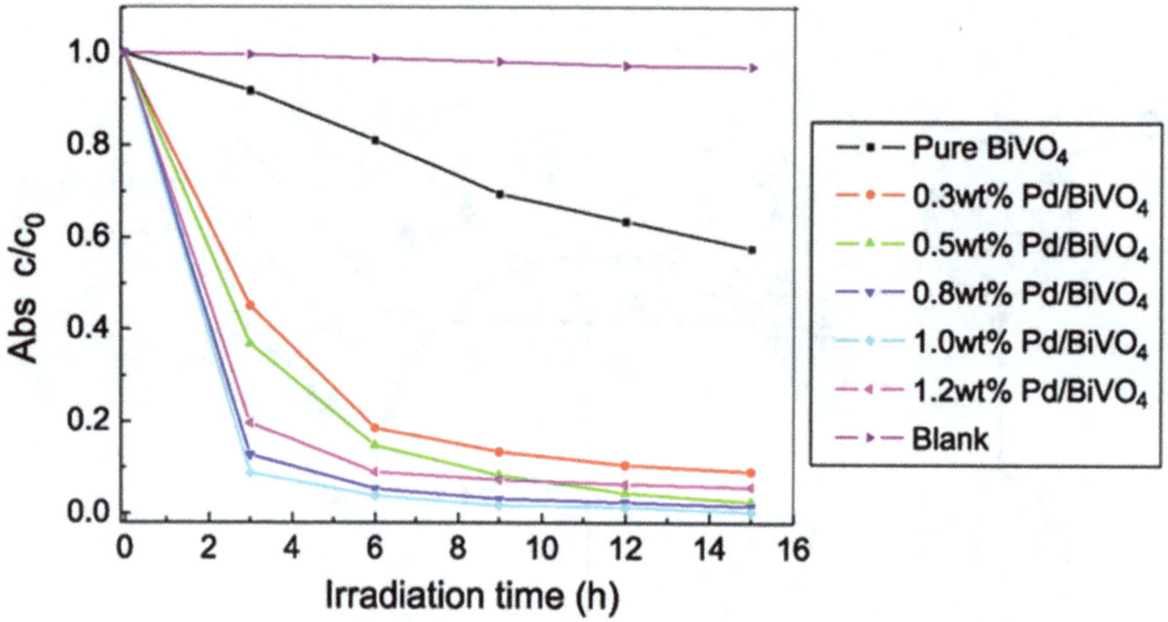

sample. The enhanced activity of the Tl-modified photocatalyst is due to the restricted recombination of photogenerated e^- and h^+ pairs. La-B/BVO photocatalysts were prepared by Min et al. (Min et al., 2013) through the sol-gel process. For photocatalytic degradation of MO a much higher rate constant i.e. 0.075 min^{-1}, was recorded for the optimized La-modified B-based BVO photocatalyst compare to the pristine BVO with a rate constant of 0.004 min^{-1} whereas only B-based BVO demonstrated the value as 0.007 min^{-1} under visible-light irradiation. The dramatic improvement in PC performance of the La-B modified BVO was assigned to the effective separation of photogenerated charge transporters due to the synergic effects of La and B in modifying the crystallites. Gao et al. (Gao et al., 2015) studied Nd-doped BVO composite with rod-like morphology, obtained via the hydrothermal method in the presence of sodium dodecyl benzenesulfonate (SDBS) as a surfactant with rod-like morphology. The authors demonstrated that although the crystallinity of BVO did not change significantly with the doping of Nd the amount of dopant controlled the photocatalytic (PC) performance of the composite. An optimal 0.8 wt% Nd-doped BVO composite was evaluated as the best desulfurization agent of the thiophene and degradation of phenol under visible light irradiation with almost 14 times higher in PC activity in comparison to pure BVO. By using the sol-gel process, Wang et al. (Wang et al., 2015) examined N and Sm co-doped BVO composite where the corn stem was used as a template. They analyzed the photodecomposition of MO under suitable illumination to optimize Sm modified N-based BVO composite. It achieved the enhancement of PCA in comparison to the any other BVO composite due to the cumulative impact of the N and Sm ions in the composite photocatalyst. Zhang et al. investigated europium-doped BVO (Eu/BVO) composite for employing it as visible-light-driven photocatalyst as prepared via hydrothermal route. The optimized 1.46 wt % Eu-based BVO composite was decolorized almost cent percent of MO solution with 180 min illumination. (Zhang & Zhang, 2010b) The light absorption in the visible region reduces

Figure 9. UV-Vis diffuses absorption spectra of photocatalysts: (a) $BiVO_4$; (b) $Pd/BiVO_4$.
Reprinted with permission from Ref. (Ge, 2008) copyright from Elsevier

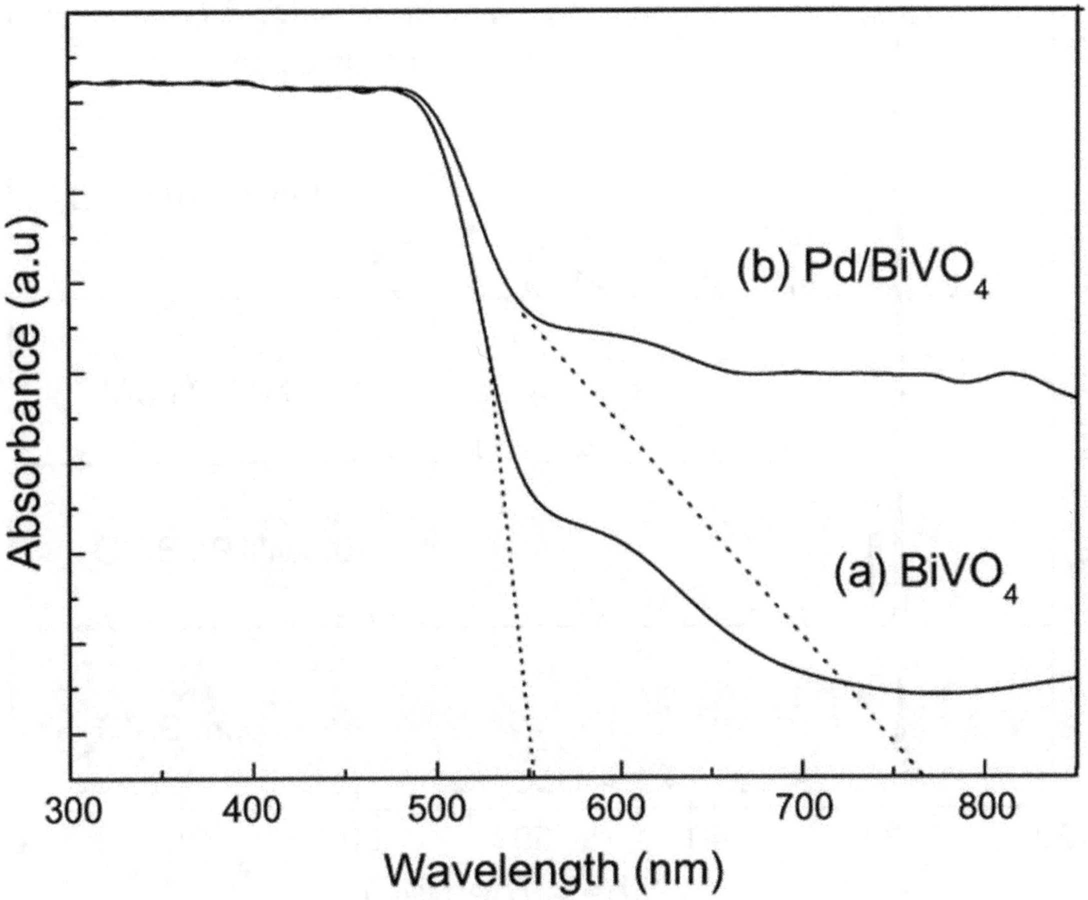

the recombination rate between the photoinduced charges to raise the PC activity of the composite. Xue et al. (Xue et al., 2017) prepared Eu-F-codoped BVO microspheres by using the hydrothermal method. This visible-light-driven photocatalyst (with a diameter of ~1-3 μm) is used for the degradation of Rh B as a noxious organic waste, as shown in Figure 11. The composite semiconductor revealed a gradual enhancement of PC performance than that of either bare BVO or individual europium-doped or fluorine-doped BVO materials. The synergistic effects of europium and fluorine co-dopant were recorded. As a result, PC performance was improved due to an increase in both surface area and separation proficiency of the photoactivated e^- and h^+.

Luo et al. (Luo et al., 2016) examined the PC efficiency of Gd-modified BVO photocatalyst as obtained through microwave-assisted hydrothermal route for use in photocatalyzed decay of Rh B under irradiation of visible light. All gadolinium doped BVO nanocomposite exhibits the tetragonal BVO phase whereas pure BVO crystallized in only monoclinic system, as confirmed by XRD analysis. After 120 min illumination, the optimal 10 at% Gd based BVO revealed ~96% degradation of Rh B. The enhanced PCA for Gd doped material could be the phase transition of BVO from monoclinic phase to tetragonal phase, as the formation of BVO tetragonal phase is facilitated through the incorporation of Gd metal.

Figure 10. XRD patterns of pure $BiVO_4$ and Pd/$BiVO_4$ composite photocatalysts.
Reprinted with permission from Ref. (Ge, 2008) copyright from Elsevier.

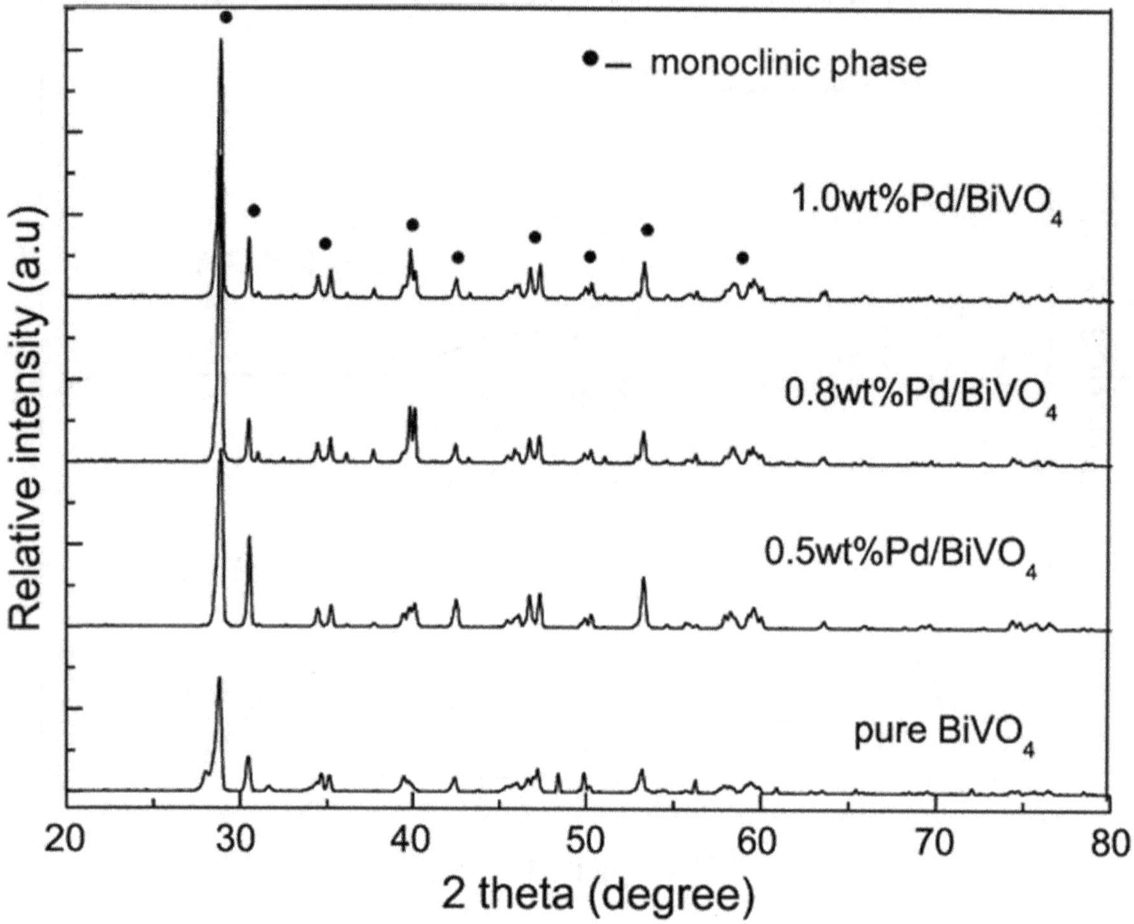

A C-based BVO with a hierarchical butterfly wing structure was fabricated by Yin et al. (Yin et al., 2013) through a sol-gel technique whereas template wings of butterfly were used for the utilization in visible-light-driven photocatalyst to degrade the MB. The improvement of photo-activated charge transporters revealed the optimized C/BVO photocatalyst as an enriched photocatalytic performer.

CONCLUSION

In this review, the photoelectrochemical water splitting process as well as the photocatalysis process for pollutant degradation of BVO photoanode has been discussed. This semiconductor attracts a lot of interest due to low band gap energy and facile synthetic routes to fabricate efficient photoexcited materials for solar renewable energy production. However, there are still remaining challenges regarding inadequate information for enhancing the stability of BVO. In this chapter, the role of surface morphology, co-catalyst, structural modification, and metal doping or heterojunction structure of BVO have been highlighted for H_2-O_2 gas production in PEC water splitting. Growth of visible light active efficient and

Figure 11. Comparisons of photocatalytic performance of bare $BiVO_4$ and doped $BiVO_4$ in degradation of Rh B under metal halide lamp irradiation. (a) Rh B concentration changes; (b) Degradation rate comparison of Rh B over different samples after 60 min light irradiation.
Reprinted with permission from Ref. (Xue et al., 2017) copyright from Elsevier.

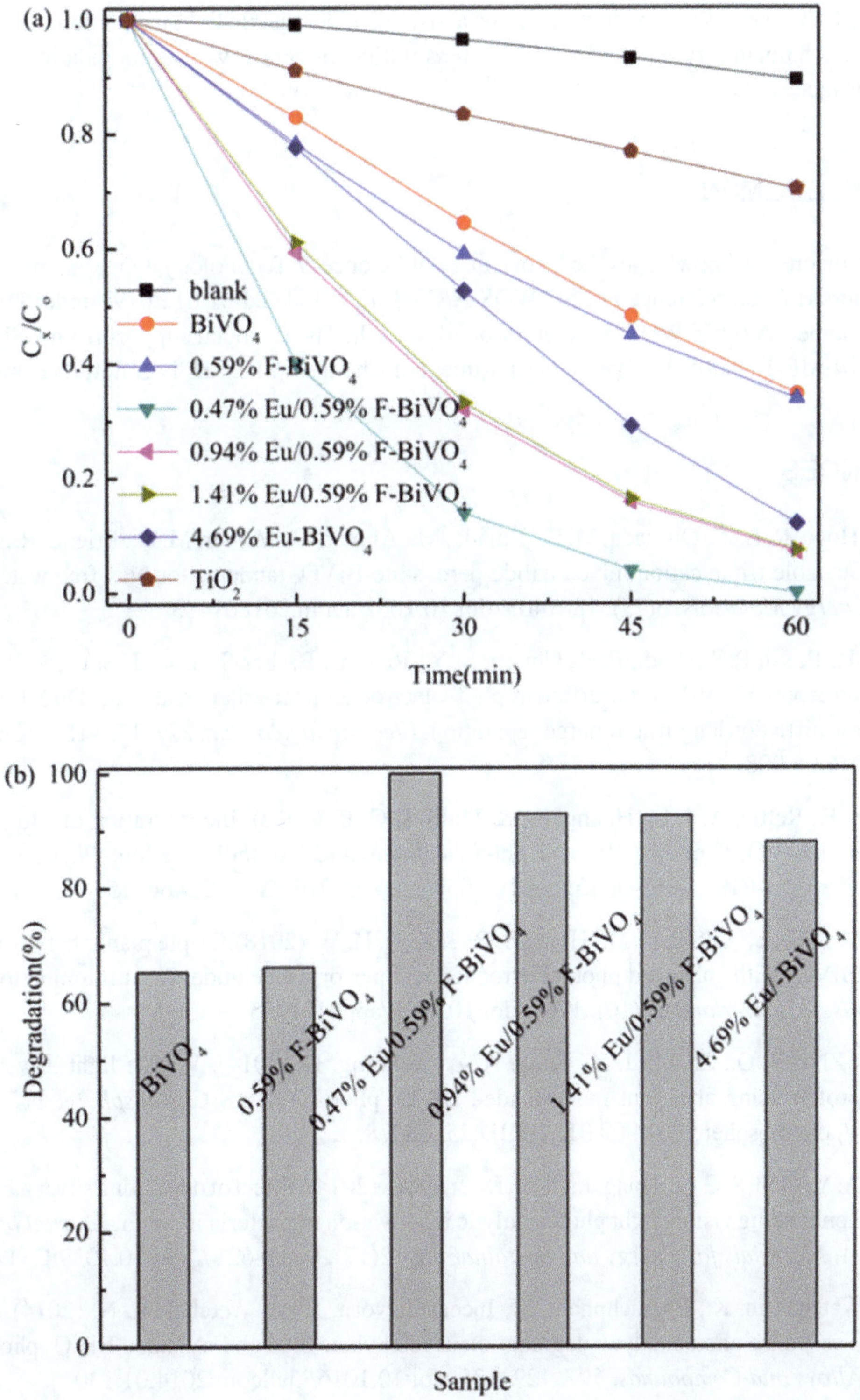

stable photocatalyst can degrade pollutants with an environmental impact. This review also indicates that further studies must be focused on the restriction of recombination between photogenerated electrons and holes to understand the better photocatalysis mechanism. A discussion over the growth of more effective morphologies, compositions, and photon absorption of BVO is also needed.

Since the BVO-based semiconductors can be activated under visible light irradiation, these materials would gain much popularity, especially for in water splitting process as well as degradation of pollutants in air and surface water.

ACKNOWLEDGMENT

The author sincerely acknowledges the Department of Science & Technology, Government of India, for financial support (vide reference no. SR/WOS-A/CS-10/2018, dated 02.01.2019) under DST-Women Scientist Scheme - A (DST-WOS-A) to carry out this work. The financial support from CSIR (File no. 01(2876)/17/EMR-II, 02.05.2017) to the Department of Chemistry, IIESTS is gratefully acknowledged.

REFERENCES

Andrei, V., Hoye, R. L. Z., Quesada, M. C., Bajada, M., Ahmad, S., Volder, M. D., Friend, R., & Reisner, E. (2018). Scalable triple cation mixed halide perovskite-$BiVO_4$ tandems for bias-free water splitting. *Advanced Energy Materials*, *8*(25), 1801403. doi:10.1002/aenm.201801403

Antony, R. P., Bassi, P. S., Abdi, F. F., Chiam, S. Y., Ren, Y., Barber, J., Loo, J. S. C., & Wong, L. H. (2016). Electrospun Mo-$BiVO_4$ for efficient photoelectrochemical water oxidation: Direct evidence of improved hole diffusion length and charge separation. *Electrochimica Acta*, *211*, 173–182. doi:10.1016/j.electacta.2016.06.008

Berglund, S. P., Rettie, A. J. E., Hoang, S., & Mullins, C. B. (2012). Incorporation of Mo and W into nanostructured $BiVO_4$ films for efficient photoelectrochemical water oxidation. *Physical Chemistry Chemical Physics*, *14*(19), 7065–7075. doi:10.1039/c2cp40807d PMID:22466715

Bhat, S. S. M., Lee, S. A., Suh, J. M., Hong, S. P., & Jang, H. W. (2018). Triple planar hetero-junction of $SnO_2/WO_3/BiVO_4$ with enhanced photoelectrochemical performance under front illumination. *Applied Sciences (Basel, Switzerland)*, *8*(10), 1765. doi:10.3390/app8101765

Bian, Z. Y., Zhu, Y. Q., Zhang, J. X., Ding, A. Z., & Wang, H. (2014). Visible-light driven degradation of ibuprofen using abundant metal-loaded $BiVO_4$ photocatalysts. *Chemosphere*, *117*, 527–531. doi:10.1016/j.chemosphere.2014.09.017 PMID:25268078

Booshehri, A. Y., Goh, S. C. K., Hong, J., Jiang, R., & Xu, R. (2014). Effect of depositing silver nanoparticles on $BiVO_4$ in enhancing visible light photocatalytic inactivation of bacteria in water. *Journal of Materials Chemistry. A, Materials for Energy and Sustainability*, *2*(17), 6209–6217. doi:10.1039/C3TA15392D

Chala, S., Wetchakun, K., Phanichphant, S., Inceesungvorn, B., & Wetchakun, N. (2014). Enhanced visible-light-response photocatalytic degradation of methylene blue on Fe-loaded $BiVO_4$ photocatalyst. *Journal of Alloys and Compounds*, *597*, 129–135. doi:10.1016/j.jallcom.2014.01.130

Chen, L., Huang, R., Ma, Y. J., Luo, S. L., Au, C. T., & Yin, S. F. (2013). Controllable synthesis of hollow and porous Ag/$BiVO_4$ composites with enhanced visible-light photocatalytic performance. *RSC Advances*, *3*(46), 24354–24361. doi:10.1039/c3ra43691h

Chen, X., Li, L., Yi, T., Zhang, W. Z., Zhang, X., & Wang, L. (2015). Microwave assisted synthesis of sheet-like Cu/$BiVO_4$ and its activities of various photocatalytic conditions. *Journal of Solid State Chemistry*, *229*, 141–149. doi:10.1016/j.jssc.2015.05.026

Cheng, J., Feng, J., & Pan, W. (2015). Enhanced Photocatalytic Activity in Electrospun Bismuth Vanadate Nano-fibers with Phase Junction. *ACS Applied Materials & Interfaces*, *7*(18), 9638–9644. doi:10.1021/acsami.5b01305 PMID:25856118

Cho, S. K., Park, H. S., Lee, H. C., Nam, K. M., & Bard, A. J. (2013). Metal doping of $BiVO_4$ by composite electrodeposition with improved photoelectrochemical water oxidation. *The Journal of Physical Chemistry C*, *117*(44), 23048–23056. doi:10.1021/jp408619u

Crabtree, G. W., Dresselhaus, M. S., & Buchanan, M. V. (2004). The hydrogen economy. *Physics Today*, *57*(12), 39–45. doi:10.1063/1.1878333

Cruz, A. M., & Pérez, U. M. G. (2010). Photocatalytic properties of $BiVO_4$ prepared by the co-precipitation method: Degradation of rhodamine B and possible reaction mechanisms under visible irradiation. *Materials Research Bulletin*, *45*(2), 135–141. doi:10.1016/j.materresbull.2009.09.029

Darwent, J. R., & Mills, A. (1982). Photo-oxidation of water sensitized by WO_3powder. *Journal of the Chemical Society. Faraday Transactions II*, *78*(2), 359–367. doi:10.1039/f29827800359

Dreyer, G., & Tillmanns, E. (1981). NeuesJahrb. Mineral. *Monatshe,* 151-154.

Drisya, K. T., López, M. S., Ramírez, J. J. R., Álvarez, J. C. D., Rousseau, A., Velumani, S., Asomoza, R., Kassiba, A., Jantrania, A., & Castaneda, H. (2020). Electronic and optical competence of TiO_2/$BiVO_4$ nanocomposites in the photocatalytic processes. *Scientific Reports*, *10*(1), 13507. doi:10.103841598-020-69032-9 PMID:32782289

Eisenberg, D., Ahn, H. S., & Bard, A. J. (2014). Enhancedphotoelectrochemical water oxidation on bismuth vanadate by electrodeposition of amorphous titanium dioxide. *Journal of the American Chemical Society*, *136*(40), 14011–14014. doi:10.1021/ja5082475 PMID:25243345

Fujishima, A., & Honda, K. (1972). Electrochemical photolysis of water at a semiconductor electrode. *Nature*, *238*(5358), 37–38. doi:10.1038/238037a0 PMID:12635268

Ganeshbabu, M., Kannan, N., Venkatesh, P. S., Paulraj, G., Jeganathan, K., & Ali, D. M. (2020). Synthesis and characterization of $BiVO_4$ nanoparticles for environmental applications. *RSC Advances*, *10*(31), 18315–18322. doi:10.1039/D0RA01065K PMID:35517221

Gao, X., Wang, Z., Zhai, X., Fu, F., & Li, W. (2015). Thesynthesize of lanthanide doped $BiVO_4$ and its enhanced photocatalytic activity. *Journal of Molecular Liquids*, *211*, 25–30. doi:10.1016/j.molliq.2015.06.058

Ge, L. (2008). Novel Pd/$BiVO_4$ composite photocatalysts for efficient degradation of methyl orange under visible light irradiation. *Materials Chemistry and Physics, 107*(2-3), 465–470. doi:10.1016/j.matchemphys.2007.08.016

Geng, Y., Zhang, P., Li, N., & Sun, Z. (2015). Synthesis of Co doped $BiVO_4$ with enhanced visible-light photocatalytic activities. *Journal of Alloys and Compounds, 651*, 744–748. doi:10.1016/j.jallcom.2015.08.123

Ginley, D., Green, M. A., & Collins, R. (2008). Solar energy conversion toward 1 terawatt. *MRS Bulletin, 33*(4), 355–364. doi:10.1557/mrs2008.71

Hernándeza, S., Gerardia, G., Bejtkab, K., Finaa, A., & Russo, N. (2016). Evaluation of the charge transfer kinetics of spin-coated $BiVO_4$ thin films for sun-driven water photoelectrolysis. *Applied Catalysis B: Environmental, 190*, 66–74. doi:10.1016/j.apcatb.2016.02.059

Ikeda, S., Tanaka, A., Shinohara, K., Hara, M., Kondo, J. N., Maruya, K., & Domen, K. (1997). Heterogeneous photocatalyst materials for water splitting. *Microporous and Mesoporous Materials, 9*, 253–258. doi:10.1016/S0927-6513(96)00112-5

Inoue, Y., Kohno, M., Kaneko, T., Ogura, S., & Sato, K. (1998). Dispersion of ruthenium oxide on barium titanates ($Ba_6Ti_{17}O_{40}$, $Ba_4Ti_{13}O_{30}$, $BaTi_4O_9$ and $Ba_2Ti_9O_{20}$) and photocatalytic activity for water decomposition. *Journal of the Chemical Society, Faraday Transactions, 94*(1), 89–94. doi:10.1039/a704947a

Intaphong, P., Phuruangrat, A., & Pookmanee, P. (2016). Synthesis and characterization of $BiVO_4$ photocatalyst by microwave method. *Integrated Ferroelectrics, 175*(1), 51–58. doi:10.1080/10584587.2016.1200910

Jiang, H., Dai, H., Meng, X., Zhang, L., Deng, J., Liu, Y., & Au, C. T. (2012). Hydrothermal fabrication and visible-light-driven photocatalytic properties of bismuth vanadate with multiple morphologies and/or porous structures for Methyl Orange degradation. *Journal of Environmental Sciences (China), 24*(3), 449–457. doi:10.1016/S1001-0742(11)60793-6 PMID:22655358

Jo, W. J., Jang, J. W., Kong, K. J., Kang, H. J., Kim, J. Y., Jun, H., Parmar, K. P. S., & Lee, J. S. (2012). Phosphatedoping into monoclinic $BiVO_4$ for enhanced photoelectrochemical water oxidation activity. *Angewandte Chemie International Edition, 51*(13), 3147–3151. doi:10.1002/anie.201108276 PMID:22344930

Joo, E. J., Park, G., Gwak, J. S., Seo, J. H., Jang, K. Y., Oh, K. H., & Nam, K. M. (2016). Efficient photoelectrochemical water oxidation by metal-doped bismuth vanadate photoanode with Iron oxyhydroxideelectrocatalyst. *Journal of Nanomaterials*, 1827151.

Kalanura, S. S., Yoo, I. L., Park, H. J., & Seo, H. (2017). Insights into the electronic bands ofWO_3/$BiVO_4$/TiO_2, revealing high solar water splitting efficiency. *Journal of Materials Chemistry. A, Materials for Energy and Sustainability, 5*(4), 1455–1461. doi:10.1039/C6TA07592D

Karunakaran, C., & Kalaivani, S. (2014). Enhanced visible light-photocatalysisby hydrothermally synthesized thallium-doped bismuth vanadate nanoparticles. *Materials Science in Semiconductor Processing, 27*, 352–361. doi:10.1016/j.mssp.2014.07.004

Kato, H., & Kudo, A. (2003). Photocatalytic water splitting into H_2 and O_2 over various tantalite photocatalysts. *Catalysis Today*, *78*(1-4), 561–569. doi:10.1016/S0920-5861(02)00355-3

Kong, D., Qi, J., Liu, D., Zhang, X., Pan, L., & Zou, J. (2019). Ni-doped $BiVO_4$ with V^{4+} species and oxygen vacancies for efficient photoelectrochemical water splitting. *Transactions of Tianjin University*, *25*(4), 340–347. doi:10.100712209-019-00202-1

Kudo, A., Omori, K., & Kato, H. (1999). A novel aqueous process for preparation of crystal form-controlled and highly crystalline $BiVO_4$ powder from layered vanadates at room temperature and its photocatalytic and photo physical properties. *Journal of the American Chemical Society*, *121*(49), 11459–11467. doi:10.1021/ja992541y

Kudo, A., Sayama, K., Tanaka, A., Asakura, K., Domen, K., Maruya, K., & Onishi, T. (1989). Nickel-loaded $K_4Nb_6O_{17}$photocatalyst in the decomposition of H_2O into H_2 and O_2: Structure and reaction mechanism. *Journal of Catalysis*, *120*(2), 337–352. doi:10.1016/0021-9517(89)90274-1

Kudo, A., Ueda, K., Kato, H., & Mikami, I. (1998). Photocatalytic O_2 evolution under visible light irradiation on $BiVO_4$ in aqueous $AgNO_3$ solution. *Catalysis Letters*, *53*(3/4), 229–230. doi:10.1023/A:1019034728816

Lei, B. X., Zeng, L. L., Zhang, P., Sun, Z. F., Sun, W., & Zhang, X. X. (2014). Hydrothermal synthesis and photocatalytic properties of visible-light induced $BiVO_4$ with different morphologies. *Advanced Powder Technology*, *25*(3), 946–951. doi:10.1016/j.apt.2014.01.014

Li, J., Chen, Y., Chen, C., & Wang, S. (2019). Solid-phase synthesis of visible-light-driven $BiVO_4$ photocatalyst and photocatalytic reduction of aqueous Cr (VI). *Bulletin of Chemical Reaction Engineering & Catalysis*, *14*(2), 336–344. doi:10.9767/bcrec.14.2.3182.336-344

Luo, W., Yang, Z., Li, Z., Zhang, J., Liu, J., Zhao, Z., Wang, Z., Yan, S., Yu, T., & Zou, Z. (2011). Solar hydrogen generation from seawater with a modified $BiVO_4$ photoanode. *Energy & Environmental Science*, *4*(10), 4046–4051. doi:10.1039/c1ee01812d

Luo, Y., Tan, G., Dong, G., Ren, H., & Xi, A. (2016). A comprehensive investigation of tetragonal Gd-doped $BiVO_4$ with enhanced photocatalytic performance under sun-light. *Applied Surface Science*, *364*, 156–165. doi:10.1016/j.apsusc.2015.12.100

Min, W., Yinsheng, C., Chao, N., Mingyan, D., & Duo, D. (2013). Lanthanum and boron co-doped $BiVO_4$ with enhanced visible light photocatalytic activity for degradation of methyl orange. *Journal of Rare Earths*, *31*(9), 878–884. doi:10.1016/S1002-0721(12)60373-1

Osterloh, F. E., & Parkinson, B. A. (2011). Recent developments in solar water splitting photocatalysis. *MRS Bulletin*, *36*(1), 17–22. doi:10.1557/mrs.2010.5

Park, H. G., & Holt, J. K. (2010). Recent advances in nano-electrode architecture for photochemical hydrogen production. *Energy & Environmental Science*, *3*(8), 1028–1036. doi:10.1039/b922057g

Park, H. S., Kweon, K. E., Ye, H., Paek, E., Hwang, G. S., & Bard, A. J. (2011). Factors in the metal doping of $BiVO_4$ for improved photoelectrocatalytic activity as studied by scanning electrochemical microscopy and first-principles density-functional calculation. *The Journal of Physical Chemistry C*, *115*(36), 17870–17879. doi:10.1021/jp204492r

Park, Y., McDonald, K. J., & Choi, K. S. (2013). Progress in bismuth vanadate photoanodes for use in solar water oxidation. *Chemical Society Reviews*, *42*(6), 2321–2337. doi:10.1039/C2CS35260E PMID:23092995

Pérez, U. M. G., Guzmán, S. S., Cruz, A. M. l., & Peral, J. (2012). Selective Synthesis of Monoclinic Bismuth Vanadate Powders by Surfactant-Assisted Co-Precipitation Method: Study of Their Electrochemical and Photocatalytic Properties. *International Journal of Electrochemical Science*, *7*, 9622–9632.

Pilli, S. K., Furtak, T. E., Brown, L. D., Deutsch, T. G., Turner, J. A., & Herring, A. M. (2011). Cobalt-phosphate (Co-Pi) catalyst modified Mo-doped $BiVO_4$photoelectrodes for solar water oxidation. *Energy & Environmental Science*, *4*(12), 5028–5034. doi:10.1039/c1ee02444b

Pookmanee, P., Kojinok, S., Punthaeod, R., Sangsricharan, S., & Phanichapat, S. (2013). Preparation and Characterization of $BiVO_4$ Powder by the Sol- gel Method. *Ferroelectrics*, *456*(1), 45–54. doi:10.1080/00150193.2013.846197

Pookmanee, P., Longchin, P., Kangwansupamonkon, W., Puntharod, R., & Phanichphant, S. (2013). Microwave-assisted synthesis Bismuth Vanadate ($BiVO_4$) Powder. *Ferroelectrics*, *455*(1), 35–42. doi:10.1080/00150193.2013.843414

Ran, R., McEvoy, J. G., & Zhang, Z. (2015). Synthesis and Optimization of Visible Light Active $BiVO_4$Photocatalysts for the Degradation of Rh B. *International Journal of Photoenergy*, 612857.

Reber, J. F., & Rusek, M. (1986). Photochemical hydrogen production with platinized suspensions of cadmium sulphide and cadmium zinc sulphide modified by silver sulphide. *Journal of Physical Chemistry*, *90*(5), 824–834. doi:10.1021/j100277a024

Regmi, C., Kim, T. H., Ray, S. K., Yamaguchi, T., & Lee, S. W. (2017). Cobalt-doped $BiVO_4$ (Co-$BiVO_4$) as a visible-light-driven photocatalyst for the degradation of malachite green and inactivation of harmful microorganisms in wastewater. *Research on Chemical Intermediates*, *43*(9), 5203–5216. doi:10.100711164-017-3036-y

Regmi, C., Kshetri, Y. K., Kim, T. H., Pandey, R. P., & Lee, S. W. (2017). Visible-light-induced Fe-doped $BiVO_4$photocatalyst for contaminated water treatment. *Mol. Catal.*, *432*, 220–231. doi:10.1016/j.mcat.2017.02.004

Regmi, C., Kshetri, Y. K., Kim, T. H., Pandey, R. P., Ray, S. K., & Lee, S. W. (2017). Fabrication of Ni-doped $BiVO_4$ semiconductors with enhanced visible-light photocatalytic performances for wastewater treatment. *Applied Surface Science*, *413*, 253–265. doi:10.1016/j.apsusc.2017.04.056

Sato, J., Saito, N., Yamada, Y., Maeda, K., Takata, T., Kondo, J. N., Hara, M., Kobayashi, H., Domen, K., & Inoue, Y. (2005). RuO_2-loaded β-Ge_3N_4 as a non-oxide photocatalyst for overall water splitting. *Journal of the American Chemical Society*, *127*(12), 4150–4151. doi:10.1021/ja042973v PMID:15783179

Sato, S. (1986). Photocatalytic activity of NO_x-doped TiO_2 in the visible light region. *Chemical Physics Letters*, *123*(1-2), 126–128. doi:10.1016/0009-2614(86)87026-9

Sayama, K., & Arakawa, H. (1994). Effect of Na_2CO_3 addition on photocatalytic decomposition of liquid water over various semiconductor catalysis. *Journal of Photochemistry and Photobiology A Chemistry*, *77*(2-3), 243–247. doi:10.1016/1010-6030(94)80049-9

Shi, L., Zhuo, S., Abulikemu, M., Mettela, G., Palaniselvam, T., Rasul, S., Tang, B., Yan, B., Saleh, N. B., & Wang, P. (2018). Annealing temperature effects on photoelectrochemical performance of bismuth vanadate thin film photoelectrodes. *RSC Advances*, *8*(51), 29179–29188. doi:10.1039/C8RA04887H PMID:35548013

Shi, W., Yan, Y., & Yan, X. (2013). Microwave- assisted synthesis of nano-scale $BiVO_4$ photocatalysts and their excellent visible-light-driven photocatalytic activity for the degradation of ciprofloxacin. *Chemical Engineering Journal*, *215*, 740–746. doi:10.1016/j.cej.2012.10.071

Sitaaraman, S. R., Shanmugapriyan, M. I., Varunkumar, K., Grace, A. N., & Sellappan, R. (2021). Synthesis of heterojunction tungsten oxide (WO_3) and Bismuth vanadate ($BiVO_4$) photoanodes by spin coating method for solar water splitting applications. *Materials Today: Proceedings*, *45*, 3920–3926. doi:10.1016/j.matpr.2020.07.601

Souza, J. S., Hirata, F. T. H., & Corio, P. (2019). Microwave-assisted synthesis of bismuth vanadate nano-flowers decorated with gold nanoparticles with enhanced photocatalytic activity. *Journal of Nanoparticle Research*, *35*, 1–9.

Tang, F., Cheng, W., Su, H., Zhao, X., & Liu, Q. (2018). Smoothing surface trapping states in 3D coral-like CoOOH-wrapped-$BiVO_4$ for efficient photoelectrochemical water oxidation. *ACS Applied Materials & Interfaces*, *10*(7), 6228–6234. doi:10.1021/acsami.7b15674 PMID:29384358

Tayyebi, A., Soltani, T., & Lee, B. K. (2019). Effect of p^H on photocatalytic and photoelectrochemical (PEC) properties of monoclinic bismuth vanadate. *Journal of Colloid and Interface Science*, *534*, 37–46. doi:10.1016/j.jcis.2018.08.095 PMID:30205253

Walter, M. G., Warren, E. L., McKone, J. R., Boettcher, S. W., Mi, Q., Santori, E. A., & Lewis, N. S. (2010). Solar water splitting cells. *Chemical Reviews*, *110*(11), 6446–6473. doi:10.1021/cr1002326 PMID:21062097

Wang, G. L., Shan, L. W., Wu, Z., & Dong, L. M. (2017). Enhanced photocatalytic properties of molybdenum-doped $BiVO_4$ prepared by sol-gel method. *J. Alloys Comp.*, *36*(2), 129–133. doi:10.100712598-015-0669-0

Wang, M., Guo, P., Chai, T., Xie, Y., Han, J., You, M., Wang, Y., & Zhu, T. (2017). Effects of Cu dopants on the structures and photocatalytic performance of cocoon-like Cu-$BiVO_4$ prepared via ethylene glycol solvothermal method. *Journal of Alloys and Compounds*, *691*, 8–14. doi:10.1016/j.jallcom.2016.08.198

Wang, M., Niu, C., Liu, J., Wang, Q., Yang, C., & Zheng, H. (2015). Effective visible light- active nitrogen and samarium co-doped $BiVO_4$ for the degradation of organic pollutants. *Journal of Alloys and Compounds*, *648*, 1109–1115. doi:10.1016/j.jallcom.2015.05.115

Wu, Y. T., Lin, L. Y., Tao, S. M., Chen, Y. S., Ma, J. S., Lee, P. Y., & Sung, Y. S. (2020). Novel in situ synthesis of $BiVO_4$photocatalyst/$Co_3(PO_4)_2$ co catalyst powder via the one step solid-state process for photoelectrochemicalcatalyzing water oxidation. *ACS Sustainable Chemistry & Engineering*, *8*(7), 2948–2956. doi:10.1021/acssuschemeng.9b07517

Xu, X., Du, M., Chen, T., Xiong, S., Wu, T., Zhao, D., & Fan, Z. (2016). New insights into Ag-doped $BiVO_4$ microspheres as visible light photocatalysts. *RSC Advances*, *6*(101), 98788–98796. doi:10.1039/C6RA20850A

Xue, S., He, H., Wu, Z., Yu, C., Fan, Q., Peng, G., & Yang, K. (2017). An interesting Eu, F-codoped $BiVO_4$ microsphere with enhanced photocatalytic performance. *Journal of Alloys and Compounds*, *694*, 989–997. doi:10.1016/j.jallcom.2016.10.146

Yan, M., Yan, Y., Wu, Y., Shi, W., & Hua, Y. (2015). Microwave-assisted synthesis of monoclinic-tetragonal $BiVO_4$ hetero-junctions with enhanced visible-light-driven photocatalytic degradation of tetracycline. *RSC Advances*, *5*(110), 90255–90264. doi:10.1039/C5RA13684A

Yao, W., Iwai, H., & Ye, J. (2008). Effects of molybdenum substitution on the photocatalyticbehaviour of $BiVO_4$. *Dalton Transactions (Cambridge, England)*, (11), 1426–1430. doi:10.1039/b713338c

Ye, K. H., Yu, X., Qiu, Z., Zhu, Y., Lu, X., & Zhang, Y. (2015). Facile synthesis of bismuth oxide/bismuth vanadate hetero-structures for efficient photoelectrochemical cells. *RSC Advances*, *5*(43), 34152–34156. doi:10.1039/C5RA03500G

Yin, C., Zhu, S., Chen, Z., Zhang, W., Gu, J., & Zhang, D. (2013). One step fabrication of C-doped $BiVO_4$ with hierarchical structures for a high-performance photocatalyst under visible-light irradiation. *Journal of Materials Chemistry. A, Materials for Energy and Sustainability*, *1*(29), 8367–8378. doi:10.1039/c3ta11833a

Yu, J., & Kudo, A. (2005). Hydrothermal Synthesis of Nano-fibrous Bismuth Vanadate. *Chemistry Letters*, *34*(6), 850–851. doi:10.1246/cl.2005.850

Zhang, A., & Zhang, J. (2010a). Characterization and photocatalytic properties of Au/$BiVO_4$ composites. *Journal of Alloys and Compounds*, *491*(1-2), 631–635. doi:10.1016/j.jallcom.2009.11.027

Zhang, A., & Zhang, J. (2010b). Effects of europium doping on the photocatalyticbehaviour of $BiVO_4$. *Journal of Hazardous Materials*, *173*(1-3), 265–272. doi:10.1016/j.jhazmat.2009.08.079 PMID:19729243

Zhong, D. K., Choi, S., & Gamelin, D. R. (2011). Near-complete suppression of surface recombination in solar photoelectrolysis by "Co-Pi" catalyst-modified W:$BiVO_4$. *Journal of the American Chemical Society*, *133*(45), 18370–18377. doi:10.1021/ja207348x PMID:21942320

Chapter 2
Photodegradation of Dyes in Visible Light by TiO2/PPy/GO Nanocomposites

Azad Kumar
https://orcid.org/0000-0001-5162-7768
M.L.K. P.G. College, India

ABSTRACT

In this research, a proficient method for the synthesis of TiO2/PPy and TiO2/PPy/GO nanocomposites is explored. These nanocomposites were prepared by one-step in situ deposition oxidative polymerization of pyrrole hydrochloride using Ammonium persulfate (APS) as an oxidant in the presence of TiO2 nanoparticles cooled in an ice bath. The obtained nanocomposites were characterized by XRD, TEM, SEM, UV-Vis, and FTIR techniques. The obtained results showed that TiO2 nanoparticles have been encapsulated by PPy with a strong effect on the morphology of TiO2/PPy and TiO2/PPy/GO nanocomposites. The photocatalytic degradation of rose Bengal, and Victoria blue dye was done under different conditions regarding concentration of dye, time of illumination, pH, and dose of the photocatalyst. The maximum photodegradation was found at 7 pH, 20 ppm concentration of VB and 25 ppm of RB dye solution, 800 mg/L for VB and 1600 mg/L for RB amount of photocatalyst, and 120 min irradiation of visible light. Kinetics of photodegradation were investigated for Victoria blue and rose Bengal dye and found first order kinetics.

INTRODUCTION

The industrial wastewater containing the various dyes are responsible for water pollution, due to their carcinogenic behaviour. Lots of investigations reported that 10-12% of dyes are used every year in textile industries such as Rose Bengal, Victoria blue, Thymol blue, Caramine, Indigo Red, Red 120, Rhodamine B, Methylene Blue, Eriochrome Black-T (EBT) (Kaur & Singhal 2014; Fox & Duxbury 1993; Mai et al., 2008; Azmi, et al., 1998) of which are major portion (20%) lost during synthesis and processing operations, which enter into water through effluents. The wastewater released from industries contains

DOI: 10.4018/978-1-6684-4553-2.ch002

highly hazardous and coloured pigments which causes serious ill effect on aquatic life and also human beings. Rose bengal is a significant xanthene dye used in textile and photochemical industries whose molecular structure as shown in Figure 1(a). It has severe toxic effects on the human health (Konstantinou & Albanis, 2004; Deng et al., 2013; Xu et al., 2010) and become dangerous contact with skin and causing itchiness, irritation, reddening and blistering. It also affects to eyes causing inflammation, eye redness, itching etc (Wang et al., 2008). Victoria blue is another triphenylmethane derivatized dye which is extensively used in the textile industry. It has been extensively used as textile dyes for silk, wool, and cotton, in the preparation of inks and in the surface-coating and dyeing of paper (Park et al., 2010; Liang et al., 2009; Ferreira & Li, 2001) as colorants in foods, drugs, cosmetics (Saquib & Muneer, 2003), as biological stains, and as anti-infective, antimicrobial, and anthelmintic agents (Li et al., 1999). The photocytotoxicity of triphenylmethane dyes, based on the production of the reactive oxygen species, is tested extensively with the regard to their photodynamic treatment (Deng et al., 2012). In the past, several physical techniques like photodegradation, coagulation, flocculation, reverse osmosis, adsorption on the activated carbon, ion exchange method ultra-filtration and chemical methods like photosensitized oxidation, adsorption (Chen et al., 2007), have been used to reduce the toxic dye effluents from wastewater (Okano et al., 1987; Wang et al., 2012; Wang et al., 2008). These methods are fairly effective in removing pollutants. However, the main drawback of these techniques is formation of secondary waste product which cannot be treated again and dumped as such (Sedlačík et al., 2012)

Titanium dioxide (TiO_2) Nanoparticles (NPs) have many excellent properties, such as low cost, simple preparation, good stability, non-toxicity, and better photodegradation ability (Chen et al., 2007), making it a good prospect for application in solar cells (Linsebigler et al., 1995) photocatalysis (Tai et al., 2007), and photocatalytic hydrogen production (Arenas et al., 2013). The wide band gap of pure TiO_2 nanoparticles have low sunlight energy conversion efficiency but also a high rate of photogenerated hole and electron recombination. Therefore, improvement of the photocatalytic properties of TiO_2 is essential.

In order to enhance the photocatalytic properties of TiO_2 number of manipulation such as metal or non-metal doping (Nag et al., 2012), compositing with other semiconductors (Vinodgopal et al., 1996), compositing with conductive materials such as graphene (Huang et al., 2015) or carbon nanotubes (Chen et al., 2007), sensitization with organic dyes and conductive polymers, such as polyaniline (Li et al., 2013), polythiophene (Deng et al., 2012), and polypyrrole (PPy) have been tested in the past (Sun et al., 2013). TiO_2 (NPs) with among these routes, the coating of conductive polymers is one of the most effective methods for the generation of good photocatalyst (Yang et al., 2013). The coating of conductive polymers can reduce the recombination rate of electron hole (e^--h^+) simultaneously (Vautier et al., 2001) act as sensitizer making effective large band gap semiconductor like TiO_2 (Mor et al., 2006) Among the various conductive polymers, PPy is one of the most promising coating agents owing to its good conductivity, high absorption coefficient in the visible part of sunlight, high charge carrier mobility and good environmental stability (Thompson et al., 2006). Therefore, PPy is suitable conducting polymer stable photosensitizer to improve the photocatalytic activity of TiO_2 in solar light. Over TiO_2 surface further, the TiO_2/PPy nanocomposite is successively used in solar cells and for the photocatalytic degradation of organic species, however its application in photodegradation of organic dyes is rarely reported.

There are many methods of preparing TiO_2/PPy nanocomposite, for example, anodic co-deposition (Kar et al., 2009), self-assembly techniques (Baran et al., 2003), photoelectrochemical polymerization (Wei et al., 2011), and hydrothermal methods (Matthews, 1988). However, in- situ chemical polymerization is promise method for preparation of TiO_2/PPy owing to its simplicity, good reproducibility, and easy scale up.

Figure 1. Molecular Structure of dyes (a) Victoria Blue (b) Rose Bengal

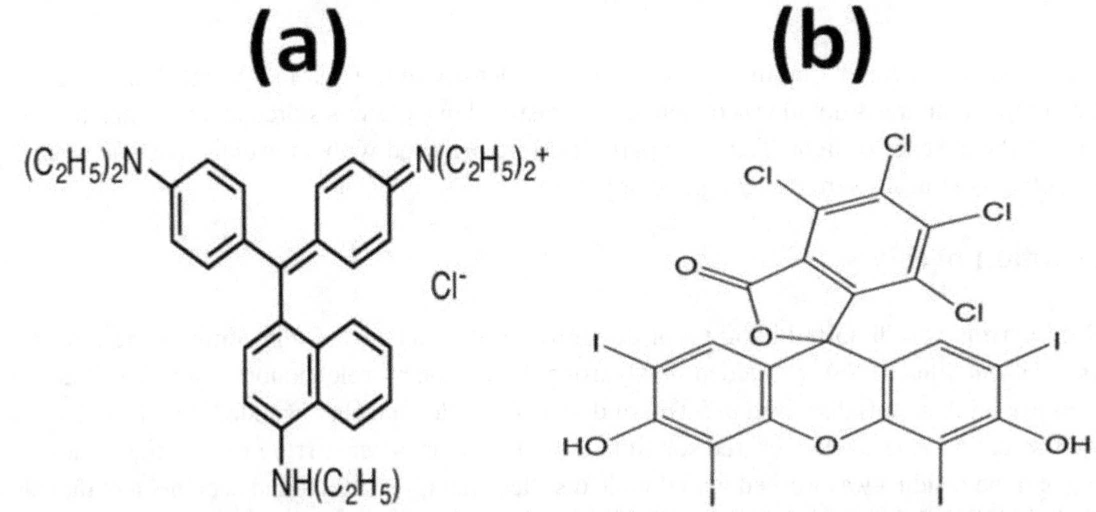

In present study, TiO_2, TiO_2/PPy and TiO_2/PPy/GO nanocomposites have been preared by one-step *in situ* polymerization of pyrrole in the reaction medium. The photocatalytic degradation of Victoria Blue and Rose Bengal have been studied in the presence of TiO_2, TiO_2/PPy and TiO_2/PPy/GO nanocomposites at different parameters i.e., concentration of dye, dose of photocatalyst, pH of reaction mixture and irradiation time. The kinetics for the photodegradation process has also been investigated in this work.

Table 1. Showing the structure and properties of dyes

Dye	Victoria Blue	Rose Bengal
Structure	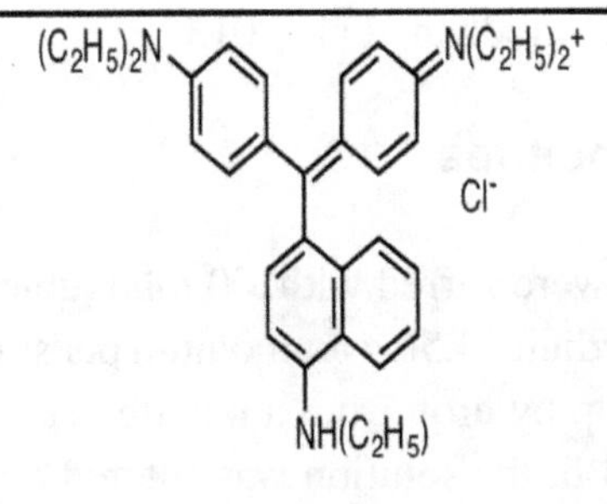	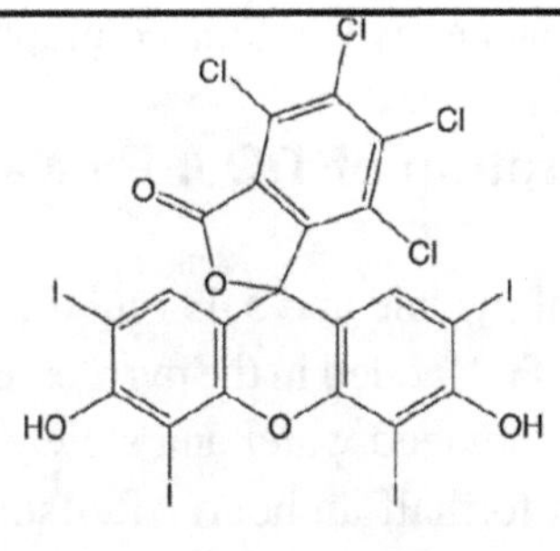
Molar mass:	$[C_{33}H_{40}N_3]Cl$. 507.5 g/mol	$C_{20}H_4Cl_4I_4O_5$ g/mol 973.67
λ_{max}	614 nm	547 nm

EXPERIMENTAL

Materials

Pyrrole monomer, having molar mass of 67 g/mol and density of 0.97 g/cm^3 (Merck India) was triply distilled until a colourless liquid was obtained. The distilled pyrrole was stored at lower than 5 °C temperature in the absence of light. TiO_2 nanoparticles were prepared with an average particle size of 50 nm. All other chemicals were reagent grade or purer.

Preparation of PPy

1.727 ml pyrrole was dissolved in 50 ml of de-ionized water and stirred for 15 min using a magnetic stirrer. 2.717 ml dilute H_2SO_4 was added slowly using drop to the pyrrole monomer solution. 2.28 g ammonium per sulphate was dissolved in 50 ml of de-ionised water and slowly added drop by drop from a burette vertically to the above prepared solution for half an hour. After stirring for 4 h, the solution was filtered and the residual was washed with double distilled water, methanol, and acetone, and then dried in an oven at 60°C. Subsequently the product is grinded to get powder of polypyrrole (Kwon et al., 2004).

Preparation of Graphene Oxide

Graphene oxide was synthesized from graphite powder using a modiðed Hummer's method. In brief, ðrst, 0.5 g of powdered ñake of graphite and 0.5 g of $NaNO_3$ were added into 24 mL of H_2SO_4 and were stirred until dissolved. Then, 3 g of $KMnO_4$ was added slowly, preventing the temperature of the suspension from exceeding 20 °C. After the mixture was stirred continuously for 1 h at 35 °C, 40 ml of distilled water was slowly added to dilute the mixture and the temperature was raised to 90 °C. To reduce the residual permanganate and manganese dioxide to colourless soluble manganese sulphate, 5 ml of 34.5% H_2O_2 was added, and the suspension was ðltered with distilled water until pH 7.0. The obtained yellow-brown suspension was exfoliated to produce single layer graphene oxide using a sonicator, and the unexfoliated precipitation was removed by centrifugation. Finally, we obtained a brown dispersion of homogeneously exfoliated graphene oxide (Pradhan et al., 2013).

Preparation of TiO_2/PPy Nanocomposites

3.454 ml pyrrole and 5.434 ml dilute H_2SO_4 were stirred with100 ml double distilled water and further 1.036 g TiO_2 added in the pyrrole reaction medium. 4.56 g ammonium per sulphate was dissolved in 100 ml of de-ionized water and slowly added drop by drop from a burette vertically to the above prepared solution for half an hour. After stirring for 4 h, the solution was filtered and the residual was washed with double distilled water, methanol and acetone, and then dried in an oven at 60 °C. This was grinded into powder (Guo et al., 2009).

Preparation of TiO_2/PPy/GO Nanocomposites

3.454 ml pyrrole and 5.434 ml dilute H_2SO_4 were stirred with double distilled water and further 1.0362 g TiO_2 and 60 mg graphene oxide added in the pyrrole reaction medium. 4.56 g ammonium per sulphate was dissolved in 100 ml of de-ionized water and slowly added drop by drop from a burette vertically to the above prepared solution for half an hour. After stirring for 4 h, the solution was filtered and the residual was washed with double distilled water, methanol and acetone, and then dried in an oven at 90°C.This was grinded into powder (Pare et al., 2009).

Characterizations

The prepared TiO_2, PPy, TiO_2/PPy and TiO_2/PPy/GO nanocomposites were characterized by x-ray diffraction (XRD) patterns in the range of $2\theta = 20–80°$ ((Cullity et al., 2001). The particle size of TiO_2 and morphology of particles was investigated with transmission electron microscope (TEM). The morphology of neat TiO_2, PPy, and TiO_2/PPy nanocomposites was investigated by scanning electron microscopy. Diffused Reflectance spectroscopy was used to determine band gap energy of prepared nanocomposites and Photoluminescence was used for the hydroxyl radical mechanism determination and e^--h^+ recombination determination.

Irradiation Procedure

The mixture of dye and photocatalysts TiO_2, PPy, TiO_2/PPy and TiO_2/PPy/GO suspension was stirred in the dark for 30 min to reach adsorption equilibrium. The desired concentration of Victoria Blue and Rose Bengal dye (20 mL) aqueous solutions in a 100 mL beaker were taken. The Irradiation of visible light on the surface of the reaction medium with constant stirring for 30 to 180 min. Degradations were performed The amount of nanocomposites material varies from 100 mg/L to 1600 mg/L at different pH values. Irradiations were carried out using one lamp (500 W). At any given irradiation time interval, the dispersion was sampled (5 mL), centrifuged, and subsequently filtered through a Millipore filter to separate the TiO_2 particles and take UV- Vis spectra to determine the residual concentration (Gole et al., 2004).

Photo-Degradation of Dyes

The photocatalytic degradation of Victoria Blue and Rose Bengal has been studied in the presence of TiO_2, TiO_2/PPy and TiO_2/PPy/GO nanocomposites at different concentration of dye solution. The solution of dye was prepared in 10:1 (V/V) ratio of water and alcohol. The known amount of photocatalyst was dispersed in the dye solution (20 ml) and the reaction mixture was irradiated with visible light with constant stirring on magnetic stirrer. After different time interval, an aliquot of solution was separated, centrifuged and absorption was recorded spectrophotometrically. The results obtained for the % degradation of Victoria Blue and Rose Bengal and Rose Bengal are shown in Figures 6-9 (Vautier et al., 2001; Reddy et al., 2002). The % degradation efficiency of dye was calculated by Equation (1).

$$\% = \frac{A_O - A_F}{A_O} * 100 \qquad (1)$$

Where η is the degradation efficiency, A_0 is the initial absorbance and A_F is the final absorbance

Figure 2. X-Ray diffraction of (a) TiO_2 (b) Pure PPy (c) TiO_2/PPy (d) TiO_2/PPy/GO

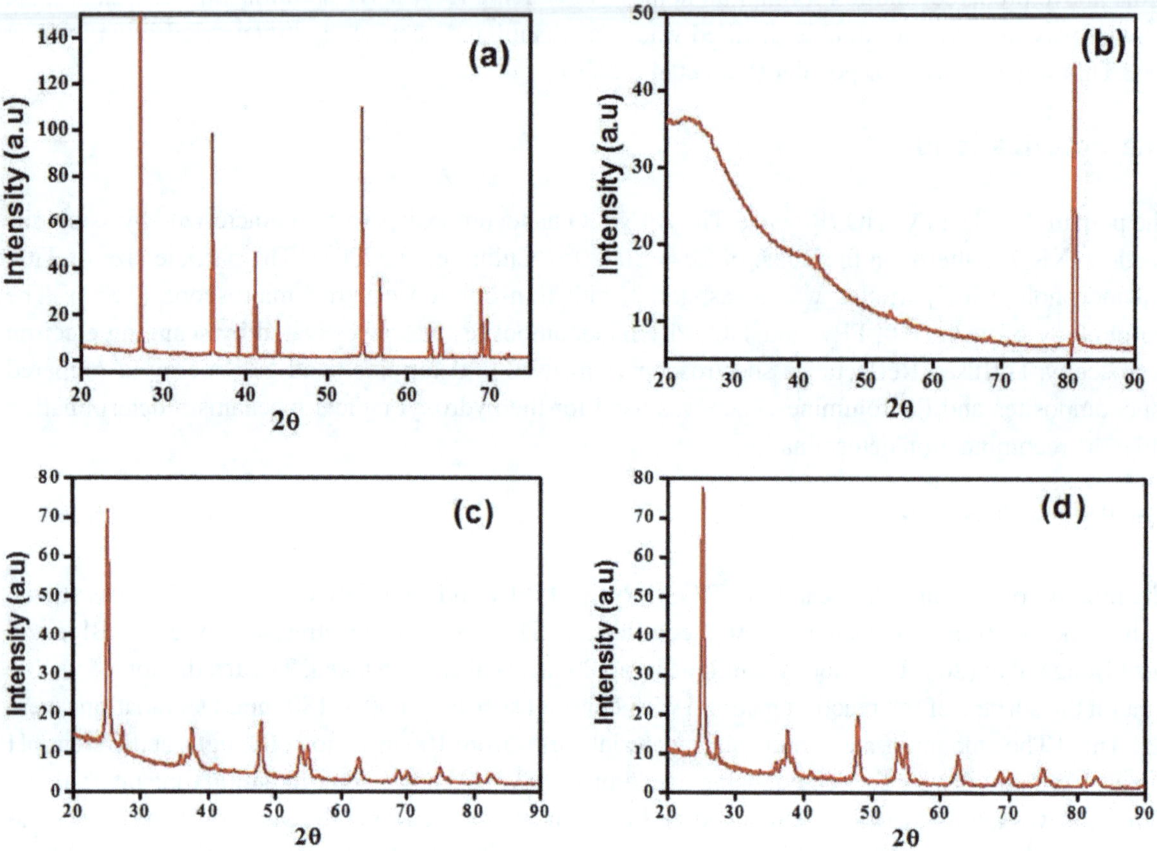

RESULTS AND DISCUSSION

XRD Patterns of Nanocomposites

XRD patterns of pure TiO_2, PPy, TiO_2/PPy and the TiO_2/PPy/GO synthesized nanocomposites are presented in Figure 2. The main diffraction peaks of pure TiO_2 nanoparticles are corresponding to 25°, 38°, and 48° (Figure 2a). The broad peak in the region of $2\theta = 10–30°$ in XRD pattern of pure PPy (Figure 2b) shows that the synthesized PPy in the absence of TiO_2 nanoparticles is amorphous. It can be seen from Figures 2c–d that the main peaks of TiO_2/PPy nanocomposites are similar to those of neat TiO_2 nanoparticles. XRD patterns of TiO_2/PPy nanocomposites show that the broad weak diffraction peak of PPy still exists, but its intensity has been decreased. It implies that when pyrrole is polymerized on TiO_2, each phase maintains his initial structure (Cullity et al., 2001, Gole et al., 2004).

Determination of Average Size of Particles/Grains in Samples

Utilizing the observed X-ray diffraction data of samples, Scherrer's calculations were attempted to know the average size of particles/grains in the samples (Sarmah and Kumar, 2011). Although, Scher-

Table 2. Average size of particles/grains in the samples of TiO_2 and TiO_2, TiO_2/PPy, TiO_2/PPy/GO

Sample	Particle Size
TiO_2	19
TiO_2/PPy	24
TiO_2/PPy/GO	30

rer's calculations are only approximate in nature, but definitely provide a first-hand idea of the average size of the particles/ grains in the samples, which may be quite accurate, provided the size of particles/ grains is below 100 nm.

$$B = \frac{0.9\lambda}{t \cos\theta} \quad (2)$$

The mean size of TiO_2, TiO_2/PPy and TiO_2/PPy/GO nanocomposites, calculated by Scherrer's Equation, are about 19, 24 nm and 30 nm, respectively. The results of Scherrer's calculations are presented in Table 2. The results suggest average size of the particles/ grains in the samples lying in nm range. The result is in good agreement with the TEM.

Transmission Electron Microscopy

TEM analysis of nanocomposites gives the size and shape of the particles on the scale of atomic diameters. Figure 3 shows the TEM image of TiO_2, PPy, TiO_2/PPy and the TiO_2/PPy/GO prepared by wet synthesis. It is observed that the particles are found in the 100 nm and are mostly of cube and hexagonal shape. Figure 3(a) shows that the particles size of TiO_2/PPy was found in 100 nm and Figure 3(b) shows that the particles size of TiO_2/PPy/graphene oxide was found in 100 nm (Kaur & Singhal, 2014).

SEM of Nanocomposites

The morphology and shape of pure TiO_2 nanoparticles, neat PPy, and TiO_2/PPy nanocomposites were characterized by SEM instrument and the obtained pictures are presented in Figure 4. As shown in Figure 4(a), TiO_2 nanoparticles were aggregated due to their high surface energy. Figure 4(c) shows that the morphology of composite at low content of TiO_2 (0.025 g TiO_2) is much like that of neat PPy and with increasing contents of TiO_2, the morphology of composites appears as nanoparticles. It indicates that the TiO_2 nanoparticles have a nucleus effect on the pyrrole polymerization and caused a homogeneous PPy shell around them. The SEM images help us draw a conclusion that the doping of TiO_2 has a strong effect on PPy's morphology and with the increase of TiO_2 contents, the composites show a transformation in morphology from typical PPy to nanoparticles (Pradhan et al., 2013).

Determination of Optical Band Gap of Nanocomposites

The band gap of polypyrrole (polypyrrole/TiO_2) was determined from absorption spectra and Tauc relation (Eq. 3)

Figure 3. TEM of (a) TiO_2 *(b) Pure PPy (c)* TiO_2*/PPy (d)* TiO_2*/PPy/GO*

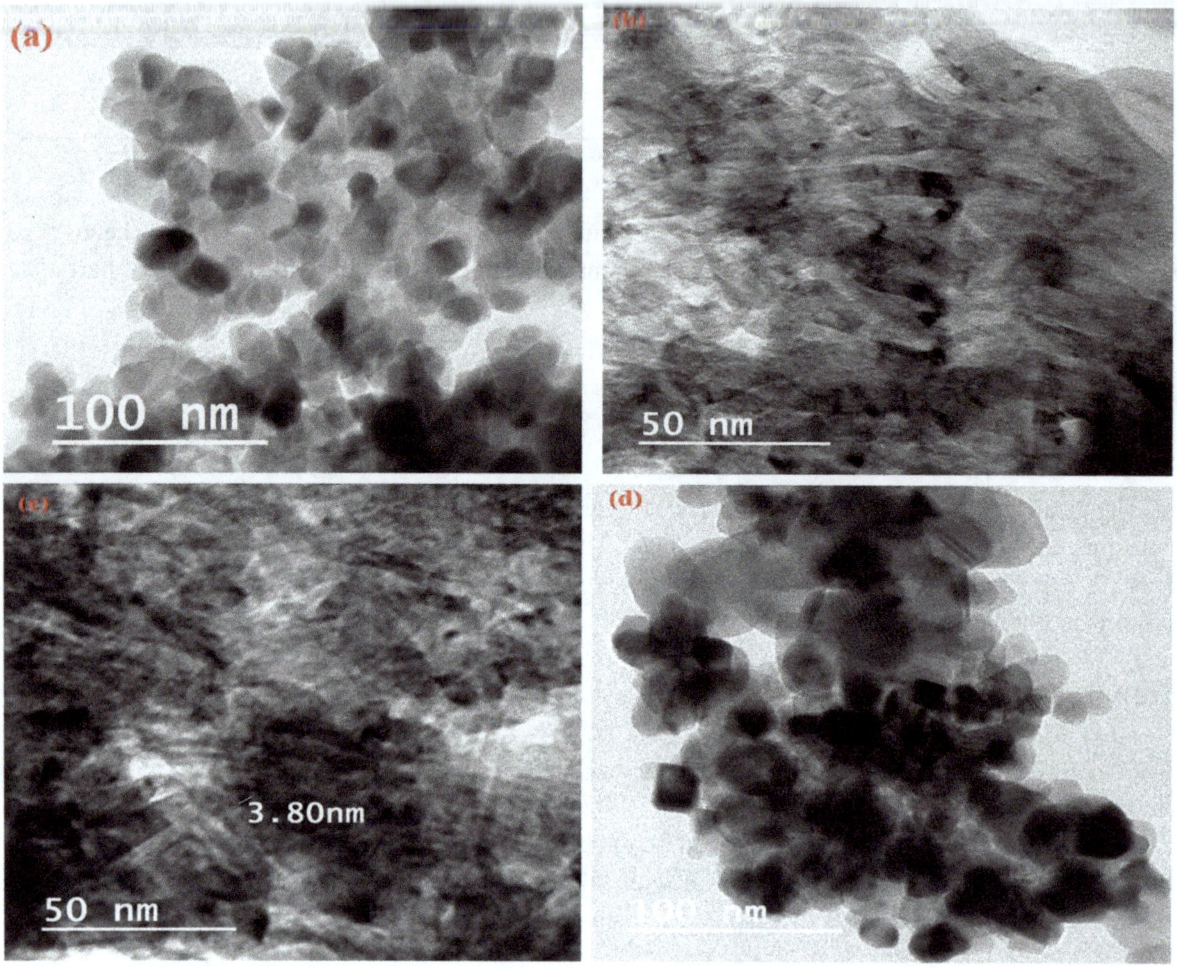

$$a.hv = B(hv - Egap)^m \tag{3}$$

where α is the absorption coefficient, hν is the photon energy, and $m = 1/2$ for direct band gap material shown in Figure 5. To describe a direct method for fitting and determination of band gap using Tauc relation (Epling & Lin 2002). After substitutions in Eq. (2), we can write that equation (4)

$$Abs.\lambda\frac{1}{m} = B\left(\frac{1}{\lambda}-\frac{1}{\lambda_{gap}}\right) \tag{4}$$

where λ is the wavelength and Abs. the corresponding value of measured absorbance. λ_{gap} can be easily obtained from curve $(Abs./\lambda)^{1/}m$ vs. $1/\lambda$ at condition $(Abs./\lambda)^{1/}m = 0$. The band gap value is obtained from relation $E_{gap} = 1239.83/\lambda_{gap.}$ The band gap of samples was calculated by extrapolation of the $(\alpha hv)^2$ versus hv plots, where α is the absorption coefficient and hv is the photon energy, $hv = (1239/\lambda)$ eV. The value of hv extrapolated to $\alpha = 0$ gives an absorption energy, which corresponds to a band gap (Eg). Figure

Figure 4. SEM images of (a) pure PPy, (b) pure TiO_2, (c) TiO_2/PPy (d) TiO_2/PPy/GO nanocomposites.

5 yields an *E*g value of 3.2 eV for TiO_2, 2.98 for pure PPy 2.65 for TiO_2 / PPy and 2.45 for TiO_2/PPy/GO (Yang et al., 2011). The slight decrease in band gap energy is due to the electrons can be excited from the HOMO to the LUMO of PPy, whereas holes were left in the HOMO of PPy. The excited-state electrons can be readily injected into the CB of TiO_2 (Hema et al., 2013).

Photodegradation

Effect of Dye Concentration

The effect of dye concentration photocatalytic degradation was studied in presence of TiO_2 TiO_2/PPy and TiO_2/PPy/GO nanocomposites materials, keeping the amount of catalyst constant. Known concentration of dye solution was prepared in water: alcohol 10:1 (V: V) ratio. The known amount of photocatalyst was dispersed in the different concentration of dye (20, 40, 60, 80 and 100 ppm for Victoria blue and 25, 50, 75, 100 and 125 ppm Rose Bengal) and reaction mixture was irradiated by visible light. The effect of photocatalytic degradation with time was measured and results are shown in Figure 6. When the concentration of solution increased, the number of dye molecule also increased therefore the effective

Figure 5. Band Gap energy of (a) pure TiO_2 (b) TiO_2/PPy, (c) pure PPy (d) TiO_2/PPy/GO nanocomposites.

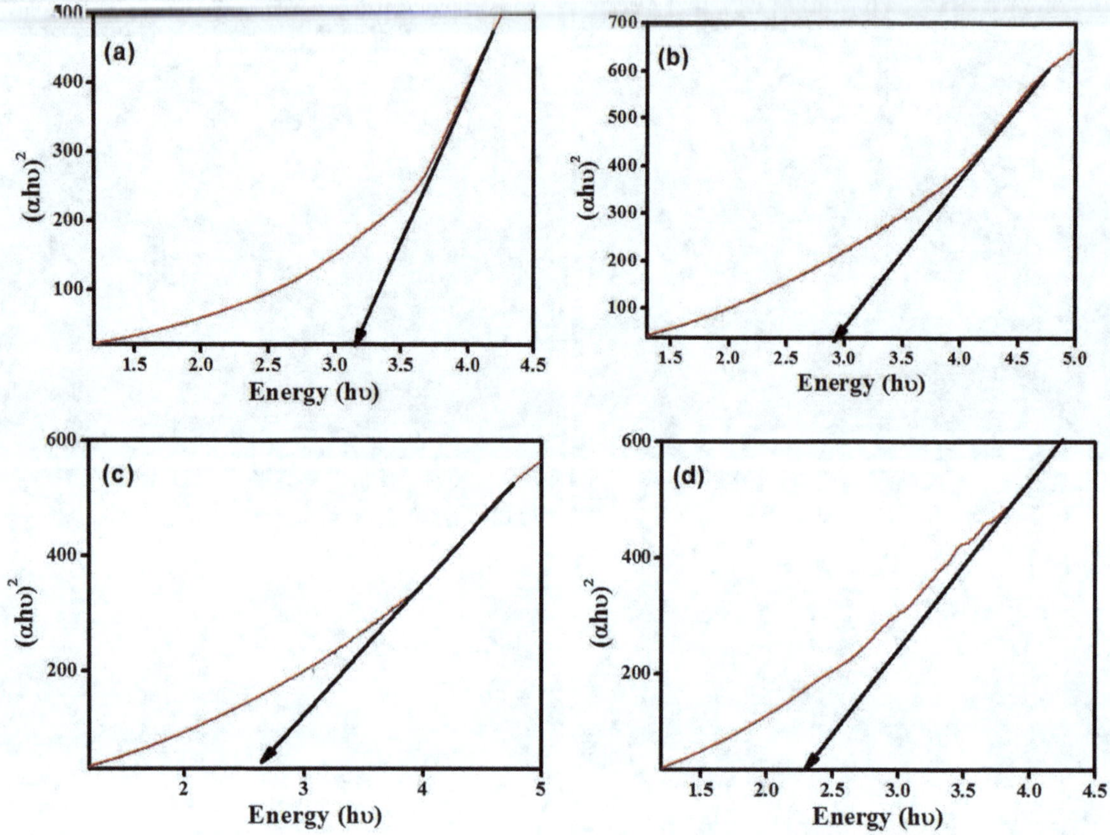

number of photons penetrating the dye reached at the catalyst surface also reduced, owing to hindrance in the path of light, thereby reducing the reactive hydroxyl and superoxide radicals and decreasing the % degradation (Guo, 2009).

Effect of pH

The photodegradation of dyes (Victoria blue and Rose Bengal) reaction was also carried out under varying pH conditions from (2 to 9), by addition of H_2SO_4 and NaOH, keeping other parameter same. The results Figure 7 shown that degradation of dye is highest in neutral medium (at pH = 7) while it decreases with increase in pH and ultimately becomes constant after pH 7. This implies that acidic condition is favourable for formation of the reactive intermediate hydroxyl radicals. This further helps in enhancing the reaction rate. On the other hand, in neutral condition the formation of reactive intermediate is relatively less favourable and hence not feasible (Vulliet et al., 2003).

Effect of Photocatalyst and Dose

It is clear from the results shown in Figure 8 that TiO_2, TiO_2/PPy and TiO_2/PPy/Go are proving as an effective photocatalyst for degradation of Victoria Blue and Rose Bengal dye; however, TiO_2/PPy/Go

Figure 6. Effect of concentration on photodegradation efficiency of dyes Victoria blue and rose bengal (a) TiO_2 (b) TiO_2/PPy and (c) TiO_2/PPy/Go

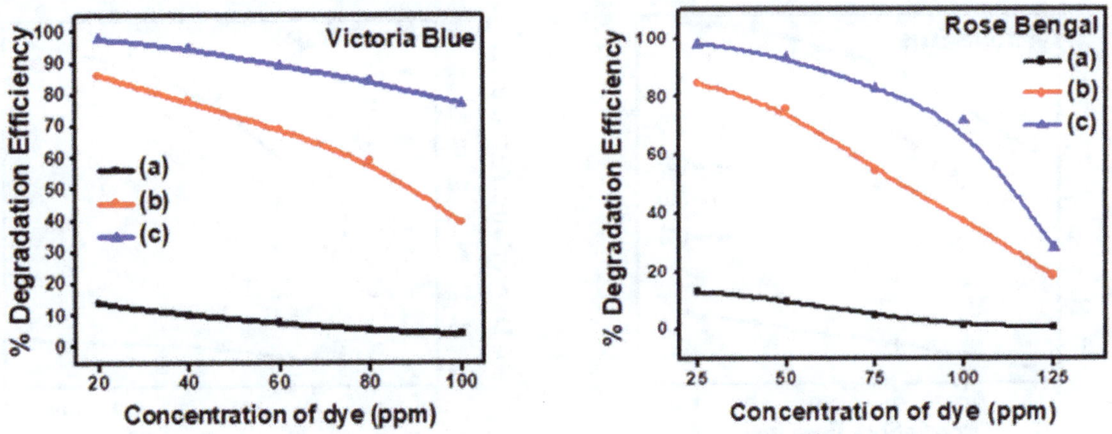

seems to be more effective photocatalyst for degradation of Victoria Blue and Rose Bengal. Effect of dose of photocatalyst on photodegradation of Victoria Blue and Rose Bengal

The effect of photocatalyst dose on the photodegradation of Victoria Blue and Rose Bengal was studied by applying different concentration (200mg/L, 100mg/L and 50mg/L) of the photocatalyst shown in Figure 9. The Degradation rate of Victoria Blue and Rose Bengal was found to increase by increasing the dose of photocatalyst from 50 mg/L to 200mg/L. this is due to the no active site increased. When the Ni is incorporated in TiO_2 the band gap energy is decreased which enhanced the photo efficiency, the surface area of photocatalyst also increased the photo efficiency of photocatalyst.

Figure 7. Effect of pH on photodegradation efficiency of dyes Victoria blue and rose bengal (a) TiO_2 (b) TiO_2/PPy and (c) TiO_2/PPy/Go

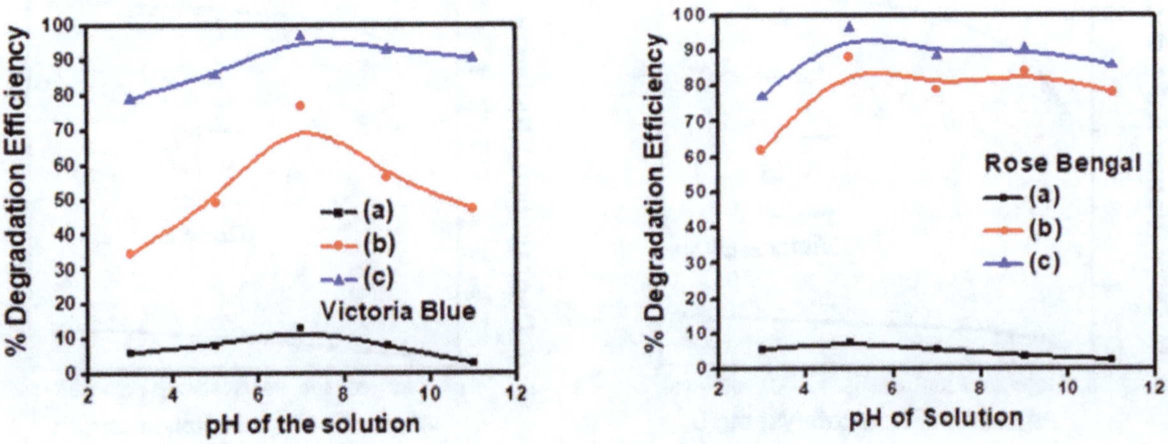

Figure 8. Effect of irradiation time on photodegradation efficiency of dyes Victoria blue and rose bengal (a) TiO_2 (b) TiO_2/PPy and (c)TiO_2/PPy/Go

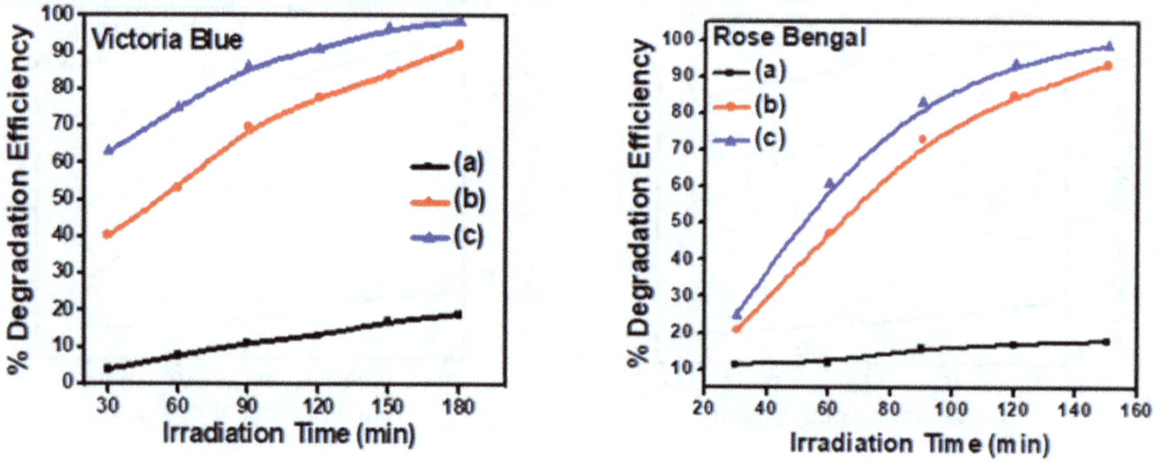

Recyclability of Photocatalyst

The photocatalyst recyclability has been studied. The photocatalyst and Victoria Blue and Rose Bengal mixture was agitated, illuminated with visible light and after desired time, the mixture was centrifuged to remove the photocatalyst. The obtained photocatalyst washed three times with distilled water and finally kept in oven for 24 h at 60 °C temperature and further it is reuse for the degradation of Victoria Blue and Rose Bengal. The photodegradation of Victoria Blue and Rose Bengal by the recyclized Photocatalyst are showing in Figures 10 and 11 for Victoria blue and rose bengal. The result shows that the recyclized photocatalyst efficiency is decreased due to the loss of some active sites and decrease of collection efficiency of photon (Guettaı & Ait Amar 2005).

Figure 9. Effect of photocatalyst amount on efficiency of dyes Victoria blue and rose bengal (a) TiO_2 (b) TiO_2/PPy and (c)TiO_2/PPy/Go

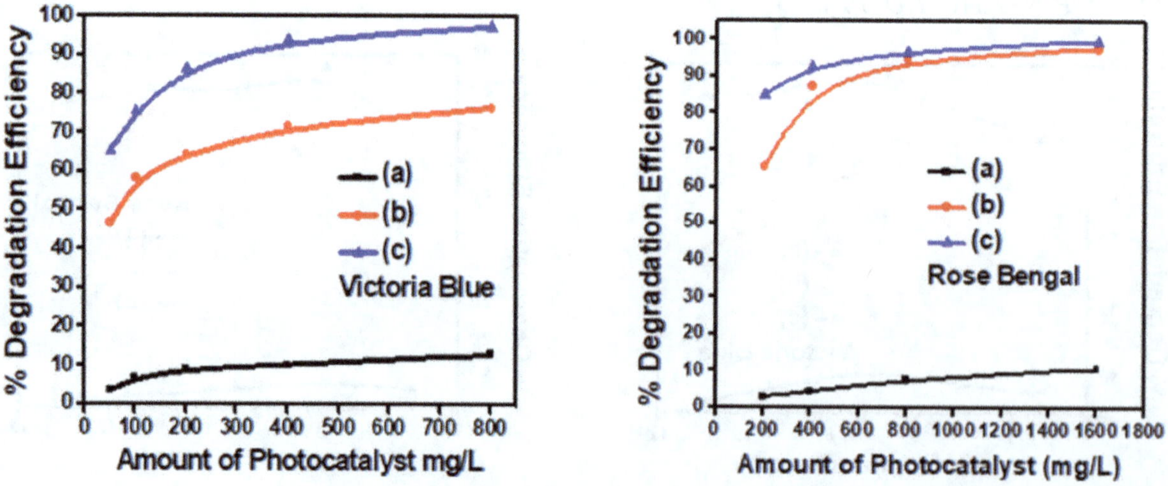

Figure 10. Photodegradation of Victoria Blue by Photocatalyst and recyclable Photocatalyst (a) TiO_2 (b) Recyclized TiO_2 (c) TiO_2/PPy (d) Recyclized TiO_2/PPy (e) TiO_2/PPy/GO (f) Recyclized TiO_2/PPy/GO

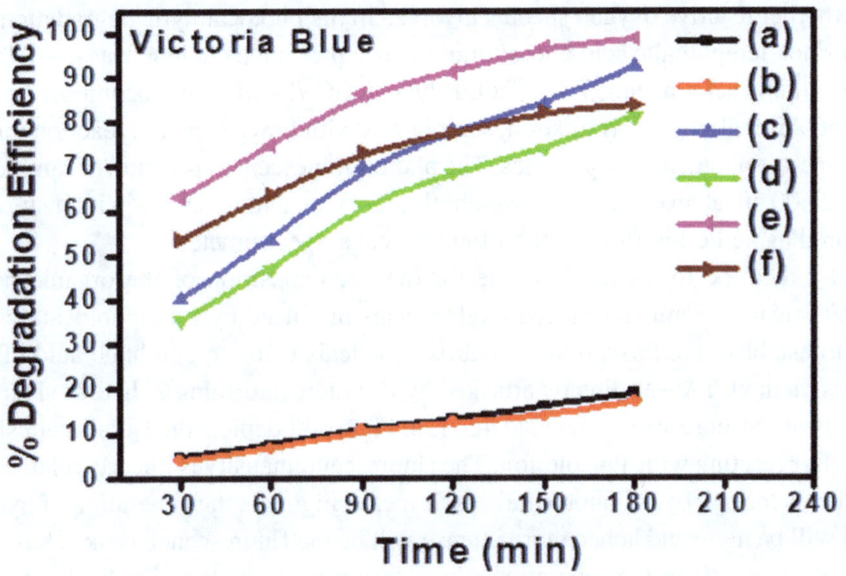

Figure 11. Photodegradation of Rose Bengal by Photocatalyst and recyclable Photocatalyst (a) TiO_2 (b) Recyclized TiO_2 (c) TiO_2/PPy (d) Recyclized TiO_2/PPy (e) TiO_2/PPy/GO (f) Recyclized TiO_2/PPy/GO

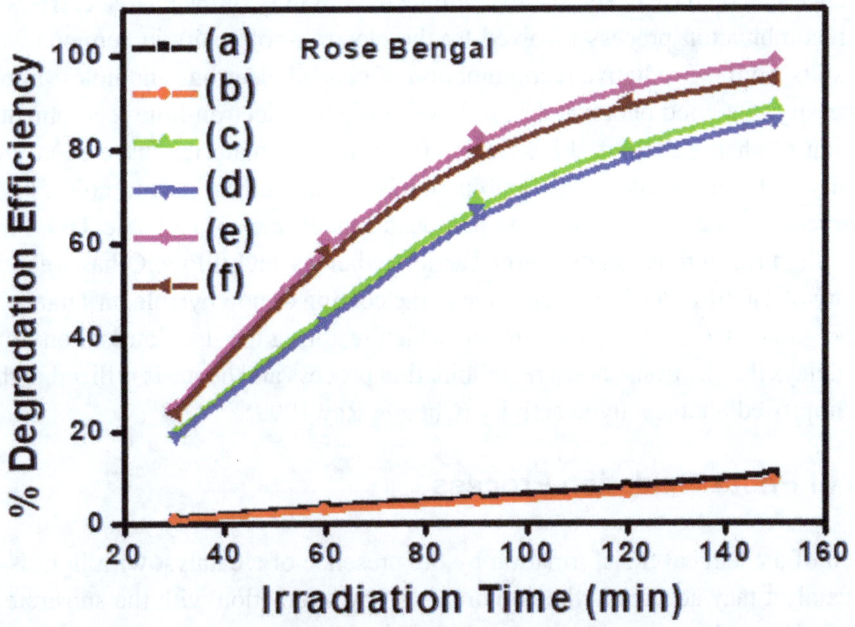

Hydroxyl Radical Formation

To determine whether reactive oxygen species involved in the photocatalytic degradation of dyes is hydroxyl radical or not, terephthalic acid photoluminescence probing technique was used. In this, alkaline solution of terephthalic acid, having TiO_2, TiO_2/PPy and TiO_2/PPy/GO nanocomposites was irradiated with visible light. After 30 min of irradiation, sample was withdrawn from the reaction mixture and was centrifuged to separate photocatalyst particles. The photoluminescence spectrum of sample was recorded between 335 and 600 nm at an excitation wavelength of 325 nm and variation in intensity of peak at 425 nm was monitored using Perkin Elmer LS 55 Fluorescence Spectrometer.

As hydroxyl radical performs the key role for the decomposition of the organic pollutants, it is necessary to investigate the amount of hydroxyl radicals produced by each photocatalyst. Thus, there is a technique to establish the formation of hydroxyl radicals using terephthalic acid (TA) as a probe molecule. In this method, TA was directly attacked by OH radical forming 2- hydroxyl terephthalic acid (TAOH) which gives a fluorescence signal at 426 nm. Figure 12 depicts the fluorescent signal of all the photocatalysts after reacting with TA solution. The fluorescent intensity is linearly related to the number of hydroxyl radicals formed by the photocatalysts. It means higher is the generation of hydroxyl radical, yield of TAOH will be more and hence more intense will be the fluorescence peak. Thus, TiO_2/PPy/GO with highest intensity confirms the generation a greater number of hydroxyl radicals compared to other photocatalysts. The fluorescence intensity follows the trend (i,e. TiO_2, < TiO_2/PPy < TiO_2/PPy/GO) of photocatalytic performance of all the photocatalyst (Pradhan et al., 2013, Chen & Ray 1999).

Lowering of Electron-Hole Recombination

Photoluminescence spectra have been used to examine the mobility of the charge carriers to the surface as well as the recombination process involved by the electron-hole pairs in semiconductor particles. PL emission results from the radiative recombination of excited electrons and holes. In other words, it is a critical necessity of a good photocatalyst to have minimum electron-hole recombination. To study the recombination of charge carriers, PL studies of synthesized materials have been undertaken. PL emission intensity is directly related to recombination of excited electrons and holes. Figure 13 shows the photoluminescence spectra of synthesized photocatalysts. It means TiO_2 and TiO_2/PPy with strong PL intensity has high recombination of charge carriers whereas TiO_2/PPy/GO has weak intensity. The weak PL intensity of TiO_2/PPy/GO may arise due to the coating of polypyrrole on Titania lattice, so that decrease in band gap of TiO_2/PPy/GO was found which resulting the decolourisation of photo excited electrons. This delays the electrons- holes recombination process and hence is utilized in the redox reaction leading to improved photocatalytic activity (Chen & Ray 1999).

Mechanism of Photo-Oxidation Process

The acceleration of a chemical transformation by the presence of a catalyst with light is called photocatalysis. The catalyst may accelerate the photoreaction by interaction with the substrate in its ground or excited state and/or with a primary photoproduct, depending upon the mechanism of the photoreaction and itself remaining unaltered at the end of each catalytic cycle. Heterogeneous photocatalysis is a process in which two active phases solid and liquid are present. The solid phase is a catalyst, usually a semiconductor. The molecular orbital of semiconductors has a band structure. The bands of interest in

Figure 12. PL spectra of photocatalyst with terephthalic acid (0.001M) TiO_2, TiO_2/PPy and TiO_2/PPy/GO

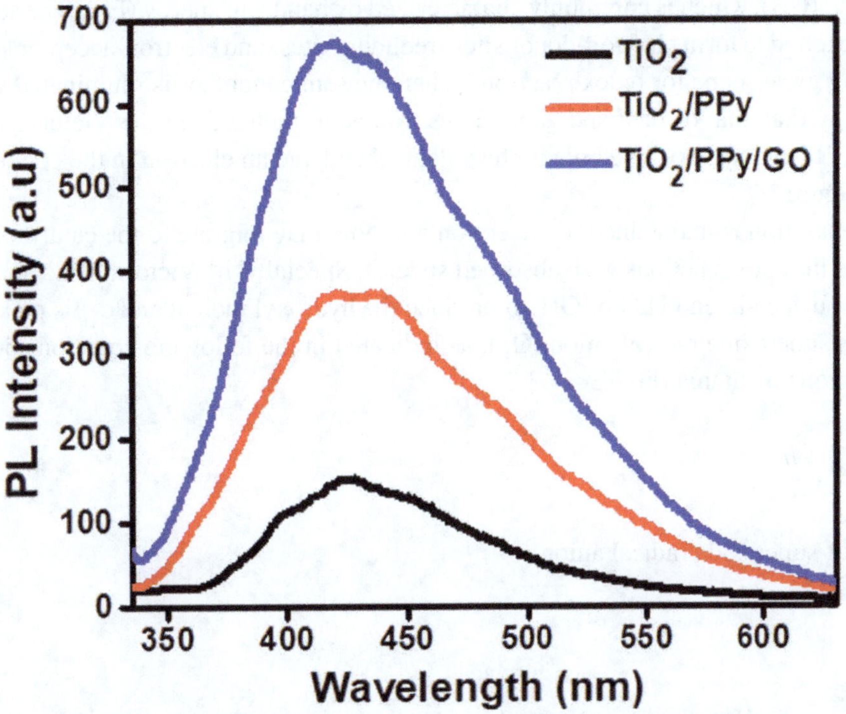

Figure 13. Photoluminescence Spectra of TiO_2, TiO_2/PPy and TiO_2/PPy/GO

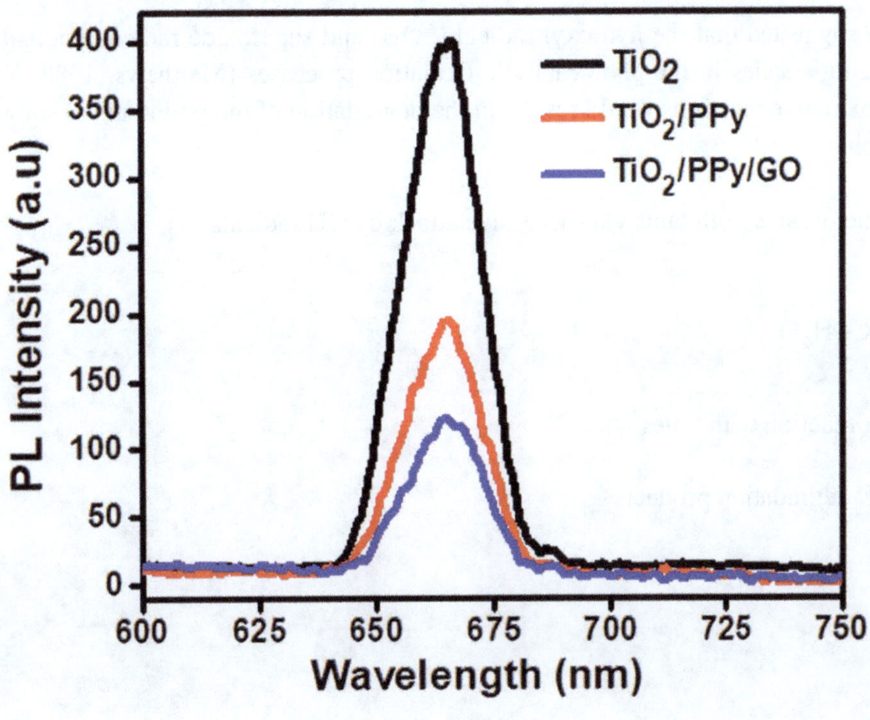

photocatalysis are the populated valence band (Victoria Blue and Rose Bengal) and its largely vacant conduction band (CB), which is commonly characterized by band gap energy (E_{bg}). The semiconductors may be photo-excited to form electron-donor sites (reducing sites) and electron-acceptor sites (oxidising sites), providing great scope for redox reaction. When the semiconductor is illuminated with light (hυ) of greater energy than that of the band gap, an electron is promoted from the Victoria Blue and Rose Bengal to the CB leaving a positive hole in the valence band and an electron in the conduction band as illustrated in Figure 14.

If charge separation is maintained, the electron and hole may migrate to the catalyst surface where they participate in redox reactions with absorbed species. Specially, h^+ Victoria Blue and Rose Bengal may react with surface-bound H_2O or OH^- to produce the hydroxyl radical and e^-_{cb} is picked up by oxygen to generate superoxide radical anion (O_2^-), as indicated in the following equations 5-7; absorption of efficient photons by titania (hυ ³ E_{bg} = 3.2 ev)

$$\mathrm{TiO_2} + h \rightarrow e^-_{cb} + h^+_{dye} \tag{5}$$

Formation of superoxide radical anion

$$\mathrm{O_2} + e^-_{cb} \rightarrow \mathrm{O_2^{-\cdot}} \tag{6}$$

Neutralization of OH^- group into OH by the hole

$$\left(\mathrm{H_2O} \Leftrightarrow \mathrm{H^+} + \mathrm{OH^-}\right) ads + h^+_{dye} \rightarrow (^{\bullet}\mathrm{OH}) + \mathrm{H^+} \tag{7}$$

It has been suggested that the hydroxyl radical (•OH) and superoxide radical anions ($O_2^{\cdot-}$) are the primary oxidizing species in the photocatalytic oxidation processes (Matthews, 1988, Vautier et al., 2001). These oxidative reactions would results in the degradation of the pollutants as shown in the following equations 8-9;

Oxidation of the organic pollutants via successive attack by OH radicals

$$\mathrm{R} + \bullet\mathrm{OH} \rightarrow \mathrm{R} + \mathrm{H_2O} \tag{8}$$

or by direct reaction with holes

R + h^+ ® R^+ ® degradation products (9)

Figure 14. Mechanism of Photodegradation by Ti encapsulated PPy

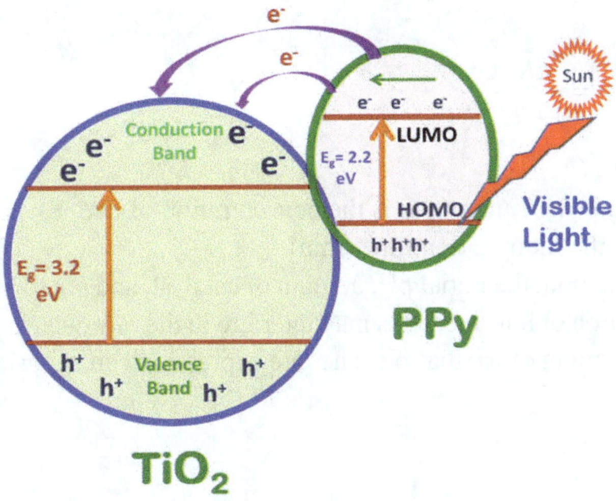

Kinetic Study of Photocatalytic Degradation

For kinetic study of photocatalytic degradation, a control experiment was first carried out under two conditions, vis (i) dye + Visible light (no catalyst) (ii) catalyst+ dye in dark without any irradiation (Figure 15). It can be seen that in under dark conditions, the amount of catalyst adsorbed becomes constant after 20 min, where adsorption equilibrium is achieved. For the kinetic study of bleaching of Victoria Blue and Rose Bengal, the initial concentration of the dyes was varied and the experiments were first conducted in dark for 20 min and then immediately followed by irradiation (Figure 15). The amount of catalyst was kept constant (0.2 g) throughout the experiment.

Figure 15. % Adsorption of Victoria Blue and Rose Bengal dye under dark condition in presence of (a) TiO_2, (b) TiO_2/PPy and (c) TiO_2/PPy/GO

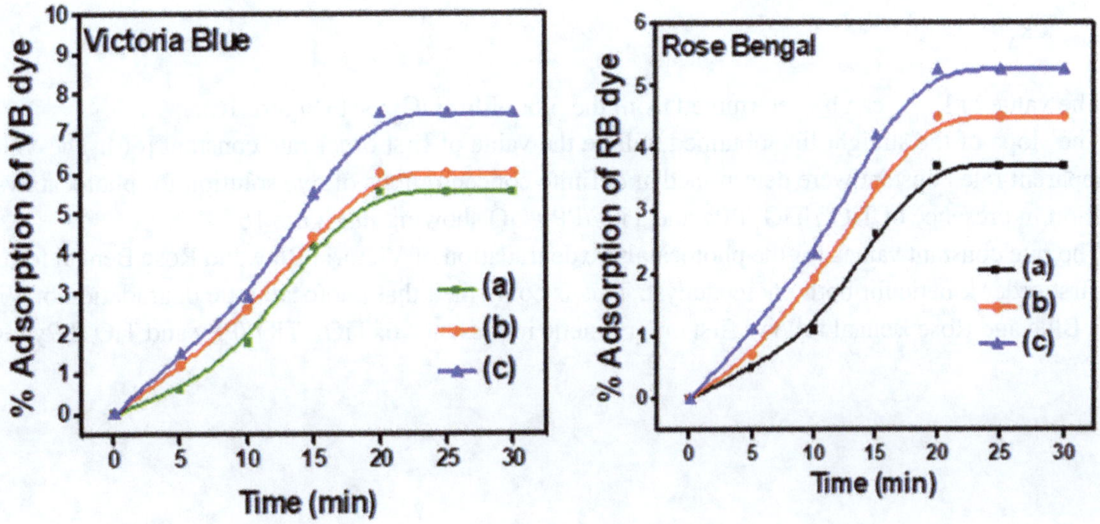

Applying the Langmuir Hinshelwood model for determining the oxidation rate of the photocatalysis of dye.

$$Rate(r) = -\frac{dC}{dt} = k\theta = \theta = \frac{kK_A C}{1 + K_A C} \tag{10}$$

Where k is the rate constant (mg/L min^{-1}), C is the concentration of dye, K_A is the adsorption constant of the dye (L/mg), and t is the illumination time (min).

During the course of reaction, the initial pH, amount of catalyst, and photo intensity were kept same. In addition to it, the formation of intermediates may interfere in the rate determination; hence the calculation was done at the beginning of irradiation. The rate expression can be written as:

$$r_o = \frac{kK_A C_o}{1 + K_A C_o} \tag{11}$$

Where r_0 is the initial rate of degradation of Victoria Blue and Rose Bengal and C_0 is the initial concentration (almost equal to Ceq). When the initial concentration $C_{initial}$ is very small, Co will also be small and Eq. (11) can be simplified as an first-order equation (Zepp & Crosby 1994; Vullie et al., 2003; Guettaı et al., 2005; Kansa et al., 2007).:

$$-\frac{dC}{dt} = kK_A C_o = \frac{\ln C_o}{C} = kK_A t \tag{12}$$

$$C = C_o e^{-k_f photo t} \tag{13}$$

Where

$k_{f,\ Photo=}\ k\ k_A$

The value of $k_{f,photo}$ can be determined from the plot of ln C_t/C_o vs. t (Figure 16).

The slope of the straight line obtained will be the value of first order rate constant [66]. The Value of apparent rate constant were determined at definite concentrations of dye solution for photocatalysis reaction in presence of TiO_2, TiO_2/PPy and TiO_2/PPy/GO showing in Figure 16.

The rate constant values for the photocatalytic degradation of Victoria Blue and Rose Bengal follow the first order kinetic for both photocatalyst. This is confirmed that photocatalytic degradation of Victoria Blue and Rose Bengal follows first order kinetic in presence of TiO_2, TiO_2/PPy and TiO_2/PPy/GO.

Figure 16. Linear first order reaction of Langmuir Hinshelwood kinetics of Victoria Blue and Rose Bengal dye vs. time (a) TiO_2 (b) TiO_2/PPy (c) TiO_2/PPy/GO

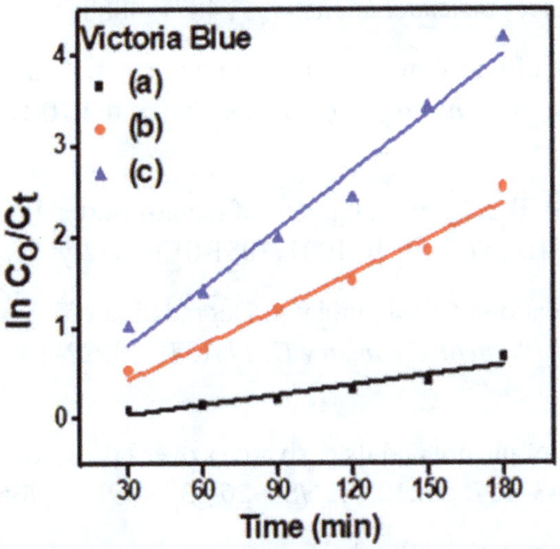

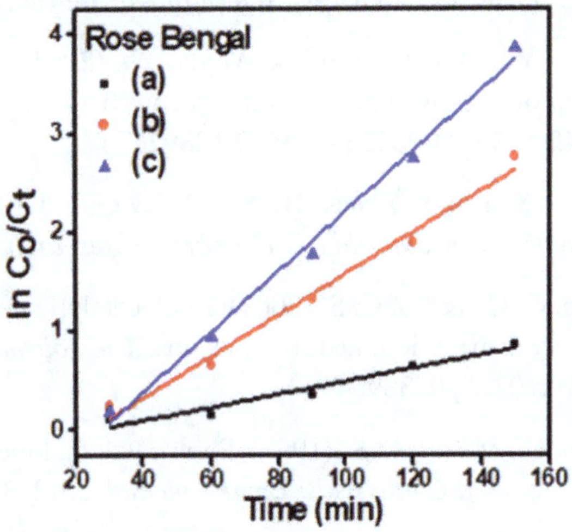

CONCLUSION

The present research work describes a proficient method for synthesis of TiO_2/PPy and TiO_2/PPy/GO nanocomposites. These nanocomposites were prepared by one-step *in situ* deposition oxidative polymerization of pyrrole hydrochloride using Ammonium per sulphate (APS) as an oxidant in the presence of ultra-fine grade powder of TiO_2 nanoparticles cooled in an ice bath. The obtained nanocomposites were characterized by XRD, TEM, SEM and UV-Vis for band gap determination. The obtained results showed that TiO_2 nanoparticles have been encapsulated by PPy with a strong effect on the morphology of TiO_2/PPy and TiO_2/PPy/GO nanocomposites. The Photocatalytic degradation of Rose Bengal and Victoria blue dye was done at different condition viz concentration of dye, time of illumination, pH and dose of photocatalyst. The maximum photodegradation were found at 7 pH, 20 ppm concentration of Victoria blue and 25 ppm of rose bengal dye solution, 800 mg/L for VB and 1600 mg/L for RB amount of photocatalyst and 120 min irradiation of visible light. Kinetics of photodegradation was investigated for Victoria blue and Rose Bengal dye and found first order kinetics. The coating of PPy and GO were enhanced the photocatalytic activity of Titania. Hence TiO_2/PPy and TiO_2/PPy/GO are the efficient photocatalyst for the degradation of Rose Bengal and Victoria Blue dye than pure TiO_2.

REFERENCES

Arenas, M. C., Rodríguez-Núnez, L. F., Rangel, D., Martínez-Álvarez, O., Martínez-Alonso, C., & Castaño, V. M. (2013). Simple one-step ultrasonic synthesis of anatase Titania/polypyrrole nanocomposites. *Ultrasonics Sonochemistry*, *20*(2), 777–784. doi:10.1016/j.ultsonch.2012.09.009 PMID:23099056

Azmi, W., Sani, R. K., & Banerjee, U. C. (1998). Biodegradation of triphenylmethane dyes. *Enzyme and Microbial Technology*, *22*(3), 185–191. doi:10.1016/S0141-0229(97)00159-2 PMID:9463944

Ba-Abbad, M. M., Kadhum, A. A. H., Mohamad, A. B., Takriff, M. S., & Sopian, K. (2012). Synthesis and catalytic activity of TiO_2 nanoparticles for photochemical oxidation of concentrated chlorophenols under direct solar radiation. *International Journal of Electrochemical Science*, *7*, 4871–4888.

Baran, W., Makowski, A., & Wardas, W. (2003). The influence of $FeCl_3$ on the photocatalytic degradation of dissolved azo dyes in aqueous TiO_2 suspensions. *Chemosphere*, *53*(1), 87–95. doi:10.1016/S0045-6535(03)00435-1 PMID:12892670

Bashir, S., Liu, J., Zhang, H., Sun, X., & Guo, J. (2013). Band gap evaluations of metal-inserted titania Nanomaterials. *Journal of Nanoparticle Research*, *15*(4), 1572. doi:10.100711051-013-1572-y

Chen, C. C., & Lu, C. S. (2007). Photocatalytic degradation of Basic violet 4: Degradation efficiency, product distribution, and mechanisms. *The Journal of Physical Chemistry C*, *111*(37), 13922–13932. doi:10.1021/jp0738964

Chen, D., & Ray, A. K. (1999). Photocatalytic kinetics of phenol and its derivatives over UV irradiated TiO_2. *Applied Catalysis B: Environmental*, *23*(2–3), 143–157. doi:10.1016/S0926-3373(99)00068-5

Chen, X., & Mao, S. S. (2007). Titanium dioxide nanomaterials: Synthesis, properties, modifications, and applications. *Chemical Reviews*, *107*(7), 2891–2959. doi:10.1021/cr0500535 PMID:17590053

Cullity, B. D., & Stock, S. R. (2001). Elements of X-ray diffraction (3rd ed.). Prentice-Hall, Inc.

Deng, F., Li, Y., Luo, X., Yang, L., & Tu, X. (2012). Preparation of conductive polypyrrole/TiO2 nanocomposite via surface molecular imprinting technique and its photocatalytic activity under simulated solar light irradiation. *Colloids and Surfaces. A, Physicochemical and Engineering Aspects*, *395*, 183–189. doi:10.1016/j.colsurfa.2011.12.029

Deng, F., Min, L., Luo, X., Wu, S., & Luo, S. (2013). Visible-light photocatalytic degradation performances and thermal stability due to the synergetic effect of TiO2 with conductive copolymers of polyaniline and polypyrrole. *Nanoscale*, *5*(18), 8703–8710. doi:10.1039/c3nr02502k PMID:23900296

Epling, G. A., & Lin, C. (2002). Photoassisted bleaching of dyes utilizing TiO2 and visible light. *Chemosphere*, *46*(4), 561–570. doi:10.1016/S0045-6535(01)00173-4 PMID:11838435

Ferreira, C. A., Domenech, S. C., & Lacaze, P. C. (2001). Synthesis and characterization of polypyrrole/TiO2 composites on mild steel. *Journal of Applied Electrochemistry*, *31*(1), 49–56. doi:10.1023/A:1004149421649

Fox, D. F., & Duxbury, D. F. (1993). The photochemistry and photophysics of triphenylmethane dyes in solid and liquid media. *Chemical Reviews*, *93*(1), 381–433. doi:10.1021/cr00017a018

Franco, A., Celma de Oliveira Lima, E., Novak, M. A., & Wells, P. R. Jr. (2007). Synthesis of nanoparticles of $Co_xFe_{(3-x)}O_4$ by combustion reaction method. *Journal of Magnetism and Magnetic Materials*, *308*(2), 198–202. doi:10.1016/j.jmmm.2006.05.022

Gole, J. L., Stout, J. D., Burda, C., Lou, Y., & Chen, X. (2004). Highly efficient formation of visible light tunable TiO2-XnX photocatalysts and their transformation at the nanoscale. *The Journal of Physical Chemistry B*, *108*(4), 1230–1240. doi:10.1021/jp030843n

Guettaı, N., & Ait Amar, H. A. (2005). Photocatalytic oxidation of methyl orange in presence of titanium dioxide in aqueous suspension. Part II: Kinetics study. *Desalination*, *185*(1–3), 439–448. doi:10.1016/j.desal.2005.04.049

Guo, J. (2009). Interface science in nanoparticles: An electronic structure view of photon-in/photon-out soft-X-ray spectroscopy. *International Journal of Quantum Chemistry*, *109*(12), 2714–2721. doi:10.1002/qua.22174

Guo, Z., Shin, K., Karki, A. B., Young, D. P., Kaner, R. B., & Hahn, H. T. (2009). Fabrication and characterization of iron oxide nanoparticles filled polypyrrole, nanocomposites. *Journal of Nanoparticle Research*, *11*(6), 1441–1452. doi:10.100711051-008-9531-8

Hema, M., Arasi, A. Y., Tamilselvi, P., & Anbarasan, R. (2013). Titania nanoparticles synthesized by sol–gel technique. *Chemical Science Transactions*, *2*(1), 239–245. doi:10.7598/cst2013.344

Huang, H., Gan, M., Ma, L., Yu, L., Hu, H., Yang, F., Li, Y., & Ge, C. (2015). Fabrication of Polyaniline/Graphen oxide/titania nanotube arrays nanocomposites and their application in supercapacitors. *Journal of Alloys and Compounds*, *630*, 214–221. doi:10.1016/j.jallcom.2015.01.059

Kansal, S. K., Singh, M., & Sud, D. (2007). Studies on photodegradation of two commercial dyes in aqueous phase using different photocatalysts. *Journal of Hazardous Materials*, *141*(3), 581–590. doi:10.1016/j.jhazmat.2006.07.035

Kar, A., Smith, Y. R., & Subramanian, V. R. (2009). Improved photocatalytic degradation of, textile dye using titanium dioxide nanotubes formed over titanium wires. *Environmental Science & Technology*, *43*(9), 3260–3265. doi:10.1021/es8031049 PMID:19534144

Kaur, J., & Singhal, S. (2014). Heterogeneous photocatalytic degradation of rose bengal: Effect of operational parameters. *Physica B, Condensed Matter*, *450*, 49–53. doi:10.1016/j.physb.2014.05.069

Konstantinou, I. K., & Albanis, T. A. (2004). TiO_2-assisted photocatalytic degradation of azo dyes in aqueous solution: Kinetic and mechanistic investigations. *Applied Catalysis B: Environmental*, *49*(1), 1–14. doi:10.1016/j.apcatb.2003.11.010

Kordouli, E., Bourikas, K., Lycourghiotis, A., & Kordulis, C. (2015). The mechanism of azo-dyes adsorption on the titanium dioxide surface and their photocatalytic degradation over samples with various anatase/rutile ratios. *Catalysis Today*, *252*, 128–135. doi:10.1016/j.cattod.2014.09.010

Kwon, J. D., Kim, P. H., Keum, J. H., & Kim, J. S. (2004). Polypyrrole/titania hybrids: Synthetic variation and test for the photovoltaic materials. *Solar Energy Materials and Solar Cells*, *83*(2–3), 311–321. doi:10.1016/j.solmat.2004.02.033

Lachheb, H., Puzenat, E., Houas, A., Ksibi, M., Elaloui, E., Guillard, C., & Herrmann, J. M. (2002). Photocatalytic degradation of various types of dyes (alizarin S, Crocein orange G, methyl Red, Congo red, methylene blue) in water by UV-irradiated titania. *Applied Catalysis B: Environmental*, *39*(1), 75–90. doi:10.1016/S0926-3373(02)00078-4

Li, X., Liu, G., & Zhao, J. (1999). Two competitive primary processes in the photodegradation of cationic triaryl methane dyes under visible irradiation in TiO_2 dispersions. *New Journal of Chemistry*, *23*(12), 1193–1196. doi:10.1039/a906765e

Li, Y., Yu, Y., Wu, L., & Zhi, J. (2013). Processable Polyaniline/Titania nanocomposites with good photocatalytic and conductivity properties prepared via peroxo-titanium complex catalyzed emulsion polymerization approach. *Applied Surface Science*, *273*, 135–143. doi:10.1016/j.apsusc.2013.01.213

Liang, H. C., & Li, X. Z. (2009). Visible-induced photocatalytic reactivity of polymer–sensitized titania nanotube films. *Applied Catalysis B: Environmental*, *86*(1–2), 8–17. doi:10.1016/j.apcatb.2008.07.015

Linsebigler, A. L., Lu, G., & Yates, J. T. Jr. (1995). Photocatalysis on TiO_2 surfaces: Principles, mechanisms, and selected results. *Chemical Reviews*, *95*(3), 735–758. doi:10.1021/cr00035a013

Mai, F. D., Lu, C. S., Wu, C. W., Huang, C. H., Chen, J. Y., & Chen, C. C. (2008). Mechanisms of photocatalytic degradation of Victoria Blue R using Nano-TiO_2. *Separation and Purification Technology*, *62*(2), 423–436. doi:10.1016/j.seppur.2008.02.006

Matthews, R. W. (1988). Kinetics of photocatalytic oxidation of organic solutes over titanium dioxide. *Journal of Catalysis*, *111*(2), 264–272. doi:10.1016/0021-9517(88)90085-1

Mor, G. K., Shankar, K., Paulose, M., Varghese, O. K., & Grimes, C. A. (2006). Use of highly-ordered TiO2 nanotube arrays in dye-sensitized solar cells. *Nano Letters*, *6*(2), 215–218. doi:10.1021/nl052099j PMID:16464037

Nabid, M. R., Golbabaee, M., Moghaddam, A. B., & Dinarvand, R., & Sedghi, R. (2008). P./TiO2 Nanocomposite: Enzymatic Synthesis and Electrochemical Properties. *International Journal of Electrochemical Science*, *3*, 1117–1126.

Nag, C. M., Chen, P. C., & Kumar, S. (2012). Hydrothermal crystallization of titania on silver nucleation sites for the synthesis of visible light nano-photocatalysts—Enhanced photoactivity using rhodamine 6G. *Applied Catalysis A*, *433–434*, 75–80.

Okano, M., Itoh, K., Fujishima, A., & Honda, K. (1987). Photoelectrochemical polymerization of pyrrole on TiO2 and its application to conducting pattern generation. *Journal of the Electrochemical Society*, *134*(4), 837–841. doi:10.1149/1.2100582

Pare, B., Singh, P., & Jonnalgadda, S. B. (2009). Degradation and mineralization of Victoria Blue B dye in a slurry photo reactor using advanced oxidation process. *Journal of Scientific and Industrial Research*, *68*, 724–729.

Park, Y., Lee, S. H., Kang, S. O., & Choi, W. (2010). Organic dye-sensitized TiO_2 for the redox conversion of water pollutants under visible light. *Chemical Communications*, *46*(14), 2477–2479. doi:10.1039/b924829c PMID:20309473

Pradhan, G. K., Padhi, D. K., & Parida, K. M. (2013). Fabrication of #-Fe2O3 nanorod/RGO composite: A novel hybrid photocatalyst for phenol degradation. *ACS Applied Materials & Interfaces*, *5*(18), 9101–9110. doi:10.1021/am402487h PMID:23962068

Reddy, K. M., Manorama, S. V., & Reddy, A. R. (2002). Bandgap studies on anatase titanium dioxide nanoparticles. *Materials Chemistry and Physics*, *78*(1), 239–245. doi:10.1016/S0254-0584(02)00343-7

Saquib, M., & Muneer, M. (2003). TiO_2-mediated photocatalytic degradation of a triphenylmethane dye (gentian violet), in aqueous suspensions. *Dyes and Pigments*, *56*(1), 37–49. doi:10.1016/S0143-7208(02)00101-8

Sarmah, S., & Kumar, A. (2011). Photocatalytic activity of polyaniline-TiO2 nanocomposites. *Indian Journal of Physics*, *85*(5), 713–726. doi:10.100712648-011-0071-1

Sedlačík, M., Mrlík, M., Pavlínek, V., Sáha, P., & Quadrat, O. (2012). Electrorheological properties of suspensions of hollow globular titanium oxide/polypyrrole particles. *Colloid & Polymer Science*, *290*(1), 41–48. doi:10.100700396-011-2521-x

Sun, L., Shi, Y., Li, B., Li, X., & Wang, Y. (2013). Preparation and characterization of polypyrrole/TiO_2 nanocomposites by reverse microemulsion polymerization and its photocatalytic activity for the degradation of methyl orange under natural light. *Polymer Composites*, *34*(7), 1076–1080. doi:10.1002/pc.22515

Tai, H., Jiang, Y., Xie, G. Y., Yu, J., & Zhao, M. (2007). Self-assembly of TiO2/polypyrrole nanocomposite ultrathin films and application for an NH3 gas sensor. *International Journal of Environmental Analytical Chemistry*, *87*(8), 539–551. doi:10.1080/03067310701272954

Thompson, T. L., & Yates, J. T. (2006). Surface science studies of the photoactivation of TiO2 new photochemical processes. *Chemical Reviews*, *106*(10), 4428–4453. doi:10.1021/cr050172k PMID:17031993

Vautier, M., Guillard, C., & Herrmann, J. M. (2001). Photocatalytic degradation of dyes in water: Case study of indigo and of indigo Carmine. *Journal of Catalysis*, *201*(1), 46–59. doi:10.1006/jcat.2001.3232

Vinodgopal, K., Wynkoop, D. E., & Kamat, P. V. (1996). Environmental photochemistry on semiconductor surfaces: Photosensitized degradation of a textile azo dye, acid orange 7, on TiO2 particles using visible light. *Environmental Science & Technology*, *30*(5), 1660–1666. doi:10.1021/es950655d

Vinuth, M., Naik, H. S. B., Vinoda, B. M., Pradeepa, S. M., Arun, K. G., & … . (2016). Rapid removal of hazardous rose bengal dye using Fe(III)–montmorillonite as an effective adsorbent in aqueous solution. *Journal of Environmental & Analytical Toxicology*, *6*(02), 355. doi:10.4172/2161-0525.1000355

Vulliet, E., Chovelon, J. M., Guillard, C., & Herrmann, J. M. (2003). Factors influencing the photocatalytic degradation of sulfonylurea herbicides by TiO_2 aqueous suspension. *Journal of Photochemistry and Photobiology A Chemistry*, *159*(1), 71–79. doi:10.1016/S1010-6030(03)00108-4

Wang, B., Li, C., Pang, J., Qing, X., Zhai, J., & Li, Q. (2012). Novel polypyrrole-sensitized hollow TiO2/fly ash cenospheres: Synthesis, characterization, and photocatalytic ability under visible light. *Applied Surface Science*, *258*(24), 9989–9996. doi:10.1016/j.apsusc.2012.06.061

Wang, D., Wang, Y., Li, X., Luo, Q., An, J., & Yue, J. (2008). Sunlight photocatalytic activity of polypyrrole–TiO_2 nanocomposites prepared by "in situ" method. *Catalysis Communications*, *9*(6), 1162–1166. doi:10.1016/j.catcom.2007.10.027

Wei, S., Mavinakuli, P., Wang, Q., Chen, D., Asapu, R., Mao, Y., Haldolaarachchige, N., Young, D. P., & Guo, Z. (2011). Polypyrrole-Titania nanocomposites derived from different oxidants. *Journal of the Electrochemical Society*, *158*(11), K205–K212. doi:10.1149/2.048111jes

Xu, S., Zhu, Y., Jiang, L., & Dan, Y. (2010). Visible light induced photocatalytic degradation of methyl orange by polythiophene/TiO_2 composite particles. *Water, Air, and Soil Pollution*, *213*(1–4), 151–159. doi:10.100711270-010-0374-4

Yang, S., Yang, X., Shao, X., Niu, R., & Wang, L. (2011). Activated carbon catalyzed persulfate oxidation of azo dye acid orange 7 at ambient temperature. *Journal of Hazardous Materials*, *186*(1), 659–666. doi:10.1016/j.jhazmat.2010.11.057 PMID:21145652

Yang, Y., Wen, J., Wei, J., Xiong, R., Shi, J., & Pan, C. (2013). Polypyrrole-decorated Ag-TiO_2 nanofibers exhibiting enhanced photocatalytic activity under visible-light illumination. *ACS Applied Materials & Interfaces*, *5*(13), 6201–6207. doi:10.1021/am401167y PMID:23767991

Zhang, L., Lv, F., Zhang, W., Li, R., Zhong, H., Zhao, Y., Zhang, Y., & Wang, X. (2009). Photo degradation of methyl orange by attapulgite-SnO_2-TiO_2 nanocomposites. *Journal of Hazardous Materials*, *171*(1–3), 294–300. doi:10.1016/j.jhazmat.2009.05.140 PMID:19577837

Chapter 3
MXene-Based Nanocomposite Photocatalysts for Wastewater Treatment

Gao Li
Dalian Institute of Chemical Physics, Chinese Academy of Sciences, China

Ali Raza
University of Sialkot, Pakistan

Sarfraz Ali
Riphah International University, Pakistan

Zhiwen Li
Dalian Institute of Chemical Physics, Chinese Academy of Sciences, China

ABSTRACT

Two-dimensional (2D) MXene has been considered as a hotspot toward environmental photocatalysis because of its outstanding structural stability, highly efficient conductivity, and versatile hydrophilicity. As an efficient photocatalytic candidate, MXenes offer rapid photogenerated charge carrier isolation, thereby providing plentiful availability for surface functional groups in respect of light-harvesting promising materials, and additionally executing a suitable foundation in favor of superior photoconversion proficiency. This chapter summarizes a comprehensive analysis of recent studies on fabrication method for MXene photocatalysts and photocatalytic performance for contaminant degradations. More significantly, MXenes are frequently employed as cocatalysts to boost the efficacy of photocatalytic activities when combined with other traditional photocatalysts such as metal oxide, metal sulfide, g-C3N4, and so on. Furthermore, in an effort to disclose the unique qualities of MXene-based nanocomposites, the stability of MXene-based nanocomposite photocatalysts is briefly examined.

DOI: 10.4018/978-1-6684-4553-2.ch003

1. INTRODUCTION

The widespread growth specifically for urban areas associated with expanding worldwide population has been considered a dynamic issue regarding the need of securing suitable and drinkable water (Shao & Chen, 2013). In addition, fast pace toward industrialization has caused to production of harmful waste spreading adverse influences on health as well as environment (Jasper et al., 2017). Severe occurrence of contaminants such as substantial metal ions, salt-like entities, medications, aromatic complexes, textile colorants, and other pollutants into drinking water and wastewater streams have posed a global eco-friendly challengeable issue caused by major impurities that are toxic in nature and therefore realized aggressive toward living organisms (Carolin et al., 2017; Gavrilescu et al., 2015). Therefore, necessarily numerous biotic and physicochemical strategies likewise aerobic/anaerobic assimilation, biochar treatment (Ye et al., 2019), membrane/sheath filtration (Bagheri & Mirbagheri, 2018), adsorption (Crini et al., 2019), and other favorable approaches have been predictably executed to remove variety of environmental impurities. Moreover, efficient nanomaterials have become frugally worthwhile and environment-friendly choices in favor of active environmentally removing agents such as adsorbents, catalysts, antibacterial mediators along with active membranes facilitating water/wastewater treatment (Das et al., 2017; Santhosh et al., 2016).

Additionally, the most promising functional nanomaterials in this respect are zeolites, metal-organic frameworks, ceramics and glasses, conductive polymers, and carbonaceous nanostructures. The architecture of functional nanomaterial-derived devices may provide an opportunity for achieving new areas of research pertaining to efficient sensitivity, suitable selectivity, high efficiency, and finally mechanically stability toward versatile ecological remediation applications (M. Sun & Li, 2018). No doubt, 2D-materials inherit atomic scale thin assemblies associated with lateral dimensions lying among ten nanometers and few micrometers. The larger lateral dimension with nanometer thickness have endowed 2D nanomaterials showing higher specific surface area that provide them anticipation for overabundance of ecological remediation applications as that of adsorption, catalysis and sensing contribution in removal activities (Ikram et al., 2020; Raza et al., 2021; Raza et al., 2020; Tan et al., 2017; Tan & Zhang, 2015; Wei et al., 2021). For environmental remediation platform, graphene-derivatives have imparted great part in the form of sorbents, membrane isolation, photocatalysts nanomaterials toward surplus sanitization, or behaving sensitive nanomaterials for contamination control as well as removal (J. Xu et al., 2018). Further, graphene-based materials and their derivatives have exhibited enormous performance like water purifying membranes connected with ideal properties likewise ultrahigh water fluidity, selectively molecules/ions filtering along with high resistance existing for biofouling (Ikram et al., 2020; Ikram et al., 2020; Ikram et al., 2020; Sun & Li, 2018). To date, GO and its derivatives are considered as the most extensive family of 2D nanomaterials regarding the removal of variety of pollutants except soils/clays (Ersan et al., 2017; Sarkar et al., 2014; Shen & Chen, 2015; Xu et al., 2018).

Rather, MXenes (Figure 1a) are introductory of newly created 2D material family emerging from diversified class attributing to transition metal carbides (TMC), transition metal nitrides (TMN), and transition metal carbonitrides (TMCN) identified with common formula $M_{n+1}X_nT_x$ (n = 1-3), denoting M as primary transition metals group, whereas X indicating carbon and/or nitrogen, T_x as significantly surface terminating groups like –OH, –O–, and/or –F respectively (Lei, Zhang, & Zhou, 2015). Up till now, nearly thirty MXenes were strongly synthesized, and the most popular among them are, $Ti_3C_2T_x$, Ti_2CT_x, $Nb_4C_3T_x$ (Ghidiu et al., 2014) Ti_3CNT_x, $Ta_4C_3T_x$ (Naguib et al., 2012), Nb_2CT_x, V_2CT_x (Lukatskaya et al., 2013), $Nb_4C_3T_x$ (Ghidiu et al., 2014). However, further variety has been theoretically anticipated

Figure 1. MAX Phase along with MXene Chemistry containing (a) Periodic table illustrating amended MAX phase configurations associated with latest reported versatile MAX phases connected with A = Co, Zn, Cu and Ni, whereas Mn as well as Fe resides at M or A location (b) Schematically illustration for etching owing to Ti_3AlC_2 MAX segment precursor over $Ti_3C_2T_x$ MXene.
Reproduced with permission from ref. (Lin, Shao, Xu, Taberna, & Simon, 2020) Copyright 2020, Elsevier Inc.

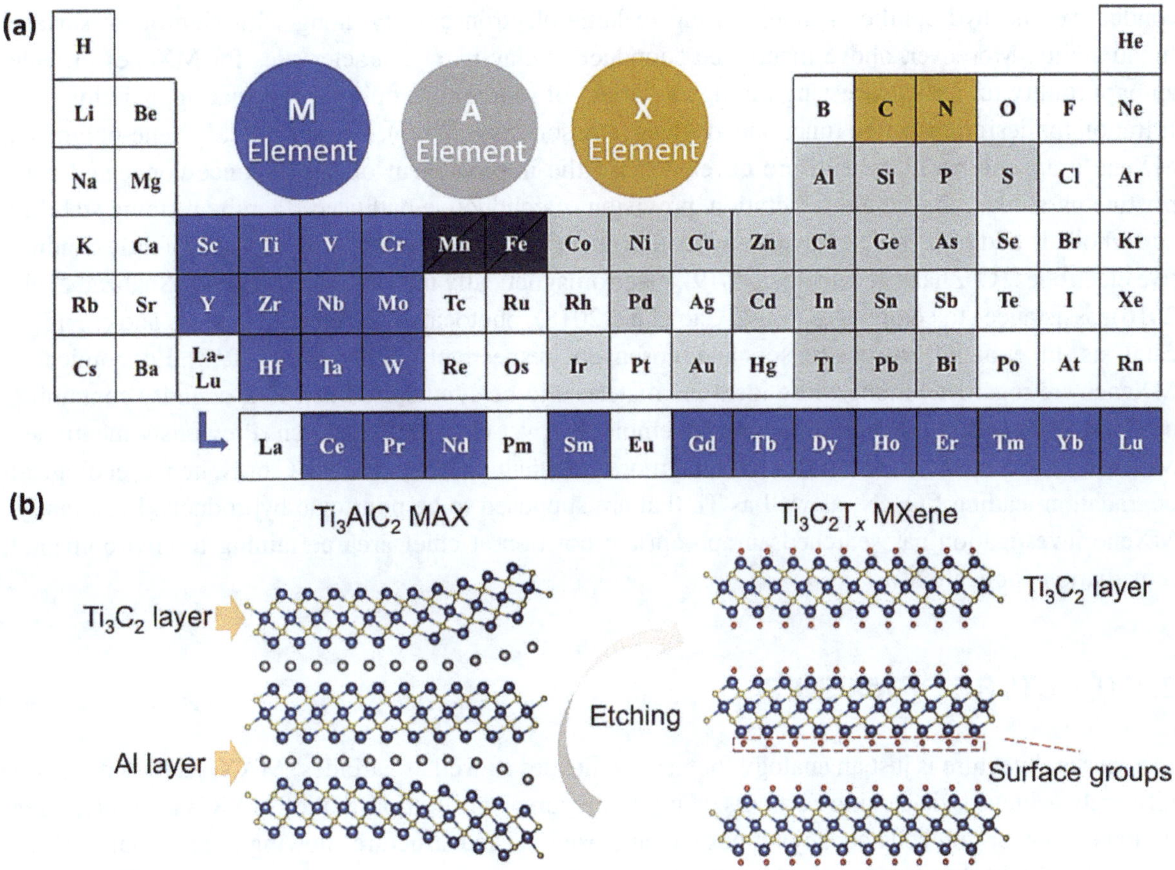

depending upon availability of MAX phase precursors (Lukatskaya Maria et al., 2013). Among those promising MXenes titanium derived MXenes like Ti_2CT_x and $Ti_3C_2T_x$ have been suggested synergetic toward eco-friendly applications caused by elemental plenty with non-piousness disintegration yields, and especially titanium carbide ($Ti_3C_2T_x$) has been extensively studied from MXene family. Additionally, various MXene configurations are added comprising more than one transition metals incorporating M layers into ordered/disordered architectures in the form of $Mo_2Ti_2C_3$, $(Ti_{0.5},Nb_{0.5})_2C$, $(V_{0.5},Cr_{0.5})_3C_2$) (Anasori, Lukatskaya, & Gogotsi, 2017; Kurtoglu, Naguib, Gogotsi, & Barsoum, 2012). Besides carbides, nitrides, and carbonitride MXene's class, there is further exploring glance in literature for other areas of research as that of electrochemical energy storage devices (Naguib et al., 2017) as well as biomedical sensing applications (K. Huang, Li, Lin, Han, & Huang, 2018), and thereby enormous forecasts related with properties. However, the remaining challengeable foundation for producing nitride MXenes via selective acid etching strategy is under discussion (Figure 1b). But an exception in this regard has been proposed as titanium carbonitride (Ti_3CN) since it has been researched toward various desired applications.

Further, it has been considered reliable and prominent limiting factor of widespread exploring nitride as well as carbonitride MXenes toward substantial ecological remediation applications. MXenes have been offered extensive interest to broaden their applications in ecological remediation because of possessing hydrophilicity, large surface area, abundant active metal hydroxide sites, associated with eco-friendly features (Zhang et al., 2018). Pristine MXenes occupy unique combination containing metallic electronic conductive and hydrophilic channels owing to large electron density along with significant surface functionality. Moreover, above mentioned "conductive clay-like" characteristics for MXenes provide an opportunity for easy processing through a variety of routes enveloping spray coating, painting, and filtration for designing active functional devices (Anasori et al., 2017). Furthermore, MXene-polymers, MXene graphene-based hybrids are developed for the improvement of performance connected with pristine parts like reduced agglomeration, provision of conductive pathways, thereby refining stability as a whole. In addition, more evolving applications in favor of MXenes encompass transparently conductive electrodes (C. Zhang & Nicolosi, 2019), electromagnetically interference shielding (Shahzad et al., 2016), assistances to composites (W.-T. Cao et al., 2018), photocatalysis, electrocatalysis and chemical catalysis (Lu et al., 2017), gas sensors and biosensors, respectively (Sinha et al., 2018). Economically, MXenes are regarded among those ideal 2D materials to be synthesized in 100 g quantity containing each batch preparing in research laboratory employing wet chemical approach. Previously mentioned MXenes have been declared as non-toxic and more abundant elements like Ti, C or N, hence, ecological degradation leading CO_2, N_2 as well as Ti, that are supposed to be non-toxic byproducts. Resultantly, MXene investigation has searched an appropriate position at chief area pertaining to environmental remediation applications.

2. STRUCTURE OF MXENES

MXenes architecture is just an analogy for ternary nitrides as well as carbides (MAX) as MXene is significantly achieved via etching A-atoms owing to precursor material as that of MAX. Generally, three various types for unit cells occupying hexagonal closed-packed structure showing space group of *P63/mmc* related with MAX (Figure 2d-f). The layered structure encompasses A layer for MAX incorporates into closed-packing M layers, whereas X atoms appear at octahedral sites existing among M layers (Barsoum & Radovic, 2011). Clearly, octahedral interstitial sites belonging MXenes occupy hexagonal closely-packed (HCP) envelope of X atoms. It has been proved that three types for crystal structure of $M_{n+1}X_n$, where n = 1, 2, 3 and are named as MXM, MXM_1XM, and $MXM_1X_1M_1XM$ respectively, and M and M_1 indicate various symmetrical sites associated with transition metal atoms, and further X and X_1 atoms represent Carbon(C) or Nitrogen (N) atoms (Jiang et al., 2020). Specific M and M_1 locations are apparently engaged by single or two varieties for elements; moreover, as for M_2X MXene indicating two various structures containing single-M element structure along with solid-solution nature M-metal structure exhibiting transition metals that are irregularly aligned in M layer of same nature. Additionally, M_3X_2 with M_4X_3 were added to two types of structures as already described in the order containing M element structure involving single or double layer owing to pristine M elements that are injected through 2nd transition metal resided with sandwich structure and is considered as a common structure (Figure 2a-c) (Anasori et al., 2017; Jiang et al., 2020). No doubt, synthetic approach shows greater impact on lattice structure for MXene, likewise ultrathin α-Mo_2C that has been synthesized employing chemical vapor deposition where Mo atoms are to some extent distorting hexagonally closed packing form along

Figure 2. (a) The Mono-M MAX phases containing M_2AX, M_3AX_2 and M_4AX_3, in group layered form of red atoms, (b) Formation of MXenes with selective etching associated with surface terminations creation (as indicating by yellow atoms) with T labelling. (c) Probable elements revealing M, A, X, and T into MAX as well as MXene phases. Unit cells along with crystal structures showing (d) 211, (e) 312, (f) 413 phases owing to MAX.

(a-c) Reproduced with permission from ref. (Hong, Wyatt, Nemani, & Anasori, 2020) Copyright 2020, Springer Nature.
(d-f) Reproduced with permission from ref. (Barsoum & Radovic, 2011) Copyright 2011, Annual Review.

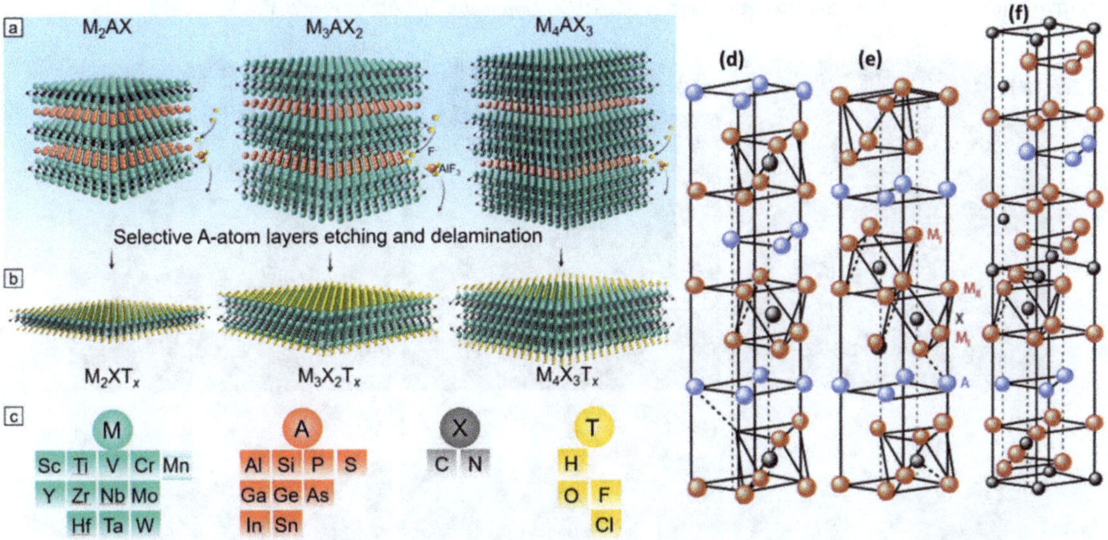

with C atoms inhabiting ordered sites behaving like octahedral voids belonging Mo lattice, thereby indicating orthorhombic crystal structure (C. Xu et al., 2015). Additionally, in case of V_2N is reproduced introducing N atoms to substitute C atoms in MXene, then V_2N structure is clearly transformed in the form of combined layered nature structure exploring trigonal geometry of V_2N whereas cubic geometry of VN (Urbankowski et al., 2017). Moreover, at the stage of alike morphology for MXenes, a large variety among them exhibits accordion nature multi-layered structure. Further, various MXenes are prepared by employing same routes but occupy variety of crystal structures.

As an example, Naguib et al. (Naguib et al., 2011) has prepared scrolled type structure of Ti_3C_2 illustrated in Figure 3a,b, as radii for few Ti_3C_2 nanosheets are measured as smaller than 20 nm that is closely related to sonication procedure followed in exfoliation process. To add, flowery nature porous structure of $Ti_3C_2T_x$ has been prepared via hydrothermal method in ethylenediamine environment at 80°C for 8 hours (Figure 3c) (Xiao et al., 2019). However, in some particular circumstances, MXene nanofibers (Figure 3d) (W. Y. Yuan et al., 2018), as well as nanoribbons (Figure 3e,f), have been successfully collected (Lian et al., 2017). Moreover, employing HF (hydrofluoric acid) treatment, exposing M atoms owing to MXenes were observed in active form, and in this way, they were likely to be binding form associated with enormously electronegatively charged surface terminating moieties, and it has been proved through verification by density functional theory as shown in Figure 2d-f. Certainly, -F, -O, and -OH are suggested well-known surface terminal species belonging MXenes. Further, various anionic moieties containing -Cl and -S may be combined excellently attached with transition metals (Liu et al., 2018). In addition, simple formula showing $M_{n+1}X_nT_x$ indicates surface termination for MXenes by

Figure 3. (a) TEM photographs for Ti_3C_2 nanoscrolling. (b) schematical appearance for MXene nanoscrolling. (c) Schematically SEM analysis showing flowery nature for $Ti_3C_2T_x$. (d) Prominently SEM results showing Ti_3C_2 nanofiber. (e) Clearly TEM analysis for alkalized Ti_3C_2 as (a-Ti_3C_2) MXene nanoribbons (MNRs) and in the last (f) Prominently HRTEM study toward a-Ti_3C_2 MNRs.

(a & b) Reproduced with permission from ref. (Michael Naguib et al., 2011) Copyright 2011, John Wiley & Sons.

(c) Reproduced with permission from ref. (Zhubing Xiao, Zhi Yang, Zhonglin Li, Pengyue Li, & Ruihu Wang, 2019) Copyright 2019, American Chemical Society.

(d) Reproduced with permission from ref. (W. Yuan, L. Cheng, Y. An, H. Wu, et al., 2018) Copyright 2018, American Chemical Society.

(e & f) Reproduced with permission from ref. (Lian et al., 2017) Copyright 2017, Elsevier B.V.

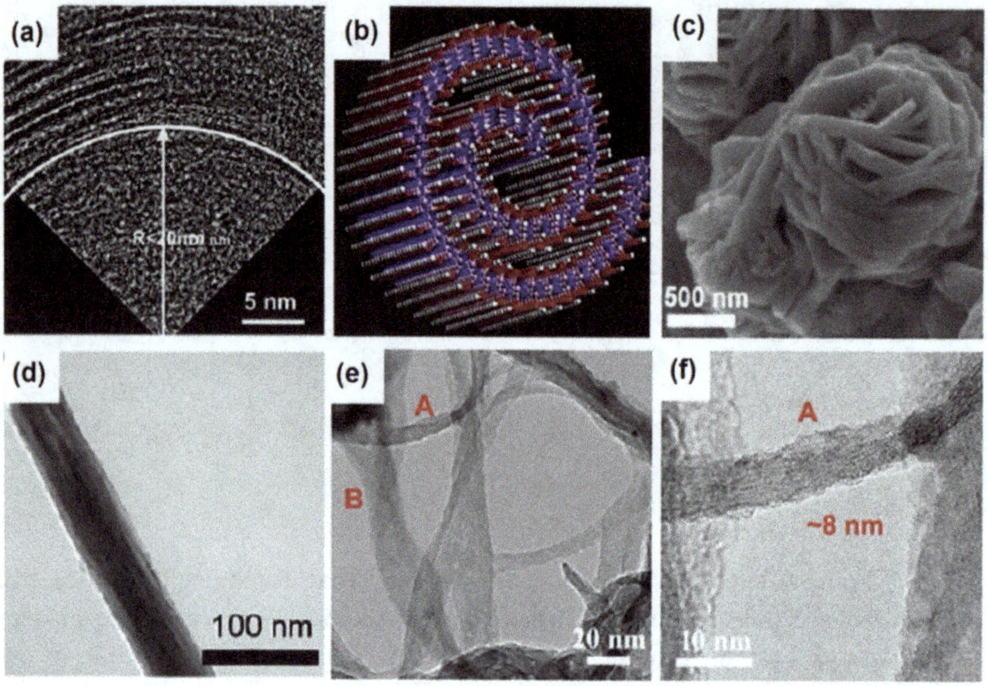

T_x, whereas exact expression is formulated by $M_{n+1}X_n(OH)_xO_yF_z$. As far as surface terminal groups are concerned their probability may be somewhere at three locations, as a first on top location of M atoms, secondly residing at hollow site among top M atoms, whereas thirdly in the form of holes lying among proceeding stacking layers for X atoms (Hu et al., 2015; Jiang et al., 2020). Generally, surface terminations attached with MXenes were arbitrarily dispersed along with mixed nature showing vital character while rendering with synthetic conditions, with transition metal elements, and finally treating after synthesis.

3. PREPARATION OF MXENES-BASED PHOTOCATALYSTS

The calcination approach, as well as hydrothermal oxidation strategy, has been extensively followed to prepare MXene-derivative photocatalysts through employing *in situ* oxidation strategy for MXene precursors. Recently, TiO_2 is studied as the most investigated semiconducting photocatalyst because of its friendly environment nature, lower cost effect, and redox potential along excellent photosensitivity. Moreover, unique TiO_2 shows certain disadvantages and difficulty to overcome large bandgap associated with highly photogenerated for carrier recombining efficiency. A lot of efforts were made for expanding light-responsive ranging and improving utility rate for photogenerated charge carriers through utilizing a sequence of approaches containing metal doping and non-metal doping, cocatalysts filling, and heterojunction architecture (Katal et al., 2020; Low et al., 2017).

3.1 Calcination Oxidation

In recent decades, 2D Ti type MXenes are supposed to be an attractive candidate to synthesize TiO_2 along with TiO_2 type photocatalysts. In this regard, Yuan et al. (Yuan et al., 2017) have very firstly published 2D layer C/TiO_2 hybrid structure that is created by oxidized CO_2 along with Ti_3C_2 (Figure 4a), whereas layered structure is successfully preserved (Figure 4b). The oxidation mechanism is rendered from CO_2 gas molecules through breaking-up of Ti-C bonding and formation of new Ti_3C_2 molecules at 700 °C, thereby making TiO_2. The synthesis method was afterward upgraded while obtaining carbon and sulfur-doped TiO_2 products of LDC-S-TiO_2/C (laminated junction composed by defect-controlled and S-doped TiO_2 with C substrate) involving sulfur intercalation owing to Ti_3C_2 associated with oxidation mechanism (Figure 4c,d) (Yuan et al., 2018). The TiO_2/C derivative of Ti_3C_2 was realized much attention in research field. Recently, Huang et al. (H. Huang et al., 2019) have achieved 2D-layered N-doped TiO_2@C by utilizing Ti_3C_2 in the form of C and Ti basis. Ti_3C_2 decorated with negative charges may be necessarily united with N-involving compounds which are regarded as positively charged melamine along with cetyltrimethylammonium bromide attained by electrostatic attraction, thereby changing composites through *in situ* method into N-doped TiO_2@C through calcination during CO_2 environment. Additionally, Kong et al. (Kong et al., 2020) have produced in order as layer-to-layer TiO_2/carbon super architecture of NPT-TiO_2/C founded by defect manufacturing for Ti_3C_2 employing nitriding-pretreatment. N atoms injection in Ti_3C_2 maintain 2D nanostructure and increase interlayer distance as 5.1 Å that becomes more beneficent toward subsequent intercalation for TiO_2 nanoplates caused by higher-temperature oxidation. An experimental study has confirmed that layer-by-layer intercalation belonging to superstructures may effectively prove efficient separation related with photogenerated carriers which in turn provide better active sites. Wu et al. (Z. Wu et al., 2020) have explored graphene/TiO_2/g-C_3N_4 through calcination of Ti_3C_2 with melamine during air environment. Resultantly, graphene decoration on TiO_2 surface behaves like an active mediator for Z-scheme favoring transformation of electrons.

To date, less amount of literature concerning non-Ti_3C_2 MXenes hybrid nano photocatalysts have been published. As an illustration, depending upon the oxidative property for Nb_2C MXene, Nb_2O_5/C/Nb_2C hybrid has been prepared following CO_2 oxidation mechanism (Su et al., 2018). Furthermore, amorphous carbon is provided among Nb_2O_5 and Nb_2C, thereby generating the Schottky barrier which in turn improves charge transformation along with carrier separation. Nb_2O_5/C/Nb_2C/g-C_3N_4 heterojunctions have also been achieved by calcination of Nb_2O_5/C/Nb_2C with melamine mixture performing at 550 °C during N_2 environment (Jiang et al., 2020). As a familiar technique, the morphology owing to nano photocatalysts performs vital role toward catalytic activity. Li et al. (2018) have successfully published Ti_3C_2- TiO_2 flowery nanoparticles through adopting many steps reaction mechanisms (Figure 4e). Firstly, $Na_2Ti_3O_7$-Ti_3C_2 hybrid products have been collected during hydrothermal atmosphere at 140 °C for 12 hrs. Afterward, $Na_2Ti_3O_7$-Ti_3C_2 nanocomposites are attained through substituting Na^+ species with H^+ ions by adding them into dilute-HCl solution. Consequently, Ti_3C_2-TiO_2 nanoflowers are significantly prepared via calcining at various temperatures (Figure 4f–h).

Figure 4. (a) Schematically pictorial for synthesis process, (b) Schematic SEM results for C/TiO$_2$. (c) Schematically illustrated synthesis process and (d) Prominently TEM photographs for LDC S TiO$_2$/C. (e) Schematically illustration for synthesis of Ti$_3$C$_2$-TiO$_2$ nanoflower-type and (f-h) The conforming morphology appearing at various temperatures.

(a&b) Reproduced with permission from ref. (Yuan et al., 2017) Copyright 2017, John Wiley & Sons.

(c&d) Reproduced with permission from ref. (W. Yuan, L. Cheng, Y. An, S. Lv, et al., 2018) Copyright 2018, John Wiley & Sons.

(e-h) Reproduced with permission from ref. (Y. Li et al., 2018) Copyright 2018, Elsevier B.V.

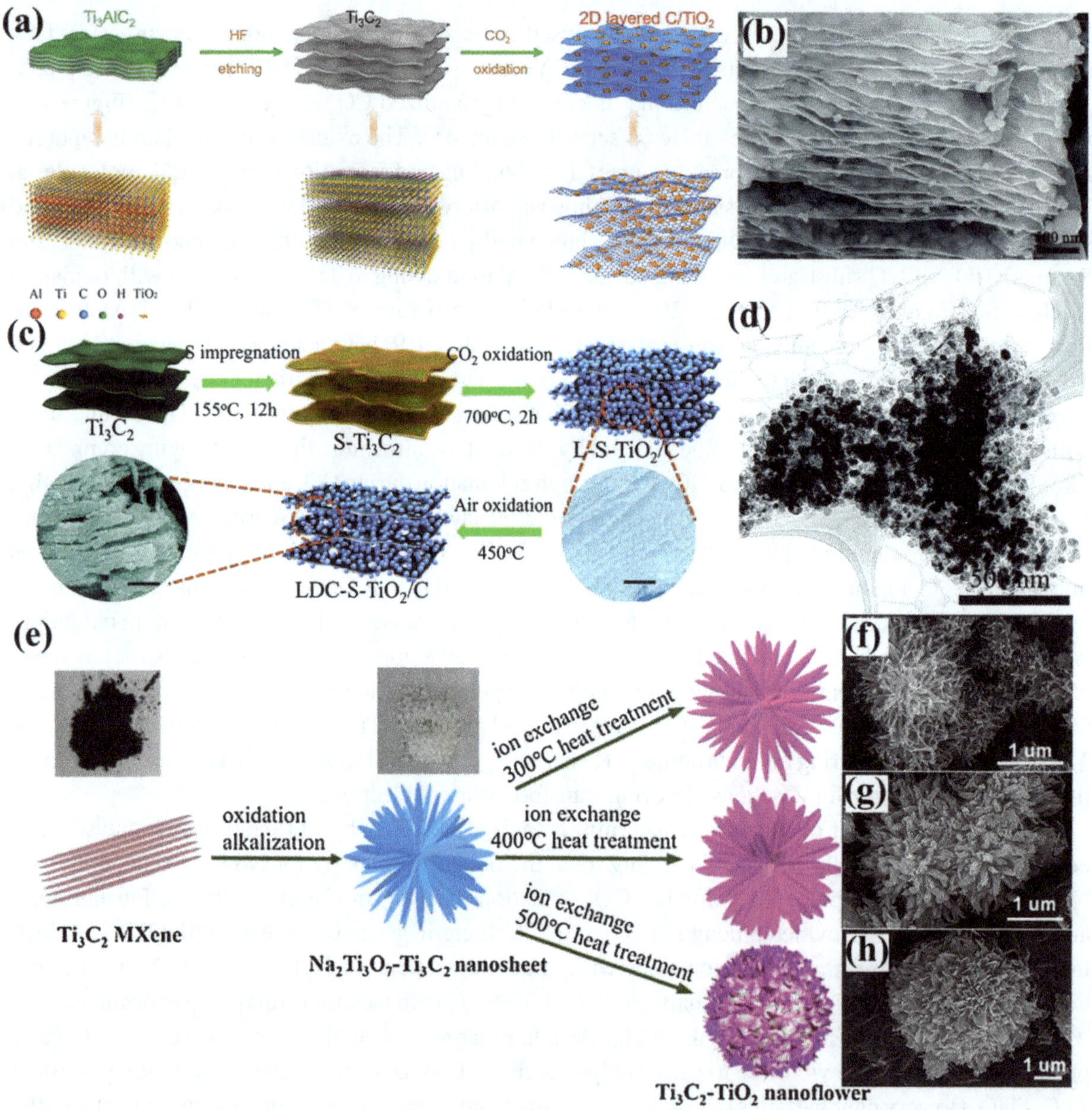

3.2 Hydrothermal Oxidation

The calcination approach and hydrothermal strategy are both considered favorable for the synthesis of MXene-derivatives as photocatalysts; rather MXenes may be entirely or partly oxidizable while incorporating it into TiO_2 or TiO_2/MXene composites. Currently, Li et al. (Li et al., 2019) have architecture 2D/2D/2D Ti_3C_2/MoS_2/TiO_2 composites following two-phase hydrothermal mechanism (Figure 5a). TiO_2 nanosheets are formed by *in situ* strategies for Ti_3C_2, whereas combined MoS_2 and TiO_2 have developed photogenerated charge carriers on TiO_2 surface, and enhanced photocatalytic activity (Figure 5b, c). Additionally, molybdenum vacancies were achieved for Mo_xS/TiO_2/Ti_3C_2 composites through hydrothermal procedure by utilizing $NaBH_4$ in the form of reducing agents. The molybdenum vacancies have increased density of active sites associated with inhibited charge carrier recombination.

Figure 5. (a) Schematically presentation for synthesis of Ti_3C_2@TiO_2@MoS_2 composites. (b, c) Schematically photocatalytic performance toward Ti_3C_2@TiO_2@MoS_2 composites irradiated with solar light. *Reproduced with permission from ref. (Y. Li, Yin, et al., 2019) Copyright 2019, Elsevier B.V.*

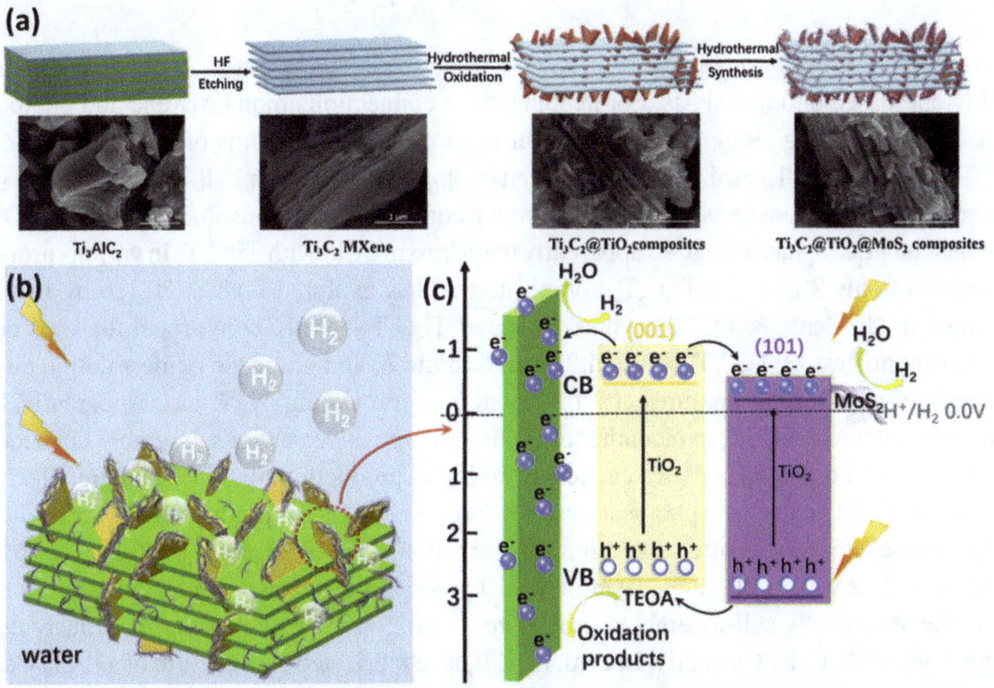

Further, two-phase hydrothermal method is utilized to design novelty into an octahedral phase of WS_2@TiO_2@Ti_3C_2 as photocatalyst, and significantly Ti_3C_2 with WS_2 have played vital role like electron acceptors (Li et al., 2019). Shao et al. (B. Shao et al., 2020) have hydrothermal approach to synthesize Ti_3C_2/TiO_2 as it was decorated by black phosphorus (BP) composites. A number of characterizations have revealed Ti_3C_2 along with Schottky barrier created among BP as well as Ti_3C_2/TiO_2 that have prominently developed photocatalytic performance. Not only calcination but also hydrothermal routes and some exceptional strategies likewise higher-energy ball-milling and wet-chemically oxidized form that were employed for preparing MXene-derivative photocatalysts. As an illustration, Cheng et al. (Cheng et al., 2018) have fabricated Ti_3C_2 nanosheets with 30% H_2O_2 addition at normal temperature for fabricating Ti_3C_2/TiO_{2-x} nanodots. Thereafter, Sun et al. (Y. Sun et al., 2019) have utilized hot water at 60 °C for oxidizing Ti_3C_2 precursors, and thereby obtaining collection of TiO_2/Ti_3C_2@amorphous nature carbon hybrids. Generally, similar preparation approaches for MXene-based photocatalysts as well as MXene-derived photocatalysts were hydrothermal along with calcination routes. The main difference between two approaches was that of synthesis for which former requires preventing oxidation for MXenes through governing particular reaction environment. On comparing between hydrothermal as well as calcination approaches, the most suggested routes were mechanical/ultrasonic mixing as well as electrostatic self-assembly as they were very straight forward for synthesizing MXene-based photocatalysts during mild environment. As far as MXene-derived photocatalysts are concerned, there was existence of typical calcination oxidation containing residual occupancy for carbon nanomaterials that may be useable in the form of cocatalyst because of enhancing visible light adsorption as well as inhibiting recombination for charge carriers (Zhong et al., 2021).

3.3 Mechanical and Electrostatic Self-Assembly

The mechanical/ultrasonic combined form approach has been suggested as the simplest strategy to prepare MXene-based photocatalysts. For keeping close connection among MXene and photocatalyst, highly mechanical stirring along with large power of ultrasonic vibration is often practiced (Zhong et al., 2021). As an example, Li et al. (2020) have been equipped with chlorophyll-a (Chl)-derived whereas multilayered $Ti_3C_2T_x$ composite as Chl/$Ti_3C_2T_x$ through combining both constituents until dry. 2D layered structure for Chl/$Ti_3C_2T_x$ has not varied apparently while mixing Chl with $Ti_3C_2T_x$ in various proportions, thereby showing only 2% of Chl/$Ti_3C_2T_x$ composite that has explored highest H_2 return. In the same way, Chen et al. (Ji, Zeng, & Li, 2019) have prepared $Ti_3C_2T_x$-based 3D hydrogel that was collected through stirring process of $Ti_3C_2T_x$ solution, Eosin Y solution, and graphene oxide solution during 0.5 hours, afterward maintaining mixtures at 70 °C along with protecting by N_2 gas. Apparently, because of electrostatic interactions, the semiconductors containing positive carrier's assembly connected with MXenes along with numerous negative carriers resulting in production for 0D/2D nature, 1D/2D form, or 2D/2D nature for MXene-based photocatalysts. Monolayer MXenes are induced to cluster collectively, during this course they may be often sonicated prior to mixing procedure. Su et al. (Su et al., 2019) have fabricated 0D P25 as well as 2D single layer of $Ti_3C_2T_x$ composites that were aided with ultrasonic mixing. In an electrostatically self-assembled procedure, the P25 units may be homogeneously dispersed over single layer of $Ti_3C_2T_x$. Currently, Li et al. (2020) have fabricated architecture of 1D/2D CdS/Ti_3C_2 composites through rendering electrostatically self-assembled procedure, as illustrated in Figure 6a. A wide variety of characterizations have proved well-combined forms of Ti_3C_2 nanosheets associated with CdS nanowires by utilizing electrostatic attraction, whereas CdS nanowires are well-decorated over Ti_3C_2 nanosheets (Figure 6b-d). Moreover, the architecture for 2D/2D photocatalysts is shown attractable attention. Further, the 2D/2D structure has presented favorable impact on photocatalytic characteristics. In addition, on comparing 0D/2D with 1D/2D nanostructures, the 2D/2D nanostructures possess largely touch areas while intimating interfacial interaction, thereby improving effective electron exploitation. Classically, 2D Ti_3C_2, as well as 2D g-C_3N_4, may be rendered through ultrasound along with mechanical stirring for set-up of 2D/2D Ti_3C_2/g-C_3N_4 composite (Figure 6e-g) (Y. Yang et al., 2019). As revealed in

Figure 6g, the connection among two constituents may be regarded as intimate due to which interfacial junction has successfully formed.

Figure 6. (a) Schematically presentation of showing fabrication for CdS/Ti_3C_2 composite. Prominently SEM photographs owing to (b) Ti_3C_2, (c) CdS and (d) CdS/Ti_3C_2 composite. Also TEM study for (e) Ti_3C_2, (f) $g\text{-}C_3N_4$ and (g) $Ti_3C_2/g\text{-}C_3N_4$.
(a-d) Reproduced with permission from ref. (J.-Y. Li et al., 2020) Copyright 2020, Elsevier B.V.
(e-g) Reproduced with permission from ref. (Yang et al., 2019) Copyright 2019, Elsevier B.V.

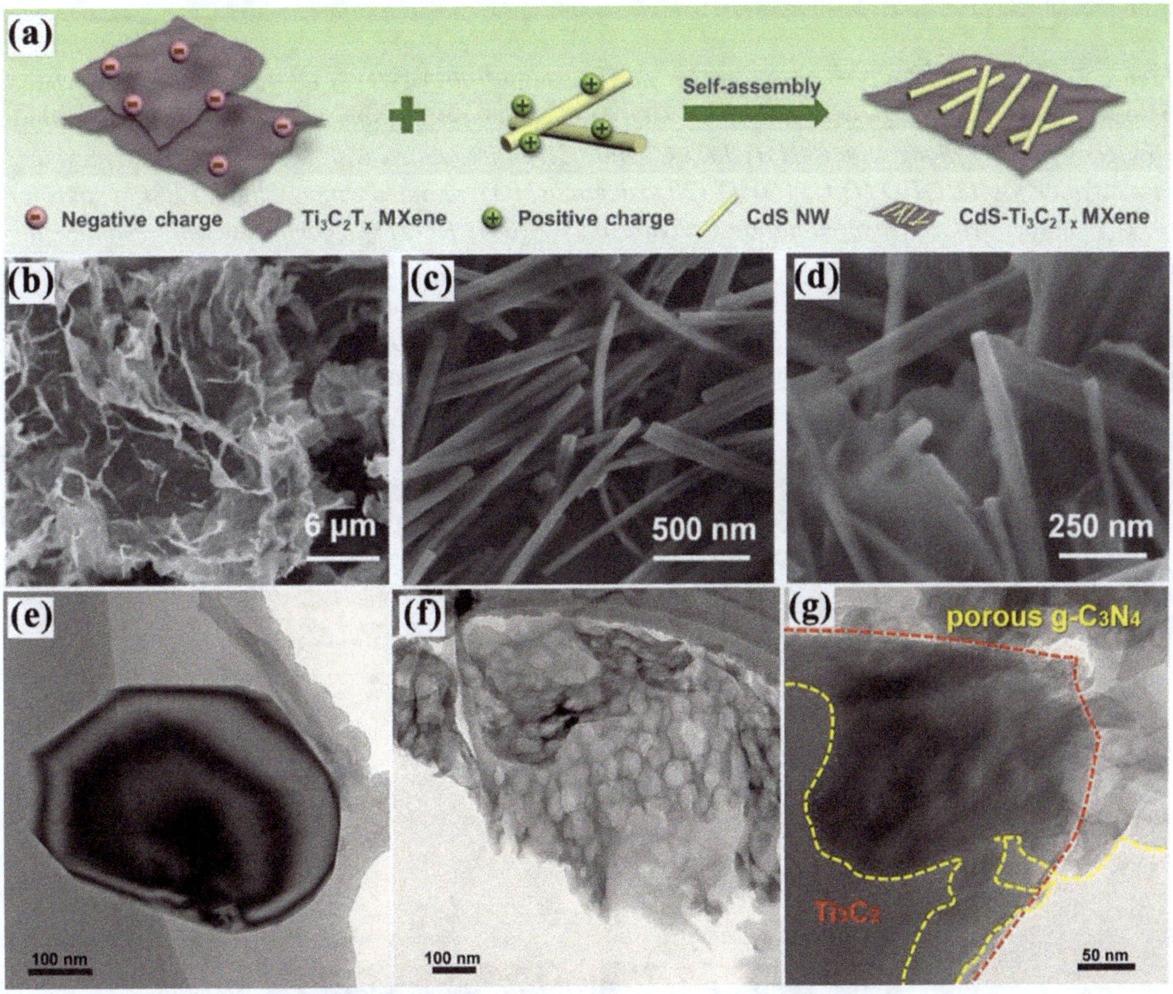

4. MXENE-BASED COMPOSITE PHOTOCATALYSTS

4.1 MXene-Metal Oxides

TiO_2 is considered the best photocatalyst as has been widely applied to break down organic as well as inorganic pollutants because of possessing high performance, abundant availability, environment-friendly type, and fine chemical stability. It is remarkable as photocatalytic characteristics belonging toTiO_2 are narrowly connected to its nanostructure caused by unlike atomic configuration as well as surface energy related to variety of crystalline facets. The (001) facets corresponding to anatase nature of TiO_2 have revealed greater photocatalytic performance as compared with (001) facets (Xiao et al., 2011). Therefore, the (001) TiO_2/Ti_3C_2 composite is synthesized by employing superficial hydrothermal oxidation route. In this way, 2D $Ti_3C_2T_x$ MXene offers an ideal platform for loading TiO_2, whereas Ti-decorated atoms on Ti_3C_2 create suitable nucleation sites. Therefore, interfacial heterojunction existing among TiO_2 and Ti_3C_2 nanosheets based on atomic-scale become conducive in nature so that reduction process related with defect-oriented recombination for photoexcited ions must be assured.

Figure 7. (a) Schematically charge-transfer mechanism on (001) TiO_2/Ti_3C_2, (b) Schematically band associations with charge flowing into {001} TiO_2-Ti_3C_2 interfaces, (c) scavenger's impact on degrading efficiency for MO along with (001)TiO_2/Ti_3C_2-160 °C-12h irradiated with UV light, (d) the recycling investigation for (001)TiO_2/Ti_3C_2-160 °C-12h to degrade MO irradiated with UV light. FESEM study for (e) Ti_3C_2 associated with layered-nature structure, and (f) (001) TiO_2/Ti_3C_2 composite while synthesizing by hydrothermal strategy at 160 °C during 12 hrs. The inset view (f) showing of enlargement as HRSEM study for TiO_2-Ti_3C_2 heterojunctions existing in area indicated by green squares, whereas schematically illustration owing to anatase-TiO_2 crystal structure corresponding to {001} as well as {101} facets.
Reproduced with permission from ref. (Peng et al., 2016) Copyright 2016, American Chemical Society.

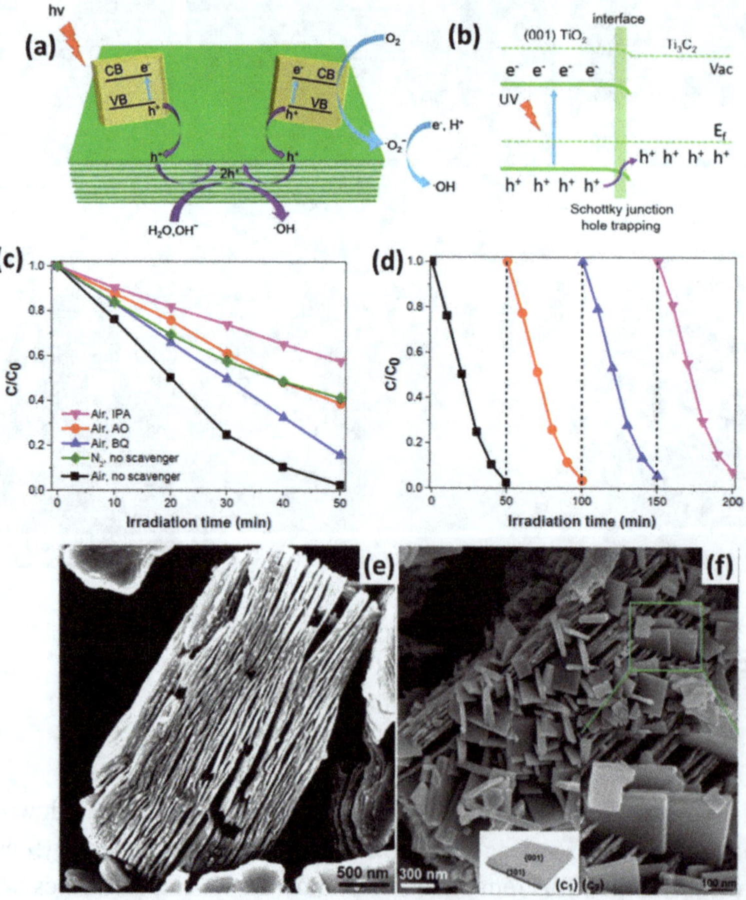

Moreover, photocatalytic oxidation mechanism for methyl orange (MO), the hydroxyl species (OH) are considered the most prominent reactive substances, that are significantly investigated with the addition of variety of scavengers. Three scavengers comprising of efficient isopropanol [IPA], most reactive ammonium oxalate [AO] with benzoquinone [BQ]) maybe impeded during reaction process revealing the inequality relation as IPA > AO > BQ have been successfully studied (Peng et al., 2016). Probably, mechanisms \concerned with photocatalytic degradation of MO illustrated in Figure 7a, b; prominently photogenerated holes rather than electrons may jump from TiO_2 toward Ti_3C_2-OH caused with work function for Ti_3C_2-OH becomes lower as compared with TiO_2 surface, that is 6.578 eV of (101) surface accordingly. Additionally, the Schottky barrier generated among Ti_3C_2 and TiO_2 inhibits reverse movement of photogenerated holes, thereby offering to prolong duration for electrons. Figure 7e,f shows the surface morphology of Ti_3C_2 and TiO_2 with 001 exposed plane. The degradation efficiency of MO drops dramatically with N_2 bubbling, as seen in Figure 7c, d, revealing that MO degradation is a photocatalytic oxidation (PCO) process. The process may be quenched by three scavengers in the following order: IPA > AO > BQ, demonstrating that ROS contributes more to MO degradation than h^+. According to these findings, the most significant reactive species in the TiO_2-catalyzed PCO process of MO is •OH, which is in agreement with Hirakawa and Nosaka (2002). The combined effect of Fe_2O_3 with Ti_3C_2 MXene significantly exhibits highly efficient performance toward photocatalytic remediation from dirty water (H. Zhang et al., 2020). The α-Fe_2O_3/$ZnFe_2O_4$@Ti_3C_2 composite has presented photocurrent feedback indicating low carrier transformation hurdles belonging to Ti_3C_2 MXene possessing fine conductivity and showing close connection among α-Fe_2O_3/$ZnFe_2O_4$ as well as Ti_3C_2, investigating via transient photocurrent response evaluations as well as electrochemical impedance spectroscopy results. In addition, the CeO_2/Ti_3C_2 MXene composite comprises of Schottky junction resulting from integrated electric field hindering photoexcited carrier recombination caused by increasing significant photocatalytic performance (Shen et al., 2019). Moreover, ZnO-nanorods/MXene product may be synthesized employing ultrasonic oscillation, thereby removing maximum rate of 98% for RhB (Liu & Chen, 2020). In this way, superior activity may be investigated toward increasing surface area influenced by ZnO nanorods whereas reducing opposition for transformation of photoexcited electron-hole pairing is greatly affected by MXene.

4.2 MXene-Metal Sulfides

Certainly, metal sulfides like that of CdS, ZnS, and In_2S_3 have played a vital role in photocatalysis field (Ran et al., 2017; H. Wang et al., 2018). Intentionally, for achieving developed photocatalytic performance for metal sulfides, a ternary Ti_3C_2-OH/ln_2S_3/CdS composite containing photocatalytic system may be fabricated by employing hydrothermal method (H. Fang et al., 2019). Figure 8a,b reveal as-prepared products in the form of spherical structure associated with great surface area for 4-TIC (Ti_3C_2-OH/ln_2S_3/CdS along with 4 wt% Ti_3C_2-OH), that is considered promising to absorb dye molecules, thereby providing suitable active sites. In addition, more favorable electrical conductivity owing to Ti_3C_2-OH whereas intimating contacts existing among above mentioned three materials likely ensured healthy degrading efficiency for Ti_3C_2-OH/ln_2S_3/CdS hybrid. Photocatalytic stuff belonging to some other ternary catalyst (In_2S_3/anatase TiO_2@ $Ti_3C_2T_x$) have been investigated to degrade MO as well as tetracycline hydrochloride (TC) (H. Wang et al., 2018). The aforesaid quasi-core-shell photocatalytic system, Ti_3C_2 MXene connected with derivative of TiO_2 has offered a robust platform for loading In_2S_3, whereas additionally the surface with edge relating toTi_3C_2 MXene may be enveloped by plentiful and irregular In_2S_3 (Figure 8e-h). Remarkably, type-II heterojunction associated with Schottky junction architecture with In_2S_3, TiO_2,

Figure 8. (a) Schematically SEM study for 4-TIC photocatalytic mechanism and (b) enlarging view for designated area in (a). (c) Prominently charge separation along with transfer, (d) the proposing mechanism owing to contaminant removal from InTi-16 system irradiated with visible light. Schematic FESEM study for (e, f) $Ti_3C_2T_x$ as well as InTi-16, (g, h) for In_2S_3/anatase TiO_2@ $Ti_3C_2T_x$, and (i) Photocatalytic activity for MO under time, (j) the pseudo-first-order rate constants toward individual samples.
(a&b) Reproduced with permission from ref. (H. Fang et al., 2019) Copyright 2019, Elsevier B.V.
(c-j) Reproduced with permission from ref. (H. Wang et al., 2018) Copyright 2018, Elsevier B.V.

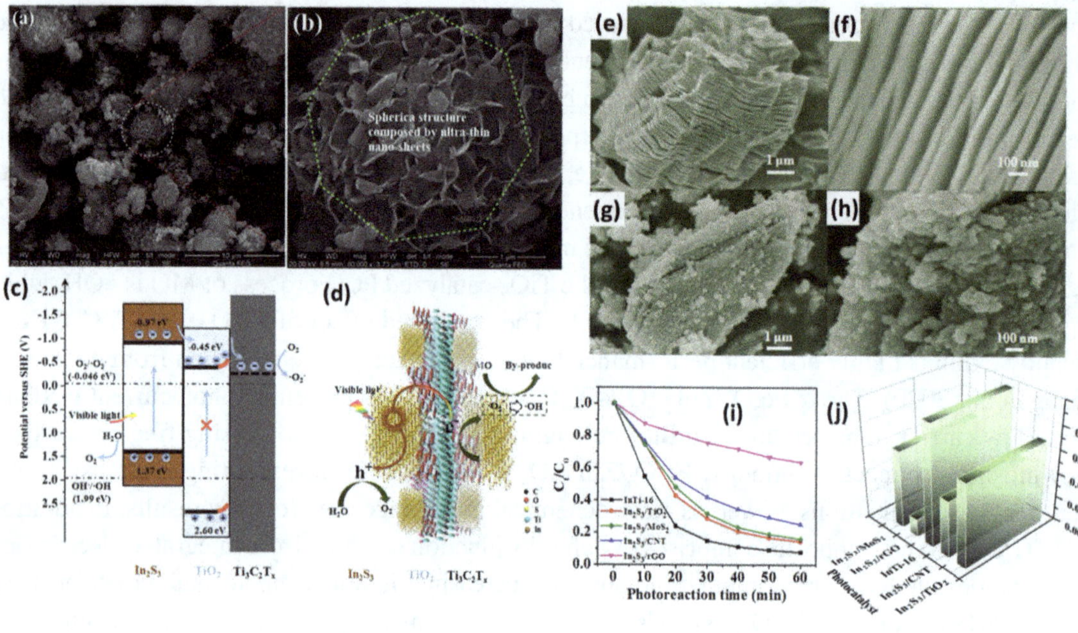

and $Ti_3C_2T_x$ provides a wide variety of channels facilitating the migration for photoexcited electrons. The charge separation mechanism along with transferring process has clearly revealed in Figure 8c, d, and exposed with visible light, electrons are significantly transferred from In_2S_3 toward TiO_2 penetrating the heterojunction. Further, the Schottky junction activation among TiO_2 and $Ti_3C_2T_x$ may boost up charge separation spatially and transferring and blocking revert flow for diffusion of electrons. Figure 8i,j shows photocatalytic degradation with time and pseudo-first order rate constants of all samples.

4.3 MXene-g-C_3N_4

The Ti_3C_2/porous g-C_3N_4 (Ti_3C_2/PCN) heterostructures as photocatalysts associated with day and night photocatalytic performance have been structured by investigating Liu et al. (Liu et al., 2020) following superficial vacuum filtering strategy. This ideal photocatalyst scheme, offer porous nature g-C_3N_4 displaying highly efficient visible light absorption with the connection of providing abundant photocatalytic active sites closely related with great specific surface area. Particularly, for 2D Ti_3C_2 MXene, the prominent van der Waals heterostructures forming among Ti_3C_2 as well as g-C_3N_4 may boost up dynamics of photoexcited charge carriers extracting from g-C_3N_4 toward MXene. During this interval but without visible light, Ti_3C_2 may exploit the retained electrons for decomposing existing pollutants. Electrons residing on conduction band for g-C_3N_4 are vastly excited toward valence band through visible light as illustrated by equation 1. Resultantly, photoexcited electrons transfer from permeable g-C_3N_4 toward Ti_3C_2 surface as indicated by equation 2, whereas electrons reacting with O_2 for producing O_2 are represented by equation 3. Moreover, additional generation of OH radicals is also given by equation 4, respectively. Further, during exposing sunlight of h^+, O_2 as well as OH involved for breaking down pollutants like that of phenol toward CO_2 and water. When the sun is set, then without light contaminants may continue to degrade via O_2 and OH, which has been formed through storage of electrons into Ti_3C_2 as illustrated by equations 3 and 4, respectively. As shown in Figure 9a, a visual illustration showing functioning process owing to Ti_3C_2/porous g-C_3N_4 as a photocatalyst during day-night reactions is given as under:

$$PCN + hv(\text{visible light}) \rightarrow PCN(e^- + h^+ \quad (1)$$

$$PCN(e^-) + Ti_3C_2 \rightarrow PCN + Ti_3C_2(e^-) \quad (2)$$

$$Ti_3C_2(e^-) + O_2 \rightarrow \Delta O_2^- \quad (3)$$

$$2{\bullet}O_2^- + 2H^+ \rightarrow \Delta OH \quad (4)$$

It has been reported that well-known ternary photocatalysts are one of those utmost utilitarian choices due to which photocatalytic efficiency may be greatly developed. As an illustration, 2D TiO_2@Ti_3C_2/g-C_3N_4 as a ternary heterojunction has also been fabricated by employing an ultrasonic-assisted calcination strategy. As a result, heterojunctions, as well as Schottky barriers, are formed into TiO_2@Ti_3C_2/g-C_3N_4 composite that becomes conducive in nature toward photocatalytic activities, as given in Figure 9b, c (Ding et al., 2019).

4.4 MXene Bi-Based

Bismuth oxyhalide being a photocatalyst has been studied as a common catalytic material, occupying an indispensable position relevant with photocatalysis having environment-friendly behavior, favorable energy band, non-toxic nature, and other peculiarities (Wu, Su, Zhu, Zhang, & Zhu, 2020). The problems associated with rapid recombination of photo-induced charge carriers may be handled inherited in naked BiOBr (Raza et al., 2021; Wei et al., 2021). In this regard Huang et al. (Q. Huang, Liu, Cai, & Xia, 2019) have fabricated BiOBr/Ti_3C_2 composite via self-assembly strategy based on electrostatically enforcement, thereby displaying outstanding photocatalytic efficiency to degrade Cr(VI), active 2,4-dinitrophenol as well as RhB, on comparing with simple BiOBr, the BiOBr/Ti_3C_2 has demonstrated highly efficient degradation capability (Huang et al., 2019). In this way, photocatalytic performance offered by BiOBr/Ti_3C_2 MXene is essentially originated from close connection as well as Schottky junction that was created among BiOBr along with Ti_3C_2 interface, as depicted in Figure 10a, b. On the other hand, one more bismuth oxyhalide (BiOCl) has been incorporated into$Ti_3C_2T_x$ MXene for achieving desired results connected with photocatalytic performance. Moreover, degradation reaction constant owing to BT-2.0 showing 2.0%wt Ti_3C_2 is significantly 3.3 folds superior to pristine BiOCl (C. Wang, Shen, Chen, Cao, & Jin, 2020). Clearly, manufacturing mechanism belonging to active species has been

Figure 9. (a) Schematically mechanism owing to day-night photocatalytic performance related with porous nature g-C_3N_4 nanolayers. (b, c) Schematically exhibition for charge transfer as well as separation through TiO_2@Ti_3C_2/g-C_3N_4 heterojunction toward RhB degradation.
(a) Reproduced with permission from ref. (N. Liu et al., 2020) Copyright 2020, Elsevier B.V.
(b&c) Reproduced with permission from ref. (Ding et al., 2019) Copyright 2019, Springer Nature.

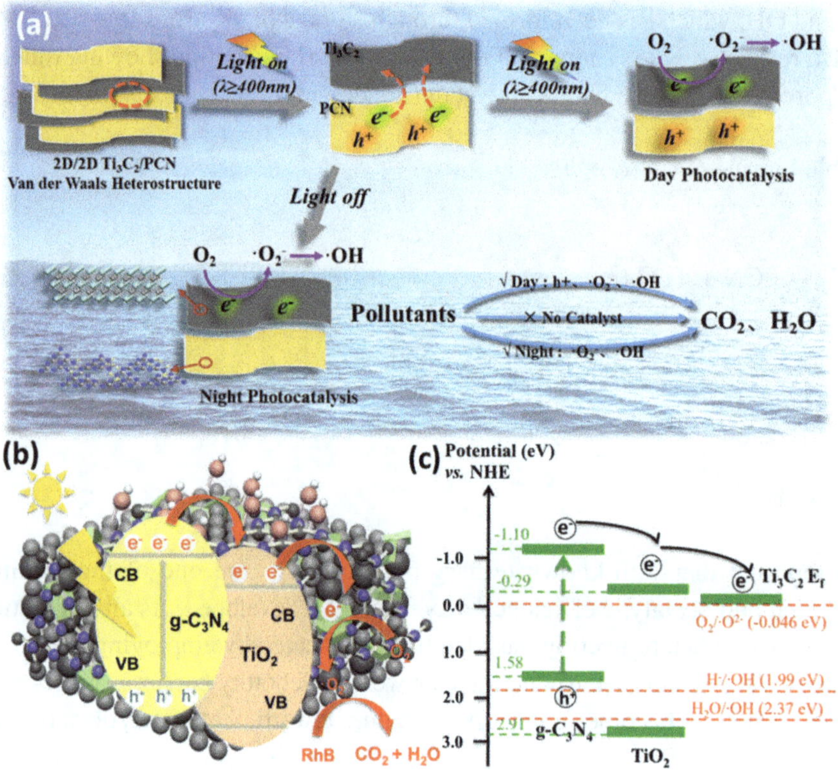

illustrated in Figure 10c. Further, Figure 10d reveals that 40% Ti_3C_2-Bi/BiOCl has offered significantly red shift toward visible light as compared with BiOCl as well as Bi/BiOCl caused by whole-spectrum absorbance for Ti_3C_2. Remarkably, the photoelectrons generated on semi-metallic bismuth which are resulted because of surface-plasmon-resonance impact, are conveyed toward Ti_3C_2 surface via direct and/or indirect channels. In addition to bismuth oxyhalides, there is another synergetic photocatalytic material Bi_2WO_6 forming Schottky junction contact with Nb_2CT_x MXene for boosting the catalytic performance. The flowery structure of Bi_2WO_6/Nb_2CT_x, as shown in Figure 10e, has presented wide BET (Brunauer–Emmett–Teller) surface-area as compared with pristine Bi_2WO_6, thereby resulting numerous active sites present in Bi_2WO_6/Nb_2CT_x hybrid (Cui et al., 2020). The optimum BET surface-area stemming with the involvement for introducing Nb_2CT_x MXene, which one reduce thickness owing to Bi_2WO_6 and forming little holes residing between Bi_2WO_6 nanosheets. Consequently, photocatalytic performance attributing to supporting nature semiconductor photocatalysts are suggested better one to degrade pollutants as indicated by Table 3.

Figure 10. Schematic diagram showing energy-band-structure belonging to Ti_3C_2 as well as BiOBr (a) prior and (b) post establishment for Schottky barrier. (c) Efficient photodegradation mechanism for BT-2.0. (d) The UV-vis DRS owing to individual as-prepared samples. (e) Clearly view for SEM image belonging to BN-2

(a&b) Reproduced with permission from ref. (Q. Huang et al., 2019) Copyright 2019, Elsevier B.V.
(c) Reproduced with permission from ref. (C. Wang et al., 2020) Copyright 2020, Elsevier B.V.
(d) Reproduced with permission from ref. (S. Wu et al., 2020) Copyright 2020, Elsevier B.V.
(e) (Cui et al., 2020) Copyright 2020, Elsevier B.V.

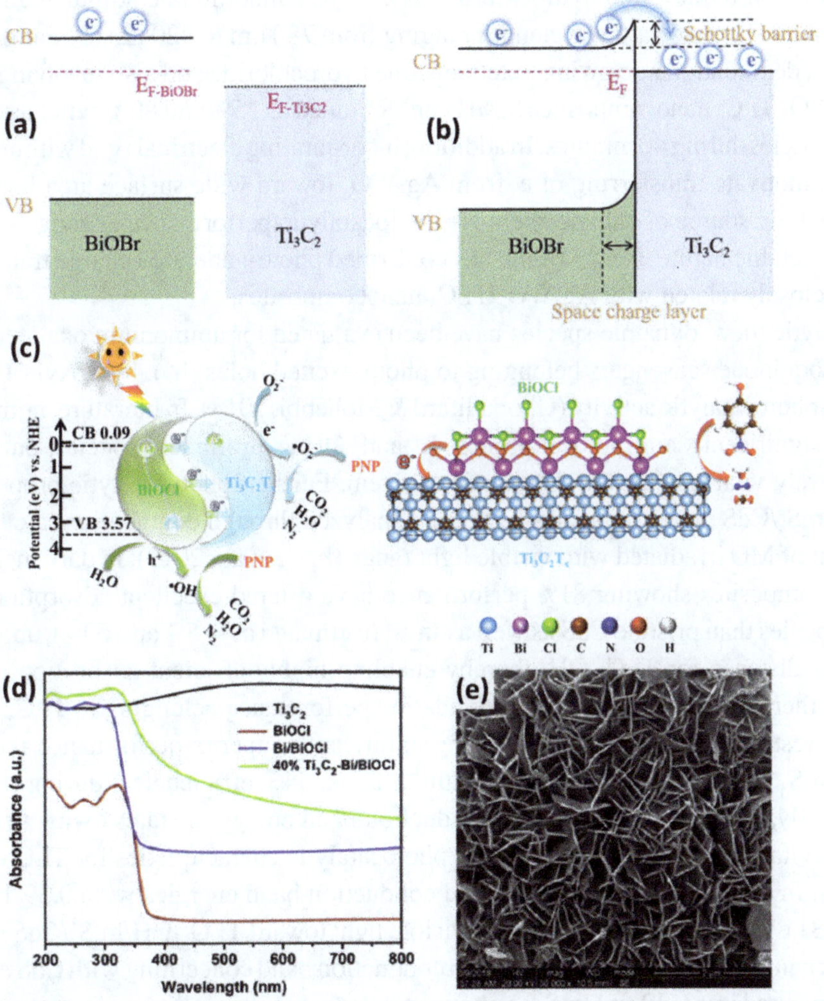

4.5 Others MXenes-Based Photocatalysts

Ag_3PO_4 was freely changed into Ag^+ species for photocatalytic reaction. In this way, Ag_3PO_4/Ti_3C_2 nanocomposite has been adopted for enhancing photocatalytic degrading activity in favor of various organic pollutants, like MO, tetracycline hydrochloride, and chloramphenicol irradiated with visible-light of 300 W for Xe lamp along with $\lambda > 420$ nm) (Cai et al., 2018). Above mentioned compounds show degradation from MXene nanocomposite in a specific order of tetracycline hydrochloride (0.32) > MO (0.094)

> chloramphenicol (0.025) depending upon rate constants (min^{-1}). Generally, a few hidden inorganic along with organic compounds exist in combined form into water surface and/or wastewater which may significantly influence photodegradation toward desired pollutants. In addition, humic acid, as well as fulvic acid, have offered clearly inhibitory effect upon degradation rate, caused by complete activity occurring among natural organic compounds as well as tetracycline hydrochloride behaving like humic acid associated with fulvic acid toward creation of photo-affected holes along with ·OH radicals (Long et al., 2017). Photocatalytic peculiarities for Ag_2WO_4/Ti_3C_2 nanocomposites have been investigated with respect to remediation tetracycline hydrochloride as well as sulfadimidine stimulated by visible light possessing 300 W Xe lamp toward wavelength ranging from 780 nm to 420 nm (Y. Fang, Cao, & Chen, 2019). Moreover, degradation rate owing to tetracycline hydrochloride and also involving sulfadimidine along with $Ag_2WO_{4/}Ti_3C_2$ nanocomposites have been examined as 25% and 80% better as compared with Ag_2WO_4 nanoparticles during 40 minutes. In addition, incorporating electrical field within $Ag_2WO_{4/}Ti_3C_2$ nanocomposites motivate transferring of e^- from Ag_2WO_4 toward wide surface area for Ti_3C_2 MXene, thereby creating a big source of enhancement for photocatalytic performance (Zhang et al., 2020). The conductivity of metallic nature Ti_3C_2 MXene has confirmed photo-generated charge transfer to enhance duration that is closely related with Ag_2WO_4/Ti_3C_2 nanocomposites.

In Photocatalytic view, dynamic species have been evaluated for ammonium oxalate, active isopropanol, and benzoquinone scavengers belonging to photo-excited holes (h^+), hydroxyl (·OH), and oxide ($\cdot O_2^-$) to endorse photocatalytic activity (Ghobadifard & Mohebbi, 2018). In literature, numerous findings have suggested significantly for removal process h^+ and ·OH as prominent species, out of which ·OH plays comparatively vital role toward degradation system. Further, photocatalytic properties inherited into Ti_3C_2-OH/ln_2S_3/CdS nanocomposites have been analyzed through degradation rate for Rhodamine B (RhB) and that of MO irradiated with visible light range (Fang et al., 2019). in dark media, Ti_3C_2-OH/ln_2S_3/CdS nanocomposites showing 81% performance have offered excellent adsorption performance owing to RhB species than pristine CdS as well as In_2S_3 (estimated as < 5% and 60%), apparently caused by suitable layer distance for Ti_3C_2-OH, thereby enabling highly efficient adsorption toward organic compounds. Furthermore, in favor of photodegradation performance belonging to Ti_3C_2-OH/ln_2S_3/CdS nanocomposites (estimated 95%) that are exploring significantly superior performance as compared with pure CdS and In_2S_3 photocatalysts (indicated with 48% and 54% efficiencies) during (8 min) interval of time. Resultantly, valence band as well as conduction band energy of states owing to photocatalysts have become inevitable parameters to determine photocatalytic characteristics for Ti_3C_2-OH/ln_2S_3/CdS nanocomposites. In this regard, valence band and conduction band energies were 0.39/1.74 eV of CdS whereas 0.89/1.31 eV of In_2S_3. Upon irradiating visible light toward Ti_3C_2-OH/ln_2S_3/CdS photocatalysts, e^- is excited which jump from valence band toward conduction band concerning with CdS as well as In_2S_3. Ti_3C_2-OH incorporated into CdS and In_2S_3 as photocatalysts may have facilitated instantaneously transfer of photo-oriented e^- toward Ti_3C_2-OH emitted from conduction band corresponding to photocatalyst, which on turn probably enhance suitable separation for photo-induced e^- and h^+ pairs (Zou et al., 2019). In literature, photocatalytic performance associated with Ti_3C_2-OH/Bi_2WO_6:Yb^{3+}/Tm^{3+} nanocomposites have been analyzed through removing RhB irradiated by visible region, whereas visible/near-infrared region and close-infrared light are also studied, respectively (Fang et al., 2020).

Out of three situations, remediation for RhB has suggested better one irradiated under visible/near-infrared limitations. Moreover, photocatalytic performance for Ti_3C_2-OH/Bi_2WO_6:Yb^{3+}/Tm^{3+} was remarkably preferred than pure Ti_3C_2-OH or Bi_2WO_6 irradiated with visible/near-infrared region. Findings were preliminarily advanced toward heterojunction configuration for Ti_3C_2-OH/Bi_2WO_6 as well as outstanding electrical conductivity in favor of Ti_3C_2-OH. Generally, a notable and ideal electrical conductivity owing to Ti_3C_2-OH into Ti_3C_2-OH/Bi_2WO_6:Yb^{3+}/Tm^{3+} nanocomposites apparently increase charge pairs separation, thereby decreasing photo-induced e-h^+ pair reoccurring to improve photocatalytic reactions for Ti_3C_2-OH/Bi_2WO_6:Yb^{3+}/Tm^{3+} nanocomposites. To add, an improvement in heterojunction formation among Ti_3C_2 - OH as well as Bi_2WO_6 may accelerate photocatalytic reactions attributing with nanocomposites (Cao et al., 2018). However, an extra concentration for Ti_3C_2-OH leads to serious scarcity of nanocomposite photocatalytic mechanism. The reason behind overabundant Ti_3C_2-OH offers light absorption toward pure Bi_2WO_6 (Zhang et al., 2018). To visible/near-infrared light, photosensitive lanthanide ions owing to Yb^{3+} may consecutively absorb near-infrared light irradiation, whereas photoactive lanthanide ions owing to Tm^{3+} may also simultaneously renovate near-infrared light toward ultraviolet region as well as visible light range (Fang et al., 2020). Consequently, the Ti_3C_2-OH/Bi_2WO_6:Yb^{3+}/Tm^{3+} photocatalysts are considered for absorbing an ultraviolet as well as visible light instantaneously to generate e^- and h^+ species into valence bands and conduction bands.

5. STABILITY OF PHOTOCATALYSTS

To find stability owing to photocatalysts has become indispensable, because of photo corrosion belonging to catalysts is severely limited toward their applications. Anti-photo corrosion investigation for Ag_3PO_4/Ti_3C_2 has been carried out by performing eight cycles upon tetracycline hydrochloride for its removal activity (Cai et al., 2018). Resultantly, photocatalytic degradation for pristine Ag_3PO_4 was decreased by 90% for eight cycles, thereby significantly offering Ag_3PO_4 decomposition caused by photo corrosion. In addition, comparatively a little amount of reduction is observed as 30% toward photocatalytic degradation that was exhibited while performing eight cycles on Ag_3PO_4/Ti_3C_2 nanomaterial, associated with exploring of improvement in anti-photo corrosion efficiency. Moreover, the accelerated factors behind enhanced photocatalytic performance as well as anti-photo corrosion efficiency owing to Ag_3PO_4/Ti_3C_2 may be explained; firstly Ti_3C_2 possesses a plenteous surface containing hydrophilic functional groups that leads to creating a solid interfacial connection along with Ag_3PO_4 in facilitation for charge carrier transfer. Secondly, electron removal reactions that are carried with proper oxidation-reduction reactions occurring on Ti surface become helpful in enhancing ·OH generation. Thirdly, Schottky junction created at interfacial of Ag_3PO_4/Ti_3C_2 composite will rapidly transfer electrons toward Ti_3C_2 surface, leading to perform role for a built-in electric field that may prohibit photo corrosion-related with Ag_3PO_4 caused by photo-induced electrons (Cai et al., 2018).

For verifying renewability owing to Ag_2WO_4/Ti_3C_2 nanocomposite, an active cycling examination has been performed toward degrading efficiency for antibiotics (Fang et al., 2019). With the completion of three cycles, reducing behavior for cyclic output has been observed caused by reducing efficient removal relevant with tetracycline hydrochloride as 9% whereas sulfadimidine as 22%, respectively. The removal decline may occur probably due to some mechanisms like photo corrosion, photolysis, and photocatalyst if again use. Generally, little concentration for Ag has been observed to be expected for creation while preliminary fabrication owing to Ag_2WO_4/Ti_3C_2 nanocomposites in the form of strong oxidation-reduction reactions occurring on terminal metal locations for Ti_3C_2 MXene (Fang et al., 2019). In another study, reusability survey owing to Ti_3C_2-OH/ln_2S_3/CdS nanocomposites has demonstrated for photocatalytic performance that is observed same after three cycles repeating degradation treatments along with RhB as > 95% (Fang et al., 2019). The photocatalytic performance stability investigation for monolayer of $TiO_2/Ti_3C_2T_x$ nanocomposites has been explored during four-cycle (duration with 16 h) reactivity, whereas photocatalytic activity toward H_2 growth rate has been curtailed ranging 2.7-2.3 mmol g^{-1} h^{-1} (Su et al., 2019). The $Ti_3C_2T_x$ is considered approximately constant while performing chemically reactive activities. Nevertheless, a very small decrement into $TiO_2/Ti_3C_2T_x$ light absorbance has also been evaluated on comparing with fresh samples, probably caused by detachment for TiO_2 particles with nanocomposites accordingly (Su et al., 2019). For further study, though facing hindrance attached with inactivation owing to $AgInS_2$ toward remarkable enhancement for interfacial carrier transfer occurring in 0D/2D $AgInS_2$-50/Ti_3C_2 Z-scheme hybrid, NH_3 generation in an average way is lessened with an estimation of 95 -70 mmol.$g^{-1}h^{-1}$ while carrying out three cycles investigation toward N_2 photo reducing circumstances as clearly depicted in Figure 11 (Qin et al., 2019).

Figure 11. (a) The comparative study revealing yield rate, and (b) the recycling investigation toward $AgInS_2$-30/Ti_3C_2. (c) The energy-band-structure situations, and (d) the pictorial view of spatial charge distance as well as conveyance during course of photocatalytic activity from $AgInS_2/Ti_3C_2$ nanosheets.
Reproduced with permission from ref. (Qin et al., 2019) Copyright 2019, Elsevier B.V.

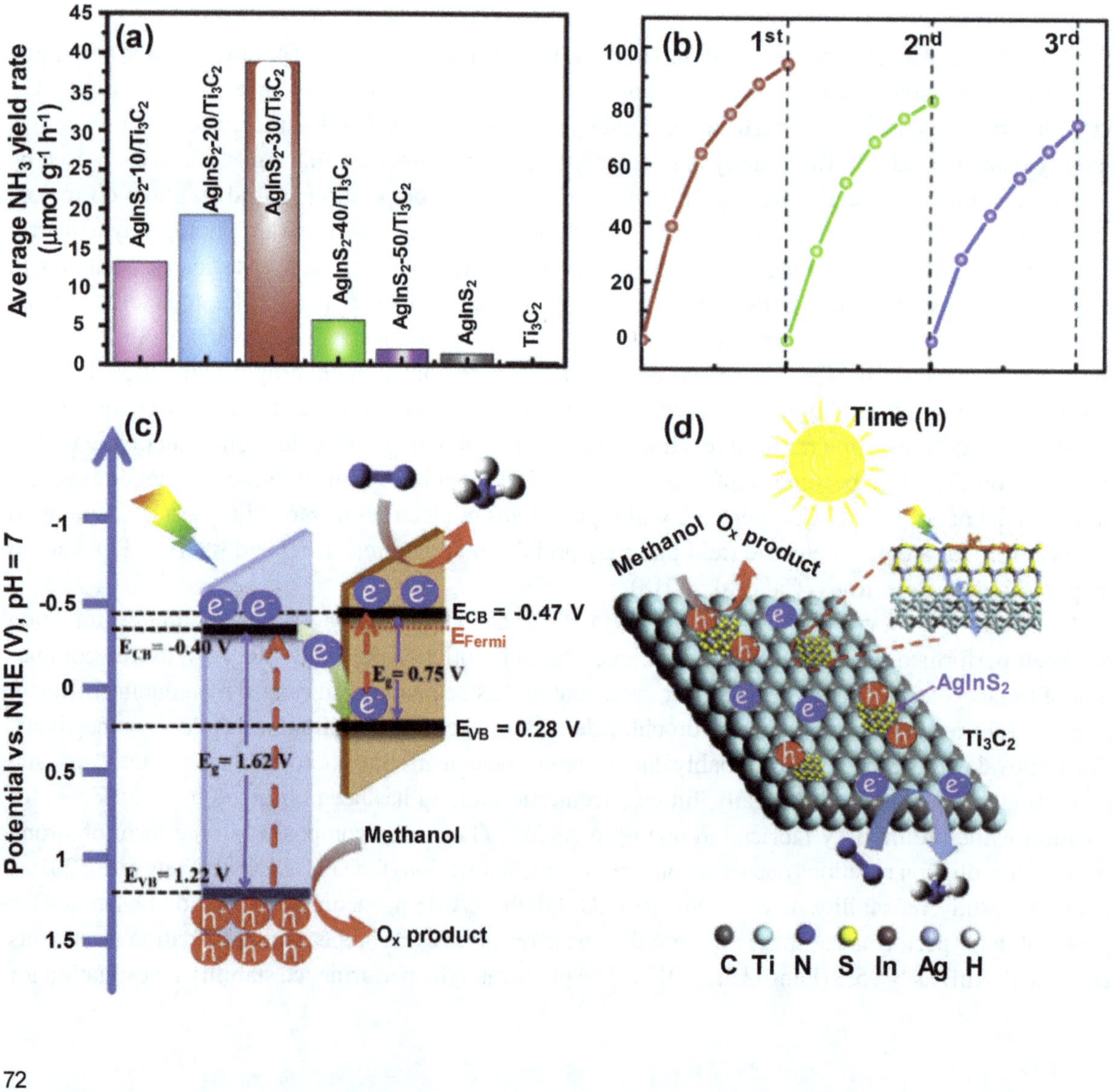

6. CONCLUSION

The technique of synthesis, structure, and photocatalytic characteristics are briefly reviewed in this chapter. This chapter presents a comprehensive review of recent research on MXene-based photocatalyst production processes and contaminant degrading photocatalytic performance. When paired with other photocatalysts including a metal oxide, metal sulfide, g-C_3N_4, and other materials covered in this chapter, MXenes are widely used as cocatalysts to increase the efficacy of photocatalytic activities. This chapter also briefly investigates the stability of MXene-based nanocomposite photocatalysts in order to reveal the distinctive properties of MXene-based nanocomposites.

ACKNOWLEDGMENT

Gao Li acknowledged the financial supported by Natural Science Foundation of China (22172167) and Liaoning Natural Science Foundation of China (2020-MS-024).

REFERENCES

Anasori, B., Lukatskaya, M. R., & Gogotsi, Y. (2017). 2D metal carbides and nitrides (MXenes) for energy storage. *Nature Reviews Materials, 2*(2), 16098. Advance online publication. doi:10.1038/natrevmats.2016.98

Bagheri, M., & Mirbagheri, S. A. (2018). Critical review of fouling mitigation strategies in membrane bioreactors treating water and wastewater. *Bioresource Technology, 258*, 318–334. doi:10.1016/j.biortech.2018.03.026 PMID:29548641

Barsoum, M. W., & Radovic, M. (2011). Elastic and Mechanical Properties of the MAX Phases. *Annual Review of Materials Research, 41*(1), 195–227. doi:10.1146/annurev-matsci-062910-100448

Cai, T., Wang, L., Liu, Y., Zhang, S., Dong, W., Chen, H., & Luo, S. (2018). Ag3PO4/Ti3C2 MXene interface materials as a Schottky catalyst with enhanced photocatalytic activities and anti-photocorrosion performance. *Applied Catalysis B: Environmental, 239*, 545–554. doi:10.1016/j.apcatb.2018.08.053

Cao, S., Shen, B., Tong, T., Fu, J., & Yu, J. (2018). 2D/2D Heterojunction of Ultrathin MXene/Bi2WO6 Nanosheets for Improved Photocatalytic CO2 Reduction. *Advanced Functional Materials, 28*(21), 1800136. doi:10.1002/adfm.201800136

Cao, W.-T., Chen, F.-F., Zhu, Y.-J., Zhang, Y.-G., Jiang, Y.-Y., Ma, M.-G., & Chen, F. (2018). Binary Strengthening and Toughening of MXene/Cellulose Nanofiber Composite Paper with Nacre-Inspired Structure and Superior Electromagnetic Interference Shielding Properties. *ACS Nano, 12*(5), 4583–4593. doi:10.1021/acsnano.8b00997 PMID:29709183

Carolin, C. F., Kumar, P. S., Saravanan, A., Joshiba, G. J., & Naushad, M. (2017). Efficient techniques for the removal of toxic heavy metals from aquatic environment: A review. *Journal of Environmental Chemical Engineering, 5*(3), 2782–2799. doi:10.1016/j.jece.2017.05.029

Cheng, X., Zu, L., Jiang, Y., Shi, D., Cai, X., Ni, Y., & Qin, Y. (2018). A titanium-based photo-Fenton bifunctional catalyst of mp MXene/TiO2−x nanodots for dramatic enhancement of catalytic efficiency in advanced oxidation processes. *Chemical Communications*, *54*(82), 11622–11625. doi:10.1039/C8CC05866K PMID:30264081

Crini, G., Lichtfouse, E., Wilson, L. D., & Morin-Crini, N. (2019). Conventional and non-conventional adsorbents for wastewater treatment. *Environmental Chemistry Letters*, *17*(1), 195–213. doi:10.100710311-018-0786-8

Cui, C., Guo, R., Xiao, H., Ren, E., Song, Q., Xiang, C., & Jiang, S. (2020). Bi2WO6/Nb2CTx MXene hybrid nanosheets with enhanced visible-light-driven photocatalytic activity for organic pollutants degradation. *Applied Surface Science*, *505*, 144595. doi:10.1016/j.apsusc.2019.144595

Das, R., Vecitis, C. D., Schulze, A., Cao, B., Ismail, A. F., Lu, X., & Ramakrishna, S. (2017). Recent advances in nanomaterials for water protection and monitoring. *Chemical Society Reviews*, *46*(22), 6946–7020. doi:10.1039/C6CS00921B PMID:28959815

Ding, X., Li, Y., Li, C., Wang, W., Wang, L., Feng, L., & Han, D. (2019). 2D visible-light-driven TiO2@ Ti3C2/g-C3N4 ternary heterostructure for high photocatalytic activity. *Journal of Materials Science*, *54*(13), 9385–9396. doi:10.100710853-018-03289-4

Ersan, G., Apul, O. G., Perreault, F., & Karanfil, T. (2017). Adsorption of organic contaminants by graphene nanosheets: A review. *Water Research*, *126*, 385–398. doi:10.1016/j.watres.2017.08.010 PMID:28987890

Fang, H., Pan, Y., Yin, M., & Pan, C. (2020). Enhanced photocatalytic activity and mechanism of Ti3C2−OH/Bi2WO6:Yb3+, Tm3+ towards degradation of RhB under visible and near infrared light irradiation. *Materials Research Bulletin*, *121*, 110618. doi:10.1016/j.materresbull.2019.110618

Fang, H., Pan, Y., Yin, M., Xu, L., Zhu, Y., & Pan, C. (2019). Facile synthesis of ternary Ti3C2–OH/ ln2S3/CdS composite with efficient adsorption and photocatalytic performance towards organic dyes. *Journal of Solid State Chemistry*, *280*, 120981. doi:10.1016/j.jssc.2019.120981

Fang, Y., Cao, Y., & Chen, Q. (2019). Synthesis of an Ag2WO4/Ti3C2 Schottky composite by electrostatic traction and its photocatalytic activity. *Ceramics International*, *45*(17, 17, Part A), 22298–22307. doi:10.1016/j.ceramint.2019.07.256

Gavrilescu, M., Demnerová, K., Aamand, J., Agathos, S., & Fava, F. (2015). Emerging pollutants in the environment: Present and future challenges in biomonitoring, ecological risks and bioremediation. *New Biotechnology*, *32*(1), 147–156. doi:10.1016/j.nbt.2014.01.001 PMID:24462777

Ghidiu, M., Naguib, M., Shi, C., Mashtalir, O., Pan, L. M., Zhang, B., & Barsoum, M. W. (2014). Synthesis and characterization of two-dimensional Nb4C3 (MXene). *Chemical Communications*, *50*(67), 9517–9520. doi:10.1039/C4CC03366C PMID:25010704

Ghobadifard, M., & Mohebbi, S. (2018). Novel nanomagnetic Ag/β-Ag2WO4/CoFe2O4 as a highly efficient photocatalyst under visible light irradiation. *New Journal of Chemistry*, *42*(12), 9530–9542. doi:10.1039/C8NJ00834E

Hirakawa, T., & Nosaka, Y. (2002). Properties of O2•- and OH• Formed in TiO2 Aqueous Suspensions by Photocatalytic Reaction and the Influence of H2O2 and Some Ions. *Langmuir*, *18*(8), 3247–3254. doi:10.1021/la015685a

Hong, W., Wyatt, B. C., Nemani, S. K., & Anasori, B. (2020). Double transition-metal MXenes: Atomistic design of two-dimensional carbides and nitrides. *MRS Bulletin*, *45*(10), 850–861. doi:10.1557/mrs.2020.251

Hu, T., Wang, J., Zhang, H., Li, Z., Hu, M., & Wang, X. (2015). Vibrational properties of Ti3C2 and Ti3C2T2 (T = O, F, OH) monosheets by first-principles calculations: A comparative study. *Physical Chemistry Chemical Physics*, *17*(15), 9997–10003. doi:10.1039/C4CP05666C PMID:25785395

Huang, H., Song, Y., Li, N., Chen, D., Xu, Q., Li, H., & Lu, J. (2019). One-step in-situ preparation of N-doped TiO2@C derived from Ti3C2 MXene for enhanced visible-light driven photodegradation. *Applied Catalysis B: Environmental*, *251*, 154–161. doi:10.1016/j.apcatb.2019.03.066

Huang, K., Li, Z., Lin, J., Han, G., & Huang, P. (2018). Two-dimensional transition metal carbides and nitrides (MXenes) for biomedical applications. *Chemical Society Reviews*, *47*(14), 5109–5124. doi:10.1039/C7CS00838D PMID:29667670

Huang, Q., Liu, Y., Cai, T., & Xia, X. (2019). Simultaneous removal of heavy metal ions and organic pollutant by BiOBr/Ti3C2 nanocomposite. *Journal of Photochemistry and Photobiology A Chemistry*, *375*, 201–208. doi:10.1016/j.jphotochem.2019.02.026

Ikram, M., Hassan, J., Imran, M., Haider, J., Ul-Hamid, A., Shahzadi, I., Ikram, M., Raza, A., Qumar, U., & Ali, S. (2020). 2D chemically exfoliated hexagonal boron nitride (hBN) nanosheets doped with Ni: Synthesis, properties and catalytic application for the treatment of industrial wastewater. *Applied Nanoscience*, *10*(9), 3525–3528. doi:10.100713204-020-01439-2

Ikram, M., Hassan, J., Raza, A., Haider, A., Naz, S., Ul-Hamid, A., Haider, J., Shahzadi, I., Qamar, U., & Ali, S. (2020). Photocatalytic and bactericidal properties and molecular docking analysis of TiO2 nanoparticles conjugated with Zr for environmental remediation. *RSC Advances*, *10*(50), 30007–30024. doi:10.1039/D0RA05862A PMID:35518250

Ikram, M., Khan, M. I., Raza, A., Imran, M., Ul-Hamid, A., & Ali, S. (2020). Outstanding performance of silver-decorated MoS2 nanopetals used as nanocatalyst for synthetic dye degradation. *Physica E, Low-Dimensional Systems and Nanostructures*, *124*, 114246. doi:10.1016/j.physe.2020.114246

Ikram, M., Umar, E., Raza, A., Haider, A., Naz, S., Ul-Hamid, A., Haider, J., Shahzadi, I., Hassan, J., & Ali, S. (2020). Dye degradation performance, bactericidal behavior and molecular docking analysis of Cu-doped TiO2 nanoparticles. *RSC Advances*, *10*(41), 24215–24233. doi:10.1039/D0RA04851H PMID:35516171

Jasper, J. T., Yang, Y., & Hoffmann, M. R. (2017). Toxic Byproduct Formation during Electrochemical Treatment of Latrine Wastewater. *Environmental Science & Technology*, *51*(12), 7111–7119. doi:10.1021/acs.est.7b01002 PMID:28538093

Ji, H., Zeng, W., & Li, Y. (2019). Gas sensing mechanisms of metal oxide semiconductors: A focus review. *Nanoscale*, *11*(47), 22664–22684. doi:10.1039/C9NR07699A PMID:31755888

Jiang, H., Zang, C., Zhang, Y., Wang, W., Yang, C., Sun, B., Shen, Y., & Bian, F. (2020). 2D MXene-derived Nb2O5/C/Nb2C/g-C3N4 heterojunctions for efficient nitrogen photofixation. *Catalysis Science & Technology*, *10*(17), 5964–5972. doi:10.1039/D0CY00656D

Jiang, X., Kuklin, A. V., Baev, A., Ge, Y., Ågren, H., Zhang, H., & Prasad, P. N. (2020). Two-dimensional MXenes: From morphological to optical, electric, and magnetic properties and applications. *Physics Reports*, *848*, 1–58. doi:10.1016/j.physrep.2019.12.006

Jiang, X. T., Kuklin, A. V., Baev, A., Ge, Y. Q., Agren, H., Zhang, H., & Prasad, P. N. (2020). Two-dimensional MXenes: From morphological to optical, electric, and magnetic properties and applications. *Physics Reports-Review Section of Physics Letters*, *848*, 1–58. doi:10.1016/j.physrep.2019.12.006

Katal, R., Masudy-Panah, S., Tanhaei, M., Farahani, M. H. D. A., & Jiangyong, H. (2020). A review on the synthesis of the various types of anatase TiO2 facets and their applications for photocatalysis. *Chemical Engineering Journal*, *384*, 123384. doi:10.1016/j.cej.2019.123384

Kong, X., Gao, P., Jiang, R., Feng, J., Yang, P., Gai, S., Chen, Y., Chi, Q., Xu, F., & Ye, W. (2020). Orderly layer-by-layered TiO2/carbon superstructures based on MXene's defect engineeringfor efficient hydrogen evolution. *Applied Catalysis A, General*, *590*, 117341. doi:10.1016/j.apcata.2019.117341

Kurtoglu, M., Naguib, M., Gogotsi, Y., & Barsoum, M. W. (2012). First principles study of two-dimensional early transition metal carbides. *MRS Communications*, *2*(4), 133–137. doi:10.1557/mrc.2012.25

Lei, J.-C., Zhang, X., & Zhou, Z. (2015). Recent advances in MXene: Preparation, properties, and applications. *Frontiers in Physics*, *10*(3), 276–286. doi:10.100711467-015-0493-x

Li, J.-Y., Li, Y.-H., Zhang, F., Tang, Z.-R., & Xu, Y.-J. (2020). Visible-light-driven integrated organic synthesis and hydrogen evolution over 1D/2D CdS-Ti3C2Tx MXene composites. *Applied Catalysis B: Environmental*, *269*, 118783. doi:10.1016/j.apcatb.2020.118783

Li, Y., Chen, X., Sun, Y., Meng, X., Dall'Agnese, Y., Chen, G., Dall'Agnese, C., Ren, H., Sasaki, S., Tamiaki, H., & Wang, X.-F. (2020). Chlorosome-Like Molecular Aggregation of Chlorophyll Derivative on Ti3C2Tx MXene Nanosheets for Efficient Noble Metal-Free Photocatalytic Hydrogen Evolution. *Advanced Materials Interfaces*, *7*(8), 1902080. doi:10.1002/admi.201902080

Li, Y., Deng, X., Tian, J., Liang, Z., & Cui, H. (2018). Ti3C2 MXene-derived Ti3C2/TiO2 nanoflowers for noble-metal-free photocatalytic overall water splitting. *Applied Materials Today*, *13*, 217–227. doi:10.1016/j.apmt.2018.09.004

Li, Y., Ding, L., Yin, S., Liang, Z., Xue, Y., Wang, X., Cui, H., & Tian, J. (2019). Photocatalytic H2 Evolution on TiO2 Assembled with Ti3C2 MXene and Metallic 1T-WS2 as Co-catalysts. *Nano-Micro Letters*, *12*(1), 6. doi:10.100740820-019-0339-0 PMID:34138075

Li, Y., Yin, Z., Ji, G., Liang, Z., Xue, Y., Guo, Y., Tian, J., Wang, X., & Cui, H. (2019). 2D/2D/2D heterojunction of Ti3C2 MXene/MoS2 nanosheets/TiO2 nanosheets with exposed (001) facets toward enhanced photocatalytic hydrogen production activity. *Applied Catalysis B: Environmental*, *246*, 12–20. doi:10.1016/j.apcatb.2019.01.051

Lian, P., Dong, Y., Wu, Z.-S., Zheng, S., Wang, X., Sen, W., ... Bao, X. (2017). Alkalized Ti3C2 MXene nanoribbons with expanded interlayer spacing for high-capacity sodium and potassium ion batteries. *Nano Energy*, *40*, 1–8. doi:10.1016/j.nanoen.2017.08.002

Lin, Z., Shao, H., Xu, K., Taberna, P.-L., & Simon, P. (2020). MXenes as High-Rate Electrodes for Energy Storage. *Trends in Chemistry*, *2*(7), 654–664. doi:10.1016/j.trechm.2020.04.010

Liu, N., Lu, N., Yu, H., Chen, S., & Quan, X. (2020). Efficient day-night photocatalysis performance of 2D/2D Ti3C2/Porous g-C3N4 nanolayers composite and its application in the degradation of organic pollutants. *Chemosphere*, *246*, 125760. doi:10.1016/j.chemosphere.2019.125760 PMID:31901663

Liu, X., & Chen, C. (2020). Mxene enhanced the photocatalytic activity of ZnO nanorods under visible light. *Materials Letters*, *261*, 127127. doi:10.1016/j.matlet.2019.127127

Liu, X., Shao, X., Li, F., & Zhao, M. (2018). Anchoring effects of S-terminated Ti2C MXene for lithium-sulfur batteries: A first-principles study. *Applied Surface Science*, *455*, 522–526. doi:10.1016/j.apsusc.2018.05.200

Long, M., Brame, J., Qin, F., Bao, J., Li, Q., & Alvarez, P. J. J. (2017). Phosphate Changes Effect of Humic Acids on TiO2 Photocatalysis: From Inhibition to Mitigation of Electron–Hole Recombination. *Environmental Science & Technology*, *51*(1), 514–521. doi:10.1021/acs.est.6b04845 PMID:27982576

Low, J., Cheng, B., & Yu, J. (2017). Surface modification and enhanced photocatalytic CO2 reduction performance of TiO2: A review. *Applied Surface Science*, *392*, 658–686. doi:10.1016/j.apsusc.2016.09.093

Lu, C., Tranca, D., Zhang, J., Rodríguez Hernández, F., Su, Y., Zhuang, X., Zhang, F., Seifert, G., & Feng, X. (2017). Molybdenum Carbide-Embedded Nitrogen-Doped Porous Carbon Nanosheets as Electrocatalysts for Water Splitting in Alkaline Media. *ACS Nano*, *11*(4), 3933–3942. doi:10.1021/acsnano.7b00365 PMID:28291319

Lukatskaya, M. R., Mashtalir, O., & Chang, R. E., Dall'Agnese, Y., Rozier, P., Taberna Pierre, L., . . . Gogotsi, Y. (2013). Cation Intercalation and High Volumetric Capacitance of Two-Dimensional Titanium Carbide. *Science*, *341*(6153), 1502–1505. Advance online publication. doi:10.1126cience.1241488 PMID:24072919

Naguib, M., Adams, R. A., Zhao, Y., Zemlyanov, D., Varma, A., Nanda, J., & Pol, V. G. (2017). Electrochemical performance of MXenes as K-ion battery anodes. *Chemical Communications*, *53*(51), 6883–6886. doi:10.1039/C7CC02026K PMID:28607970

Naguib, M., Kurtoglu, M., Presser, V., Lu, J., Niu, J., Heon, M., Hultman, L., Gogotsi, Y., & Barsoum, M. W. (2011). Two-dimensional nanocrystals produced by exfoliation of Ti3 AlC2. *Advanced Materials*, *23*(37), 4248–4253. doi:10.1002/adma.201102306 PMID:21861270

Naguib, M., Kurtoglu, M., Presser, V., Lu, J., Niu, J., Heon, M., Hultman, L., Gogotsi, Y., & Barsoum, M. W. (2011). Two-Dimensional Nanocrystals Produced by Exfoliation of Ti3AlC2. *Advanced Materials*, *23*(37), 4248–4253. doi:10.1002/adma.201102306 PMID:21861270

Naguib, M., Mashtalir, O., Carle, J., Presser, V., Lu, J., Hultman, L., Gogotsi, Y., & Barsoum, M. W. (2012). Two Dimensional Transition Metal Carbides. *ACS Nano*, *6*(2), 1322–1331. doi:10.1021/nn204153h PMID:22279971

Peng, C., Yang, X., Li, Y., Yu, H., Wang, H., & Peng, F. (2016). Hybrids of Two-Dimensional Ti3C2 and TiO2 Exposing {001} Facets toward Enhanced Photocatalytic Activity. *ACS Applied Materials & Interfaces*, *8*(9), 6051–6060. doi:10.1021/acsami.5b11973 PMID:26859317

Qin, J., Hu, X., Li, X., Yin, Z., Liu, B., & Lam, K. (2019). 0D/2D AgInS2/MXene Z-scheme heterojunction nanosheets for improved ammonia photosynthesis of N2. *Nano Energy*, *61*, 27–35. doi:10.1016/j.nanoen.2019.04.028

Ran, J., Gao, G., Li, F.-T., Ma, T.-Y., Du, A., & Qiao, S.-Z. (2017). Ti3C2 MXene co-catalyst on metal sulfide photo-absorbers for enhanced visible-light photocatalytic hydrogen production. *Nature Communications*, *8*(1), 13907. doi:10.1038/ncomms13907 PMID:28045015

Raza, A., Qin, Z., Ahmad, S. O. A., Ikram, M., & Li, G. (2021). Recent advances in structural tailoring of BiOX-based 2D composites for solar energy harvesting. *Journal of Environmental Chemical Engineering*, *9*(6), 106569. doi:10.1016/j.jece.2021.106569

Raza, A., Qumar, U., Haider, A., Naz, S., Haider, J., Ul-Hamid, A., Ikram, M., Ali, S., Goumri-Said, S., & Kanoun, M. B. (2021). Liquid-phase exfoliated MoS2 nanosheets doped with p-type transition metals: A comparative analysis of photocatalytic and antimicrobial potential combined with density functional theory. *Dalton Transactions (Cambridge, England)*, *50*(19), 6598–6619. doi:10.1039/D1DT00236H PMID:33899890

Raza, A., Qumar, U., Hassan, J., Ikram, M., Ul-Hamid, A., Haider, J., Imran, M., & Ali, S. (2020). A comparative study of dirac 2D materials, TMDCs and 2D insulators with regard to their structures and photocatalytic/sonophotocatalytic behavior. *Applied Nanoscience*, *10*(10), 3875–3899. doi:10.100713204-020-01475-y

Santhosh, C., Velmurugan, V., Jacob, G., Jeong, S. K., Grace, A. N., & Bhatnagar, A. (2016). Role of nanomaterials in water treatment applications: A review. *Chemical Engineering Journal*, *306*, 1116–1137. doi:10.1016/j.cej.2016.08.053

Sarkar, C., Bora, C., & Dolui, S. K. (2014). Selective Dye Adsorption by pH Modulation on Amine-Functionalized Reduced Graphene Oxide–Carbon Nanotube Hybrid. *Industrial & Engineering Chemistry Research*, *53*(42), 16148–16155. doi:10.1021/ie502653t

Shahzad, F., Alhabeb, M., Hatter Christine, B., Anasori, B., Man Hong, S., Koo Chong, M., & Gogotsi, Y. (2016). Electromagnetic interference shielding with 2D transition metal carbides (MXenes). *Science*, *353*(6304), 1137–1140. doi:10.1126cience.aag2421 PMID:27609888

Shao, B., Wang, J., Liu, Z., Zeng, G., Tang, L., Liang, Q., He, Q., Wu, T., Liu, Y., & Yuan, X. (2020). Ti3C2Tx MXene decorated black phosphorus nanosheets with improved visible-light photocatalytic activity: Experimental and theoretical studies. *Journal of Materials Chemistry. A, Materials for Energy and Sustainability*, *8*(10), 5171–5185. doi:10.1039/C9TA13610J

Shao, L., & Chen, G. Q. (2013). Water footprint assessment for wastewater treatment: Method, indicator, and application. *Environmental Science & Technology*, *47*(14), 7787–7794. doi:10.1021/es402013t PMID:23777208

Shen, J., Shen, J., Zhang, W., Yu, X., Tang, H., Zhang, M., Zulfiqar, & Liu, Q. (2019). Built-in electric field induced CeO2/Ti3C2-MXene Schottky-junction for coupled photocatalytic tetracycline degradation and CO2 reduction. *Ceramics International*, *45*(18), 24146–24153. doi:10.1016/j.ceramint.2019.08.123

Shen, Y., & Chen, B. (2015). Sulfonated Graphene Nanosheets as a Superb Adsorbent for Various Environmental Pollutants in Water. *Environmental Science & Technology*, *49*(12), 7364–7372. doi:10.1021/acs.est.5b01057 PMID:26008607

Sinha, A., Dhanjai, Zhao, H., Huang, Y., Lu, X., Chen, J., & Jain, R. (2018). MXene: An emerging material for sensing and biosensing. *Trends in Analytical Chemistry*, *105*, 424–435. doi:10.1016/j.trac.2018.05.021

Su, T., Hood, Z. D., Naguib, M., Bai, L., Luo, S., Rouleau, C. M., Ivanov, I. N., Ji, H., Qin, Z., & Wu, Z. (2019). Monolayer Ti3C2Tx as an Effective Co-catalyst for Enhanced Photocatalytic Hydrogen Production over TiO2. *ACS Applied Energy Materials*, *2*(7), 4640–4651. doi:10.1021/acsaem.8b02268

Su, T., Peng, R., Hood, Z. D., Naguib, M., Ivanov, I. N., Keum, J. K., Qin, Z., Guo, Z., & Wu, Z. (2018). One-Step Synthesis of Nb2O5/C/Nb2C (MXene) Composites and Their Use as Photocatalysts for Hydrogen Evolution. *ChemSusChem*, *11*(4), 688–699. doi:10.1002/cssc.201702317 PMID:29281767

Sun, M., & Li, J. (2018). Graphene oxide membranes: Functional structures, preparation and environmental applications. *Nano Today*, *20*, 121–137. doi:10.1016/j.nantod.2018.04.007

Sun, Y., Sun, Y., Meng, X., Gao, Y., Dall'Agnese, Y., Chen, G., Dall'Agnese, C., & Wang, X.-F. (2019). Eosin Y-sensitized partially oxidized Ti3C2 MXene for photocatalytic hydrogen evolution. *Catalysis Science & Technology*, *9*(2), 310–315. doi:10.1039/C8CY02240B

Tan, C., Cao, X., Wu, X.-J., He, Q., Yang, J., Zhang, X., Chen, J., Zhao, W., Han, S., Nam, G.-H., Sindoro, M., & Zhang, H. (2017). Recent Advances in Ultrathin Two-Dimensional Nanomaterials. *Chemical Reviews*, *117*(9), 6225–6331. doi:10.1021/acs.chemrev.6b00558 PMID:28306244

Tan, C., & Zhang, H. (2015). Wet-chemical synthesis and applications of non-layer structured two-dimensional nanomaterials. *Nature Communications*, *6*(1), 7873. doi:10.1038/ncomms8873 PMID:26303763

Urbankowski, P., Anasori, B., Hantanasirisakul, K., Yang, L., Zhang, L., Haines, B., May, S. J., Billinge, S. J. L., & Gogotsi, Y. (2017). 2D molybdenum and vanadium nitrides synthesized by ammoniation of 2D transition metal carbides (MXenes). *Nanoscale*, *9*(45), 17722–17730. doi:10.1039/C7NR06721F PMID:29134998

Wang, C., Shen, J., Chen, R., Cao, F., & Jin, B. (2020). Self-assembled BiOCl/Ti3C2Tx composites with efficient photo-induced charge separation activity for photocatalytic degradation of p-nitrophenol. *Applied Surface Science*, *519*, 146175. doi:10.1016/j.apsusc.2020.146175

Wang, H., Wu, Y., Xiao, T., Yuan, X., Zeng, G., Tu, W., Wu, S., Lee, H. Y., Tan, Y. Z., & Chew, J. W. (2018). Formation of quasi-core-shell In2S3/anatase TiO2@metallic Ti3C2Tx hybrids with favorable charge transfer channels for excellent visible-light-photocatalytic performance. *Applied Catalysis B: Environmental*, *233*, 213–225. doi:10.1016/j.apcatb.2018.04.012

Wei, X., Akbar, M. U., Raza, A., & Li, G. (2021). A review on bismuth oxyhalide based materials for photocatalysis. *Nanoscale Advances*, *3*(12), 3353–3372. doi:10.1039/D1NA00223F

Wu, S., Su, Y., Zhu, Y., Zhang, Y., & Zhu, M. (2020). In-situ growing Bi/BiOCl microspheres on Ti3C2 nanosheets for upgrading visible-light-driven photocatalytic activity. *Applied Surface Science*, *520*, 146339. doi:10.1016/j.apsusc.2020.146339

Wu, Z., Liang, Y., Yuan, X., Zou, D., Fang, J., Jiang, L., zhang, J., Yang, H., & Xiao, Z. (2020). MXene Ti3C2 derived Z–scheme photocatalyst of graphene layers anchored TiO2/g–C3N4 for visible light photocatalytic degradation of refractory organic pollutants. *Chemical Engineering Journal*, *394*, 124921. doi:10.1016/j.cej.2020.124921

Xiao, Q. K., Wang, H. Y., Li, F., & Gao, Y. (2011). 3D object retrieval based on a graph model descriptor. *Neurocomputing*, *74*(17), 3486–3493. doi:10.1016/j.neucom.2011.06.002

Xiao, Z., Yang, Z., Li, Z., Li, P., & Wang, R. (2019). Synchronous Gains of Areal and Volumetric Capacities in Lithium-Sulfur Batteries Promised by Flower-like Porous Ti3C2T x Matrix. *ACS Nano*, *13*(3), 3404–3412. doi:10.1021/acsnano.8b09296 PMID:30790514

Xu, C., Wang, L., Liu, Z., Chen, L., Guo, J., Kang, N., Ma, X.-L., Cheng, H.-M., & Ren, W. (2015). Large-area high-quality 2D ultrathin Mo2C superconducting crystals. *Nature Materials*, *14*(11), 1135–1141. doi:10.1038/nmat4374 PMID:26280223

Xu, J., Cao, Z., Zhang, Y., Yuan, Z., Lou, Z., Xu, X., & Wang, X. (2018). A review of functionalized carbon nanotubes and graphene for heavy metal adsorption from water: Preparation, application, and mechanism. *Chemosphere*, *195*, 351–364. doi:10.1016/j.chemosphere.2017.12.061 PMID:29272803

Yang, X. H., Yang, H. G., & Li, C. (2011). Controllable Nanocarving of Anatase TiO2 Single Crystals with Reactive {001} Facets. *Chemistry (Weinheim an der Bergstrasse, Germany)*, *17*(24), 6615–6619. doi:10.1002/chem.201100134 PMID:21538611

Yang, Y., Zeng, Z., Zeng, G., Huang, D., Xiao, R., Zhang, C., Zhou, C., Xiong, W., Wang, W., Cheng, M., Xue, W., Guo, H., Tang, X., & He, D. (2019). Ti3C2 Mxene/porous g-C3N4 interfacial Schottky junction for boosting spatial charge separation in photocatalytic H2O2 production. *Applied Catalysis B: Environmental*, *258*, 117956. doi:10.1016/j.apcatb.2019.117956

Ye, S., Zeng, G., Wu, H., Liang, J., Zhang, C., Dai, J., Xiong, W., Song, B., Wu, S., & Yu, J. (2019). The effects of activated biochar addition on remediation efficiency of co-composting with contaminated wetland soil. *Resources, Conservation and Recycling*, *140*, 278–285. doi:10.1016/j.resconrec.2018.10.004

Yuan, W., Cheng, L., An, Y., Lv, S., Wu, H., Fan, X., Zhang, Y., Guo, X., & Tang, J. (2018). Laminated Hybrid Junction of Sulfur-Doped TiO2 and a Carbon Substrate Derived from Ti3C2 MXenes: Toward Highly Visible Light-Driven Photocatalytic Hydrogen Evolution. *Advancement of Science*, *5*(6), 1700870. doi:10.1002/advs.201700870 PMID:29938169

Yuan, W., Cheng, L., An, Y., Wu, H., Yao, N., Fan, X., & Guo, X. (2018). MXene Nanofibers as Highly Active Catalysts for Hydrogen Evolution Reaction. *ACS Sustainable Chemistry & Engineering*, *6*(7), 8976–8982. doi:10.1021/acssuschemeng.8b01348

Yuan, W., Cheng, L., Zhang, Y., Wu, H., Lv, S., Chai, L., Guo, X., & Zheng, L. (2017). 2D-Layered Carbon/TiO2 Hybrids Derived from Ti3C2MXenes for Photocatalytic Hydrogen Evolution under Visible Light Irradiation. *Advanced Materials Interfaces*, *4*(20), 1700577. doi:10.1002/admi.201700577

Yuan, W. Y., Cheng, L. F., An, Y. R., Wu, H., Yao, N., Fan, X. L., & Guo, X. H. (2018). MXene Nanofibers as Highly Active Catalysts for Hydrogen Evolution Reaction. *ACS Sustainable Chemistry & Engineering*, *6*(7), 8976–8982. doi:10.1021/acssuschemeng.8b01348

Zhang, C., & Nicolosi, V. (2019). Graphene and MXene-based transparent conductive electrodes and supercapacitors. *Energy Storage Materials*, *16*, 102–125. doi:10.1016/j.ensm.2018.05.003

Zhang, G., Wu, S., Li, Y., & Zhang, Q. (2020). Significant improvement in activity, durability, and light-to-fuel efficiency of Ni nanoparticles by La2O3 cluster modification for photothermocatalytic CO2 reduction. *Applied Catalysis B: Environmental*, *264*, 118544. doi:10.1016/j.apcatb.2019.118544

Zhang, H., Li, M., Zhu, C., Tang, Q., Kang, P., & Cao, J. (2020). Preparation of magnetic α-Fe2O3/ZnFe2O4@Ti3C2 MXene with excellent photocatalytic performance. *Ceramics International*, *46*(1), 81–88. doi:10.1016/j.ceramint.2019.08.236

Zhang, M., Qin, J., Rajendran, S., Zhang, X., & Liu, R. (2018). Heterostructured d-Ti3C2/TiO2/g-C3N4 Nanocomposites with Enhanced Visible-Light Photocatalytic Hydrogen Production Activity. *ChemSusChem*, *11*(24), 4226–4236. doi:10.1002/cssc.201802284 PMID:30334348

Zhang, Y., Wang, L., Zhang, N., & Zhou, Z. (2018). Adsorptive environmental applications of MXene nanomaterials: A review. *RSC Advances*, *8*(36), 19895–19905. doi:10.1039/C8RA03077D PMID:35541640

Zhong, Q., Li, Y., & Zhang, G. (2021). Two-dimensional MXene-based and MXene-derived photocatalysts: Recent developments and perspectives. *Chemical Engineering Journal*, *409*, 128099. doi:10.1016/j.cej.2020.128099

Zou, Y., Shi, J.-W., Sun, L., Ma, D., Mao, S., Lv, Y., & Cheng, Y. (2019). Energy-band-controlled ZnxCd1−xIn2S4 solid solution coupled with g-C3N4 nanosheets as 2D/2D heterostructure toward efficient photocatalytic H2 evolution. *Chemical Engineering Journal*, *378*, 122192. doi:10.1016/j.cej.2019.122192

Chapter 4
Designing and Structural Modification of Novel Biomaterials to Increase Their Bioefficacy Towards Decontamination of Cadmium

Nirupama Singh
RCCV Girls College, Ghaziabad, India

ABSTRACT

Heavy metals have come up as a threatening pollutant in aqueous media, leading to life threatening consequences. Biomaterial has been a novel and innovative wing of Green Chemistry, eradicating the threats in a cost effective and clean manner. The present study has been focused on the successful removal of a life-threatening heavy metal Cd (II) from aqueous solution using biosorbent created using selected plants. The present study establishes the fact that carboxylic acid group plays an important role in the metal binding process using protection of COOH group by propylamination and esterification. We could also conclude that the enrichment of COOH group onto the biomaterial using synthetic modifications succination leads to the increase in the sorption efficiency.

INTRODUCTION

The problem originates with the tremendous increase in quantity of various toxic pollutants, encountered in wastewater (Brostlap et al.,1988), toxic metals are very imperative associate of filthy dozen club of pollutants. Thus, the decontamination of metal bearing waste water seems to be a critical environmental concern. The conventional methods utilized for the subsiding of toxic metals are often inadequate due to the technical and economic constraints (Delvin et al., 2002). For that reason, the search has brought newly emerging conception of Green Chemistry which is being considered as a novel opinion guiding the upcoming generation products and proceedings (Gardea-Torresdey et al., 1996, Ahalya et al., 2003).

DOI: 10.4018/978-1-6684-4553-2.ch004

Biomaterial generated via green path has acquired much significance for decontamination of water (Adrehold et al., 1996, Baig et al., 1999). Bioremediation incorporates procedures that trim down complete treatment charge via the relevant use of aboriginal agricultural byproducts and wastes (Isabel et al., 2004), which appears to be a wise choice as they reduce direct dependence on expensive water purifying chemicals, insignificant transportation necessities and propose indisputable local resources, as suitable solutions to deal with local affair of water standards. Rejuvenation of the biosorbent escalates the cost efficiency of the process thus warrants its future success (Kar et al., 1992; Khalid et al., 1998). However, biomaterials have been found to be linked with drawbacks associated to constancy and simultaneous sorption range of lethal metals, limiting their viable utilizations (Stenberg et.al 2002; Adesola et al., 2006). Consequently, biosynthetic improvisation improves the binding capability selectivity and steadiness of natural biomaterials has fascinated the eager attention of scientific community (Kobya et al., 2004; Manju et al., 1997). The use of green methods for structural modifications on biomaterials to maximize their bioefficacy adds further important dimension to the popularity of biosorption process for decontamination of water (Martins et al., 2004; Meunier et al., 2003; Nicholas et al., 2003)

Heavy Metals: Environmental Threat

Heavy metals being the significant affiliate of filthy dozen toxic pollutants encounter various ecosystem environment (Tiwari et al., 1999). As they are biologically imperishable, their danger multiplies by their agglomeration resulting into serious ecological concern. Heavy metals, although are found in trace amount in the earth's shell, anthropogenic actions viz, manufacturing processing and utilization of metals, amalgam and metallic compounds largely summing up their natural environmental levels. Land dumping of solid wastes, indiscriminate use and disposal of pesticides, contamination of water, all badly insult the natural sink and disperse a large amount of various toxic metals into all the three components of environment viz: atmosphere, hydrosphere lithosphere. The toxic metals enter the human organization by two prominent ways (Hima et al., 2003). First, the metals impose bad effects due to their direct consumption as potable water and through inhalation. Secondly, circuitous uptake of lethal metals via consumption of food, depending upon the water composition of the ecosystem metals enters in to the human body. Based on ecological pollution viewpoint, metals can be divided into categories viz: Toxic and accessible, Nontoxic but accessible and Toxic but non accessible (Sure et al., 1953; Tegbe et al., 2006).

It is the first category of potential toxic and comparatively accessible metals which has attracted the eager attention of the researchers (Tiemann et al., 2002). These toxic metals are called rare metal as they occur in extremely minute quantity in the earth's crust. Based on their densities they are further subdivided, those with densities more than 5 g/cm^3 are elected as heavy metals (IARC and Namasivayam et.al). Present study has been focused on the decontamination of Cadmium from aqueous medium.

Decontamination Techniques

The use of available water sources are being strongly impacted due to the constant increase in the degrees of industrialization and increasing standards of livelihood. Mastering toxic metals release and removing lethal heavy metals from aqueous solutions has turn out to be a confront for the present time. Most commonly utilized measures for eradicating metal ions from aqueous streams involve methods like, precipitation, ion exchange, reverse osmosis, electro dialysis, coagulation and flocculation etc. (Coupal et al., 1976; Gardea-Torresdey et al., 1990). The basic principle, procedural details and commercially accessible instrumentation for the above discussed phenomenon has been described in short.

Reverse Osmosis (RO)

Being a separation process pressure reverse osmosis uses pressure technique it forces a solution via a membrane which retains the solute particles on one side and allows the passage of pure solvent to other side. In other words, it can be described as the process of flowing or passage of a solvent from a section of high solute concentration via a membrane towards a region having low solute concentration by applying a pressure more than the osmotic pressure (Ajmal et al., 2000; Asheh et al., 1998). Reverse osmosis is basically the opposite of the usual osmosis process, here the solvent goes from a low solute concentration, high solute concentration area via a membrane when no external pressure is exerted. The semipermeable membrane present here allows only the passage of solvent but does not allows the solute to pass through it. But the technique is associated with several demerits which reduces its utilization, it need periodic replacement of filters, not successful in removing bacteria, the semipermeable membrane gets clogged resulting into the sludge trouble, large amount of water get wasted in removing the sludge hence the technique proves out to be an expensive technique (Banat et al., 2003, Azab et al., 1989).

Electrodialysis (ED)

Electrodialysis is a technique which is utilized to carry salt ions through ion exchange membranes from one solution to another solution, beneath the impact of applied galvanic potential variance (Basso et al., 2002, Gardea-Torresdey et al., 2004). The process is carried out in an electrodialysis cell. The cell comprises of a feed cubicle and another compartment is concentrate compartment which is devised by a cation swap layer and an anion swap membrane which is located in-between the two electrodes. Approximately in all the electrodialysis procedures, manifold electrodialysis compartments are organized into electrodialysis stack, which alternates the cation and anion exchange membranes developing numerous electrodialysis cells.

Ultrafiltration (UF)

Ultrafiltration is variety of water decontamination technique where liquid is forced against a semi permeable membrane via hydrostatic pressure. The semipermiable membrane retains the solutes of high moleculer weight and suspended particles, whereas molecules of low molecular weight and water easily pass via the membrane. Ultrafiltration separation procedure is utilized in the industries, purification research work and for concentrating macromolecular solutions, specially protein solutions (Krishnan et al., 2003, Basu et al., 2003). Fundamentally this technique is not very distinct from reverse osmosis or micro filtration else than it can retain size range of molecules.

Ion Exchange

In this technique ions are exchanged between a solution of electrolyte and complex or among two electrolytes. In the majority of cases, the term ion exchange is utilized to denote the processes of decontamination, purification and separation of aqueous solutions with the help of mineral ion exchangers or solid polymeric exchangers. Distinctive ion exchangers are viz: gel polymers or functionalized porous ion exchange resins, clay and humus and zeolites etc. The ion exchangers are either cation or anion exchangers, positively charged ions (cations) are exchanged in cation exchangers, negatively charged ions are exchanged in anion exchangers (anions) (Basso et al., 2002, EES). Certain amphoteric exchangers are also available they are capable of exchanging together cations and anions. The process is associated with certain demerits which restrict its usage viz, restricted accessibility; time to time resin needs to be replaced and the process does not inactivate microbes.

Chemical Precipitation

In the process of chemical precipitation during chemical reaction solid is formed in the solution. When the chemical reaction takes place and precipitation happens precipitate is formed the liquid retained on top of the solid is the supernate. Chemical precipitation reactions can be utilized for qualitative chemical analysis, making pigments and for removing salts from water. The technique requires a specialized handling and generates a large amount of sludge which is difficult to separate (Low et al., 1991; Paknikar et al., 2003).

However, all these prior mentioned methods have disadvantages like incomplete metal removal, high reagent and energy requirements, very expensive and none of the stabilizers are persistent, Boddu et al., 2003.

Biosorption: An Emerging Pathway

Biosorption is comparatively a novel practice that came to existence during 1980's and procured substantial consideration as it has emerged out to be very potential in the elimination of taint from effluvium in an environmental affable manner (Gardea-Torresdey et al., 1996a; Volesky 2001). For the cost effective remediation of lethal metal ions from aqueous system ample amount of attention has been provided to the use of living and non-living biological resources leading to the cost-effective decontamination. (Iqbal et al., 2002; Regin et al., 2000). Lifeless organisms provide various benefits over living organisms because no maintenance is required by the lifeless resources neither it ever gets affected by the high concentrations of pollutants. On the contrary living creatures always require dietetic care and supply and they usually get affected by elevated concentration of contaminants (Gardea-Torresdey et al., 1996b). As compared to the artificial ion exchange resins, plant based biosorbing materials never need toxic chemicals for their preparation. Hence, the plant originated biosorbent materials are measured additionally effective for removing toxic heavy metal ions out of the aquatic system in a sustainable, cost effective and eco-friendly manner (Goodwin et al., 1975; Ghimire et al., 2002; Gonzalez et al., 2006).

PLANTS UNDER STUDY

Saraca indica (Ashoka) leaf Powder, the plant Saraca indica is a naturally Occurring herb. With dark green foliage and very bright and fragrant yellow and orange flowers is an evergreen plant. It is a small tree having height up to 10 m, with reddish brown wood and almost black bark leaves are paripinnate, stipules intra petiolar, united, scarious, oblong, lanceolate, and glabrous; flowers orange to scarlet. Flowering time is December to May and fruiting time is June to July. It occurs up to the altitudes 600 meters. The Leaves are membranous, alternate, paripinnate, 15-20 cm long, leaf-lets 6-12 oblong or oblong lanceolate, 1.25 cm in width. The shoots: young shoots are pendent. Flowers are orange or yellow eventually turning vermillion, very fragrant scents at night, 2.5-3.8 cm across (Pagnelli et al., 2002, 2003). Pedicles slender small pubescent, sepals 4 mm long. Seeds are 4-8 ellipsoid, oblong 38 cm and packed together.

Ficus religiosa (Pipal) leaf Powder, Pipal exhibits immense therapeutic importance. The leaves of pipal serve as a magnificent laxative as well as a wonderful stimulant for the body. It proves to be very helpful for the patients suffering from Jaundice. It assists in controlling the excessive urine released during jaundice. Leaves of Pipal are extremely efficient in the treatment of heart related disorders (Sawalha et al., 2005; Nagarnaik et al., 2005). It also helps in managing the palpitation of heart and thereby helps in fighting cardiac weakness. Pipal the holy tree is also known by variety of names in different languages and regions like Bodhi in Sanskrit, Piplo in Gujrati, Al or Aryal in Malayalam etc. It is a belief that this tree protects humans from the negative energies, evil eye and also keeps away dreadful dreams. The Pipal-bark has light grey color and is smooth. The leaves have a distinctive shape of heart and have long and tapering tips, colors of flower is red. Flowers in the month of February and Fruits in May / June, widely found in uplands and plain area.

Azadirachta indica (Neem) leaf Powder, Neem is a tree of the mahogany family Meliaceae. Neem twigs are utilized for brushing teeth in India, Bangladesh and Pakistan this practice is almost the oldest and most valuable forms of dental care. Several parts of the tree viz, seeds, leaves, flowers and bark are utilized for preparing several diverse therapeutic preparations. It is a very fast-growing tree which reaches a height of 15-20 m, and sometimes even 35-45 m. Neem is evergreen tree but it may not survive in severe drought condition (Shukla et al., 2004, Rosa et al., 2003). Neem tree has wide spread branches, the dense top of the tree use to be roundish or oval and may acquire a diameter of 15-20 m in mature and free standing specimen. The roots are sturdy taproot and well developed lateral roots. Leaves of neem are usually 25-45 cm long, the terminal leaflet is frequently missing. The trunk is comparatively short and straight and can reach a diameter of 1.2 m. The bark use to be very hard, fissured or scaly, the color varies from whitish-grey to reddish-brown. The flowers are white and have fragrance, flowers are usually arranged axillary, in general relatively drooping panicles usually 25 cm long. Individual flower use to be 5-6 mm long and 8-12 mm wide.

Objectives

Objectives of sorption studies involve evaluation of efficiency of individual and compost biomaterial [ashoka, pipal and neem] for abatement of heavy metal, sorption efficiency (%) of Cadmium from aqueous system using biosorbent. Determination of equilibrium isotherms, followed by mechanistic modeling using adsorption isotherms and modification of the composite for enhancing sorption efficiency.

Material and Methods

Biosorbent Preparation

The leaves of different plants: Neem, Ashoka and Pipal has been collected in the month of January and washed with water to get rid of the sticked dirt, dried at 65°C for 24 hours, crushed and sieved via (105 um, 210 um) mesh copper sieves. Prior to the implementation of the adsorption experiment no physical or chemical treatments have been provided to the biosorbent.

Sorption Studies

Sorption study consists of series of experiments as a function of single biosorbent material (Ashoka, Pipal and Neem leaf powder] individually (1.0, 2.0, 4.0, 6.0 g) as well as mixed or ternary biosorbent dosage (0.5, 1.0g, 2.0, 4.0g), contact time 40 minutes, metal concentration (1, 5, 10, 25 and 50 mg/L), particle size (105 µm) all the study has been performed in different batch of experiments. The solution containing Cd (II) as Cadmium Nitrate has been taken into separate Erlenmeyer flasks. After pH adjustment, required quantity of biosorbent (single as well as mixed biosorbent) has been introduced and as a final point metal bearing suspension was permitted to settle down. The remaining/residual biomaterial absorbed/adsorbed with metal ion has been filtered using Whatman filter paper no. 42. Filtrate has been collected and subjected to metal ion estimation using Atomic Absorption Spectroscopy (Perkin Elmer-A analyst 100). Total Sorption of percent metal by the sorbent has been computed utilizing the following equation:

$$\% \text{ Sorption} = (Co-Ce) / Co \times 100$$

Where, Co is initial concentration of metal ions in the solution and Ce is the final concentration of metal ions in the solution.

Adsorption Studies

Cadmium sorption by biosorbent has been calculated (Vankar et al., 2002, Volesky 2001) utilizing the mass balance equation for the biosorbent:

$$q= [V (C_i-C_f)] / S$$

Where, q is the cadmium metal uptake (mg metal/g dry weight), V represents the volume of metal containing solution contacted (batch) with the biosorbent (L), Ci is the initial concentration of metal in solution (mg/L), C_f is the final concentration of target metal in solution (mg/L), S representing dry weight of biosorbent added (g)

Sorption Isotherm

The metal biosorption equilibrium onto biomaterial has been interpreted utilizing Freundlich and Langmuir isotherms. The classical Freundlich equation has been depicted below:

$$q = K_f C_e^{1/n}$$

Where, K_f and n are the characteristic constants, C_e, is the final concentration of metal (mg/L) in the solution and q is the heavy metal adsorbed on the biosorbent (mg/g dry weight).

The classical Langmuir equation can be expressed as:

$C_e/q_e = (1/Q_0 b) + (C_e/Q_b)$

Where, q_e is the quantity of Cadmium adsorbed at equilibrium, C_e is the equilibrium concentration, Q_0 and b are the Langmuir constants linked with adsorption capacity and energy of adsorption respectively.

The biosorption capacity of the sorbent i.e K_f and Q_o and biosorption intensity/energy (1/n and b) has been estimated utilizing the slope and intercept of the Langmuir and Freundlich isotherms.

Chemical Modification onto Biomaterial

Chemical modification has been done for identification of the functional groups responsible for the sorption. All the three biomaterials (Neem, Pipal and Ashoka) under study have been reported contain free amino acids as constituent of their chemical composition. (Becker et. Al 1997, Jose et al 1999, Ray et al 1995) Efforts has been made to establish the role of carboxylic acids for the sorption process.

Identification of the Functional Groups Responsible for the Sorption

1. **Esterification:** The plant biomaterial has been subjected to esterification of carboxylic groups in a similar manner explained by Gardea-Torresdey et al, 1998. 10.0 gm of biomaterial sorbent material has been suspended in 640.0 ml of methanol. 5.4 ml of conc. HCI has been mixed in the suspension to compose the mixture of 0.1 M in HCl. Periodically over a two days period three 3.0 ml aliquots has been removed (Abdel-Ghani et al., 2007, Karta et al., 2005). All these samples and final products have been washed several times with deionized water to get rid of excess HCI and CH_3OH. The final product has been freeze-dried and further stored for consequent experiments.
2. **Propylamination:** in this process 4.0 g of biomaterial has been washed with acetone, filtered using Whatman 42 filter paper, again washing has been done with water and finally washed with acetone (Ashkenazy et al., 1996, Poon et al., 1986). This acetone washed biomaterial further has been washed with propylamine and kept over the magnetic stirrer for approximately 30 min. Biomaterial obtained has been then centrifuged. The obtained biomaterial was then used for further IR studies and sorption studies.

Modifications for Increasing the Sorption Efficiency of Biomaterial

1. **Acetylation:** Amino group on the biomaterial has been acetylated via washing 12.0 gm of biomaterial first in 0.1 M HCl to remove any debris. Followed by washing in sodium phosphate/sodium acetate buffer (0.1M Na_3PO_4/1.0M $NaC_2H_3O_2$) at pH 7.2. The biomaterial has been reacted with 64.0 ml of acetic anhydride and stirred while maintaining pH of 7.2 for 1 hr. The acetylated biomaterial has been then centrifuged for 5 min at 3000 rpm. After removing the supernatant, biomaterial was then resuspended in 1M hydroxylamine for removing O-acetyl groups. The biomaterial was then washed with 0.1M HCI to remove any more soluble materials and finally washed de-ionised water.
2. **Succination:** The amino groups on the biomaterial have been succinated by washing 12.0 gm of biomaterial first in 0.1 M HCI to remove any debris, followed by washing in 0.1M Sodium acetate at pH 8.0. The biomaterial was then resuspended 500 ml of 1M $NaC_2H_30_2.H_2O$ at 8.0. An additional 16.0 gm anhydride has been added after 15 minute interval for the next 1.5 hrs. leading to 6 additions of 16.0 g succinic anhydride to the biomaterial. Then biomaterial has been washed with

0.1 M HCI centrifuged and then washed repeatedly with de-ionised water. Now, the amino group is neutralized, but forms additional carboxyl group. This must lead to enhanced metal binding for those metals which bind to carboxyl ligands.

Results and Discussion

Table 1 represents soluble Cd (II) concentration in percentage, after sorption on powdered, *Azadiracta Indica (Neem), Ficus Religiosa* (Pipal) and *Saraca Indica* (Ashoka), leaf powder individually as a function of biomass dosage (1-6 g) and metal concentration (5-50 ppm) at particles size 105 (μm), pH 6.5, volume (200 ml) respectively.

Table 2 represents soluble Cd (II) ion concentration (%) after sorption on mixed powdered biosorbents under same experimental conditions mentioned above.

A scrutiny of the tables clearly depicts the fact that the biomaterial *Azadiracta Indica* shows maximum affinity (97.60%) for sorption of Cd (II), while the minimum value (92.46%) has been shown by *Saraca indica.* It is also evident from the tables, that when the mixed biosorbent (ternary) used the percentage of sorption for Cd (II) ions increases from 97.60% to 98% for Cd (II). Apparently overall increase in sorption is not very clearly visible but the significant increase in sorption can be clearly observed by reducing the amount of biomaterial dosage up to just half.

Identification of Sorption Responsible Functional Groups

Synthetic approach for the protection of COOH group by esterification and propylamination has been considered. The esterification of carboxylic acids was accomplished by methanol. The protection of COOH group leading to the development of ester has resulted into the decreased sorption efficiency and thus role of COOH group in metal binding has got highlighted.

$$RCOOH + CH_3OH \longrightarrow RCOOCH_3 + H_2O$$

In a very similar way the carboxylic acid group has also been transformed into amide by reaction with propylamine. Propylamination of the biomaterial lead to the decreased sorption efficiency supporting the function of carboxylic acid for metal sorption Reacting the carboxylic groups with propylamine, neutralizes these anions, considerably decreasing the metal ion uptake, demonstrating that negatively charged carboxylic group has an important role to play in biosorption process because of the electrostatic attraction.

$$CH_3CH_2CH_2NH_2 + RCOOH \longrightarrow RCOONH_2 + CH_3CH_2CH_3$$

Table 1. Sorption efficiency (%) of Saraca Insica, Ficus religiosa and Azadirachta indica for Cd (II) as afunction of metal concentration and biomass dosage, at volume (200 ml), pH 6.5 and particle size (105 μM)

Initial Conc. mg/L	Soluble Cd (II) concentration on Saraca indica leaf powder			
	Biomass dosage 1.0 g	Biomass dosage 2.0 g	Biomass dosage 4.0 g	Biomass dosage 6.0 g
5	43.19	56.71	69.81	69.97
10	54.12	72.56	83.32	83.46
25	66.83	81.28	92.46	92.63
50	66.99	81.39	92.71	92.79
Initial Conc. mg/L	Soluble Cd (II) concentration on Ficus religiosa leaf powder			
	Biomass dosage 1.0 g	Biomass dosage 2.0 g	Biomass dosage 4.0 g	Biomass dosage 6.0 g
5	43.78	62.49	72.83	72.93
10	56.18	71.63	80.71	80.83
25	63.32	82.63	94.82	94.93
50	63.49	82.79	94.98	95.04
Initial Conc. mg/L	Soluble Cd (II) concentration on Azadirachta indica leaf powder			
	Biomass dosage 1.0 g	Biomass dosage 2.0 g	Biomass dosage 4.0 g	Biomass dosage 6.0 g
5	48.56	62.19	74.33	74.42
10	52.39	76.39	82.92	83.01
25	73.83	92.32	97.60	97.78
50	73.99	92.41	97.63	97.72

Table 2. Sorption efficiency (%) of composite biomaterial for Cd (II) as a function of metal concentration and biomass dosage, at volume 200 ml, pH 6.5 and particle size (105 μM)

Initial Conc. mg/L	Soluble Cd (II) concentration on composite biomaterial			
	Biomass dosage 0.5 g	Biomass dosage 1.0 g	Biomass dosage 2.0 g	Biomass dosage 4.0 g
5	53.19	63.83	76.27	76.36
10	62.89	73.15	84.67	84.76
25	71.83	82.56	98.17	98.31
50	71.97	82.67	98.29	98.39

Table 3. Freundlich & Langmuir isotherm constants for sorption of Cadmium onto plant biomaterial

	Kf	1/n	R2	Q0	b	R2
Pipal	**0.76**	**0.69**	**0.93**	**2.63**	**0.24**	**0.93**
Ashoka	**0.71**	**0.74**	**0.95**	**2.56**	**0.29**	**0.92**
Neem	**0.65**	**0.82**	**0.95**	**2.51**	**0.34**	**0.95**
Mixture	**0.58**	**0.91**	**0.98**	**2.46**	**0.43**	**0.98**

Figure 1. Succination of novel biomaterial has been done to add a carboxyl group on to nitrogen ligand.

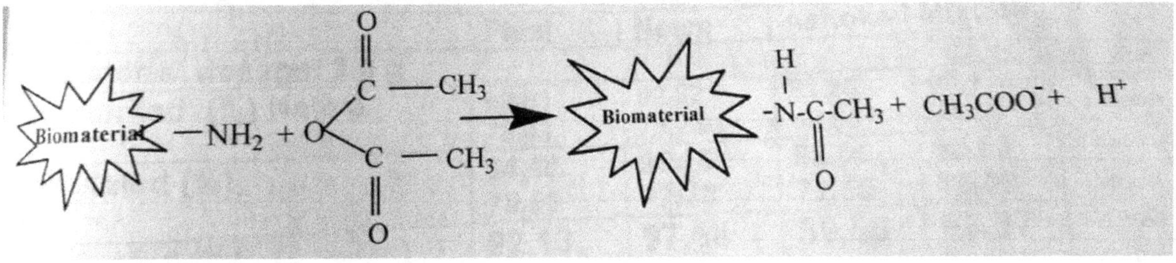

CHEMICAL MODIFICATIONS ONTO BIOMATERIAL: IMPROVING SORPTION EFFICIENCY

Based on our studies, it was determined that carboxyl ligands play an important role in binding metal ions to the sorbents. Thus, escalating the quantity of these groups should enhance the biomaterial metal binding capability. This has been achieved via acetylation and succination of the biomaterial.

In the process of acetylation of the biomaterial, available amino ligands are obstructed by acetic anhydride leading to amplify metal binding capacity. Acetylation leads into the neutralization of amino groups. It reduces the number of positively charged sites on biomaterial surface that leads to the modest enhancement in metal uptake capacity.

Biomaterials thus modified by above two possible ways and sorption efficiency was found to be increased using the same amount of biomaterial (4.0 g) for the metal under study.

Sorption effectiveness of all the three modified biomaterial for Cadmium metal was also monitored at biomaterial dose of 2.0 g. more interestingly almost equally good percentage of sorption was obtained at considerably lower biomaterial dosage [2.0 g] under the same experimental conditions.

Table 4. Enhancement of sorption efficiency of different chemically modified biomaterial in case of Cd (II) metal solution (Biomaterial dosage 4.0 gm).

	Pipal	Neem	Ashoka	Mixture
Biomaterial dosage 4.0 gm				
Unmodified	94.82	98.60	92.46	98.17
Acetylated Biomass	96.12	98.37	94.69	98.79
Succinaed Biomass	98.06	99.19	97.23	99.79

Table 5. Enhancement of sorption efficiency of different chemically modified biomaterial in case of Cd (II) metal solution (Biomaterial dosage 2.0 gm).

	Pipal	Neem	Ashoka	Mixture
Biomaterial dosage 2.0 gm				
Unmodified	82.63	92.32	81.28	98.17
Acetylated Biomass	84.48	95.16	83.56	98.59
Succinaed Biomass	92.13	97.68	89.56	99.37

Figure 2a. Freundlich isotherm plot for the adsorption of Cd (II) ion on Ficus religiosa leaf powder

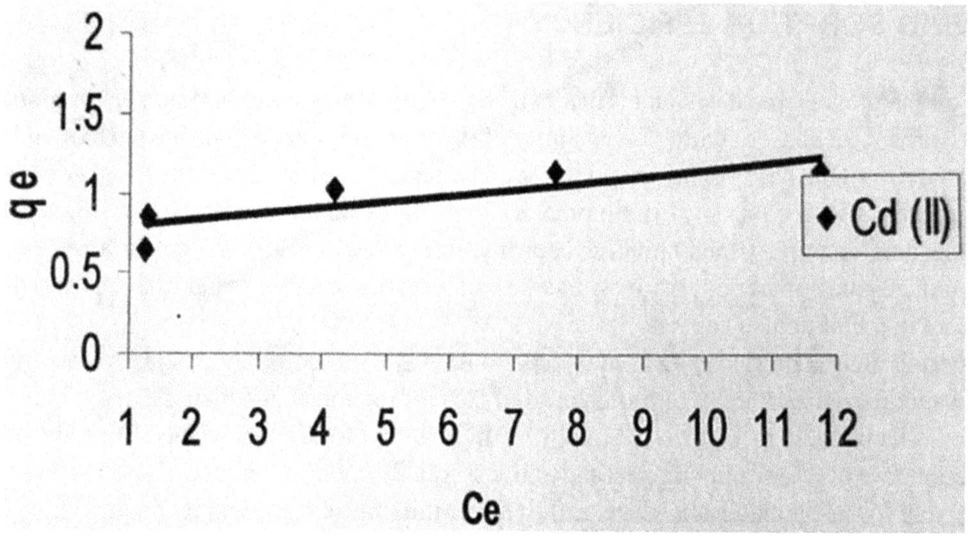

Figure 2b. Freundlich isotherm plot for the adsorption of Cd (II) ion on Saraca indica leaf powder

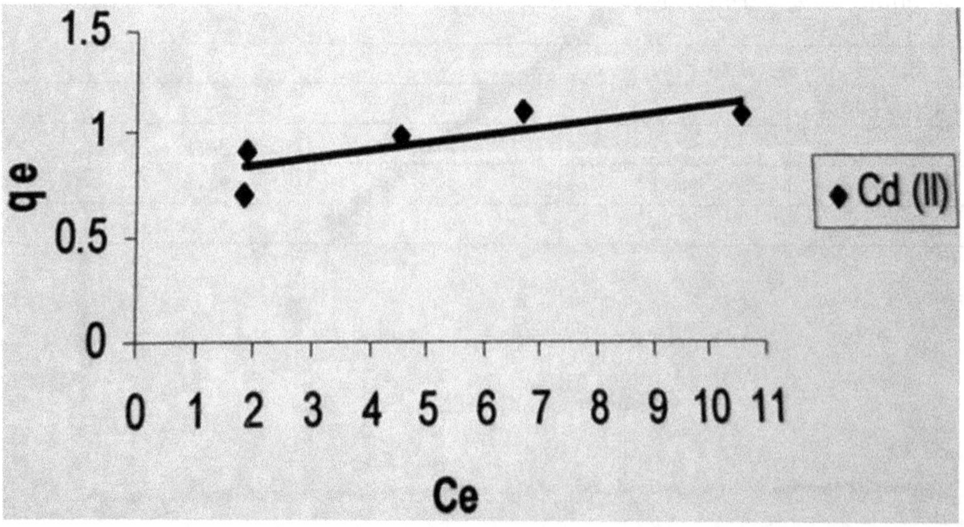

Figure 2c. Freundlich isotherm plot for the adsorption of Cd (II) ion on Azadirachta indica leaf powder

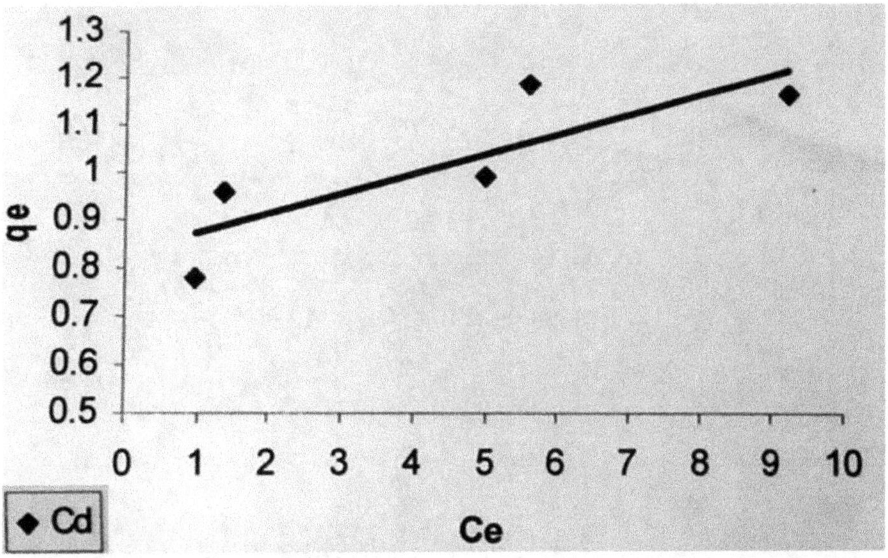

Figure 2d. Freundlich isotherm plot for the adsorption of Cd (II) ion composite (ternary) mixture

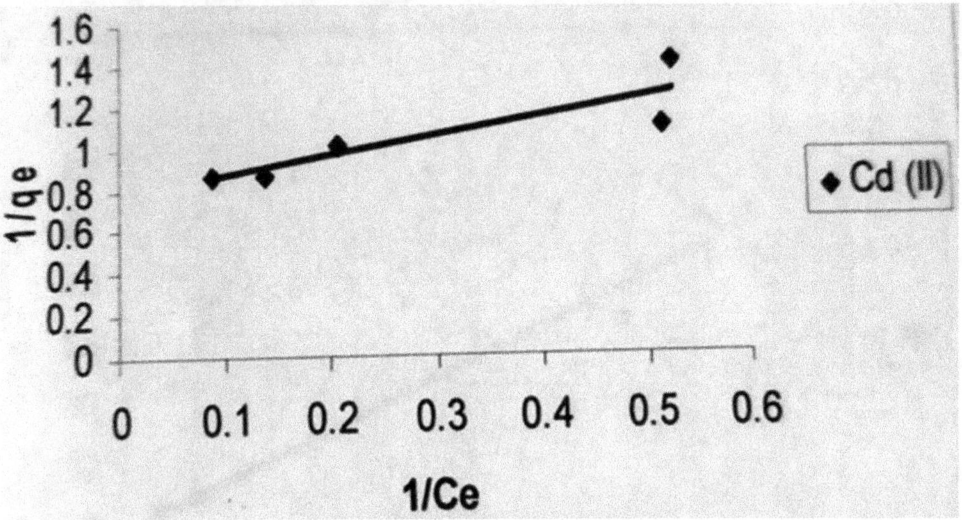

Figure 3a. Langmuir isotherm plot for the adsorption of Cd (II) ion on Ficus religiosa leaf powder

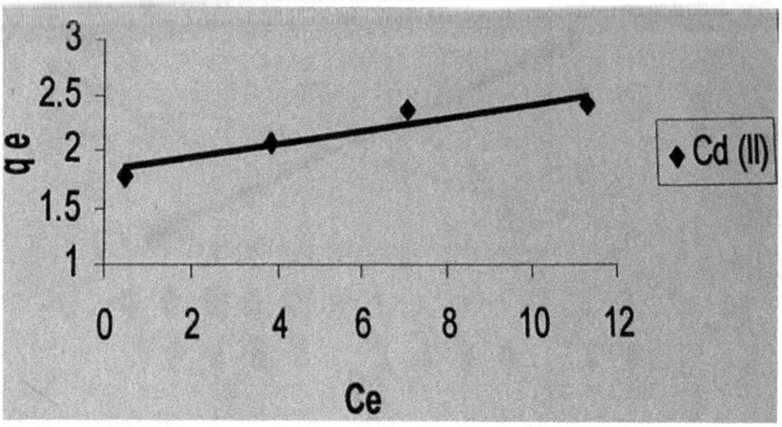

Figure 3b. Langmuir isotherm plot for the adsorption of Cd (II) ion on Saraca indica leaf powder

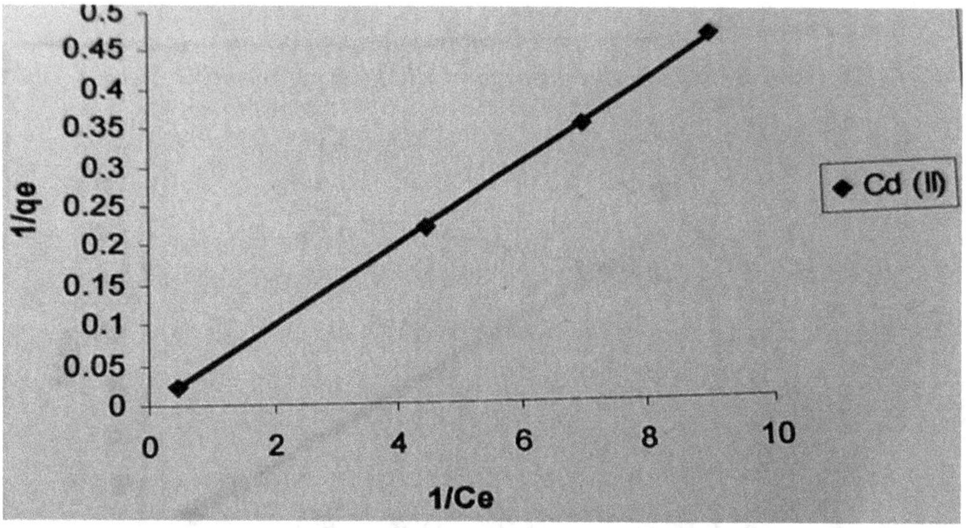

Fig. 3c. Langmuir isotherm plot for the adsorption of Cd (II) ion on Azadirachta indica leaf powder

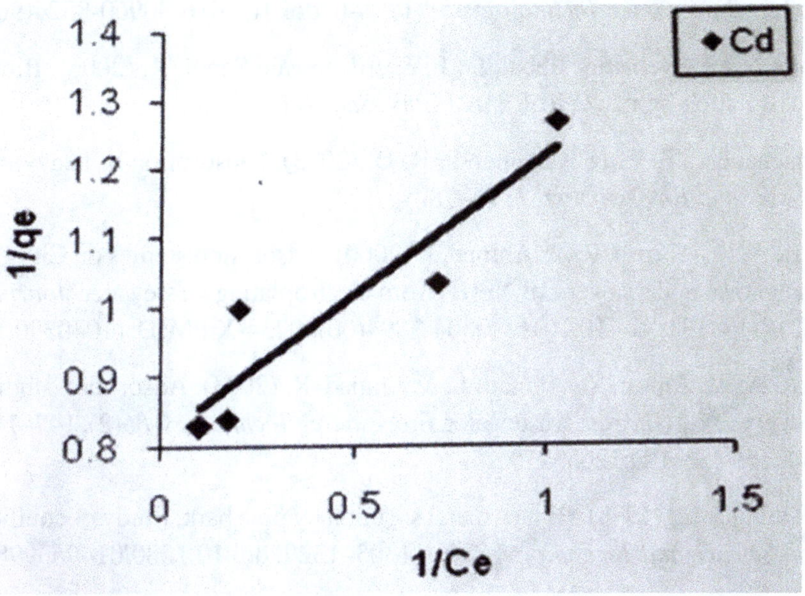

Figure 3d. Langmuir isotherm plot for the adsorption of Cd (II) ion on Azadirachta indica leaf powder

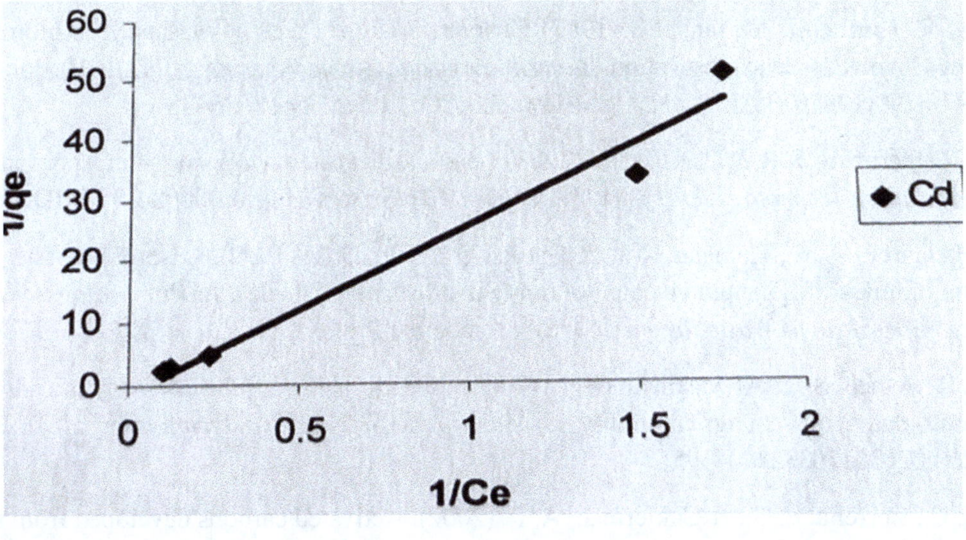

REFERENCES

Abdel-Ghani, N. T., Hefny, M., & El-Chaghaby, G. A. F. (2007). Removal of lead from aqueous solution using low cost abundantly available adsorbents. *International Journal of Environmental Science and Technology*, *4*(1), 67–73. doi:10.1007/BF03325963

Aderhold, D., Williams, C. J., & Edyvean, R. G. J. (1996). The removal of heavy-metal ions by seaweeds and their derivatives. *Bioresource Technology*, *58*(1), 1–6. doi:10.1016/S0960-8524(96)00072-7

Adesola Babarinde N.A., Oyebamiji Babalola, J., & Adebowale Sanni, R. (2006). Biosorption of lead ions from aqueous solution by maize leaf. *Int. J. Phy. Sci.*, *1*(1), 23-26.

Ahalya, N., Ramachandra, T. V., & Kanamandi, R. D. (2003). Biosorption of heavy metals. *Research Journal of Chemistry and Environment*, *7*, 71–78.

Ajmal, M., Rao, R. A. K., Ahmad, R., & Ahmad, J. (2000). Adsorption studies on Citrus reticulata (fruit peel of orange): Removal and recovery of Ni (II) from electroplating wastewater. *Journal of Hazardous Materials*, *79*(1-2), 117–131. doi:10.1016/S0304-3894(00)00234-X PMID:11040390

Ajmal, M., Rao, R. A. K., Anwar, S., Ahmad, J., & Ahmad, R. (2003). Adsorption studies on rice husk: Removal and recovery of Cd (II) from wastewater. *Bioresource Technology*, *86*(2), 147–149. doi:10.1016/S0960-8524(02)00159-1 PMID:12653279

Al-Asheh, S., & Duvnjak, Z. (1998). Binary metal sorption by pine bark: Study of equilibria and mechanisms. *Separation Science and Technology*, *33*(9), 1303–1329. doi:10.1080/01496399808544985

Alluri, H. K., Ronda, S. R., Settalluri, V. S., Bondili, J. S., Suryanarayana, V., & Venkateshwar, P. (2007). Biosorption: An eco-friendly alternative for heavy metal removal. *African Journal of Biotechnology*, *6*(25).

An, H. K., Park, B. Y., & Kim, D. S. (2001). Crab shell for the removal of heavy metals from aqueous solution. *Water Research*, *35*(15), 3551–3556. doi:10.1016/S0043-1354(01)00099-9 PMID:11561614

Ashkenazy, R., Gottlieb, L., & Yannai, S. (1997). Characterization of acetone-washed yeast biomass functional groups involved in lead biosorption. *Biotechnology and Bioengineering*, *55*(1), 1–10. doi:10.1002/(SICI)1097-0290(19970705)55:1<1::AID-BIT1>3.0.CO;2-H PMID:18636438

Axtell, N. R., Sternberg, S. P., & Claussen, K. (2003). Lead and nickel removal using Microspora and Lemna minor. *Bioresource Technology*, *89*(1), 41–48. doi:10.1016/S0960-8524(03)00034-8 PMID:12676499

Baig, T. H., Garcia, A. E., Tiemann, K. J., & Gardea-Torresdey, J. L. (1999). Adsorption of heavy metal ions by the biomass of Solanum elaeagnifolium (Silverleaf night-shade). In *Proceedings of the 1999 conference on Hazardous Waste Research, (131)*. Academic Press.

Banat, F., Al-Asheh, S., & Al-Makhadmeh, L. (2003). Kinetics and equilibrium study of cadmium ion sorption onto date pits—An agricultural waste. *Adsorption Science and Technology*, *21*(3), 245–260. doi:10.1260/026361703322404395

Basso, M. C., Cerrella, E. G., & Cukierman, A. L. (2002). Activated carbons developed from a rapidly renewable biosource for removal of cadmium (II) and nickel (II) ions from dilute aqueous solutions. *Industrial & Engineering Chemistry Research*, *41*(2), 180–189. doi:10.1021/ie010664x

Basu, A., Kumar, S., & Mukherjee, S. (2003). Arsenic reduction from aqueous environment by water lettuce (Pistia stratiotes L.). *Indian Journal of Environmental Health*, *45*(2), 143–150. PMID:15270347

Boddu, V. M., Abburi, K., Talbott, J. L., & Smith, E. D. (2003). Removal of hexavalent chromium from wastewater using a new composite chitosan biosorbent. *Environmental Science & Technology*, *37*(19), 4449–4456. doi:10.1021/es021013a PMID:14572099

Borstlap, A. C., & Schuurmans, J. (1988). Kinetics of L-valine uptake in tobacco leaf discs. Comparison of wild-type, the digenic mutant Valr-2, and its monogenic derivatives. *Planta*, *176*(1), 42–50. doi:10.1007/BF00392478 PMID:24220733

Clausen, C. A. (2000). Isolating metal-tolerant bacteria capable of removing copper, chromium, and arsenic from treated wood. *Waste Management & Research*, *18*(3), 264–268. doi:10.1177/0734242X0001800308

Coupal, B., & Lalancette, J. M. (1976). The treatment of waste waters with peat moss. *Water Research*, *10*(12), 1071–1076. doi:10.1016/0043-1354(76)90038-5

De la Rosa, G., Gardea-Torresdey, J. L., Peralta-Videa, J. R., Herrera, I., & Contreras, C. (2003). Use of silica-immobilized humin for heavy metal removal from aqueous solution under flow conditions. *Bioresource Technology*, *90*(1), 11–17. doi:10.1016/S0960-8524(03)00099-3 PMID:12835051

Gardea-Torresdey, J. L., Becker-Hapak, M. K., Hosea, J. M., & Darnall, D. W. (1990). Effect of chemical modification of algal carboxyl groups on metal ion binding. *Environmental Science & Technology*, *24*(9), 1372–1378. doi:10.1021/es00079a011

Gardea-Torresdey, J. L., De La Rosa, G., & Peralta-Videa, J. R. (2004). Use of phytofiltration technologies in the removal of heavy metals: A review. *Pure and Applied Chemistry*, *76*(4), 801–813. doi:10.1351/pac200476040801

Gardea-Torresdey, J. L., Tang, L., & Salvador, J. M. (1996a). Copper adsorption by esterified and unesterified fractions of Sphagnum peat moss and its different humic substances. *Journal of Hazardous Materials*, *48*(1-3), 191–206. doi:10.1016/0304-3894(95)00156-5

Gardea-Torresdey, J. L., Tiemann, K. J., Dokken, K., & Gamez, G. (1998). Investigation of metal binding in alfalfa biomass through chemical modification of amino and sulfhydryl ligands. In *Proceedings of the 1998 Conference on Hazardous Waste Research* (pp. 111-121). Snowbird.

Gardea-Torresdey, J. L., Tiemann, K. J., Gonzalez, J. H., Henning, J. A., & Townsend, M. S. (1996b). Ability of silica-immobilized Medicago sativa (alfalfa) to remove copper ions from solution. *Journal of Hazardous Materials*, *48*(1-3), 181–190. doi:10.1016/0304-3894(95)00155-7

Ghimire, K. N., Inoue, K., Makino, K., & Miyajima, T. (2002). Adsorptive removal of arsenic using orange juice residue. *Separation Science and Technology*, *37*(12), 2785–2799. doi:10.1081/SS-120005466

Goodwin, T. W., & Mercer, E. I. (1972). *Introduction to plant biochemistry*. Academic Press.

Goyal, P., Sharma, P., Srivastava, S., & Srivastava, M. M. (2008). Saraca indica leaf powder for decontamination of Pb: Removal, recovery, adsorbent characterization and equilibrium modeling. *International Journal of Environmental Science and Technology*, *5*(1), 27–34. doi:10.1007/BF03325994

International Agency of Research on Cancer. (1976). *Cadmium and Cadmium Compounds. Monographs on the Evaluation of Carcinogenic Risk of Chemicals to Man* (Vol. 11). International Agency for Research on Cancer.

Iqbal, M., Saeed, A., & Akhtar, N. (2002). Petiolar felt-sheath of palm: A new biosorbent for the removal of heavy metals from contaminated water. *Bioresource Technology*, *81*(2), 151–153. doi:10.1016/S0960-8524(01)00126-2 PMID:11762907

Jordao, C. P., Pereira, M. D. G., Einloft, R., Santana, M. B., Bellato, C. R., & Vargas de Mello, J. W. (2002). Removal of Cu, Cr, Ni, Zn, and Cd from electroplating wastes and synthetic solutions by vermicompost of cattle manure. *Journal of Environmental Science and Health. Part A, Toxic/Hazardous Substances & Environmental Engineering*, *37*(5), 875–892. doi:10.1081/ESE-120003594 PMID:12049122

Kar, R. N., Sahoo, B. N., & Sukla, C. B. (1992). Removal of heavy metals from pure water using sulphate-reducing bacteria (SRB). *Pollution Research*, *11*, 1–13.

Kartal, S. N., & Imamura, Y. (2005). Removal of copper, chromium, and arsenic from CCA-treated wood onto chitin and chitosan. *Bioresource Technology*, *96*(3), 389–392. doi:10.1016/j.biortech.2004.03.004 PMID:15474943

Khalid, N., Rahman, A., Ahmad, S., Kiani, S. N., & Ahmed, J. (1998). Adsorption of cadmium from aqueous solutions on rice husk. *Radiochimica Acta*, *83*(3), 157–162. doi:10.1524/ract.1998.83.3.157

Kobya, M. (2004). Removal of Cr (VI) from aqueous solutions by adsorption onto hazelnut shell activated carbon: Kinetic and equilibrium studies. *Bioresource Technology*, *91*(3), 317–321. doi:10.1016/j.biortech.2003.07.001 PMID:14607493

Kratochvil, D., Pimentel, P., & Volesky, B. (1998). Removal of trivalent and hexavalent chromium by seaweed biosorbent. *Environmental Science & Technology*, *32*(18), 2693–2698. doi:10.1021/es971073u

Krishnan, K. A., & Anirudhan, T. S. (2003). Removal of cadmium (II) from aqueous solutions by steam-activated sulphurised carbon prepared from sugar-cane bagasse pith: Kinetics and equilibrium studies. *Water S.A.*, *29*(2), 147–156. doi:10.4314/wsa.v29i2.4849

Low, K. S., & Lee, C. K. (1991). Cadmium uptake by the moss, Calymperes delessertii, Besch. *Bioresource Technology*, *38*(1), 1–6. doi:10.1016/0960-8524(91)90214-5

Makkar, H. P. S., & Becker, K. (1997). Nutrients and antiquality factors in different morphological parts of the Moringa oleifera tree. *The Journal of Agricultural Science*, *128*(3), 311–322. doi:10.1017/S0021859697004292

Manju, G. N., & Anirudhan, T. S. (1997). Use of coconut fibre pith-based pseudo-activated carbon for chromium (VI) removal. *Indian Journal of Environmental Health*, *39*, 289–298.

Martins, R. J., Pardo, R., & Boaventura, R. A. (2004). Cadmium (II) and zinc (II) adsorption by the aquatic moss Fontinalis antipyretica: Effect of temperature, pH and water hardness. *Water Research*, *38*(3), 693–699. doi:10.1016/j.watres.2003.10.013 PMID:14723939

Meunier, N., Laroulandie, J., Blais, J. F., & Tyagi, R. D. (2003). Cocoa shells for heavy metal removal from acidic solutions. *Bioresource Technology*, *90*(3), 255–263. doi:10.1016/S0960-8524(03)00129-9 PMID:14575948

Nagarnaik, P. B., Bhole, A. G., & Natarajan, G. S. (2003). Adsorption of arsenic on flyash. *Indian Journal of Environmental Health*, *45*(1), 1–4. PMID:14723275

Namasivayam, C., & Ranganathan, K. (1995). Removal of Cd (II) from wastewater by adsorption on "waste" Fe (III) Cr (III) hydroxide. *Water Research, 29*(7), 1737–1744. doi:10.1016/0043-1354(94)00320-7

Niu, H., & Volesky, B. (2003). Characteristics of anionic metal species biosorption with waste crab shells. *Hydrometallurgy, 71*(1-2), 209–215. doi:10.1016/S0304-386X(03)00158-0

Oliveira, J. T. A., Silveira, S. B., Vasconcelos, I. M., Cavada, B. S., & Moreira, R. A. (1999). Compositional and nutritional attributes of seeds from the multiple purpose tree Moringa oleifera Lamarck. *Journal of the Science of Food and Agriculture, 79*(6), 815–820. doi:10.1002/(SICI)1097-0010(19990501)79:6<815::AID-JSFA290>3.0.CO;2-P

Pagnanelli, F., Mainelli, S., Vegliò, F., & Toro, L. (2003). Heavy metal removal by olive pomace: Biosorbent characterisation and equilibrium modelling. *Chemical Engineering Science, 58*(20), 4709–4717. doi:10.1016/j.ces.2003.08.001

Pagnanelli, F., Toro, L., & Sara, M. (2002). Removal and Recovery of Ni (II), Pb (II) and Cr (III) by Olive Mill Residue. *Environmental Science & Technology, 15*, 47–59.

Paknikar, K. M., Pethkar, A. V., & Puranik, P. R. (2003). *Bioremediation of metalliferous wastes and products using inactivated microbial biomass*. Academic Press.

Poon, C. P. C. (1986). Removal of cadmium from wastewaters. In Cadmium in the Environment (pp. 46-55). doi:10.1007/978-3-0348-7238-6_7

Raji, C., Manju, G. N., & Anirudhan, T. S. (1997). *Removal of heavy metal ions from water using sawdust-based activated carbon*. Academic Press.

Ray, S., & Chatterjee, B. P. (1995). Saracin: A lectin from Saraca indica seed integument recognizes complex carbohydrates. *Phytochemistry, 40*(3), 643–649. doi:10.1016/0031-9422(95)98168-G PMID:7576454

Romero-Gonzalez, J., Peralta-Videa, J. R., Rodriguez, E., Delgado, M., & Gardea-Torresdey, J. L. (2006). Potential of Agave lechuguilla biomass for Cr (III) removal from aqueous solutions: Thermodynamic studies. *Bioresource Technology, 97*(1), 178–182. doi:10.1016/j.biortech.2005.01.037 PMID:16154514

Salah Azab, M., & Peterson, P. J. (1989). The removal of cadmium from water by the use of biological sorbents. *Water Science and Technology, 21*(12), 1705–1706. doi:10.2166/wst.1989.0149

Sawalha, M. F., Gardea-Torresdey, J. L., Parsons, J. G., Saupe, G., & Peralta-Videa, J. R. (2005). Determination of adsorption and speciation of chromium species by saltbush (Atriplex canescens) biomass using a combination of XAS and ICP–OES. *Microchemical Journal, 81*(1), 122–132. doi:10.1016/j.microc.2005.01.008

Seki, K., Saito, N., & Aoyama, M. (1997). Removal of heavy metal ions from solutions by coniferous barks. *Wood Science and Technology, 31*(6), 441–447. doi:10.1007/BF00702566

Shukla, S. R., & Pai, R. S. (2005). Adsorption of Cu (II), Ni (II) and Zn (II) on modified jute fibres. *Bioresource Technology, 96*(13), 1430–1438. doi:10.1016/j.biortech.2004.12.010 PMID:15939269

Son, B. C., Park, K., Song, S. H., & Yoo, Y. J. (2004). Selective biosorption of mixed heavy metal ions using polysaccharides. *Korean Journal of Chemical Engineering*, *21*(6), 1168–1172. doi:10.1007/BF02719489

Sternberg, S. P., & Dorn, R. W. (2002). Cadmium removal using Cladophora in batch, semi-batch and flow reactors. *Bioresource Technology*, *81*(3), 249–255. doi:10.1016/S0960-8524(01)00131-6 PMID:11806406

Sun, G., & Shi, W. (1998). Sunflower stalks as adsorbents for the removal of metal ions from wastewater. *Industrial & Engineering Chemistry Research*, *37*(4), 1324–1328. doi:10.1021/ie970468j

Sure, B., Easterling, L., Dowell, J., Crudup, M., & Scrimshaw, N. S. (1953). Protein Efficiency, Improvement in Whole Yellow Corn with Lysine, Tryptophan, and Threonine. *Journal of Agricultural and Food Chemistry*, *1*(9), 626–629. doi:10.1021/jf60009a007

Tegbe, T. S. B., Adeyinka, I. A., Baye, K. D., & Alawa, J. P. (2006). Evaluation of Feeding Graded Levels of Dried and Milled FicusthonningiiLeaves on Growth Performance, Carcass Characteristics and Organs of Weaner Rabbits. *Pakistan Journal of Nutrition*, *5*(6), 548–550. doi:10.3923/pjn.2006.548.550

Tiemann, K. J., Gamez, G., Dokken, K., Parsons, J. G., & Gardea-Torresdey, J. L. (2002). Chemical modification and X-ray absorption studies for lead (II) binding by Medicago sativa (alfalfa) biomass. *Microchemical Journal*, *71*(2-3), 287–293. doi:10.1016/S0026-265X(02)00021-8

Tiwari, D., Mishra, S. P., Mishra, M., & Dubey, R. S. (1999). Biosorptive behaviour of Mango (Mangifera indica) and Neem (Azadirachta indica) bark for Hg2+, Cr3+ and Cd2+ toxic ions from aqueous solutions: A radiotracer study. *Applied Radiation and Isotopes*, *50*(4), 631–642. doi:10.1016/S0969-8043(98)00104-3

Vankar, P. S., Tiwari, V., & Srivastava, J. (2007). Antioxidants from supercritical carbon dioxide fluid extracts (SCFE) of bark-peel of Eucalyptus globulus. *EJEAF Chem.*, *6*(11), 2550–2556.

Vieira, R. H., & Volesky, B. (2000). Biosorption: A solution to pollution? *International Microbiology*, *3*(1), 17–24. PMID:10963329

Villaescusa, I., Fiol, N., Martínez, M., Miralles, N., Poch, J., & Serarols, J. (2004). Removal of copper and nickel ions from aqueous solutions by grape stalks wastes. *Water Research*, *38*(4), 992–1002. doi:10.1016/j.watres.2003.10.040 PMID:14769419

Volesky, B. (2001). Detoxification of metal-bearing effluents: Biosorption for the next century. *Hydrometallurgy*, *59*(2-3), 203–216. doi:10.1016/S0304-386X(00)00160-2

Chapter 5
Hybrid 2D Nanomaterials for Photocatalytic Degradation of Wastewater Pollutants

Ali Raza
University of Sialkot, Pakistan

Muhammad Ikram
GC University, Lahore, Pakistan

Usman Qumar
GC University, Lahore, Pakistan

Asma Rafiq
University of the Punjab, Lahore, Pakistan

ABSTRACT

Developing innovative technologies for the effective treatment of organic contaminants comprising agricultural wastes, industrial dyes, and chemicals is gaining extraordinary importance across the globe. In the last few years, photocatalytic degradation has become an effective and established route to eliminate these pollutants from aqueous solution relative to simple adsorption. 2D nanomaterials exhibit great potential as an effectual photocatalyst in degradation of contaminants, especially hybridization with other functional components due to wide-ranging band structures, sufficient active sites, and significant specific surface area. Herein, the unique hybridization of 2D nanomaterials with numerous functional species is reviewed comprehensively by highlighting their improved photocatalytic performances and remarkable environmentally friendly activity. The chapter outlines the mechanism of photocatalytic degradation to explore the advantages/disadvantages of regular 2D materials and discover the significance of developing hybrid 2D photocatalysts.

DOI: 10.4018/978-1-6684-4553-2.ch005

INTRODUCTION

The need for industrial and agricultural products has been increasing substantially with a rise in population across the globe, which results in release of organic contaminants in environment. Numerous environmental factors pose huge impact on quality of life because of poisonousness and prolonged persistence of the pollutants produced by pesticide production pharmaceutical production, textile industries, and other organic compound production (Hasija et al., 2019; Perreault et al., 2015; Suyana et al., 2017). Though, organic dyes are hazardous even at minute amounts and can bring adverse effects to ecosystem and human health. Recently, the elimination of dyes and other contaminants from water has become a challenging task (Rafiq et al., 2021). Hence, several technologies have been adopted to eliminate water pollutants, comprising biodegradation, adsorption, catalytic degradation, sedimentation, membrane filtering, and coagulation (Ahmed et al., 2018; Ge et al., 2018). The photodegradation of impurities has gained noteworthy importance in the past several years. The photocatalytic reaction involves heterogeneous catalysis, where a semiconductor photocatalyst absorbs UV light to reduce numerous environmental pollutants. Photodegradation offers advantages over the traditional wastewater treatment approaches. For instance, the complete degradation of contaminants using photocatalysts can take place within a few hours at room temperature (Rafiq et al., 2021).

Functional 2D materials are being widely studied for photocatalysis owing to exceptional properties ascribed to structure properties. 2D nanostructures exhibit remarkable advantages in charge separation, architecture modulation, light-harvesting, and tunability relative to 0D and 1D nanomaterials (Ali et al., 2021; Y. Liu et al., 2019; Raza, Qumar, et al., 2022; Yasir et al., 2022). The significant lateral size of 2D nanosheets provides adequate active sites for adsorption of contaminant and surface reaction (Sun et al., 2015). Additionally, the ultrathin thickness prefers charge diffusion to material surface via reducing the transportation distance (Deng et al., 2016), which is a unique aspect of 2D materials. Several efforts have been carried out to design effective photocatalysts with 2D nanomaterials due to exclusive electronic, optical, and physicochemical features of 2D nanosheets, in addition to the assessment of their potential for water decontamination (Qumar et al., 2022; Raza, Altaf, et al., 2022; Raza, Rafiq, et al., 2022; Su et al., 2018; Tan et al., 2017).

In the hybrid photocatalyst, graphene enhances electron transport to separate photogenerated electrons from holes, even though attaining a greater affinity for adsorption due to its larger surface area (Huang et al., 2012; Li et al., 2016; Perreault et al., 2015; Raza, Zhang, et al., 2022). When employed as a photocatalyst or co-photocatalyst for the elimination of impurities, MoS_2 nanosheets with a bandgap of 1.8 eV are normally utilized (Wang & Mi, 2017). Other photocatalysts like Bi_2O_3 (2.5 eV), g-C_3N_4 (2.6 eV), WO_3 (2.7 eV), and BiOX (3.0 eV) have also been widely employed as visible light-stimulated photocatalysts (Haque et al., 2017; Jiang et al., 2017; Patnaik et al., 2016). Keeping in view the importance of numerous 2D materials for photocatalytic degradation of harmful chemicals and dyes here, this chapter is specifically focused on a brief review and discussion of various hybrid 2D photocatalysts and to enhance their activity by defect and hybridization engineering.

GRAPHENE-BASED PHOTOCATALYSTS

Graphene exhibit 2D π-conjugated nanostructure by covalent bonding (σ bonds) among adjacent carbons, hence exploring higher charge mobility and thermal conductivity, huge specific surface area, and outstanding mechanical strength (Guan & Han, 2019). Recently, extensive research has been carried out on graphene-based hybrid materials as unique sorbent or photocatalyst for purification of contaminants (Huang et al., 2012; Perreault et al., 2015). Normally, graphene is cited as graphene oxide (GO) or reduced graphene oxide (rGO) in photocatalytic system and is capable of tuning the photocatalysts band structure and repress the electron-hole pairs recombination for a substantial photocatalytic enhancement (Zhang et al., 2012). In this chapter, we summarize current developments on the hybrid photocatalysts synthesis consisting of graphene and various structural materials, and their practices in photocatalysis.

Graphene Hybridized with g-C_3N_4 Nanosheets

The g-C_3N_4 nanosheets displaying the monoatomic layer such as graphene, exhibiting absorption in visible range and considered supreme to synthesize hybrid photocatalysts using graphene for ecological applications. Up to now, numerous effectual approaches are being designed to prepare 2D/2D nanocomposites exhibiting various combined formats, which considerably increase their photocatalytic activity. Figure 1a reveals the fabrication of a 3D g-C_3N_4/GO structure by the hydrothermal coassembly for designing an effectual photocatalyst of g-C_3N_4 with an enhanced transfer of electrons using GO (Tong et al., 2015). The elimination of methyl orange (MO) by employing the highly interconnected porous aerogel reached up to 92% within 4 h, whereas the pristine g-C_3N_4 value was only 12%. On the other hand, g-C_3N_4-rGO hybrids had been immobilized on 3D nickel foam via facile dip-coating technique. After absorption of visible light, a successive hydrazine hydrate reduction method for tetracycline and MO decolorization was observed (Wang et al., 2018), maximum photocatalytic performance was monitored when the rGO/g-C_3N_4 ratio was 1:9 in weight. Both nitrogen-doped carbon nanotubes (N-CNT) and oxygen modified monolayer g-C_3N_4 had been hybridized using GO by electrostatic contact to attain ample coupling heterointerfaces, hence a membrane heterostructured was formed after vacuum filtration (Figure 1b) (Córdoba et al., 2018). The GO and N-CNT presence under the visible light irradiation can effectively decrease the photogenerated charge carriers' recombination in 94.30% tetracycline hydrochloride degradation. Another study was carried out to prepare 3D hybrid photocatalyst with extraordinarily advanced photocatalytic performance for removal of MB in which 1D Ag@$AgVO_3$ nanowires were penetrated via graphene and protonated g-C_3N_4 nanosheets. The resultant increment in photocatalytic activity was accredited to numerous synergistic effects from the enhanced surface area of unique composition, outstanding graphene conductivity, and strong Ag@$AgVO_3$ absorption for visible light (Zhang et al., 2015). Zeng's et al employed Ag@Ag_3PO_4/g-C_3N_4/rGO to prepare an indirect reproductive Z-scheme network for eliminating pollutant 2,2¢,4,4¢ -tetrabrominated diphenyl ether (BDE-47) within 93.4% efficacy in visible light exposure in 120 min, displaying an improvement of »174 times relative to pristine g-C_3N_4 sheets (Liang, Zhang, et al., 2019). Additionally, several research groups reported the various 2D nanomaterials combination into g-C_3N_4/graphene nanocomposite to fabricate ternary photocatalysts using 2D/2D/2D configuration. For instance, an in situ adsorption technique was employed for insetting g-C_3N_4 sheets in the center of MoS_2 sheets and graphene to display an unusual photocatalytic response resulting from close contacted surface among two adjacent nanosheets, that was ~4.8 times greater relative to pure g-C_3N_4 nanosheets for the decolorization of RhB (Tian et al., 2018). Tonda's and his colleagues combined CoAl-layered double hydroxide, g-C_3N_4, and rGO employing a one-step hydrothermal approach to design innovative heterostructures with upgraded photocatalytic activity in the decolorization of Congo red (CR) and tetracycline (Jo & Tonda, 2019). To increase the interfacial charge transfer for

Figure 1. Hybridization of g-C_3N_4 with GO sheets for photocatalytic decolorization. (a) Depiction for the preparation of 3D g-C_3N_4/GO aerogel and its enhanced properties for photodegradation. (b) Synthesis of O-g-C_3N_4/GO/N-CNT nanocomposites by electrostatic interaction and mechanism for photocatalytic degradation of impurities.

(a) Reproduced with permission from ref. (Tong et al., 2015) Copyright 2015, American Chemical Society.
(b) Reproduced with permission from ref. (Qu et al., 2018) Copyright 2018, American Chemical Society.

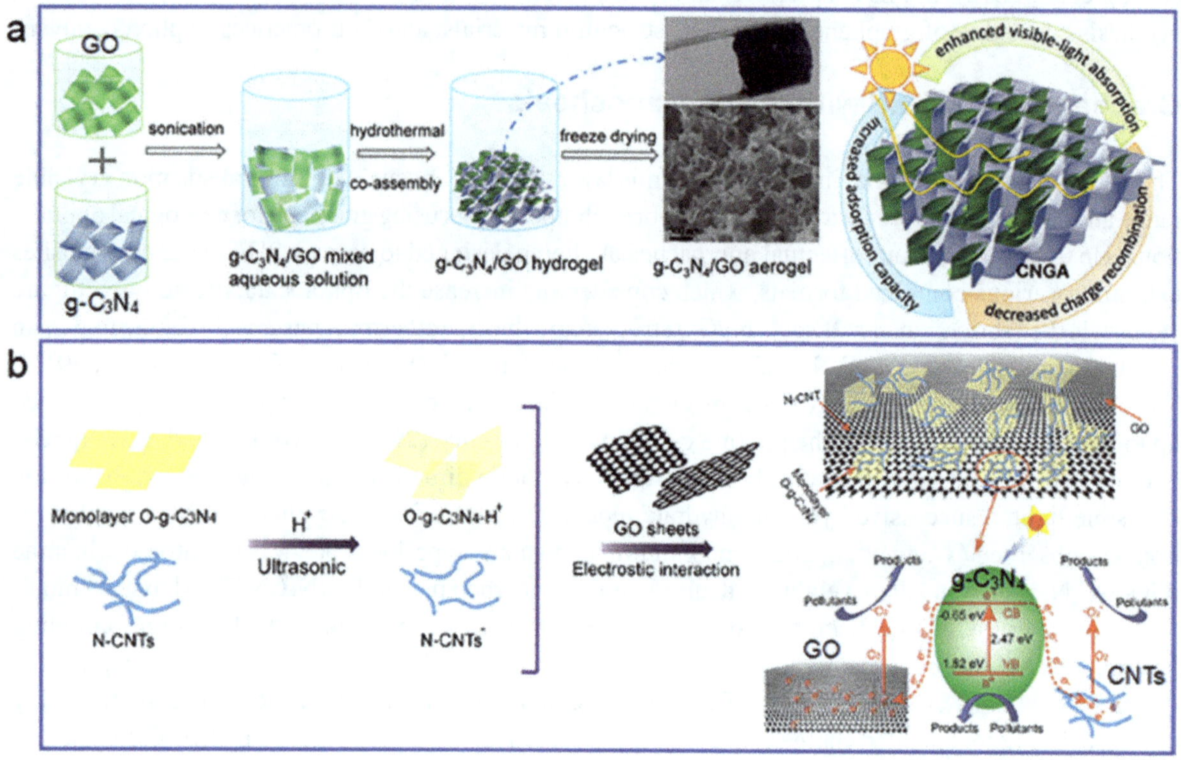

substantial improvement of photocatalytic performance, the 2D/2D/2D arrangement was employed to provide a large intimate interfacial contact.

Hybridized with Metal Oxides

Graphene exhibits unusual conductivity and remarkable electron mobility, whereas metal oxides (MOs) display significant light-harvesting aptitude in visible and UV regions. Therefore, their combination is favorable for designing hybrid photocatalysts exhibiting notable photocatalytic activity. Furthermore, the TiO_2/rGO composites also displayed a considerably significant aptitude for persulfate activation under exposure of in to generate more hydroxyl radicals and sulfate radicals ($SO_4^{\bullet-}$) for whole elimination of micro contaminants like SMX, bisphenol A (BPA), acetaminophen and phenol (Yang et al., 2019). Wang's et al studied the electronic properties and chemical structure of TiO_2–graphene composites by employing density functional theory (DFT) simulation and explored that TiO_2–graphene (Akola et al., 2008) heterostructures retained exceptional charge carrier separation and oxidation capabilities, and the time-consuming lifetimes of photogenerated radicals relative to other nanocomposites comprising TiO_2 with numerous planes (Yang et al., 2013). The coupling of rGO sheets improved harvesting potential for visible light, lessened its BG, increased charge carrier separation, and repressed the accumulation of TiO_{2-x} nanoparticles (NPs) to absorb more organic contaminants for Ti^{3+} and O_2 vacancies self-doped TiO_2 (Xu et al., 2018). The hybrid TiO_{2-x}/rGO photocatalysts TiO_2/rGO, and TiO_{2-x} samples, revealed a rapid reaction rate relative to pure TiO_2 for BPA degradation. TiO_2 and Ni were affixed in Shin's work onto the surface of GO sheets via microwave and the resultant Ni-integrated TiO_2/GO photocatalysts displayed outstanding adsorption skill for target impurities in dark and enhanced photodegradation rate under exposure of visible light (Pham et al., 2016). Likewise, Ag and Pt NPs have been improved onto/into GO-TiO_2 mesocrystals using in situ reducing approach (Sreeja et al., 2019). The variation of Pt and rGO abridged the BG of TiO_2 and considerably increased the absorption potential for visible light illumination and electron-hole separation rate, which offered an increased MO photo decolorization relative to pure TiO_2, Pt/TiO_2, and rGO/TiO_2, respectively (Huo et al., 2019). TiO_2–rGO/MoS_2 is an additional significant ternary photocatalyst where the interfacial charge moved from the TiO_2 to the MoS_2 via middle rGO nanosheets for efficiently constraining the recombination of comparatively advantageous electrons-holes (Nimbalkar et al., 2016). The electron-hole pair recombination had suppressed collaboratively in the ternary photocatalyst with the S-scheme heterostructure creation and Schottky junction among rGO and TiO_2 nanosheets. For superior photocatalytic performances, Fe-based compounds are also commonly used to hybridize with graphene. For example, with the addition of 10% rGO loading in magnetite (Fe_3O_4) the decolorization efficacy for 2-methylisoborneol (MIB) was improved from 22.5% to 99% (Moztahida et al., 2019). Moreover, silver-magnetite/graphene nanomaterials had been employed as a broadspectrum catalyst exhibiting excellent proficiency for removal of pollutants under illumination of ultraviolet source (Saleh & Taufik, 2019). Additionally, rGO hybridization with silver and magnetite NPs might be utilized for activating peroxydisulfate (PDS) in the effectual photodegradation of endocrine-disrupting compounds and pharmaceuticals (Park et al., 2018). The preliminary pseudo-first-order rate of phenol elimination was recorded as »8 times greater compared to magnetite NPs. The rGO nanosheets can improve the mobility of photoexcited charge carriers and construct various heterojunctions for the separation of photoinduced electrons and holes in the Z-scheme rGO–Fe_2O_3–MoS_2 composites in 3D configuration (Chen et al., 2017). Under the exposure of visible light, the ternary nanocomposites revealed considerably upgraded photocatalytic performance and exceptional stability in degradation of MB and RhB relative to pure Fe_2O_3 NPs. Currently, rGO and $NiFe_{0.7}Co_{1.3}O_4$ has been hybridized simultaneously as an outstanding photocatalyst to stimulate persulfate for degrading BPA,(Xu et al., 2019) whereas magnetically separated Co–Fe Prussian blue analogs and rGO have joined into nanocomposites for increasing the photocatalytic activity to levofloxacin hydrochloride after peroxymonosulfate (PMS) activation (Pi et al., 2018). Other MOs are also used for synthesizing hybrid photocatalysts using graphene in addition to TiO_2 and Fe-based compounds. Normally, a WO_3 and rGO nanocomposite was developed by employing a hydrothermal route in the RhB photodegradation (Ahmed et al., 2018). For elimination of neutral red and ciprofloxacin photo catalytically from wastewater, manganese dioxide (MnO_2) nanorods have been loaded on rGO sheets using facile hydrothermal method to attain closed interaction of 1D/2D composite with sufficient reaction sites and large surface area. (Chhabra et al., 2019) Meanwhile, MnO_2 sheet/N-doped graphene aerogel has been controllably synthesized to boost up PMS for IBP degradation (Dong et al., 2019). For effectual removal of 4-nitrophenol by improving light absorption, Park et al (Zhou et al., 2019) fabricated a ternary composite of rGO/zirconium dioxide(ZrO_2)/Ag_3PO_4 exhibiting band gap of about 2.3 eV.

Hybridization with Other Compounds

MoS_2 nanomaterials exhibiting BG (»1.8 eV) may rapidly supply photogenerated electrons-holes, whereas graphene having excellent conductivity are advantageous in faster separation of photogenerated carriers. Hence, MoS_2–GO composite hydrogel has been synthesized using one-step hydrothermal technique in decolorization of MB with enhanced photocatalytic activity (Ding et al., 2015). To attain a better photocatalytic performance of 99% within 28 min for MB, MoS_2/graphene composites have been fabricated (Zhao et al., 2017). On the other hand, the rGO/MoS_2 hybrid nanostructures have been fabricated using ultrasound-driven mixed MoS_2 and GO exfoliation in an ethanol/water combination and successive GO reduction (Cravanzola et al., 2016). Currently, ZnS–MoS_2 hybrid was attached over rGO sheets using solvent thermal approach as a unique photocatalyst exhibiting enhanced absorption in visible region (Hu et al., 2018) rGO/ZnS–MoS_2 photocatalyst effectually removes industrial discharge due to their effective charge separation and modified band structure. Similarly, Ag_3PO_4 microcrystals have been grown on a MoS_2 and graphene hybrid using two-step hydrothermal method for removal of phenol impurities and served as a favorable photocatalyst as shown in Figure 2 (Peng et al., 2014). Various other metal sulfides have also described in literature to hybrid on rGO sheets other than MoS_2 in organic dyes photodegradation. Normally, CdS NPs (10 nm-sized) have been anchored on rGO using facile condensation method in dimethylformamide to prepare CdS/rGO composites exhibiting superior activity by »3 times relative to pure CdS in RhB decolorization (Meng et al., 2016). By employing surface modification approach, CdS NPs, nanosheets, and nanorods are grafted on rGO sheets using potential electronic reactions (Bera et al., 2015). Consequently, the 2D/2D heterostructure photodegradation was observed to be »4 and »3.4 times greater compared to 0D/2D and 1D/2D hybrid nanocomposite, respectively. Furthermore, it has also been observed that the 2D/2D photocatalyst displayed significant degradation response of »2.5 times that of pristine CdS nanosheets. Likewise, CdS sheet–rGO nanomaterials were employed to decrease heteroaromatic, aromatic, aliphatic, and sulfonyl azides photocatalytically to the analogous amines in the occurrence of hydrazine hydrate (Singha et al., 2018). On the other hand, Cu_2SnS_3 and Cu_2SnSe_3 quantum dots (QDs) had been hybridized onto rGO sheets using one-pot colloidal fabrication, resulting in extraordinarily improved photocatalytic responses for MO and RhB degradation (Han et al., 2018). The CdS/$ZnIn_2S_4$/rGO ternary photocatalyst was also prepared by uniform growth of ultrathin $ZnIn_2S_4$ sheets perpendicular to the CdS nanowires surface using facile solvothermal method, followed by successive incorporation of rGO acting as a cocatalyst for considerably enhancing their stability and photoactivity (Tian et al., 2017). To utilize the adsorptive nature and strengthened charge transfer of rGO and silver surface plasmon resonance effect, Ag nanomaterials (8 nm-sized) have been dispersed on wrinkled rGO nanosheets to obtain a maximum degradation response in textile discharges (V et al., 2019). Though, GO enwrapped Ag phosphate (GO/Ag_3PO_4) nanomaterials showed excellent activities in the elimination of several polycyclic aromatic hydrocarbons photocatalytically (Yang et al., 2018) Ma et al. utilized commercial $K_4Fe(CN)_6$ to prepare Fe_3C@N-CNT/graphene nanomaterials, which showed outstanding photocatalytic activity in PMS enhancement for BPA elimination (Ma et al., 2019). On the other hand, Niu's and his colleagues employed a surface charge-mediated self-assembly to develop Bi@Bi_5O_7I/rGO 2D/2D heterostructure for excellent degradation efficiency in levofloxacin decomposition (Liang, Niu, et al., 2019).

Figure 2. The synthesis and photocatalytic activity of Ag_3PO_4–MoS_2/graphene (GR) photocatalyst. (a) Synthesis technique, (b) photocatalytic dichlorophenol degradation by various photocatalysts, (c) degradation of numerous organic phenols via hybrid photocatalyst.
Reproduced with permission from ref. (Peng et al., 2014) Copyright 2014, Royal Society of Chemistry.

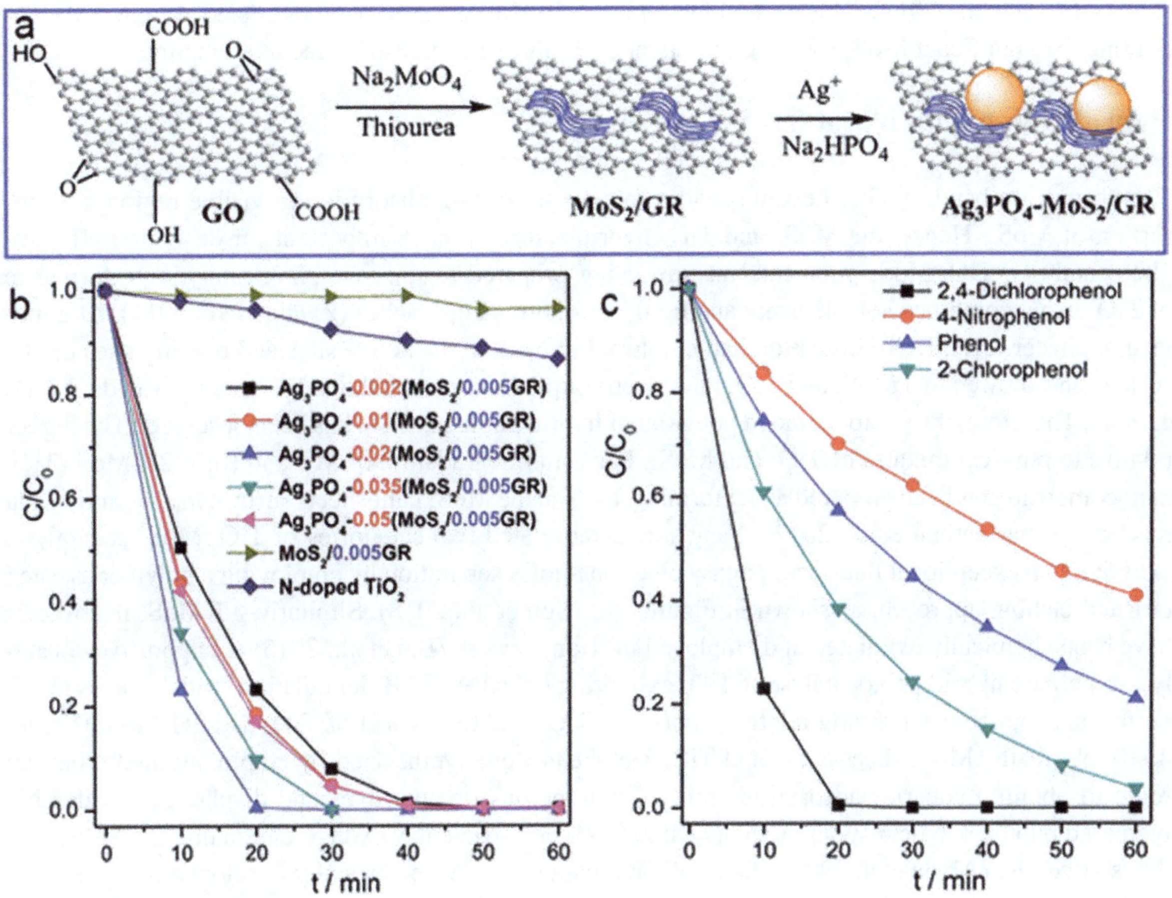

MoS_2-BASED PHOTOCATALYSTS

MoS_2 monolayers with three layers of atoms are composed by sandwiching a Mo among two S layers. MoS_2 crystals comprise of three crystalline stages: metallic 1T level, semiconductor 2H level, and 3R stage, that will display diverse features in various hybridized photocatalysts (Guan & Han, 2019). MoS_2 materials have been regarded as outstanding photocatalysts or cocatalysts for the photocatalytic degradation of contaminants due to their strong light adsorption, high photocatalytic capability, and cost-effectiveness (Wu et al., 2018). Especially, MoS_2 thin films with perpendicularly aligned layers showed several 2D sites because of their developed dangling bonds that presented excellent chemical reactivity for effectual degradation of harmful compounds (Islam et al., 2017). Concurrently, TiO_2/MoS_2 nanocomposites have been prepared to impede charge recombination and improve absorption in visible region (Rahmanian et al., 2018). Furthermore, CDs and Ag_3PO_4 NPs have been deposited on MoS_2 sheets for enhancing their photodegraded capability by separating photoproduced electrons-holes effectually (Li et al., 2019). The incorporation of 1T-MoS_2 sheets using MIL-53(Fe) generated needle-shaped nanocomposites exhibiting mesopores and micropores, that enhanced the IBP photocatalytic rate relative to pure MIL-53(Fe) and1T-MoS_2, respectively (N. Liu et al., 2019). Heterojunctions can be prepared by modifying MoS_2 as cocatalysts to stimulate the photoproduced electrons transfer and hence increase their photocatalytic activity (Peng et al., 2017). This part will summarize the MoS_2-based photodegraded systems by highlighting the significant MoS_2 role as catalysts or cocatalysts for industrial decontamination.

Hybridization with Metal Oxides

The absorption band of TiO_2 becomes wider to have maximum absorption in visible region after insertion of MoS_2. Hence, the MoS_2 and TiO_2 hybridization is much important for designing effective photocatalysts. TiO_2/MoS_2 photocatalysts are being prepared to improve photocatalytic performance of TiO_2 using combined solvothermal and liquid exfoliation approaches (Zhang et al., 2016). At a molecular cluster level, DFT simulation had employed to explore the active sites and binding sites on the surface and showed that MoS_2 and TiO_2 have been coupled via chemical bonds instead of van der Waals contact. Therefore, the photodegraded potential of hybrid photocatalysts has been observed to be higher relative to physical mixture of TiO_2 and MoS_2. For better degradation of MB and RhB, 2H MoS_2/TiO_2 nanocomposite has been prepared using forming TiO_2 on the MoS_2 nanosheets surface in a hydrothermal reaction (Sabarinathan et al., 2017). Yang's et al fabricated two categories of TiO_2/MoS_2 cocatalysts exhibiting an exceptional interface for easy electrons-holes separation by employing a polymer assisted targeted-etching approach, as shown in Figure 3a-f (Sun et al., 2018). Similarly, 1T MoS_2 nanosheets have been chemically exfoliated and employed in Xiong's work (Bai et al., 2015) as supportive cocatalyst to prepare hybrid photocatalyst of TiO_2 exhibiting enhanced RhB degradation, which shows better performance photocatalytically relative to pristine TiO_2 and the hybrid of TiO_2 and 2H-MoS_2 (Figure 4a-d). Normally, MoS_2 sheet-decorated TiO_2 nanobelts were synthesized by employing hydrothermal route to obtain a superior adsorption ability for numerous organic dyes and display a considerably increased photocatalytic response (Zhou et al., 2013). For efficient MO degradation under UV and visible source, the (Akola et al., 2008) faces of TiO_2 nanosheets were vertically developed in Cui's work on the graphite fibers and then MoS_2 sheets were further fabricated at TiO_2 nanosheets interface (Lu et al., 2017). Liu et al. effectively prepared a sandwich-like TiO_2/MoS_2/TiO_2 photocatalyst by integrating the in situ chemical reduction, mechanochemical process, and extreme calcination temperature in argon atmosphere (e.g., 350 °C) (Liu et al., 2016). The rate of degradation of MO was up to 89.86% by using the sandwich-like photocatalyst owing to development of heterojunctions.

Hybridization with Noble Metals, BiOI and Black Phosphorus

The degradation efficacy over MoS_2 nanosheets can be considerably enhanced via integrating noble metals by localized interface plasmon and rapid photoexcited electrons transferring (Islam et al., 2017). For example, 1T@2H-MoS_2 nanosheets decorated with Au NPs showed enhanced photocatalytic response in 91.2% MB degradation in 2 h (Lin et al., 2019). The 2H phase modification into the 1T was stimulated by loading of Au NPs and intraparticle charge transferring arise from 2H to 1T MoS_2 for increasing the photogenerated electron-hole pair separation (Lee et al., 2008). Other research findings were reported on BSA-coated Au nanoclusters and explored that these nanoclusters can be employed as an exfoliating agent for MoS_2 nanosheets along with precursor in epitaxially growing Au NPs on MoS_2 in the exfolia-

Figure 3. TEM micrographs of (a) monodispersed TiO_2 microspheres and (b) MoS_2/TiO_2 core-shell microspheres, (c) HRTEM of inset portion of (b), (d) HRTEM of core–shell microspheres. Elemental mapping micrographs of (e) Mo/S and (f) Ti/O.
Reproduced with permission from ref. (Sun et al., 2018) Copyright 2017, Royal Society of Chemistry.

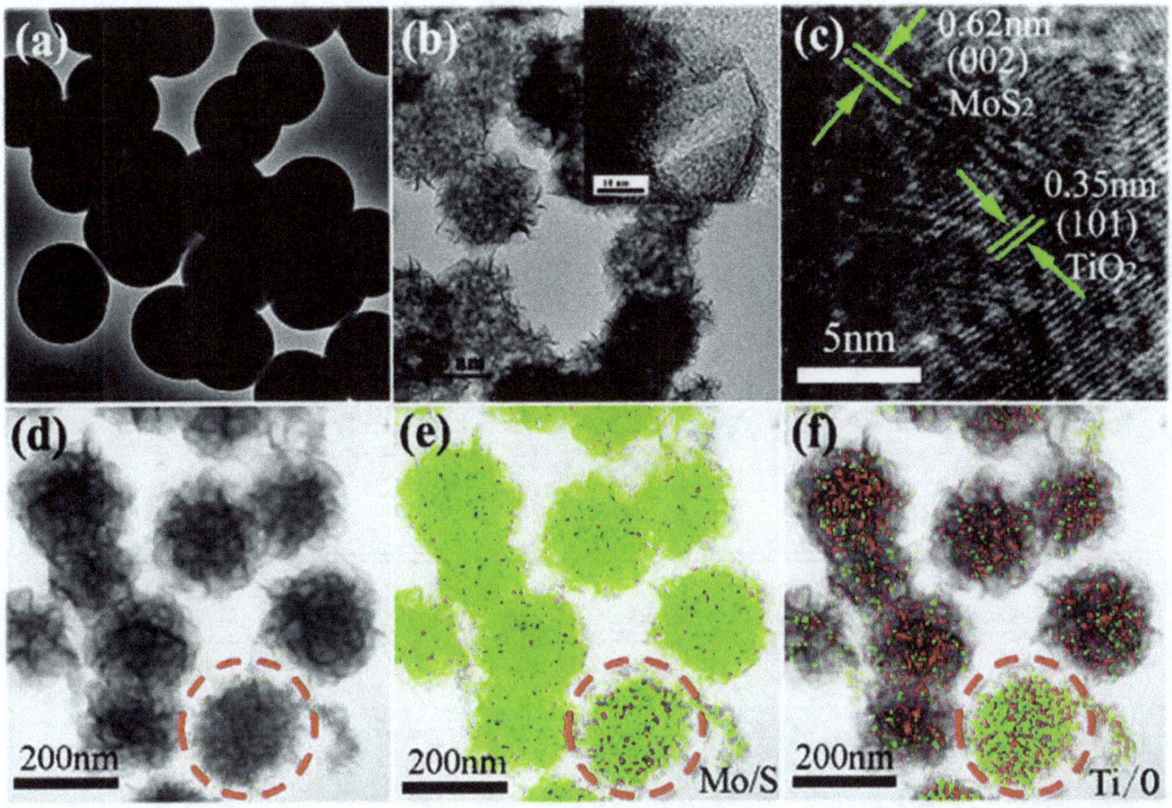

tion method (Figure 4e-h) (Guan et al., 2018). The Au-decorated MoS_2 nanosheets showed an improved degradation of organic dyes photocatalytically by rapid transfer of photogenerated electrons (Figure 4i, j). MoS_2 nanosheets may also control the Pd catalyst's electronic structure via interfacial electronic metal–support reaction (Shi et al., 2019). Furthermore, a Z-scheme ternary MoS_2/BiOI/AgI hierarchical system was attained by precipitation method in mixed water solution and ethylene glycol (Jahurul Islam et al., 2016). In BiOI samples, the presence of O-vacancies and the 2D alignment of MoS_2 materials extended the lifetime of charge carrier and enhanced the RhB degradation for 7 and 16 times relative to BiOI/AgI and BiOI, respectively. Liu's group affixed black phosphorus QDs (BPQDs) onto MoS_2 nanosheets for consuming maximum sunlight to generate 0D/2D nanohybrids using grinding and sonicating method (Feng et al., 2018). The photoactivity was observed to be increased up to 3×10^{-2} min^{-1}, due to increased light absorption and accelerated charge separation, which was appeared to be 13 and 27 folds higher compared to pristine BPQDs and MoS_2, respectively. WO_3 has been gaining importance to remediate environmental pollution due to exhibiting friendly environment, earth richness and photoresponse in visible range (Haque et al., 2017). WO_3 can be synthesized into numerous nanostructural materials to promote an increment in photocatalytic degradation. Normally, 2D WO_3 nanomaterials have been fabricated recently to deliver more active sites for improving photocatalytic degradation (Guan et al., 2017).

Figure 4. Hybridization of MoS_2 nanosheets. (a) Illustration for fabricating TiO_2–MoS_2(1T) and TiO_2–MoS_2(2H) hybrid structures. TEM micrographs of (b) TiO_2–MoS_2(1T) and (c) TiO_2–MoS_2(2H) structure, (d) photocatalytic decolorization curves of RhB with various products. (e) Schematic exfoliation of MoS_2 nanosheets and surface growth of Au25 clusters into Au_m NPs, (f) TEM and (g) HRTEM micrographs of Au_m/MoS_2 nanosheets, (h) TEM micrograph of Au_{m+n}/MoS_2 nanosheets attained after reaction with H_2O_2 for 4 h, (i) depiction for photocatalytic degradation on Au_{m+n}/MoS_2 nanosheets, (j) photocatalytic performances of several samples.

(a-d) Reproduced with permission from ref. (Bai et al., 2015) Copyright 2015, Tsinghua University Press.
(e-j) Reproduced with permission from ref. (Guan et al., 2018) Copyright 2018, Royal Society of Chemistry.

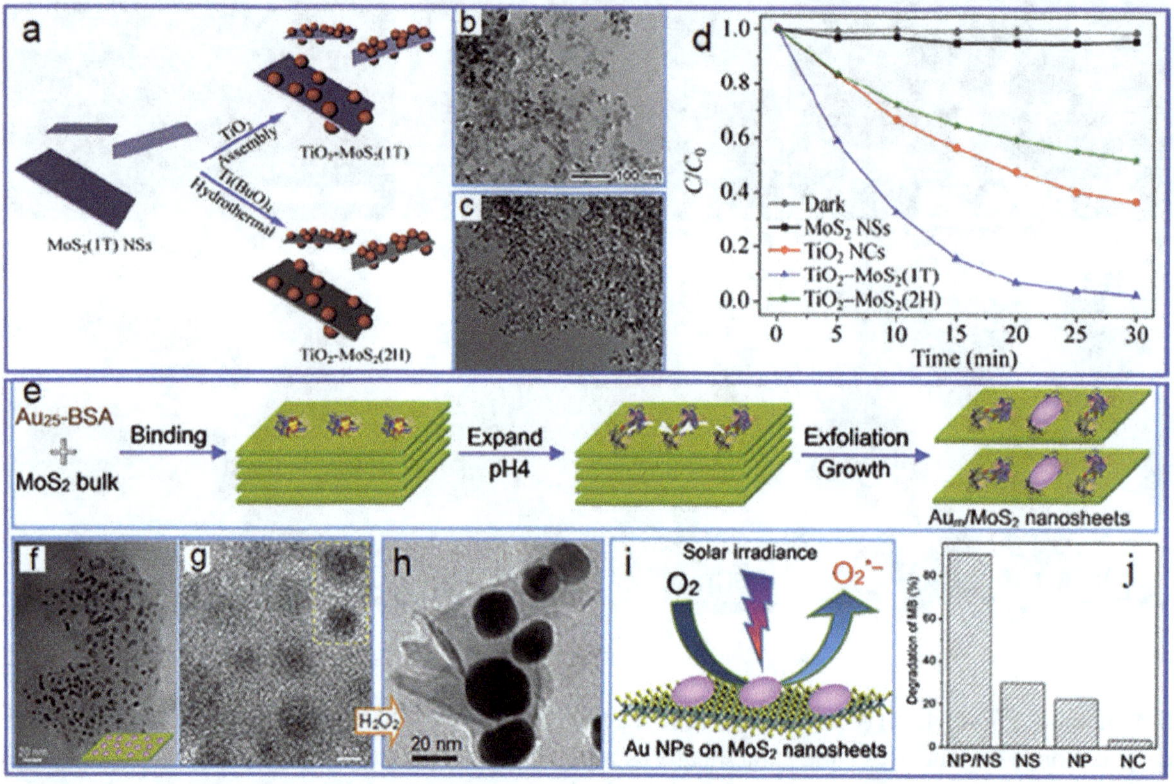

WO_3 nanoplates displayed the paramount degradation of RhB photocatalytically among various microstructures of WO_3 (i.e., nanoplate, nanorod and nanosphere) (Farhadian et al., 2015). Excellent catalytic activity was also observed for WO_3 in CR and MB dyes degradation relative to nanorods (Parthibavarman et al., 2018). Moreover, the hybridization process for different functional structures could considerably enhance the degradation potential of WO_3 materials using synergistic effect (Liu et al., 2014). In this unit, WO_3 nanosheets applications for dye degradation will be described photocatalytically to highlight the innovative hybrid structures comprising WO_3 nanosheets and their enhanced photocatalytic activity.

C_3N_4-BASED PHOTOCATALYST

Hybridization with Metal Oxides

For attaining enhanced photocatalytic degradation for elimination of pollutants, the hybridization of sheets of g-C_3N_4 with oxides of metals (e.g., ZnO, V_2O_5, WO_3, TiO_2, and Fe_3O_4) is also productive via effectual interface contact. Calculations based on DFT have recommended that the TiO_2–C_3N_4 heterojunction interface can deliver rapid separation of charge and transfer by both C–O and bridges of Ti–N. Hence, easing the quick separation/transfer of photoinduced electrons and holes (W.-K. Wang et al., 2019). A photocatalyst that is heterojunction comprising of layers of g-C_3N_4 and particles of TiO_2 was synthesized for the enhanced degradation of CBZ and diclofenac (Hu et al., 2019) Superoxide holes and radicals were certified as vital oxidative species by the scrounger experiments under exposure of visible light. Likewise, anatase TiO_{2-x} particles were fabricated on g-C_3N_4 sheets for developing heterostructures that are Z-scheme with an enhanced rate of separation of photoproduced electron–hole pairs (Tan et al., 2018), which detached electrons from TiO_2 and holes from g-C_3N_4 for considerably decreasing intraparticle recombination of electron–hole (Maldonado et al., 2019). Moreover, a ternary carbon plane/C_3N_4/TiO_2 photocatalyst was fabricated by calcining tetracycline/melamine mixture and then combination with TiO_2, which displayed extraordinary degradation activity in effectual elimination of numerous pollutants: norfloxacin (95.3%), MB (98.6%), and tetracycline (94.0%) (Liu et al., 2019). A heterogeneous $ZnTiO_3$ photocatalyst nanofibers and sheets of C_3N_4 was prepared in Lee's work by employing an electrospinning and sonication technique, and the photocatalyst that is hybrid showed a substantial improvement in removal of 4-nitrophenol, 4-chlorophenol, MB, and phenol photocatalytically under irradiation of visible light (Pawar et al., 2017). Ag NPs were decorated on the $Bi_5FeTi_3O_{15}$ interface layers and g-C_3N_4 sheets in Zhang's work to prepare Ag-bridged $Bi_5FeTi_3O_{15}$/C_3N_4 heterojunctions for enhanced tetracycline dilapidation (Wang et al., 2019). DFT calculations confirmed that g-C_3N_4 and $Bi_5FeTi_3O_{15}$ had matched band structures to make an extra fast Z-scheme transfer route for charge carriers. The presence of Ag NPs increased the absorption of light region and reduced the recombination of photoproduced electron–hole pairs. g-C_3N_4 layers were formed in the interlayers of N-$KTiNbO_5$ nanosheets by space-confined effect for enhancing the contact area between N-$KTiNbO_5$ and g-C_3N_4 nanosheets in N-doped $KTiNbO_5$/C_3N_4 heterostructures, producing an effectual charge transfer across ample interfaces and a considerably enhanced degradation activity of BPA and RhB (Liu et al., 2018). V_2O_5 nanorods were adapted onto the surface of g-C_3N_4 nanosheets in Guttena's work for developing 1D/2D hybrid photocatalysts exhibiting higher degradation activity relative to pristine V_2O_5 and g-C_3N_4 for elimination of CR (Dadigala et al., 2019). WO_3 materials were explored to design C_3N_4 hybrid as effective photocatalysts due to their outstanding properties and 2D structures. By synthesis of direct Z-scheme heterostructures, a heterogeneous WO_3–C_3N_4 composite was prepared using hydrothermal approach to improve degradation of SMX (Zhu et al., 2017). Heterostructure and oxygen vacancy was concurrently formed using one-step calcination method to prepare a WO_{3-x}/C_3N_4 photocatalyst for increasing degradation of tetracycline, RhB, and S. aureus under exposure to imitation light (Zhang et al., 2019). Conversely, the Z-scheme photocatalysts of WO_3/C_3N_4 were prepared by anchoring WO_3 nanoplates at the surface of g-C_3N_4 using in-situ acidic precipitation and following calcination process (Chai et al., 2018). This close interaction between C_3N_4 and WO_3 was not only enhanced interfacial reaction areas but also assisted the transferring and separation of photoproduced carriers, generating a considerably increased photocatalytic performance. The photocatalytic performance of hybrid photocatalysts was increased up to 4.70 times relative to one of pristine C_3N_4 in the decolorization of RhB (Deng et al., 2019). Ag NPs have been incorporated in WO_3/C_3N_4 for employing surface plasmon resonance effect to enhance absorption in visible range for effectual decolorization of tetracycline and RhB. ZnO nanomaterials were also dropped on g-C_3N_4 effectively to formulate Z-scheme heterojunctions just like WO_3 (Figure 5a), which revealed 0.0735 min^{-1} interaction rate constant for photodegradation of cephalexin, ~5.4 & ~8.1 times superior than pure g-C_3N_4 and ZnO, respectively (Figure 5b-d) (Li et al., 2018). On the other hand, Co_3O_4 NPs have been formed uniformly in the network of g-C_3N_4 sheets by one-pot melamine pyrolysis and cobalt nitrate, that considerably increased the photocatalytic activity of MB and tetracycline because of interactive role of Co_3O_4 NPs in g-C_3N_4 network (Suyana et al., 2017). Literature also reported the fabrication of Co_3O_4 QD–C_3N_4 heterostructures by attaching Co_3O_4 QDs onto the g-C_3N_4 surface by employing a facile chemically and subsequently annealing reaction in air (Figure 5e) (Gao et al., 2018). Exceptional enhancement in photocatalytic activity was attained by employing the 0D/2D heterostructures as catalysts in numerous dye solutions on account of the synergistic effect (Figure 5f). Furthermore, a $ZnFe_2O_4$–C_3N_4 photocatalyst through magnetic property was prepared after reflux treating the mixed $ZnFe_2O_4$ and C_3N_4 in methanol at 90 °C, which were employed to design the photo-Fenton system with H_2O_2 for decolorization of Orange II (Yao et al., 2014). Consequently, a higher 0.012 min^{-1} degradation rate was attained to reveal »2.4 times improvement than physical mixture of $ZnFe_2O_4$ and C_3N_4 NPs., the recyclability of hybrid catalysts had been examined to display outstanding efficiency with five successive runs by magnetic separation method. 0D Cu-doped FeOOH clusters have been incorporated onto the g-C_3N_4 surface by Wang and his colleagues to fabricate 0D/2D heterostructures, which displayed improvement in MB reduction relative to pristine C_3N_4 nanosheets (Zhang et al., 2018). However, 0D/2D photocatalyst showed exceptional degradation activity to specify their high stability and good robustness after being utilized for ten cycles at various pH (4.8–10.1).

ENGINEERING PROTOCOLS

Defect Engineering in 2D Photocatalysts

Anion Vacancies

Due to the low formation energy of oxygen vacancies, they have been considered as more prevalent and extensively studied defects in transition-MOs (Yan et al., 2017). Due to the atomic thickness of the O vacancies (V_O), the electronic structure and physiochemical properties of 2DMs are being effectively customized, affecting the photocatalytic efficiency (Hou et al., 2017). Along with altering the carrier concentration and electronic structure, engineered V_o can aid in the molecule activation, including O_2, N_2, and CO_2, as well as enhanced photocatalytic efficiency. Zhang and colleagues discovered that the V_O in BiOBr contains clustered electrons for -back-donation, which can cause changes in the adsorbed N_2 molecule, lengthening the bond between N-atoms from 1.078 to 1.133 for free molecular nitrogen (Li et al., 2015). Regarding that, the N_2 molecule can also be effectively reduced to NH_3 using electrons transmitted through the interface from its excited BiOBr. Comparable to N_2 activation, it has been shown that V_O in ultra-thin ZnAl-LDH is conducive to CO_2 activation (Zhao et al., 2015). An increase in the density of V_O was observed upon gradual decrement in thickness of prepared samples (210 to 2.7 nm) due to the emergence of various unsaturated coordinate Zn ions adjacent to the O_2 vacancies.

Figure 5. Typical hybridized systems of g-C_3N_4 and metal oxides. (a) HRTEM micrograph of g-C_3N_4@ZnO hybrids, (b) mechanism for photocatalytic reaction of antibiotic over g-C_3N_4@ZnO system. (c) photocatalytic degradation and (d) pseudo-first-order kinetic fitting of cephalexin vis different samples. (e) TEM image of Co_3O_4 QDs on g-C_3N_4 nanosheets, (f) concentration ratio of numerous dye solution before and after degradation.

(a-d) Reproduced with permission from ref. (Li et al., 2018) Copyright 2018, Elsevier B.V.
(e-f) Reproduced with permission from ref. (Gao et al., 2018) Copyright 2018, Wiley-VCH.

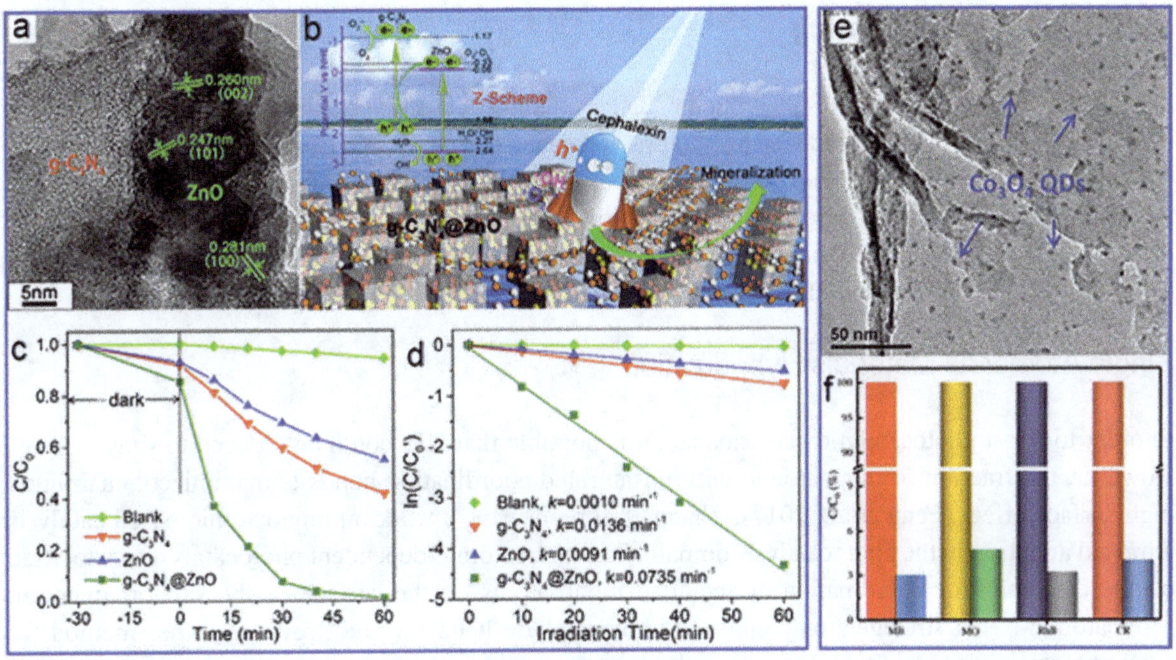

Kong and group adopted the plasma engraving method to produce defects of O-vacancy and Ti^{3+} on TiO_2 nanosheets' surface (Kong et al., 2018). The electronic configuration of the 2D TiO_2 nanosheets undergoes considerable variation as a result of the manufactured defects, with BG decreasing (3.13-2.88 eV), along with upshifting of CB and VB edges forming a defective state in the forbidden gap. Due to the formation of this defective state, the H_2 generation activity reached dramatically twofold as compared to pure TiO_2. Besides facilitating the forming of intermediate bands in the BG, the VO in WO_3 atomic layers promotes the adsorption and activation of CO_2 into radical COOH• species (Liang et al., 2020). The critical function of V_O in WO_3 layers enables the creation of more CO and O_2 in the infrared region. Fengcai Lei and colleagues achieved totally controlled formation of Vo rich and Vo deficient In_2O_3 NSs by rapidly heating $In(OH)_3$ NSs in the presence of oxygen or air (Lei et al., 2014). Vo was detected in the ESR and XPS spectra. The observed 531.4 eV peak indicated the formation of ultra-thin In_2O_3 NSs enriched with V_O and a peak maxima region, indicating that more V_O rich In_2O_3 NSs were produced than Vo poor In_2O_3 NSs or their analogous bulk forms. Additionally, a strong V_o signal at g = 2.004 was detected in the ESR, indicating that the Vo-rich In_2O_3 contains the largest amount of oxygen vacancies. Vo innovation significantly altered the electronic structure of In_2O_3 NSs with high Vo material. As shown by DRS and XPS examination, Vo-doped In_2O_3 sample exhibited a narrower energy gap and an upshift was observed in VB tip. DFT calculations clearly demonstrated that abundant density of states was produced at valence band maxima, and a new concentration of defects revealed that Vo rich In_2O_3 NSs were more abundant than Vo poor In_2O_3 NSs. As a result, the Vo-doped In_2O_3 produced a stronger electric field and a higher carrier level. Irradiation further excited the electrons into CB. Thus, Vo rich In_2O_3 NSs outperformed Vo weak In_2O_3 NSs and bulk In_2O_3 by 2.5 and 15 times, respectively, for H_2O oxidation. These findings substantiated the efficacy of anion vacancy in electronic configuration engineering.

Hybridization Engineering

2DMs have an extremely high SSA, which enhances the importance of the surface state relative to the bulk inside. Charge carriers produced by photons are dispersed at the surface to take part in the oxidation/reduction reactions. Thus, hybridization of surfaces to boost the effective ingestion of excitons is enviable in the absence of a 2D structure. In this section, consistent with surface hybridization, various 2D hybridization techniques with robust case studies are added, including QDS/2DMs, single atoms/2DMs, molecular/2DMs, and layered 2D/2D hybridization.

Single Atoms/2D Materials Hybridization

In order to boost photocatalytic performance, it is possible that NPs could be reduced to single atoms. However, the fraction of monoatomic with unsaturated coordination bonds is maximized, enabling a high surface effect (Feng et al., 2017). Zhang et al pioneering's work on monoatomic-based catalysis attracted attention in the photocatalysis domain. The monoatomic-dependent photocatalyst was focused on the dispersion or coordination of secluded mono-atoms on the surface of the support material. Monoatomic-based strategies may enhance photocatalytic behavior and provide another method for adjusting selectivity. Additionally, active single atoms, chemical bonding between single atoms, and 2DMs based supports have developed into a robust and straightforward charge transfer process. Thus, it is highly desirable to build a single atom/2DMs hybridization in order to achieve a superior photocatalytic response (Chen et al., 2016). Via calcination, protonation, and coupled exfoliation, single Rh atoms have been scattered on uniform ultra-thin 2D TiO_2 NSs (Ida et al., 2015). In the HAADF STEM image, distinct brightest spots indicated Rh atoms, while moderate brightness spots indicated Ti atoms. In addition to that, the EXAFS analysis revealed that Rh atoms in prepared samples exhibited a chemical environment similar to that of Rh_2O_3, which exhibited the bonding with O atoms and undergoing oxidation. Fabricated co-catalysts were used as the reaction center for photocatalytic H_2 generation, as predicted by DFT simulations. Thus, the rate of H_2 evolution was increased tenfold as compared with pure TiO_2 NSs. Wu and colleagues investigated the use of single Pt atoms as co-catalysts in order to enhance the hydrogen generation behavior of C_3N_4 NSs under irradiation (Xin et al., 2014). To form Pt single atoms/C_3N_4, a basic liquid phase reaction with H_2PtCl_6 and C_3N_4 was used in conjunction with low-temperature annealing. HAADF STEM technique was used to determine the dispersion and structure of Pt. Specific transparent spots have been observed to be uniformly scattered on graphitic-C_3N_4 sheets, with 99.4% of Pt having a size greater than 0.2 nm, indicating that Pt exists entirely as monoatomic. As the doping concentration of Pt exceeded 0.38%, dispersion of Pt atoms became denser and numerous nanoclusters were formed. The local atomic configuration of the Pt/C_3N_4 has been investigated using extended EXAFS spectroscopy. The coordination number was evaluated to be approximately 5 for the Pt atoms which confirmed the decoration of monoatomic on the top of g-C_3N_4 surface having BG value of

2.03 eV. The photocatalytic H_2 evolution behavior was significantly enhanced regarding the fabrication of a single Pt atom/C_3N_4 structure. Pt/C_3N_4 (0.16 wt % Pt loading) reached a production rate of nearly 318 mol h^{-1}, about 50 times relative to pristine C_3N_4. Concurrently, prepared structures demonstrated remarkable stability during the photocatalytic H_2 generation after several cycles. The advantageous quality of UTAS, surface trap states of C_3N_4 was largely altered by the secluded single Pt atom, which increases the exciton life period and increases the possibility of e^-s engaging in H^+ reduction. The observation that secluded metal atoms possess high surface energy was shown, as well as the possibility that these atoms cooperate closely with the surface of the supports. Through the interaction of affected metal atoms with available defects on the surface of support, the hybridization energy scheme can become a local minimum. Hence, these atoms could be secured as well as maintained in their stable state. Surface defects are more likely to form in ultrathin 2DMs due to the extremely high SSA and minute atomic flee radiation. Thus, a monoatomic-anchored surface DR ultrathin 2D structure can be constructed to enhance photocatalytic behavior (Di et al., 2018).

CONCLUSION

Regarding the field of 2D materials, intensive efforts have been devoted in recent years. The daunting quest for unique 2D materials remains ongoing and is primarily intended to discover novel 2D materials and their remarkable properties. In this chapter, we aimed to represent a thorough analysis of the latest innovations made in the field of photocatalytic degradation by 2D materials. In addition to current progress in photocatalysis, a throwback of basic knowledge is outlined. Various combination of fabrication methods employed for preparing novel 2D NMs is also illustrated. It is widely believed that 2D materials exhibit excellent photocatalytic performance. The potential for various 2D nanomaterials has been reported at length to remediate aqueous systems contaminated with dyes.

REFERENCES

Ahmed, B., Ojha, A. K., Singh, A., Hirsch, F., Fischer, I., Patrice, D., & Materny, A. (2018). Well-controlled in-situ growth of 2D WO3 rectangular sheets on reduced graphene oxide with strong photocatalytic and antibacterial properties. *Journal of Hazardous Materials*, *347*, 266–278. doi:10.1016/j.jhazmat.2017.12.069 PMID:29329009

Akola, J., Walter, M., Whetten, R. L., Hakkinen, H., & Gronbeck, H. (2008). On the structure of thiolate-protected Au25. *Journal of the American Chemical Society*, *130*(12), 3756–3757. doi:10.1021/ja800594p PMID:18321117

Ali, H., Guler, A. C., Masar, M., Urbanek, P., Urbanek, M., Skoda, D., Suly, P., Machovsky, M., Galusek, D., & Kuritka, I. (2021). Solid-State Synthesis of Direct Z-Scheme Cu2O/WO3 Nanocomposites with Enhanced Visible-Light Photocatalytic Performance. *Catalysts*, *11*(2), 293. doi:10.3390/catal11020293

Bai, S., Wang, L., Chen, X., Du, J., & Xiong, Y. (2015). Chemically exfoliated metallic MoS2 nanosheets: A promising supporting co-catalyst for enhancing the photocatalytic performance of TiO2 nanocrystals. *Nano Research*, *8*(1), 175–183. doi:10.100712274-014-0606-9

Bera, R., Kundu, S., & Patra, A. (2015). 2D Hybrid Nanostructure of Reduced Graphene Oxide–CdS Nanosheet for Enhanced Photocatalysis. *ACS Applied Materials & Interfaces*, *7*(24), 13251–13259. doi:10.1021/acsami.5b03800 PMID:26029992

Chai, B., Liu, C., Yan, J., Ren, Z., & Wang, Z. (2018). In-situ synthesis of WO3 nanoplates anchored on g-C3N4 Z-scheme photocatalysts for significantly enhanced photocatalytic activity. *Applied Surface Science*, *448*, 1–8. doi:10.1016/j.apsusc.2018.04.116

Chen, L., He, F., Zhao, N., & Guo, R. (2017). Fabrication of 3D quasi-hierarchical Z-scheme RGO-Fe2O3-MoS2 nanoheterostructures for highly enhanced visible-light-driven photocatalytic degradation. *Applied Surface Science*, *420*, 669–680. doi:10.1016/j.apsusc.2017.05.099

Chen, Z., Pronkin, S., Fellinger, T.-P., Kailasam, K., Vilé, G., Albani, D., Krumeich, F., Leary, R., Barnard, J., Thomas, J. M., Pérez-Ramírez, J., Antonietti, M., & Dontsova, D. (2016). Merging Single-Atom-Dispersed Silver and Carbon Nitride to a Joint Electronic System via Copolymerization with Silver Tricyanomethanide. *ACS Nano*, *10*(3), 3166–3175. doi:10.1021/acsnano.5b04210 PMID:26863408

Chhabra, T., Kumar, A., Bahuguna, A., & Krishnan, V. (2019). Reduced graphene oxide supported MnO2 nanorods as recyclable and efficient adsorptive photocatalysts for pollutants removal. *Vacuum*, *160*, 333–346. doi:10.1016/j.vacuum.2018.11.053

Córdoba, R., Ibarra, A., Mailly, D., & De Teresa, J. M. (2018). Vertical Growth of Superconducting Crystalline Hollow Nanowires by He+ Focused Ion Beam Induced Deposition. *Nano Letters*, *18*(2), 1379–1386. doi:10.1021/acs.nanolett.7b05103 PMID:29357248

Cravanzola, S., Cesano, F., Magnacca, G., Zecchina, A., & Scarano, D. (2016). Designing rGO/MoS2 hybrid nanostructures for photocatalytic applications. *RSC Advances*, *6*(64), 59001–59008. doi:10.1039/C6RA08633K

Dadigala, R., Bandi, R., Gangapuram, B. R., Dasari, A., Belay, H. H., & Guttena, V. (2019). Fabrication of novel 1D/2D V2O5/g-C3N4 composites as Z-scheme photocatalysts for CR degradation and Cr (VI) reduction under sunlight irradiation. *Journal of Environmental Chemical Engineering*, *7*(1), 102822. doi:10.1016/j.jece.2018.102822

Deng, D., Novoselov, K. S., Fu, Q., Zheng, N., Tian, Z., & Bao, X. (2016). Catalysis with two-dimensional materials and their heterostructures. *Nature Nanotechnology*, *11*(3), 218–230. doi:10.1038/nnano.2015.340 PMID:26936816

Deng, S., Yang, Z., Lv, G., Zhu, Y., Li, H., Wang, F., & Zhang, X. (2019). WO3 nanosheets/g-C3N4 nanosheets' nanocomposite as an effective photocatalyst for degradation of rhodamine B. *Applied Physics. A, Materials Science & Processing*, *125*(1), 44. doi:10.100700339-018-2331-9

Di, J., Xiong, J., Li, H., & Liu, Z. (2018). Ultrathin 2D Photocatalysts: Electronic-Structure Tailoring, Hybridization, and Applications. *Advanced Materials*, *30*(1), 1704548. doi:10.1002/adma.201704548 PMID:29178550

Ding, Y., Zhou, Y., Nie, W., & Chen, P. (2015). MoS2–GO nanocomposites synthesized via a hydrothermal hydrogel method for solar light photocatalytic degradation of methylene blue. *Applied Surface Science*, *357*, 1606–1612. doi:10.1016/j.apsusc.2015.10.030

Dong, Q., Wang, J., Duan, X., Tan, X., Liu, S., & Wang, S. (2019). Self-assembly of 3D MnO2/N-doped graphene hybrid aerogel for catalytic degradation of water pollutants: Structure-dependent activity. *Chemical Engineering Journal*, *369*, 1049–1058. doi:10.1016/j.cej.2019.03.139

Farhadian, M., Sangpour, P., & Hosseinzadeh, G. (2015). Morphology dependent photocatalytic activity of WO3 nanostructures. *Journal of Energy Chemistry*, *24*(2), 171–177. doi:10.1016/S2095-4956(15)60297-2

Feng, R., Lei, W., Sui, X., Liu, X., Qi, X., Tang, K., Liu, G., & Liu, M. (2018). Anchoring black phosphorus quantum dots on molybdenum disulfide nanosheets: A 0D/2D nanohybrid with enhanced visible–and NIR –light photoactivity. *Applied Catalysis B: Environmental*, *238*, 444–453. doi:10.1016/j.apcatb.2018.07.052

Feng, X., Sheng, N., Liu, Y., Chen, X., Chen, D., Yang, C., & Zhou, X. (2017). Simultaneously Enhanced Stability and Selectivity for Propene Epoxidation with H2 and O2 on Au Catalysts Supported on Nano-Crystalline Mesoporous TS-1. *ACS Catalysis*, *7*(4), 2668–2675. doi:10.1021/acscatal.6b03498

Gao, H., Yang, H., Xu, J., Zhang, S., & Li, J. (2018). Strongly Coupled g-C3N4 Nanosheets-Co3O4 Quantum Dots as 2D/0D Heterostructure Composite for Peroxymonosulfate Activation. *Small, 14*(31), 1801353.

Ge, L., Peng, Z., Wang, W., Tan, F., Wang, X., Su, B., Qiao, X., & Wong, P. K. (2018). g-C3N4/MgO nanosheets: Light-independent, metal-poisoning-free catalysts for the activation of hydrogen peroxide to degrade organics. *Journal of Materials Chemistry. A, Materials for Energy and Sustainability*, *6*(34), 16421–16429. doi:10.1039/C8TA05488F

Guan, G., & Han, M.-Y. (2019). Functionalized Hybridization of 2D Nanomaterials. *Advancement of Science*, *6*(23), 1901837. PMID:31832321

Guan, G., Liu, S., Cheng, Y., Zhang, Y.-W., & Han, M.-Y. (2018). BSA-caged metal clusters to exfoliate MoS2 nanosheets towards their hybridized functionalization. *Nanoscale*, *10*(23), 10911–10917. doi:10.1039/C8NR02121J PMID:29850713

Guan, G., Xia, J., Liu, S., Cheng, Y., Bai, S., Tee, S. Y., Zhang, Y.-W., & Han, M.-Y. (2017). Electrostatic-Driven Exfoliation and Hybridization of 2D Nanomaterials. *Advanced Materials*, *29*(32), 1700326. doi:10.1002/adma.201700326 PMID:28640388

Han, Y., Yang, Y., Zhao, J., Yin, X., & Que, W. (2018). Facile One-Pot Synthesis of Ternary Copper-Tin-Chalcogenide Quantum Dots on Reduced Graphene Oxide for Enhanced Photocatalytic Activity. *Catalysis Letters*, *148*(10), 3112–3118. doi:10.100710562-018-2525-y

Haque, F., Daeneke, T., Kalantar-zadeh, K., & Ou, J. Z. (2017). Two-Dimensional Transition Metal Oxide and Chalcogenide-Based Photocatalysts. *Nano-Micro Letters*, *10*(2), 23. doi:10.100740820-017-0176-y PMID:30393672

Hasija, V., Raizada, P., Sudhaik, A., Sharma, K., Kumar, A., Singh, P., Jonnalagadda, S. B., & Thakur, V. K. (2019). Recent advances in noble metal free doped graphitic carbon nitride based nanohybrids for photocatalysis of organic contaminants in water: A review. *Applied Materials Today*, *15*, 494–524. doi:10.1016/j.apmt.2019.04.003

Hou, J., Cao, S., Wu, Y., Liang, F., Sun, Y., Lin, Z., & Sun, L. (2017). Simultaneously efficient light absorption and charge transport of phosphate and oxygen-vacancy confined in bismuth tungstate atomic layers triggering robust solar CO2 reduction. *Nano Energy, 32*, 359–366. doi:10.1016/j.nanoen.2016.12.054

Hu, X., Deng, F., Huang, W., Zeng, G., Luo, X., & Dionysiou, D. D. (2018). The band structure control of visible-light-driven rGO/ZnS-MoS2 for excellent photocatalytic degradation performance and long-term stability. *Chemical Engineering Journal, 350*, 248–256. doi:10.1016/j.cej.2018.05.182

Hu, Z., Cai, X., Wang, Z., Li, S., Wang, Z., & Xie, X. (2019). Construction of carbon-doped supramolecule-based g-C3N4/TiO2 composites for removal of diclofenac and carbamazepine: A comparative study of operating parameters, mechanisms, degradation pathways. *Journal of Hazardous Materials, 380*, 120812. doi:10.1016/j.jhazmat.2019.120812 PMID:31326838

Huang, C., Li, C., & Shi, G. (2012). Graphene based catalysts. *Energy & Environmental Science, 5*(10), 8848–8868. doi:10.1039/c2ee22238h

Huo, J., Yuan, C., & Wang, Y. (2019). Nanocomposites of Three-Dimensionally Ordered Porous TiO2 Decorated with Pt and Reduced Graphene Oxide for the Visible-Light Photocatalytic Degradation of Waterborne Pollutants. *ACS Applied Nano Materials, 2*(5), 2713–2724. doi:10.1021/acsanm.9b00215

Ida, S., Kim, N., Ertekin, E., Takenaka, S., & Ishihara, T. (2015). Photocatalytic Reaction Centers in Two-Dimensional Titanium Oxide Crystals. *Journal of the American Chemical Society, 137*(1), 239–244. doi:10.1021/ja509970z PMID:25479408

Islam, M. A., Church, J., Han, C., Chung, H.-S., Ji, E., Kim, J. H., Choudhary, N., Lee, G.-H., Lee, W. H., & Jung, Y. (2017). Noble metal-coated MoS2 nanofilms with vertically-aligned 2D layers for visible light-driven photocatalytic degradation of emerging water contaminants. *Scientific Reports, 7*(1), 14944. doi:10.103841598-017-14816-9 PMID:29097721

Jahurul Islam, M., Amaranatha Reddy, D., Han, N. S., Choi, J., Song, J. K., & Kim, T. K. (2016). An oxygen-vacancy rich 3D novel hierarchical MoS2/BiOI/AgI ternary nanocomposite: Enhanced photocatalytic activity through photogenerated electron shuttling in a Z-scheme manner. *Physical Chemistry Chemical Physics, 18*(36), 24984–24993. doi:10.1039/C6CP02246D PMID:27722571

Jiang, L., Yuan, X., Pan, Y., Liang, J., Zeng, G., Wu, Z., & Wang, H. (2017). Doping of graphitic carbon nitride for photocatalysis: A reveiw. *Applied Catalysis B: Environmental, 217*, 388–406. doi:10.1016/j.apcatb.2017.06.003

Jo, W.-K., & Tonda, S. (2019). Novel CoAl-LDH/g-C3N4/RGO ternary heterojunction with notable 2D/2D/2D configuration for highly efficient visible-light-induced photocatalytic elimination of dye and antibiotic pollutants. *Journal of Hazardous Materials, 368*, 778–787. doi:10.1016/j.jhazmat.2019.01.114 PMID:30739031

Kong, X., Xu, Y., Cui, Z., Li, Z., Liang, Y., Gao, Z., Zhu, S., & Yang, X. (2018). Defect enhances photocatalytic activity of ultrathin TiO2 (B) nanosheets for hydrogen production by plasma engraving method. *Applied Catalysis B: Environmental, 230*, 11–17. doi:10.1016/j.apcatb.2018.02.019

Lee, J.-S., Shevchenko, E. V., & Talapin, D. V. (2008). Au–PbS Core–Shell Nanocrystals: Plasmonic Absorption Enhancement and Electrical Doping via Intra-particle Charge Transfer. *Journal of the American Chemical Society*, *130*(30), 9673–9675. doi:10.1021/ja802890f PMID:18597463

Lei, F., Sun, Y., Liu, K., Gao, S., Liang, L., Pan, B., & Xie, Y. (2014). Oxygen Vacancies Confined in Ultrathin Indium Oxide Porous Sheets for Promoted Visible-Light Water Splitting. *Journal of the American Chemical Society*, *136*(19), 6826–6829. doi:10.1021/ja501866r PMID:24773473

Li, H., Shang, J., Ai, Z., & Zhang, L. (2015). Efficient Visible Light Nitrogen Fixation with BiOBr Nanosheets of Oxygen Vacancies on the Exposed {001} Facets. *Journal of the American Chemical Society*, *137*(19), 6393–6399. doi:10.1021/jacs.5b03105 PMID:25874655

Li, N., Liu, Z., Liu, M., Xue, C., Chang, Q., Wang, H., Li, Y., Song, Z., & Hu, S. (2019). Facile Synthesis of Carbon Dots@2D MoS2 Heterostructure with Enhanced Photocatalytic Properties. *Inorganic Chemistry*, *58*(9), 5746–5752. doi:10.1021/acs.inorgchem.9b00111 PMID:30950600

Li, N., Tian, Y., Zhao, J., Zhang, J., Zuo, W., Kong, L., & Cui, H. (2018). Z-scheme 2D/3D g-C3N4@ ZnO with enhanced photocatalytic activity for cephalexin oxidation under solar light. *Chemical Engineering Journal*, *352*, 412–422. doi:10.1016/j.cej.2018.07.038

Li, X., Yu, J., Wageh, S., Al-Ghamdi, A. A., & Xie, J. (2016). Graphene in Photocatalysis: A Review. *Small, 12*(48), 6640-6696.

Liang, C., Niu, C.-G., Zhang, L., Wen, X.-J., Yang, S.-F., Guo, H., & Zeng, G.-M. (2019). Construction of 2D heterojunction system with enhanced photocatalytic performance: Plasmonic Bi and reduced graphene oxide co-modified Bi5O7I with high-speed charge transfer channels. *Journal of Hazardous Materials*, *361*, 245–258. doi:10.1016/j.jhazmat.2018.08.099 PMID:30199824

Liang, C., Zhang, L., Guo, H., Niu, C.-G., Wen, X.-J., Tang, N., Liu, H.-Y., Yang, Y.-Y., Shao, B.-B., & Zeng, G.-M. (2019). Photo-removal of 2,2¢4,4¢-tetrabromodiphenyl ether in liquid medium by reduced graphene oxide bridged artificial Z-scheme system of Ag@Ag3PO4/g-C3N4. *Chemical Engineering Journal*, *361*, 373–386. doi:10.1016/j.cej.2018.12.092

Liang, L., Li, X., Zhang, J., Ling, P., Sun, Y., Wang, C., Zhang, Q., Pan, Y., Xu, Q., Zhu, J., Luo, Y., & Xie, Y. (2020). Efficient infrared light induced CO2 reduction with nearly 100% CO selectivity enabled by metallic CoN porous atomic layers. *Nano Energy*, *69*, 104421. doi:10.1016/j.nanoen.2019.104421

Lin, Y., Tian, Z., Zhang, L., Ma, J., Jiang, Z., Deibert, B. J., Ge, R., & Chen, L. (2019). Chromium-ruthenium oxide solid solution electrocatalyst for highly efficient oxygen evolution reaction in acidic media. *Nature Communications*, *10*(1), 162. doi:10.103841467-018-08144-3 PMID:30635581

Liu, C., Dong, S., & Chen, Y. (2019). Enhancement of visible-light-driven photocatalytic activity of carbon plane/g-C3N4/TiO2 nanocomposite by improving heterojunction contact. *Chemical Engineering Journal*, *371*, 706–718. doi:10.1016/j.cej.2019.04.089

Liu, C., Zhu, H., Zhu, Y., Dong, P., Hou, H., Xu, Q., Chen, X., Xi, X., & Hou, W. (2018). Ordered layered N-doped KTiNbO5/g-C3N4 heterojunction with enhanced visible light photocatalytic activity. *Applied Catalysis B: Environmental*, *228*, 54–63. doi:10.1016/j.apcatb.2018.01.074

Liu, N., Huang, W., Tang, M., Yin, C., Gao, B., Li, Z., Tang, L., Lei, J., Cui, L., & Zhang, X. (2019). In-situ fabrication of needle-shaped MIL-53(Fe) with 1T-MoS2 and study on its enhanced photocatalytic mechanism of ibuprofen. *Chemical Engineering Journal*, *359*, 254–264. doi:10.1016/j.cej.2018.11.143

Liu, X., Xing, Z., Zhang, H., Wang, W., Zhang, Y., Li, Z., Wu, X., Yu, X., & Zhou, W. (2016). Fabrication of 3D Mesoporous Black TiO2/MoS2/TiO2 Nanosheets for Visible-Light-Driven Photocatalysis. *ChemSusChem*, *9*(10), 1118–1124. doi:10.1002/cssc.201600170 PMID:27111114

Liu, X., Yan, Y., Da, Z., Shi, W., Ma, C., Lv, P., Tang, Y., Yao, G., Wu, Y., Huo, P., & Yan, Y. (2014). Significantly enhanced photocatalytic performance of CdS coupled WO3 nanosheets and the mechanism study. *Chemical Engineering Journal*, *241*, 243–250. doi:10.1016/j.cej.2013.12.058

Liu, Y., Zeng, X., Hu, X., Hu, J., & Zhang, X. (2019). Two-dimensional nanomaterials for photocatalytic water disinfection: Recent progress and future challenges. *Journal of Chemical Technology and Biotechnology (Oxford, Oxfordshire)*, *94*(1), 22–37. doi:10.1002/jctb.5779

Lu, W., Li, J., Sheng, Y., Zhang, X., You, J., & Chen, L. (2017). One-pot synthesis of magnetic iron oxide nanoparticle-multiwalled carbon nanotube composites for enhanced removal of Cr(VI) from aqueous solution. *Journal of Colloid and Interface Science*, *505*, 1134–1146. doi:10.1016/j.jcis.2017.07.013 PMID:28709373

Ma, W., Wang, N., Du, Y., Tong, T., Zhang, L., Andrew Lin, K.-Y., & Han, X. (2019). One-step synthesis of novel Fe3C@nitrogen-doped carbon nanotubes/graphene nanosheets for catalytic degradation of Bisphenol A in the presence of peroxymonosulfate. *Chemical Engineering Journal*, *356*, 1022–1031. doi:10.1016/j.cej.2018.09.093

Maldonado, M. I., Saggioro, E., Peral, J., Rodríguez-Castellón, E., Jiménez-Jiménez, J., & Malato, S. (2019). Hydrogen generation by irradiation of commercial CuO + TiO2 mixtures at solar pilot plant scale and in presence of organic electron donors. *Applied Catalysis B: Environmental*, *257*, 117890. doi:10.1016/j.apcatb.2019.117890

Meng, N., Zhou, Y., Nie, W., & Chen, P. (2016). Synthesis of CdS-decorated RGO nanocomposites by reflux condensation method and its improved photocatalytic activity. *Journal of Nanoparticle Research*, *18*(8), 241. doi:10.100711051-016-3522-y

Moztahida, M., Nawaz, M., Kim, J., Shahzad, A., Kim, S., Jang, J., & Lee, D. S. (2019). Reduced graphene oxide-loaded-magnetite: A Fenton-like heterogeneous catalyst for photocatalytic degradation of 2-methylisoborneol. *Chemical Engineering Journal*, *370*, 855–865. doi:10.1016/j.cej.2019.03.214

Nimbalkar, D. B., Lo, H.-H., Ramacharyulu, P. V. R. K., & Ke, S.-C. (2016). Improved photocatalytic activity of RGO/MoS2 nanosheets decorated on TiO2 nanoparticles. *RSC Advances*, *6*(38), 31661–31667. doi:10.1039/C6RA01591C

Park, C. M., Heo, J., Wang, D., Su, C., & Yoon, Y. (2018). Heterogeneous activation of persulfate by reduced graphene oxide–elemental silver/magnetite nanohybrids for the oxidative degradation of pharmaceuticals and endocrine disrupting compounds in water. *Applied Catalysis B: Environmental*, *225*, 91–99. doi:10.1016/j.apcatb.2017.11.058 PMID:32704206

Parthibavarman, M., Karthik, M., & Prabhakaran, S. (2018). Facile and one step synthesis of WO3 nanorods and nanosheets as an efficient photocatalyst and humidity sensing material. *Vacuum*, *155*, 224–232. doi:10.1016/j.vacuum.2018.06.021

Patnaik, S., Martha, S., Acharya, S., & Parida, K. M. (2016). An overview of the modification of g-C3N4 with high carbon containing materials for photocatalytic applications. *Inorganic Chemistry Frontiers*, *3*(3), 336–347. doi:10.1039/C5QI00255A

Pawar, R. C., Kang, S., Park, J. H., Kim, J., Ahn, S., & Lee, C. S. (2017). Evaluation of a multi-dimensional hybrid photocatalyst for enrichment of H2 evolution and elimination of dye/non-dye pollutants. *Catalysis Science & Technology*, *7*(12), 2579–2590. doi:10.1039/C7CY00466D

Peng, W., Li, Y., Zhang, F., Zhang, G., & Fan, X. (2017). Roles of Two-Dimensional Transition Metal Dichalcogenides as Cocatalysts in Photocatalytic Hydrogen Evolution and Environmental Remediation. *Industrial & Engineering Chemistry Research*, *56*(16), 4611–4626. doi:10.1021/acs.iecr.7b00371

Peng, W., Wang, X., & Li, X.-y. (2014). The synergetic effect of MoS2 and graphene on Ag3PO4 for its ultra-enhanced photocatalytic activity in phenol degradation under visible light. *Nanoscale*, *6*(14), 8311–8317. doi:10.1039/c4nr01654h PMID:24933179

Perreault, F., Fonseca de Faria, A., & Elimelech, M. (2015). Environmental applications of graphene-based nanomaterials. *Chemical Society Reviews*, *44*(16), 5861–5896. doi:10.1039/C5CS00021A PMID:25812036

Pham, T.-T., Nguyen-Huy, C., & Shin, E. W. (2016). Facile one-pot synthesis of nickel-incorporated titanium dioxide/graphene oxide composites: Enhancement of photodegradation under visible-irradiation. *Applied Surface Science*, *377*, 301–310. doi:10.1016/j.apsusc.2016.03.144

Pi, Y., Ma, L., Zhao, P., Cao, Y., Gao, H., Wang, C., Li, Q., Dong, S., & Sun, J. (2018). Facile green synthetic graphene-based Co-Fe Prussian blue analogues as an activator of peroxymonosulfate for the degradation of levofloxacin hydrochloride. *Journal of Colloid and Interface Science*, *526*, 18–27. doi:10.1016/j.jcis.2018.04.070 PMID:29709668

Qu, L., Zhu, G., Ji, J., Yadav, T. P., Chen, Y., Yang, G., Xu, H., & Li, H. (2018). Recyclable Visible Light-Driven O-g-C3N4/Graphene Oxide/N-Carbon Nanotube Membrane for Efficient Removal of Organic Pollutants. *ACS Applied Materials & Interfaces*, *10*(49), 42427–42435. doi:10.1021/acsami.8b15905 PMID:30444339

Qumar, U., Hassan, J. Z., Bhatti, R. A., Raza, A., Nazir, G., Nabgan, W., & Ikram, M. (2022). Photocatalysis vs adsorption by metal oxide nanoparticles. *Journal of Materials Science and Technology*, *131*, 122–166. doi:10.1016/j.jmst.2022.05.020

Rafiq, A., Ikram, M., Ali, S., Niaz, F., Khan, M., Khan, Q., & Maqbool, M. (2021). Photocatalytic degradation of dyes using semiconductor photocatalysts to clean industrial water pollution. *Journal of Industrial and Engineering Chemistry*, *97*, 111–128. doi:10.1016/j.jiec.2021.02.017

Rahmanian, E., Malekfar, R., & Pumera, M. (2018). Nanohybrids of Two-Dimensional Transition-Metal Dichalcogenides and Titanium Dioxide for Photocatalytic Applications. *Chemistry (Weinheim an der Bergstrasse, Germany)*, *24*(1), 18–31. doi:10.1002/chem.201703434 PMID:28872715

Raza, A., Altaf, S., Ali, S., Ikram, M., & Li, G. (2022). Recent advances in carbonaceous sustainable nanomaterials for wastewater treatments. *Sustainable Materials and Technologies*, *32*, e00406. doi:10.1016/j.susmat.2022.e00406

Raza, A., Qumar, U., Rafi, A. A., & Ikram, M. (2022). MXene-based nanocomposites for solar energy harvesting. *Sustainable Materials and Technologies*, *33*, e00462. doi:10.1016/j.susmat.2022.e00462

Raza, A., Rafiq, A., Qumar, U., & Hassan, J. Z. (2022). 2D hybrid photocatalysts for solar energy harvesting. *Sustainable Materials and Technologies*, *33*, e00469. doi:10.1016/j.susmat.2022.e00469

Raza, A., Zhang, X., Ali, S., Cao, C., Rafi, A. A., & Li, G. (2022). Photoelectrochemical Energy Conversion over 2D Materials. *Photochem, 2*(2).

Sabarinathan, M., Harish, S., Archana, J., Navaneethan, M., Ikeda, H., & Hayakawa, Y. (2017). Highly efficient visible-light photocatalytic activity of MoS2–TiO2 mixtures hybrid photocatalyst and functional properties. *RSC Advances*, *7*(40), 24754–24763. doi:10.1039/C7RA03633G

Saleh, R., & Taufik, A. (2019). Ultraviolet-light-assisted heterogeneous Fenton reaction of Ag-Fe3O4/graphene composites for the degradation of organic dyes. *Journal of Environmental Chemical Engineering*, *7*(1), 102895. doi:10.1016/j.jece.2019.102895

Shi, Y., Huang, X.-K., Wang, Y., Zhou, Y., Yang, D.-R., Wang, F.-B., Gao, W., & Xia, X.-H. (2019). Electronic Metal–Support Interaction To Modulate MoS2-Supported Pd Nanoparticles for the Degradation of Organic Dyes. *ACS Applied Nano Materials*, *2*(6), 3385–3393. doi:10.1021/acsanm.9b00297

Singha, K., Mondal, A., Ghosh, S. C., & Panda, A. B. (2018). Visible-light-driven Efficient Photocatalytic Reduction of Organic Azides to Amines over CdS Sheet–rGO Nanocomposite. *Chemistry, an Asian Journal*, *13*(3), 255–260. doi:10.1002/asia.201701614 PMID:29265682

Sreeja, V. G., Vinitha, G., Reshmi, R., Jayaraj, M. K., & Anila, E. I. (2019). Structural, Spectral, Electrical and Nonlinear Optical Characterizations of rGO-PANI Composites. *Materials Today: Proceedings*, *10*, 456–465. doi:10.1016/j.matpr.2019.03.010

Su, T., Shao, Q., Qin, Z., Guo, Z., & Wu, Z. (2018). Role of Interfaces in Two-Dimensional Photocatalyst for Water Splitting. *ACS Catalysis*, *8*(3), 2253–2276. doi:10.1021/acscatal.7b03437

Sun, Y., Gao, S., Lei, F., & Xie, Y. (2015). Atomically-thin two-dimensional sheets for understanding active sites in catalysis. *Chemical Society Reviews*, *44*(3), 623–636. doi:10.1039/C4CS00236A PMID:25382246

Sun, Y., Lin, H., Wang, C., Wu, Q., Wang, X., & Yang, M. (2018). Morphology-controlled synthesis of TiO2/MoS2 nanocomposites with enhanced visible-light photocatalytic activity. *Inorganic Chemistry Frontiers*, *5*(1), 145–152. doi:10.1039/C7QI00491E

Suyana, P., Ganguly, P., Nair, B. N., Mohamed, A. P., Warrier, K. G. K., & Hareesh, U. S. (2017). Co3O4–C3N4 p–n nano-heterojunctions for the simultaneous degradation of a mixture of pollutants under solar irradiation. *Environmental Science. Nano*, *4*(1), 212–221. doi:10.1039/C6EN00410E

Tan, B., Ye, X., Li, Y., Ma, X., Wang, Y., & Ye, J. (2018). Defective Anatase TiO2−x Mesocrystal Growth In Situ on g-C3N4 Nanosheets: Construction of 3D/2D Z-Scheme Heterostructures for Highly Efficient Visible-Light Photocatalysis. *Chemistry (Weinheim an der Bergstrasse, Germany)*, *24*(50), 13311–13321. doi:10.1002/chem.201802366 PMID:29957872

Tan, C., Lai, Z., & Zhang, H. (2017). Ultrathin Two-Dimensional Multinary Layered Metal Chalcogenide Nanomaterials. *Advanced Materials*, *29*(37), 1701392. doi:10.1002/adma.201701392 PMID:28752578

Tian, H., Liu, M., & Zheng, W. (2018). Constructing 2D graphitic carbon nitride nanosheets/layered MoS2/graphene ternary nanojunction with enhanced photocatalytic activity. *Applied Catalysis B: Environmental*, *225*, 468–476. doi:10.1016/j.apcatb.2017.12.019

Tian, Q., Wu, W., Liu, J., Wu, Z., Yao, W., Ding, J., & Jiang, C. (2017). Dimensional heterostructures of 1D CdS/2D ZnIn2S4 composited with 2D graphene: Designed synthesis and superior photocatalytic performance. *Dalton Transactions (Cambridge, England)*, *46*(9), 2770–2777. doi:10.1039/C7DT00018A PMID:28168251

Tong, Z., Yang, D., Shi, J., Nan, Y., Sun, Y., & Jiang, Z. (2015). Three-Dimensional Porous Aerogel Constructed by g-C3N4 and Graphene Oxide Nanosheets with Excellent Visible-Light Photocatalytic Performance. *ACS Applied Materials & Interfaces*, *7*(46), 25693–25701. doi:10.1021/acsami.5b09503 PMID:26545166

V, P. P., R, P., Sathe, V., & Mahalingam, U. (2019). Graphene boosted silver nanoparticles as surface enhanced Raman spectroscopic sensors and photocatalysts for removal of standard and industrial dye contaminants. *Sensors and Actuators B: Chemical, 281*, 679-688.

Wang, K., Li, J., & Zhang, G. (2019). Ag-Bridged Z-Scheme 2D/2D Bi5FeTi3O15/g-C3N4 Heterojunction for Enhanced Photocatalysis: Mediator-Induced Interfacial Charge Transfer and Mechanism Insights. *ACS Applied Materials & Interfaces*, *11*(31), 27686–27696. doi:10.1021/acsami.9b05074 PMID:31282639

Wang, W.-K., Zhu, W., Mao, L., Zhang, J., Zhou, Z., & Zhao, G. (2019). Two-dimensional TiO2-g-C3N4 with both TiN and CO bridges with excellent conductivity for synergistic photoelectrocatalytic degradation of bisphenol A. *Journal of Colloid and Interface Science*, *557*, 227–235. doi:10.1016/j.jcis.2019.08.088 PMID:31521972

Wang, X., Wang, H., Yu, K., & Hu, X. (2018). Immobilization of 2D/2D structured g-C3N4 nanosheet/reduced graphene oxide hybrids on 3D nickel foam and its photocatalytic performance. *Materials Research Bulletin*, *97*, 306–313. doi:10.1016/j.materresbull.2017.09.024

Wang, Z., & Mi, B. (2017). Environmental Applications of 2D Molybdenum Disulfide (MoS2) Nanosheets. *Environmental Science & Technology*, *51*(15), 8229–8244. doi:10.1021/acs.est.7b01466 PMID:28661657

Wu, M., Li, L., Liu, N., Wang, D., Xue, Y., & Tang, L. (2018). Molybdenum disulfide (MoS2) as a co-catalyst for photocatalytic degradation of organic contaminants: A review. *Process Safety and Environmental Protection*, *118*, 40–58. doi:10.1016/j.psep.2018.06.025

Xin, S., Yin, Y.-X., Guo, Y.-G., & Wan, L.-J. (2014). A High-Energy Room-Temperature Sodium-Sulfur Battery. *Advanced Materials*, *26*(8), 1261–1265. doi:10.1002/adma.201304126 PMID:24338949

Xu, L., Yang, L., Johansson, E. M. J., Wang, Y., & Jin, P. (2018). Photocatalytic activity and mechanism of bisphenol a removal over TiO2−x/rGO nanocomposite driven by visible light. *Chemical Engineering Journal*, *350*, 1043–1055. doi:10.1016/j.cej.2018.06.046

Xu, X., Qin, J., Wei, Y., Ye, S., Shen, J., Yao, Y., Ding, B., Shu, Y., He, G., & Chen, H. (2019). Heterogeneous activation of persulfate by NiFe2−xCoxO4-RGO for oxidative degradation of bisphenol A in water. *Chemical Engineering Journal*, *365*, 259–269. doi:10.1016/j.cej.2019.02.019

Yan, D., Li, Y., Huo, J., Chen, R., Dai, L., & Wang, S. (2017). Defect Chemistry of Nonprecious-Metal Electrocatalysts for Oxygen Reactions. *Advanced Materials*, *29*(48), 1606459. doi:10.1002/adma.201606459 PMID:28508469

Yang, L., Xu, L., Bai, X., & Jin, P. (2019). Enhanced visible-light activation of persulfate by Ti3+ self-doped TiO2/graphene nanocomposite for the rapid and efficient degradation of micropollutants in water. *Journal of Hazardous Materials*, *365*, 107–117. doi:10.1016/j.jhazmat.2018.10.090 PMID:30412807

Yang, N., Liu, Y., Wen, H., Tang, Z., Zhao, H., Li, Y., & Wang, D. (2013). Photocatalytic Properties of Graphdiyne and Graphene Modified TiO2: From Theory to Experiment. *ACS Nano*, *7*(2), 1504–1512. doi:10.1021/nn305288z PMID:23350627

Yang, X., Cai, H., Bao, M., Yu, J., Lu, J., & Li, Y. (2018). Insight into the highly efficient degradation of PAHs in water over graphene oxide/Ag3PO4 composites under visible light irradiation. *Chemical Engineering Journal*, *334*, 355–376. doi:10.1016/j.cej.2017.09.104

Yao, Y., Lu, F., Qin, J., Wei, F., Xu, C., & Wang, S. (2014). Magnetic ZnFe2O4–C3N4 Hybrid for Photocatalytic Degradation of Aqueous Organic Pollutants by Visible Light. *Industrial & Engineering Chemistry Research*, *53*(44), 17294–17302. doi:10.1021/ie503437z

Yasir, M., Masar, M., Sopik, T., Ali, H., Urbanek, M., Antos, J., Machovsky, M., & Kuritka, I. (2022). ZnO nanowires and nanorods based ZnO/WO3/Pt heterojunction for efficient photocatalytic degradation of estriol (E3) hormone. *Materials Letters*, *319*, 132291. doi:10.1016/j.matlet.2022.132291

Zhang, F., Huang, L., Ding, P., Wang, C., Wang, Q., Wang, H., Li, Y., Xu, H., & Li, H. (2019). One-step oxygen vacancy engineering of WO3-x/2D g-C3N4 heterostructure: Triple effects for sustaining photoactivity. *Journal of Alloys and Compounds*, *795*, 426–435. doi:10.1016/j.jallcom.2019.04.297

Zhang, S., Gao, H., Huang, Y., Wang, X., Hayat, T., Li, J., Xu, X., & Wang, X. (2018). Ultrathin g-C3N4 nanosheets coupled with amorphous Cu-doped FeOOH nanoclusters as 2D/0D heterogeneous catalysts for water remediation. *Environmental Science. Nano*, *5*(5), 1179–1190. doi:10.1039/C8EN00124C

Zhang, S., Li, J., Wang, X., Huang, Y., Zeng, M., & Xu, J. (2015). Rationally designed 1D Ag@AgVO3 nanowire/graphene/protonated g-C3N4 nanosheet heterojunctions for enhanced photocatalysis via electrostatic self-assembly and photochemical reduction methods. *Journal of Materials Chemistry. A, Materials for Energy and Sustainability*, *3*(18), 10119–10126. doi:10.1039/C5TA00635J

Zhang, W., Xiao, X., Li, Y., Zeng, X., Zheng, L., & Wan, C. (2016). Liquid exfoliation of layered metal sulphide for enhanced photocatalytic activity of TiO2 nanoclusters and DFT study. *RSC Advances*, *6*(40), 33705–33712. doi:10.1039/C6RA03534E

Zhang, Y., Zhang, N., Tang, Z.-R., & Xu, Y.-J. (2012). Graphene Transforms Wide Band Gap ZnS to a Visible Light Photocatalyst. The New Role of Graphene as a Macromolecular Photosensitizer. *ACS Nano*, *6*(11), 9777–9789. doi:10.1021/nn304154s PMID:23106763

Zhao, Y., Chen, G., Bian, T., Zhou, C., Waterhouse, G. I. N., Wu, L.-Z., Tung, C.-H., Smith, L. J., O'Hare, D., & Zhang, T. (2015). Defect-Rich Ultrathin ZnAl-Layered Double Hydroxide Nanosheets for Efficient Photoreduction of CO2 to CO with Water. *Advanced Materials*, *27*(47), 7824–7831. doi:10.1002/adma.201503730 PMID:26509528

Zhao, Y., Zhang, X., Wang, C., Zhao, Y., Zhou, H., Li, J., & Jin, H. (2017). The synthesis of hierarchical nanostructured MoS2/Graphene composites with enhanced visible-light photo-degradation property. *Applied Surface Science*, *412*, 207–213. doi:10.1016/j.apsusc.2017.03.181

Zhou, W., Yin, Z., Du, Y., Huang, X., Zeng, Z., Fan, Z., Liu, H., Wang, J., & Zhang, H. (2013). Synthesis of Few-Layer MoS2 Nanosheet-Coated TiO2 Nanobelt Heterostructures for Enhanced Photocatalytic Activities. *Small*, *9*(1), 140–147. doi:10.1002mll.201201161 PMID:23034984

Zhou, X., Wang, J., Sheng, N., Cui, R., Deng, Y., & Dai, J. (2019). Corrigendum to "Subchronic reproductive effects of 6:2 chlorinated polyfluorinated ether sulfonate (6:2 Cl-PFAES), an alternative to PFOS, on adult male mice". *Journal of Hazardous Materials*, *365*, 972. doi:10.1016/j.jhazmat.2018.10.085 PMID:30448035

Zhu, W., Sun, F., Goei, R., & Zhou, Y. (2017). Construction of WO3–g-C3N4 composites as efficient photocatalysts for pharmaceutical degradation under visible light. *Catalysis Science & Technology*, *7*(12), 2591–2600. doi:10.1039/C7CY00529F PMID:29129990

Chapter 6
Metal Oxide–Cellulosic Nanocomposite for the Removal of Dyes From Wastewater

Suneeta Bhandari
https://orcid.org/0000-0001-7351-4561
G.B. Pant University of Agriculture and Technology, India

ABSTRACT

Water is a vital component of life, and its availability is critical for all living things. Due to rising water demand, traditional water/wastewater treatment methods are inefficient in supplying adequate safe water. The leaching of harmful compounds into the process water is a problem with most commercial and chemically manufactured materials for water treatment. As a result of research into developing better materials that could achieve high efficiency without posing a health concern, non-toxic composite materials made of cellulose and metal oxides were investigated. Due to its great physical, chemical, and mechanical qualities, cellulose is one of the materials gaining popularity. Nanocomposites have been approved as a solution for water purification that avoids the issues associated with using simply metal oxides. The purpose of this study is to review the potential applications of cellulose integrated with metal oxides for wastewater treatment and harmful metal removal from dyes via industrial waste.

INTRODUCTION

Rapid industrialization, urbanization, and population growth have put a major strain on water supplies, which has resulted in a significant increase in the demand for clean water (Velusamy et al., 2021). Because the use of chemicals is unavoidable owing to economic expansion, it is critical to ensure their safe use and disposal (Pai et al., 2021). Our ocean, rivers, reservoirs, lakes, and ocean are all drowning. As a result of research into developing better materials that could achieve high efficiency without posing a health concern, non-toxic composite materials made of cellulose and metal oxides were investigated (Oyewo et al., 2020). The dumping of industrial effluents into our waterways is a major cause of water pollution, which is a serious environmental issue (Oyewo et al., 2020). Contaminants such as viruses,

DOI: 10.4018/978-1-6684-4553-2.ch006

Figure 1. Shows the water pollution creates by industrial effluents.

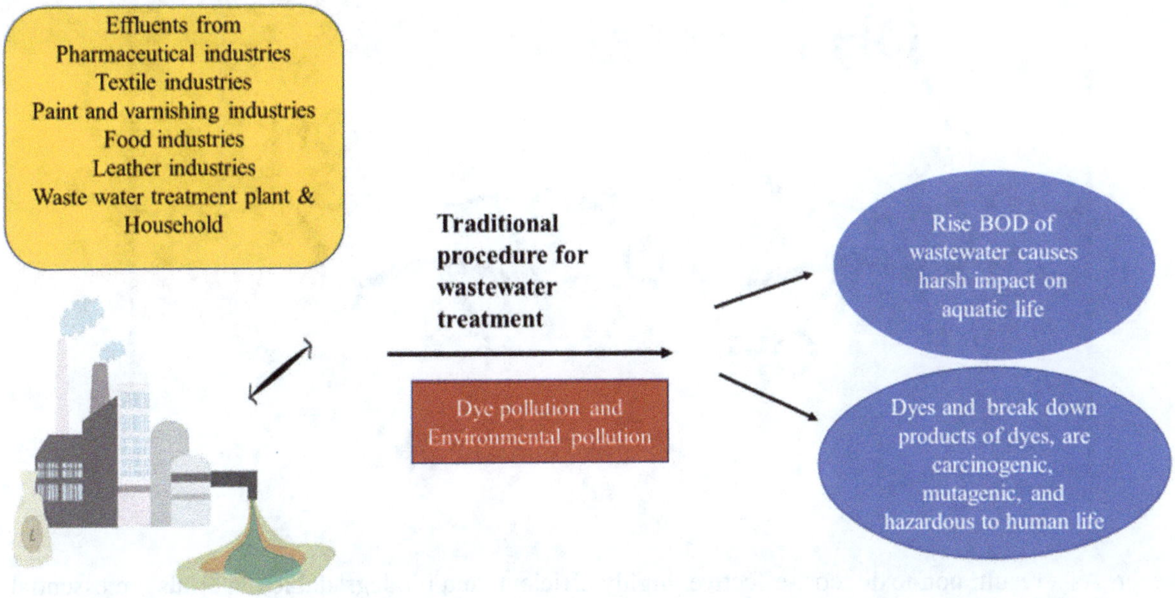

heavy metals, dyes, and pigments may be present in these discharges. According to the World Health Organization's (WHO) most recent study, over 844 million people worldwide do not have access to safe drinking water (Wutich et al., 2019). Many diseases are transferred by waterborne pathogens in the form of disease-causing bacteria, viruses, or protozoa, Cholera, typhoid, hepatitis, and COVID-19 (Sharma et al., 2020) are some of the most common diseases (Sarkar et al., 2020). Industrial effluents can be absorbed or recycled but most of the wastewater percolates through the soil or directly into the water system, changing the concentration of the aquifers beneath the earth's surface.

As a result, the main targets are effluent from the printing, textile, paints, cosmetics, food, and mining industries. Despite the damage to the environment, the continued demand for these toxic metals and dyes cannot be overlooked. Some of the metals are utilized in medicine, neutron radiography, mineral identification, and steel manufacture as an additive. Arsenic, lead, chromium, cadmium, iron, and vanadium are heavy metals that have recently been linked to water pollution. In India's biggest cities, an estimated 38354 million liters per day (MLD) of sewage is created, but the sewage treatment capacity is only 11786 MLD (Velusamy et al., 2021). Similarly, just 60% of industrial wastewater is treated, largely from large-scale companies in india. Even at trace levels, the presence of heavy metals in residential water can cause unthinkable health problems such as brain damage, cancer, and system diseases. Furthermore, 20 percent of individuals die each year as a result of using contaminated drinking water (Hunge et al., 2015). Textile industries, power plants, and chemical industries discharge wastewater that contains hazardous and non-biodegradable recalcitrant organic compounds and heavy metals (Abdel Aal et al., 2009; Kansal et al., 2007). Textile wastewater contamination poses a significant hazard to water resources and the economy. Hence, textile industries stand out among diverse industries because they consume a significant amount of water, energy, and chemicals (Akarslan et al., 2018)

Chemically synthesized water treatment solutions have also been used, although they have been linked to a number of drawbacks, including the leaching of harmful compounds back into the process

Figure 2. Shows cellulose polymer chemical structure

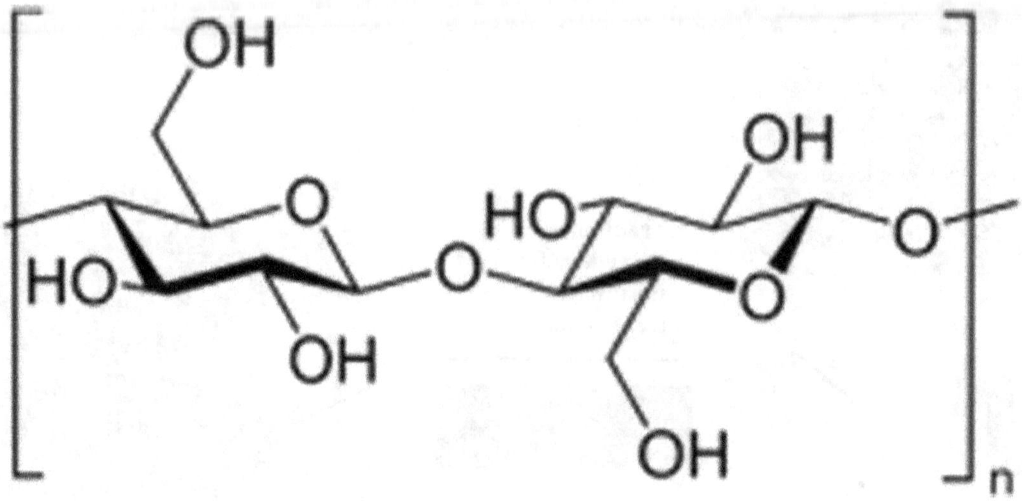

water. As a result, non-toxic, cost-effective, highly efficient, and biodegradable materials are essential for complete removal of these contaminants from water; cellulose is a preferable substitute. Figure 2 Shows cellulose polymer chemical structure.

Cellulose is mostly derived from plants (e.g., wood and cotton), although it can also be made by bacteria, tunicates, and fungi (Eichhorn et al., 2010). Among the agricultural byproducts investigated as sources of cellulose, such as sugar beet, sisal fibres, and barley straw, the rice husk and straw have gotten a lot of attention. Cellulose is a glycosidic bond-linked linear homopolymer of -D, 1,4 glucose units. It is made up of nano-sized microfibrils that are surrounded by lignin and hemicellulose (Habibi et al., 2010). As shown in Figure 3, around 36 glucose chains are attached to each other by van der Waals forces and intra- and intermolecular hydrogen bonds to form protofibrils, which stack together to form long fibres called microfibrils, which then join together to form cellulosic fibres (Habibi, 2014; Rojas et al., 2015). The diameter of single cellulose microfibrils varies from 2 to 20 nanometers (Azizi et al., 2005).

Cellulose is biodegradable, non-toxic, low-cost, and plentiful in a variety of natural resources and agricultural wastes. cellulose could be used in a variety of water treatment methods to remove a variety of impurities, including harmful metals and dyes. As a result, the use of cellulose-metal oxide composite photocatalysts in water treatment could be more appealing due to their high surface area, light stability, and low toxicity (Noohpisheh et al., 2020; Jing et al., 2018). As a result, the importance of cellulose and metal oxide chemistry, as well as their interactions, is rigorously examined in this study (Oyewo et al., 2020).

DYES

Dyes are one of many compounds found in industrial wastewater that are considered significant contaminants. Because of their excellent thermal and photostability, dyes can last a long time in the environment. Textile industries utilize 80% of all dye produced, making them the largest dye consumers. Dyes are difficult to remove from wastewater using traditional procedures due to their high-water solubility.

Figure 3. Shows the cellulose fibers and microfibers' structural arrangement (Rojas et al., 2015).

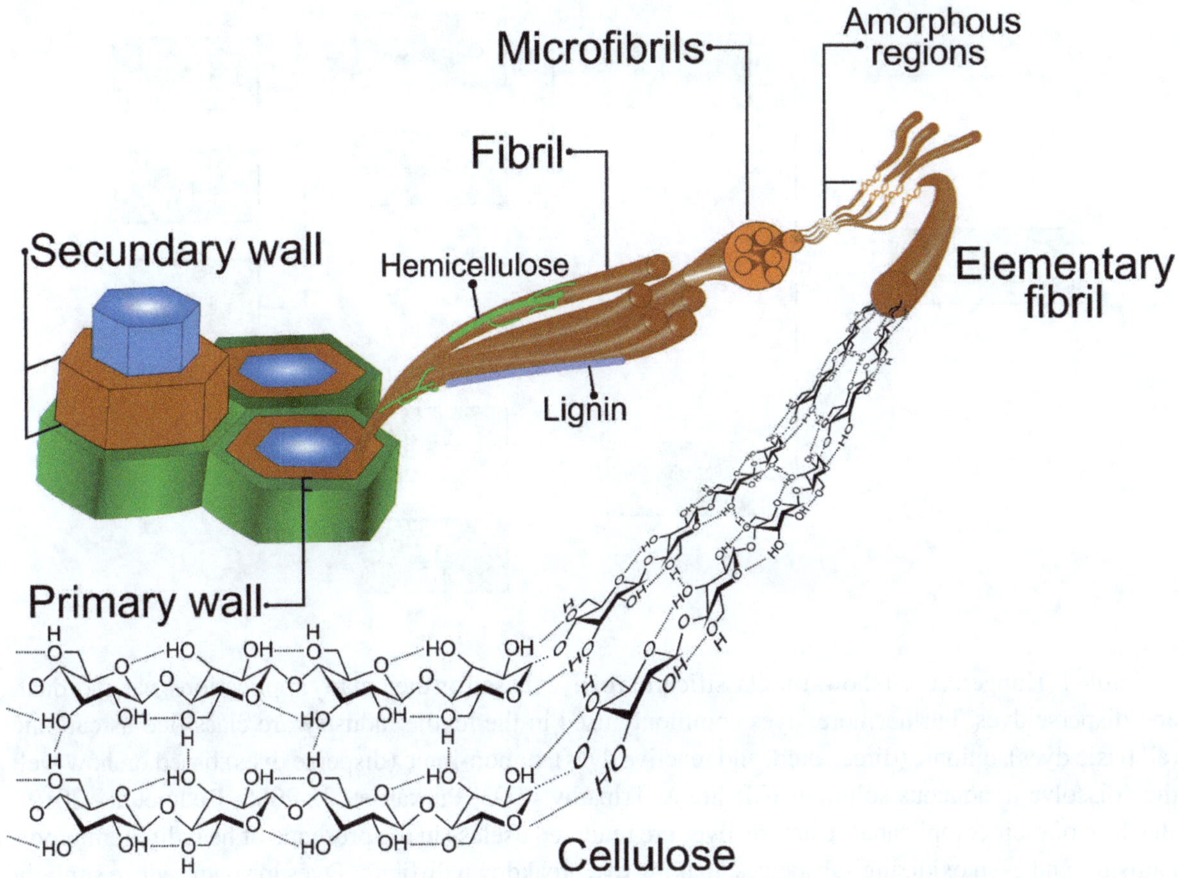

Many dyes, as well as the breakdown products of dyes, are carcinogenic, mutagenic, and hazardous to living things. Dyes raise the BOD of wastewater, which has an impact on aquatic life. They are not easily degradable, and wastewater treatment systems and traditional procedures such as adsorption, ultrafiltration, chemicals, and electrochemical approaches rarely remove them from water (Subha et al., 2018).

By hydrogen bonding, van der Waals forces, or electrostatic interactions, the dye molecule attaches to the fiber's surface. Excess dyestuffs mix with water and are released as effluent as a result of the incorrect coloring process. If these dyestuffs are released into the environment, they can pose a major threat. Much research has looked into the toxicity of various dyes, with the conclusion that dyes are extremely genotoxic and carcinogenic. The dye can enter the body through the pores of the skin, inhalation, or ingestion (Pai et al., 2021).

Classification of Dyes

Dyes are classified by their ingredients, colors, and applications, with application being the most often used method for dye categorization (Clarke & Anliker, 1980; Gupta, 2000). Dyes, on the other hand, can be cationic or anionic and can be used in a variety of sectors, including textiles, leather, pulp & paper, and paint. Figure 4 Shows the dye classification based on ionic charge.

Figure 4. Dye classification based on ionic charge

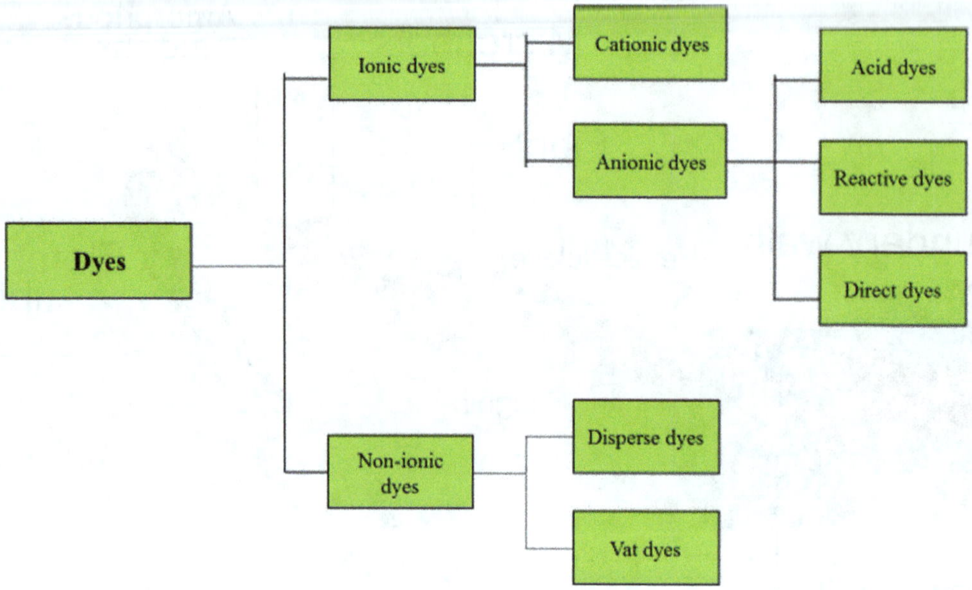

Table 1 (Hunger, 2003) shows the classification of dyes based on their many applications: vat, mordant, and disperse dyes. Furthermore, dyes commonly used in the textile industry are classified as cationic (all basic dyes), anionic (direct, acid, and reactive dyes), or non-ionic (disperse dyes) based on how well they dissolve in aqueous solutions (Mishra & Tripathy, 1993; Purkait et al., 2005; Ruan et al., 2019). Because of their complicated structure, dyes are rendered useless in the presence of heat, light, microorganisms, and even oxidizing substances, making dye breakdown difficult. Dyes in water reduce sunlight penetration and oxygen solubility, lowering water quality and impeding photosynthesis in aquatic plants and wildlife (Clarke & Anliker, 1980; Rafatullah et al., 2010). It can also cause serious health problems in humans, such as an increase in heart rate, vomiting, shock, the formation of Heinz bodies, cyanosis, jaundice, quadriplegia, and tissue necrosis (Bell et al., 2000). Heavy metals are classed as inorganic water pollutants, while dye molecules are categorized as organic water pollutants (Oyewo et al., 2020).

Methods for Dye Removal

For the scavenging of dyes, a variety of technologies have been developed and applied, which can be grouped into three categories: biological approaches, chemical methods, and physical methods (Kobya et al., 2005). Their practical applications are limited by high capital costs, low efficiency, and the formation of surplus sludge. Some of these procedures have been proven to be more adaptable and superior to others, and they may be used to remove a wide spectrum of colors from wastewater (Foo et al., 2010). Adsorption and photocatalysis are better alternatives due to their cheap initial cost, produces nontoxic by-products, and removes colors fully even from dilute solutions (Srinivasan and Viraraghavan, 2010).

A stage in wastewater treatment called physicochemical fractionation separates hydrophilic and hydrophobic components (Kim & Yu, 2005). The methodologies and adsorbents used in wastewater treatment have recently been compared (Crini & Lichtfouse, 2019; Crini et al., 2019). Some of the technologies used to remediate dye-contaminated waters include reverse osmosis, coagulation, flocculation, ion

Table 1. Dye classification, examples, applications, water solubility, and toxicity.

Type of Dyes	Chemical Structure of Dye	Example	Application	Toxicity	Solubility	Reference
Direct Dyes	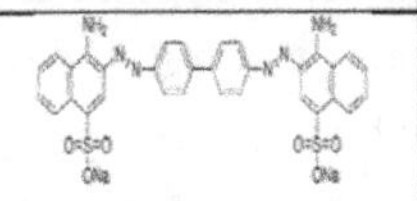 Congo red (CR)	Congo red (CR), Direct red	Cotton, leather, Coloring paper products	Bladder cancer	Water soluble, anionic compounds	(Shaban et al., 2017; Hunger, 2003)
Acid Dyes	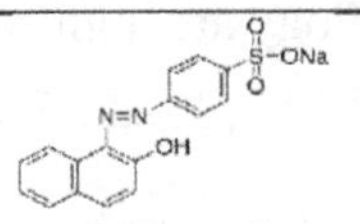 Acid orange 7	Acid orange 7,Acid yellow 36	Nylon, wool, silk, textile, leather, and pharmaceutical sectors.	carcinogenic and mutagenic	Water soluble	Dutta et al., 2021; Ruan et al., 2019)
Basic Dyes	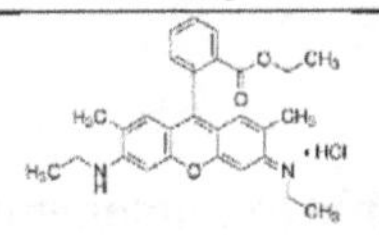 rhodamine 6G	Basic red 1 or rhodamine 6G, Methylene blue	Polyesters, Paper	Changing the chemical and physical qualities of water bodies, as well as having negative consequences for the flora and fauna	Water soluble	Benkhaya et al., 2020; Dutta et al., 2021)
Vat Dyes	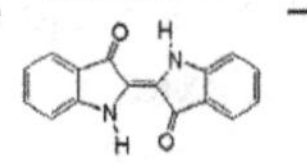 Vat blue 1	Vat blue 1, Vat orange 15	cellulosic fibers	Allergic, rhinitis, dermatitis, conjunctivitis	Water-insoluble	(Ruan et al., 2019; Dutta et al., 2021; Yamjala et al., 2016)
Disperse Dyes	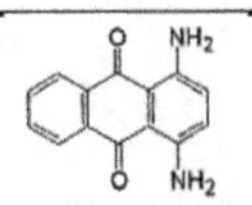 Disperse violet 1	Disperse violet 1, Disperse orange 3	nylon, acrylic fiber	Carcinogenic and allergic to skin	Water-insoluble	(Dutta et al., 2021; Ruan et al., 2019)
Reactive Dye	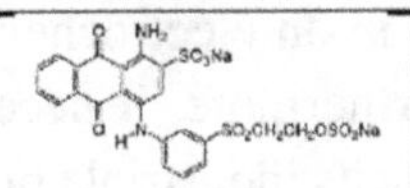 reactive blue 19	reactive blue 19, reactive red 120	Cotton, nylon and wool	Allergic reaction, dermatitis, conjunctivitis	Water soluble	(Benkhaya et al., 2020; Dutta et al., 2021)

exchange, activated carbon adsorption, advanced oxidation, ozonation, photocatalysis, Fenton process, photo-Fenton, electrochemical oxidation (Lade et al., 2015), and filtration (Singh et al., 2015). These procedures are frequently expensive, and amine remnants in sludges may be detected after treatment. Some methods for dye removal are given below.

Coagulation

Coagulant is a substance used in coagulation to improve the density of colloidal form. The efficiency of the coagulant is determined by the coagulant's kind, dosage, pH, temperature, alkalinity, and mixing circumstances. Although coagulation is effective at removing heavy metals from wastewater, it has the drawback of generating secondary pollutants that are detrimental to both humans and the environment. As a result, it must be subjected to additional secondary treatment for complete removal.

Ion Exchange

Ion exchange is a separation method that replaces ions with one another in order to achieve high removal efficiency. This approach produces less sludge than other methods, which is its biggest advantage. The sole drawback to this method is that it causes matrix fouling at high concentration.

Membrane Filtration

Membrane Filtration is a type of filtration that uses a membrane is becoming more popular in water purification; depending on the driving power and ionic and molecular flow modes through the membrane, the membrane can be characterized as dense or porous. Toxic metal ions have been successfully removed from wastewater using a variety of membranes, including ultrafiltration, reverse osmosis, microfiltration, electrodialysis and nanofiltration. The main disadvantages of this technology are high-power costs, membrane fouling and handling rejection, while the benefits include relatively low energy requirements, high efficiency, ease of operation and reliability (Eriksson, 1988).

Chemical Precipitation

This is used to treat wastewater effluent because it is cost-effective and simple to implement (Matlock et al., 2002; Izadi et al., 2017). Sulphide precipitation and hydroxide precipitation are also included. To reduce metals to an acceptable level for discharge into the environment, the hydroxide approach necessitates the use of numerous chemicals (Jüttner et al., 2000).

Electrochemical Method

Electrochemical wastewater treatment technologies are very successful at removing heavy metal ions from industrial wastewater effluent. Using anodic and cathodic processes in an electrochemical cell, this approach recovers heavy metals in their elemental metallic state. Furthermore, it necessitates a significant financial investment as well as increased power supply, restricting its industrial applications.

Bioremediation

For the elimination of industrial wastewater effluents, bioremediation employs microorganisms. This microbe-assisted approach is also commonly used to remove harmful heavy metal contaminants, although conventional wastewater treatment is not cost-effective. It also pollutes the aquatic environment with harmful secondary contaminants.

Adsorption

Metal ions such as Cd, Pb, Cr, Zn, Cu, Co, Hg, and As are widespread contaminants in an aquatic environment, and adsorption using graphene-based materials in wastewater treatment is a severe complication worldwide.

Nanocomposite

Nanocomposites are multi-phase materials with dimensions in the nano range (10–100 nm) in at least one phase. Nanocomposite materials have recently emerged as viable options for overcoming the limits of various engineering materials. The dispersed matrix and dispersion phase materials of nanocomposite materials can be categorized (Pandey et al., 2017). Nanocomposites' unique features have brought in a revolution in bioremediation (Mohanraj et al., 2020). Several research groups have recently experimented with various biopolymeric nanocomposites for enhanced degradation of synthetic azo dyes, such as metal/ metal oxide/biopolymer or metal oxide/conducting polymer. Chitosan/ZnO/AgCl nanocomposite based on hydrogel beads is an example of a biopolymer/metal oxide nanocomposite that allows for complete photocatalytic degradation of methylene blue (Taghizadeh et al., 2020; Adnan et al., 2020; Sathiyavimal et al., 2020).

Metal Oxide-cellulose (MOC) Nanocomposites Chemistry

Metal oxide-cellulose (MOC) nanocomposites are a new field that allows inorganic nanoparticles to be incorporated into multifunctional natural-based materials. To avoid aggregation, the synthesis of MOC nanocomposites should be done with care to ensure that the metal oxide nanoparticles are well dispersed within the cellulose matrices. Cellulose is a natural polymeric substance with extended OH groups that can be changed to improve its characteristics by adding other functional groups. Carboxymethylcellulose (CMC), ethylhydroxypropylcellulose (EHPC), hydroxylpropylcellulose (HPC), cellulose carbamate, cellulose acetate and amino cellulose have all been developed in this area. cellulose microfiber (CMF), cellulose microcrystal (CMC), cellulose nanofiber (CNF) and cellulose nanocrystal (CNC), are some of the nanostructures found in nanocellulose (Oyewo et al., 2020).

Metal oxide nanoparticles provide efficient surface area for photocatalysis and other processes in the composite. Biodegradability, nontoxicity, high thermal strength, ion-adsorption capacity, biocompatibility, stability and sensitivity are some of the features of nanocellulose components (Ali et al., 2016).

Preparation Methods for Metal Oxide-cellulose Nanocomposites

The synthesis of metal-oxide cellulose nanocomposites can be done in a variety of ways, including hydrothermal or wet chemical, solvothermal, wet spinning, sol-gel, coprecipitation, and microwave (Ramesh et al., 2018; Li et al., 2013; Fu et al., 2015). Some methods for metal oxide-cellulose nanocomposites are given below.

Hydrothermal

A high-porosity combination of natural cellulose fibers and manganese dioxide nanosheets was created using a simple hydrothermal technique. The cellulose matrix was impregnated with aqueous potassium permanganate (KMnO4) solution and oleic acid, which served as a reducing agent (Biliuta & Coseri, 2019). The reaction conditions might also be tweaked to get varied forms of integrated MnO2 nanoparticles (Oyewo et al., 2020).

Sol-gel Method

In the synthesis of organic–inorganic nanocomposite, polar and hydrophilic polymers, such as polyvinyl alcohol, are used. Biocompatibility, ease of processing, and high-water swelling ability are all advantages of this approach. A sol-gel method of synthesizing cellulose-PVA-TiO2 nanocomposites was published by Ramesh et al. (Ramesh et al. 2018). After heating and stirring a cellulose/LiCl/N, N-dimethylacatamide solution, tetra ethylene glycol and PVA were added to the cellulose sol. Finally, titanium tetraisopropoxide was added, and the entire solution was heated for 10 hours at 95 degrees Celsius. To make the composite, the solution was rinsed with ethanol and dried in the oven. Martins et al. made NFC/ZnO nanocomposites by heating nano fibrillated cellulose with poly (diallyldimethylammonium chloride) (PDDA) solution and then adding ZnO nanoparticle solution (Martins et al., 2013; Oyewo et al., 2020).

Wet Spinning Method

In a NaOH/urea aqueous solution, cellulose-Fe2O3 nanocomposite fibres with super active characteristics were generated by wet spinning (Yu et al., 2013). The presence of reactive hydroxyl groups on cellulose was allowed to impact the nature of composite formation or the growth strategy of ZnO nanostructures in the synthesis of ZnO-cellulose (Ma et al., 2016; Bakhsh et al., 2018).

If anhydrides, acetates, or epoxides were added to the cellulose, they might replace some of the reactive OH, causing ZnO to grow and agglomerate differently on the cellulose structures. All of the highlighted preparation methods depict cellulose as a good platform for inorganic nanoparticle development (Ma et al., 2016).

Microwave Method

The microwave oven approach was used to make ZnO-mediated cellulose carbamate nanocomposites (Yu et al., 2013). To make cellulose carbamate, the cellulose was first treated with urea in aqueous solution. The cellulose carbamate was then rinsed after being distributed in a NaOH/ZnO aqueous solution. Because it produces a high yield, completes the reaction quickly, has control over the heating process, and does not come into contact with the materials, the microwave approach has several benefits over the traditional heating method (Mahmoud et al., 2015).

Liquid-phase Deposition

SnO_2 was produced by dissolving SnF_2 in aqueous medium and heating it to a low temperature. SnO_2 was consistently deposited onto the flexible, non-exfoliable cellulose surface (Mahadeva & Kim, 2011).

Classification of Metal Oxide-cellulose Nanocomposites

Metal-oxide cellulose nanocomposites come in a variety of forms and have been made in a variety of techniques. Metal oxide-cellulose nanocomposites (for example, Ag-cellulose), metal oxide-modified cellulose nanocomposites (for example, Fe2O3-cellulose acetate), and metal oxide PVA-cellulose nanocomposites are among them (e.g. TiO2-PVA-cellulose) (Oyewo et al., 2020).

- metal oxide-cellulose nanocomposites
- metal oxide-modified cellulose nanocomposites
- metal oxide PVA- cellulose nanocomposites
- bacterial cellulose nanocomposites

Bacterial cellulose nanocomposites that can be changed for regeneration utilizing N-methyl-morpholine-N-oxide monohydrate is bacterial cellulose combined with metal oxides. The metal oxides were appropriately adhered to the regenerated bacterial cellulose's surface and internal matrix. They are said to have good thermal, antibacterial properties, and mechanical as well as reasonable biocompatibility and low toxicity (Ullah et al., 2013).

Application of Metal Oxide-cellulose (MOC) Nanocomposites for Removal of Dyes from Wastewater

The organic and inorganic hybrid effects in MOC nanocomposites result in new optical, mechanical, and thermal functions (Mahadeva & Kim, 2011). The nanoparticles integrated into the cellulose matrix have a high surface area, which provides it a stable nature and retention ability that avoids leaching or contamination. The number of nanoparticles on the surface of a nanocomposite material contributes to the extent of dye degradation because the surface area of the material is a significant element in photocatalytic reactions.

ZnO-cellulose, iron oxide-cellulose, and TiO2-cellulose nanocomposites have all been developed and employed in the photodegradation of dye components in water. Incorporating metal oxides or inorganic minerals into cellulose is a novel technique to improve material qualities, and it could also serve as a foundation for water purification. It has been reported that a bead cellulose integrated iron oxyhydroxide was used to remove arsenic ions from aqueous solutions, which may include chelation or ion exchange (Guo et al., 2009; Yu et al., 2013). The production of MOC composites sorbents for water treatment reduces the dangers associated with metal oxide nanoparticles (Guo et al., 2007). The functional groups in cellulose, or their modified forms, have a role in the formation of complexes with metal ions in wastewater.

Nanohybrids are receiving a lot of attention for multifunctional water purification systems, and their size-dependent features are different in their aspect towards photocatalysis and water purification (Krivoshapkin et al., 2019). Low-cost nanoparticles such as TiO2, ZnO, and Fe2O3 are the most studied in composite creation with cellulose materials. They are photosensitive, have a high quantum efficiency, have a high absorption power, are largely non-toxic, and have a wide bandgap. The breakdown of methylene blue dye using ZnO/cellulose nanocomposites, and the nanocomposites were more efficient than ZnO nanoparticles alone (Lefatshe et al., 2015).

Polyamide-amine-epichlorohydrin (PAE) was used to cross-link the cellulosic fibres and stable the TiO2 NPs on the nanomatrix, resulting in microfibrillated cellulose (MFC)-TiO2 composites (Garusinghe et al., 2018). The nanocomposite was utilised to degrade methyl orange dye and did not desorb into the methyl orange solution, indicating that it is a stable and suitable water purification material. MFC is recyclable, biocompatible, has a high retention capacity, is stable, and has a low cost. The photocatalytic efficiency is also determined by the amount and kind of metal oxide in the cellulose matrix (Oyewo et al., 2020).

CONCLUSION

Cellulose is the most abundant, renewable, and environmentally friendly biopolymer on the earth and may be derived in vast quantities from plants, crops and forest trash. Cellulose and some environmentally friendly metal oxides materials are readily available, and their composites are effective waste-water treatment materials. Cellulose is a biodegradable, non-toxic, low-cost material that can be found in a wide range of natural resources and agricultural waste. To remove a variety of impurities, including harmful metals and dyes, cellulose could be used in a variety of water treatment methods. As a result, due to their high surface area, light stability, and low toxicity, the application of cellulose metal oxide composite as an efficient dye adsorption and photo degradation in water. Incorporation of metal oxide into cellulose improves the stability of the material, prevent desorption of the nanoparticles into the water system, reduce toxicity effects and also helps the material to be long-lasting. Moreover, multi-metal or polymetal oxides can be used with the cellulose instead of pure metal oxides in the composites. Overall, when all procedures are relatively safe, would exhibit less aggregation and offer advance dye removal from wastewater purification process.

REFERENCES

Abdel Aal, A., Mahmoud, S. A., & Aboul-Gheit, A. K. (2009). Sol–Gel and Thermally Evaporated Nanostructured Thin ZnO Films for Photocatalytic Degradation of Trichlorophenol. *Nanoscale Research Letters*, *4*, 627.

Adnan, M. A. M., Phoon, B. L., & Julkapli, N. M. (2020). Mitigation of pollutants by chitosan/metallic oxide photocatalyst: A review. *Journal of Cleaner Production*, *261*, 121190. Advance online publication. doi:10.1016/j.jclepro.2020.121190

Akarslan, F., & Demiralay, H. (2018). Effective of textile materials harmful to human health. *Acta Physica Polonica*, *128*, B-407–B-408.

Ali, A., Ambreen, S., Maqbool, Q., Naz, S., Shams, M. F., Ahmad, M., Phull, A. R., & Zia, M. (2016). Zinc impregnated cellulose nanocomposites: Synthesis, characterization and applications. *Journal of Physics and Chemistry of Solids*, *98*, 174–182.

Azizi, S., Alloin, F., & Dufresne, A. (2005). Review of recent research into cellulosic whiskers, their properties and their application in nanocomposite field. *Biomacromolecules*, *6*, 612–626.

Bakhsh, E. M., Khan, S. A., Marwani, H. M., Danish, E. Y., Asiri, A. M., & Khan, S. B. (2018). Performance of cellulose acetate-ferric oxide nanocomposite supported metal catalysts toward the reduction of environmental pollutants. *International Journal of Biological Macromolecules*, *107*, 668–677.

Bell, J., Plumb, J. J., Buckley, C. A., & Stuckey, D. C. (2000). Treatment and decolorization of dyes in an anaerobic baffled reactor. *Journal of the Environmental Engineering Division*, *126*, 1026–1032.

Benkhaya, S., M'rabet, S., & El Harfi, A. (2020). A review on classifications, recent synthesis and applications of textile dyes. *Inorganic Chemistry Communications*, *115*, 107891.

Biliuta, G., & Coseri, S. (2019). Cellulose: A ubiquitous platform for ecofriendly metal nanoparticles preparation. *Coordination Chemistry Reviews*, *383*, 155–173.

Clarke, E., & Anliker, R. (1980). Organic Dyes and Pigments. The Handbook of Environmental Chemistry, 3, 181-215.

Crini, G., & Lichtfouse, E. (2019). Advantages and disadvantages of techniques used in wastewater treatment. *Environmental Chemistry Letters*, *17*, 145–155.

Crini, G., Lichtfouse, E., Wilson, L. D., & Morin-Crini, N. (2019a). Conventional and non-conventional adsorbents for wastewater treatment. *Environmental Chemistry Letters*, *17*, 195–213.

Dutta, S., Gupta, B., Srivastava, S. K., & Gupta, A. K. (2021). Recent advances on the removal of dyes from wastewater using various adsorbents: A critical review. *Materials Advances*, *2*, 4497–4531.

Eichhorn, S. J., Dufresne, A., & Aranguren, M. (2010). Review: Current international research into cellulose nanofibres and nanocomposites. *Journal of Materials Science*, *45*, 1–33.

Eriksson, P. (1988). Nanofiltration extends the range of membrane ultrafiltration. *Environment and Progress*, *7*, 58–62.

Foo, K. Y., & Hameed, B. H. (2010). An overview of dye removal via activated carbon adsorption process. *Desalination and Water Treatment*, *19*, 255–274.

Garusinghe, U. M., Raghuwanshi, V. S., Batchelor, W., & Garnier, G. (2018). Water Resistant Cellulose – Titanium Dioxide Composites for Photocatalysis. *Scientific Reports*, *8*, 2306.

Guo, X., Du, Y., Chen, F., Park, H. S., & Xie, Y. (2007). Mechanism of removal of arsenic by bead cellulose loaded with iron oxyhydroxide (β-FeOOH): EXAFS study. *Journal of Colloid and Interface Science*, *314*, 427–433.

Guo, Y., Zhou, J., Song, Y., & Zhang, L. (n.d.). An efficient and environmentally friendly method for the synthesis of cellulose carbamate by microwave heating. *Macromolecular Rapid Communications*, *30*, 1504–1508.

Gupta, V. (2009). Application of Low-Cost Adsorbents for Dye Removal: A Review. *Journal of* Environmental Management, 90, 2313–2342.

Habibi, Y. (2014). Key advances in the chemical modification of nanocelluloses. *Macromolecules*, *45*, 9220.

Habibi, Y., Lucia, L. A., & Rojas, O. J. (2010). Cellulose nanocrystals: Chemistry, self-assembly, and applications. *Chemical Reviews*, 3479–3500.

Hunge, Y. M., Mohite, V. S., Kumbhar, S. S., Rajpure, K. V., Moholkar, A. V., & Bhosale, C. H. (2015). Photoelctrocatalytic degradation of methyl red using sprayed WO_3 thin films under visible light irradiation. *Journal of Materials Science*, *26*, 8404–8412.

Hunger, K. (2003). *Industrial Dyes-Chemistry, Properties, Application*. Wiley-VCH.

Izadi, A., Mohebbi, A., Amiri, M., & Izadi, N. (2017). Removal of iron ions from industrial copper raffinate and electrowinning electrolyte solutions by chemical precipitation and ion exchange. *Minerals Engineering*, *113*, 23–35.

Jing, S., Cao, X., Zhong, L., Peng, X., Sun, R., & Liu, J. (2018). Effectively enhancing conversion of cellulose to HMF by combining in-situ carbonic acid from CO2 and metal oxides. *Industrial Crops and Products*, *126*, 151–157.

Jüttner, K., Galla, U., & Schmieder, H. (2000). Electrochemical approaches to environmental problems in the process industry. *Electrochimica Acta*, *45*, 2575–2594.

Kansal, S. K., Singh, M., & Sud, D. (2007). Studies on photo degradation of two commercial dyes in aqueous phase using different photocatalyst. *Journal of Hazardous Materials*, *141*, 581–590.

Kim, H. C., & Yu, M. J. (2005). Characterization of natural organic matter in conventional water treatment processes for selection of treatment processes focused on DBPs control. *Water Research*, *39*, 4779–4789.

Kobya, M., Demirbas, E., Senturk, E., & Ince, M. (2005). Adsorption of Heavy Metal Ions from Aqueous Solutions by Activated Carbon Prepared from Apricot Stone. *Bioresource Technology*, *96*, 1518–1521.

Krivoshapkin, P. V., Ivanets, A. I., Torlopov, M. A., Mikhaylov, V. I., Srivastava, V., Sillanpää, M., Prozorovich, V. G., Kouznetsova, T. F., Koshevaya, E. D., & Krivoshapkina, E. F. (2019). Nanochitin/manganese oxide-biodegradable hybrid sorbent for heavy metal ions. *Carbohydrate Polymers*, *210*, 135–143.

Lade, H., Kadam, A., Paul, D., & Govindwar, S. (2015). Biodegradation and detoxification of textile azo dyes by bacterial consortium under sequential microaerophilic/aerobic processes. *EXCLI Journal*, *14*, 158.

Lefatshe, K., Muiva, C. M., & Kebaabetswe, L. P. (2017). Extraction of nanocellulose and in-situ casting of ZnO/cellulose nanocomposite with enhanced photocatalytic and antibacterial activity. *Polymers*, *164*, 301–308.

Li, S. M., Dong, Y. Y., Ma, M. G., Fu, L. H., Sun, R. C., & Xu, F. (2013). Hydrothermal synthesis, characterization, and bactericidal activities of hybrid from cellulose and TiO_2. *Carbohydrate Polymers*, *96*, 15–20.

Ma, J., Sun, Z., Wang, Z., & Zhou, X. (2016). Preparation of ZnO–cellulose nanocomposites by different cellulose solution systems with a colloid mill. *Cellulose (London, England)*, *23*, 3703–3715.

Mahadeva, S. K., & Kim, J. (2009). Hybrid nanocomposite based on cellulose and tin oxide: Growth, structure, tensile and electrical characteristics. *Science and Technology of Advanced Materials*, *12*, 055006.

Mahmoud, A. M., Ibrahim, F. A., Shaban, S. A., & Youssef, N. A. (2015). Adsorption of heavy metal ion from aqueous solution by nickel oxide nano catalyst prepared by different methods. *Egyptian Journal of Petroleum*, *24*, 27–35.

Martins, N. C. T., Freire, C. S. R., Neto, C. P., Silvestre, A. J. D., Causio, J., Baldi, G., Sadocco, P., & Trindade, T. (2013). Antibacterial paper based on composite coatings of nanofibrillated cellulose and ZnO. *Colloids and Surfaces. A, Physicochemical and Engineering Aspects*, *417*, 111–119.

Matlock, M. M., Howerton, B. S., & Atwood, D. A. (2002). Chemical precipitation of heavy metals from acid mine drainage. *Water Research*, *36*, 4757–4764.

Mishra, G., & Tripathy, M. (1993). A critical review of the treatment. for decolorization of textile effluent. *Colourage*, *40*, 35–38.

Mohanraj, J., Durgalakshmi, D., Rakkesh, R. A., Balakumar, S., Rajendran, S., & Karimi-Maleh, H. (2020). Facile synthesis of paper based graphene electrodes for point of care devices: A double stranded DNA (dsDNA) biosensor. *Journal of Colloid and Interface Science*, *566*, 463–472. doi:10.1016/j.jcis.2020.01.089 PMID:32032811

Noohpisheh, Z., Amiri, H., Farhadi, S., & Mohammadi-gholami, A. (2020). Green synthesis of Ag-ZnO nanocomposites using Trigonella foenum-graecum leaf extract and their antibacterial, antifungal, antioxidant and photocatalytic properties. *Spectrochimica Acta. Part A: Molecular and Biomolecular Spectroscopy*, *240*, 118595.

Oyewo, O. A., Elemike, E. E., Onwudiwe, D. C., & Onyango, M. S. (2020). Metal oxide-cellulose nanocomposites for the removal of toxic metals and dyes from wastewater. *International Journal of Biological Macromolecules*. j.ijbiomac.2020.08.074 doi:10.1016/

Pai, S., Kini, M. S., & Selvaraj, R. (2021). A review on adsorptive removal of dyes from wastewater by hydroxyapatite nanocomposites. *Environmental Science and Pollution Research International*, *28*, 11835–11849.

Pai, S., Kini, M. S., & Selvaraj, R. (2021). A review on adsorptive removal of dyes from wastewater by hydroxyapatite nanocomposites. *Environmental Science and Pollution Research International*, *28*, 11835–11849.

Pandey, N., Shukla, S. K., & Singh, N. B. (2017). Water purification by polymer nanocomposites: An overview. *Nanocomposites*, *3*(2), 47–66. doi:10.1080/20550324.2017.1329983

Purkait, M. K., DasGupta, S., & De, S. (2005). Adsorption of eosin dye on activated carbon and its surfactant based desorption. *Journal of Environmental Management*, *76*, 135.

Rafatullah, M., Sulaiman, O., Hashim, R., & Ahmad, A. (2010). Adsorption of methylene blue on low-cost adsorbents: A review. *Journal of Hazardous Materials*, *177*, 70–80.

Ramesh, S., Kim, H. S., & Kim, J. H. (2018). Cellulose–Polyvinyl Alcohol–Nano-TiO_2 Hybrid Nano-composite: Thermal, Optical, and Antimicrobial Properties against Pathogenic Bacteria. *Polymer-Plastics Technology and Engineering*, *57*, 669–681.

Rojas, J., Bedoya, M., & Ciro, Y. (2015). Current Trends in the Production of Cellulose Nanoparticles and Nanocomposites for Biomedical Applications. In M. Poletto, & H. L. Ornaghi (Eds.), *Cellulose - Fundamental Aspects and Current Trends*. IntechOpen. doi:10.5772/61334

Ruan, W., Hu, J., Qi, J., Hou, Y., Zhou, C., & Wei, X. (2019). Wenqian Ruan Removal of Dyes from Wastewater by. Nanomaterials: A Review. Jiwei Hu Jimei Qi Yu Hou Chao Zhou Xionghui Wei*Advanced. Materials Letters*, *10*, 9–20.

Sarkar, S., Ponce, N. T., Banerjee, A., Bandopadhyay, R., Rajendran, S., & Lichtfouse, E. (2020). Green polymeric nanomaterials for the photocatalytic degradation of dyes: A review. *Environmental Chemistry Letters*, *18*, 1569–1580.

Sathiyavimal, S., Vasantharaj, S., Kaliannan, T., & Pugazhendhi, A. (2020). Eco-biocompatibility of chitosan coated biosynthesized copper oxide nanocomposite for enhanced industrial (Azo) dye removal from aqueous solution and antibacterial properties. *Carbohydrate Polymers*, *241*, 116243. Advance online publication. doi:10.1016/j.carbpol.2020.116243 PMID:32507166

Shaban, M., Abukhadra, M. R., Hamd, A., Amin, R. R., & Abdel Khalek, A. (2017). Photocatalytic removal of Congo red dye using MCM-48/Ni_2O_3 composite synthesized based on silica gel extracted from rice husk ash; fabrication and application. *Journal of Environmental Management*, *204*, 189–199.

Sharma, V. K., Jinadatha, C., & Lichtfouse, E. (2020). Environmental chemistry is most relevant to study coronavirus pandemics. *Environmental Chemistry Letters*. doi:10.1007/s10311-020-01017-6

Singh, R. L., Singh, P. K., & Singh, R. P. (2015). Enzymatic decolorization and degradation of azo dyes—A review. *International Biodeterioration & Biodegradation*, *104*, 21–31. doi:10.1016/j.ibiod.2015.04.027

Srinivasan, A., & Viraraghavan, T. (2010). Decolorization of Dye Wastewaters by Biosorbents, A Review. *Journal of Environmental Management*, *91*, 1915–1929.

Subha, V., Divya, K., Gayathri, S., Jagan Mohan, E., Keerthanaa, N., Vinitha, M., Kirubanandan, S., & Renganathan, S. (2018). Applications of iron oxide nano composite in waste water treatment–dye decolourisation and anti–microbial activity. *MOJ Drug Design Development & Therapy*, *2*, 178–184.

Taghizadeh, M. T., Siyahi, V., Ashassi-Sorkhabi, H., & Zarrini, G. (2020). ZnO, AgCl and AgCl/ZnO nanocomposites incorporated chitosan in the form of hydrogel beads for photocatalytic degradation of MB, E. coli and S. aureus. *International Journal of Biological Macromolecules*, *147*, 1018–1028. doi:10.1016/j.ijbiomac.2019.10.070 PMID:31739064

Ullah, M. W., Park, J. K., Ul-Islam, M., Khan, S., & Khattak, W. A. (2013). Synthesis of regenerated bacterial cellulose-zinc oxide nanocomposite films for biomedical applications. *Cellulose (London, England)*, *21*, 433–447.

Velusamy, S., Roy, A., Sundaram, S., & Kumar Mallick, T. (2021). A Review on Heavy Metal Ions and Containing Dyes Removal Through Graphene Oxide-Based Adsorption Strategies for Textile Wastewater Treatment. *Chemical Record (New York, N.Y.)*, *21*, 1570–1610.

Wutich, A., Rosinger, A. Y., Stoler, J., Jepson, W., & Brewis, A. (2019). Measuring human water needs. *American Journal of Human Biology*, *3*, e23350. https://doi.org/10.1002/ajhb.23350

Yamjala, K., Nainar, M. S., & Ramisetti, N. R. (2016). Methods for the analysis of azo dyes employed in food industry: A review. *Food Chemistry*, *192*, 813–824.

Yu, X., Tong, S., Ge, M., Zuo, J., Cao, C., & Song, W. (2013). One-step synthesis of magnetic composites of cellulose@iron oxide nanoparticles for arsenic removal. *Journal of Materials Chemistry. A, Materials for Energy and Sustainability*, *1*, 959–965.

Chapter 7
Biopolymer-Based Nanocomposite Materials for Detection and Removal of Pollutants in Wastewater

Ratnesh Das
Dr. Harisingh Gour Central University, India

Arunesh Kumar Mishra
Dr. Harisingh Gour Central University, India

Pratibha Mishra
Dr. Harisingh Gour Central University, India

Megha Das
Indira Gandhi National Tribal University, India

ABSTRACT

Biopolymer-based nanocomposites, particularly chitosan, cellulose, alginate, starch, and carrageenan, are increasingly being employed as reinforcements for composite materials because they are biodegradable, recyclable, renewable, abundant, conveniently available, cost-effective, and non-abrasive to processing equipment. These biopolymer nanocomposite materials are also lightweight, stiff, and have good mechanical properties. Biopolymer nanocomposites have interfacial limitations because all nanocomposite biopolymers are hydrophilic. Water recycling has been made possible by biopolymer-based nanocomposite materials, which have a variety of applications for cleaning wastewater, making it a viable and cost-effective solution to water scarcity. The growing concern about heavy metal contamination has necessitated the development of new and better-suited sorbent materials for effective detoxification.

DOI: 10.4018/978-1-6684-4553-2.ch007

Figure 1. Different types of water pollutants

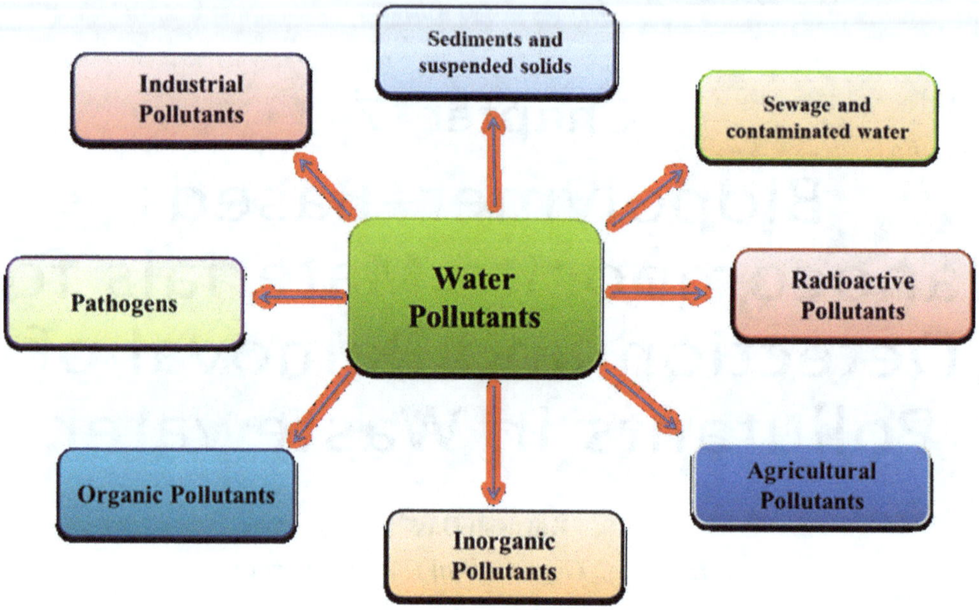

INTRODUCTION

Water treatment technology that is cleaner and more efficient is needed to reduce pollution from several common industrial and household activities. Water treatment refers to the procedures that are used to make water more suitable for a specific purpose. It can be used for a variety of purposes, including residential, industrial, medical, and many more. All water treatment methods aim to remove impurities from the water and make it drinkable (Ahmad et al., 2015; Zhang et al., 2016). The treatment of water before it is used and the treatment of sewage, household, agricultural, and industrially produced after it has been utilized are included in water treatment. For the aforementioned sorts of water treatment techniques, there are numerous possibilities. All have advantages and disadvantages, as well as convenience of use, cost-effectiveness, and end usage, which all impact their choice (Ritter et al., 2002; Rajasekhar et al., 2018; Jassby et al., 2018; Rodriguez-Proteau et al., 2006). New self-assembly technologies allow nanoparticles to self-assemble in a controlled manner, resulting in novel materials with novel transport, optical, mechanical, magnetic, electrical, and electronic properties that could be used to remove pollutants from aqueous systems at low cost or even at low concentration (Khan et al., 2019; Kumar et al., 2014). Nowadays, the major problem is the contamination of the drinking water with pollutants such as sediments and suspended solids, industrial, agricultural, inorganic, organic, biological, and radioactive pollutants like strontium, lanthanides, and actinides (Bharagava et al., 2018; Saxena et al., 2019). The main source of these radioactive pollutants is the operation of nuclear power plants, research facilities, and the use of radioisotopes in the industry as well as diagnostic medicine (Lytle et al., 2014). The development of industrialization and urbanization has resulted in a daily worsening of the water situation. Pharmaceuticals, textile, dyes, fertilizer, smelting processes, mining, and others all produce enormous amounts of wastewater (Bharagava and Chandra, 2010; Kishor et al., 2019; Houa et al., 2020).

Figure 2. Distribution of water on earth

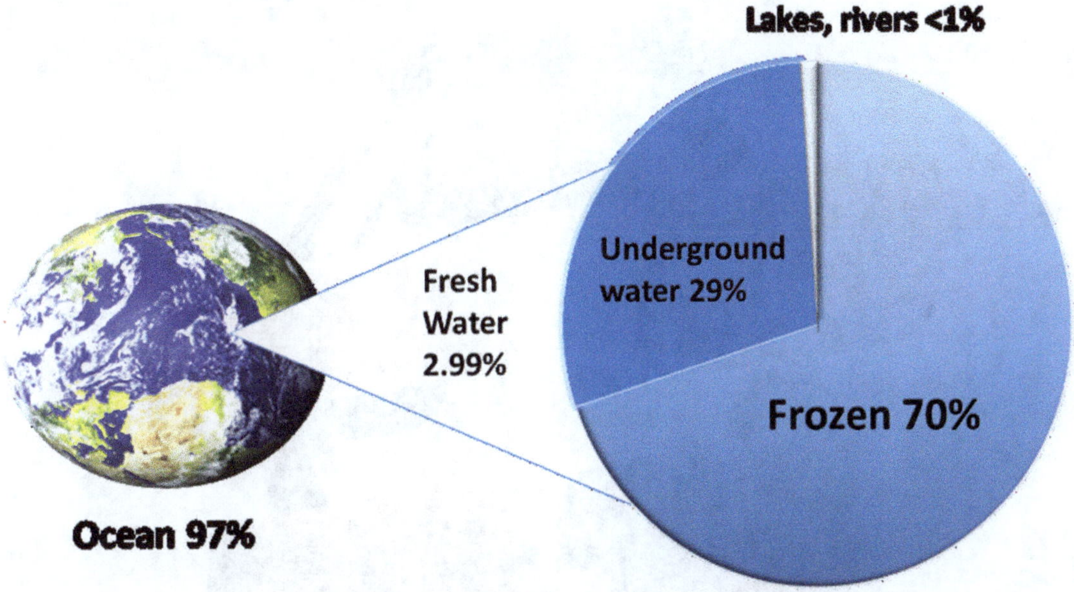

Water covers about 70% of our earth's surface is covered with water and out of this 97.5% of water is present in seas and oceans and 1.73% is present in the form of glaciers. Only 0.77% of fresh water is available on the earth's surface for human and agricultural use (Ahuja, S., 2015). According to the World Health Organization, about 1.7 million people have died as a result of water pollution, and four billion cases of various health conditions have been documented annually owing to waterborne infections (Briggs et al., 2016).

Biopolymers are biodegradable polymers that are either manufactured fully by living organisms or are derived from natural sources. according to the International Union of Pure and Applied Chemists (IUPAC), A biopolymer is a substance made up of a single type of biomacromolecule (Vert et al., 2012). A biopolymer-based nanocomposite has just entered the water treatment field. Many different types of biopolymer-based nanocomposite materials are being researched and used in water treatment systems (Crini, 2005). Water recycling has been made possible by biopolymer-based nanocomposite materials, which have a variety of applications for cleaning wastewater, making it a viable and cost-effective solution to water scarcity. The growing concern about heavy metal contamination has necessitated the development of new and better-suited sorbent materials for effective detoxification. Biopolymers like cellulose, chitosan, alginate, and carrageenan are effective in the treatment of wastewater (Nasrollahzadeh et al., 2020). Polymeric materials become more effective when they are chemically altered with grafting functionalization or chelating agents (Beaugeard et al., 2020; Gao et al., 2018; Rivas et al., 2018). This work will give an overview of current advances in biopolymer nanocomposites and their use in wastewater treatment. Due to their high toxicity and bioaccumulation through the food chain and hence in the human body, the presence of heavy metals such as Cd, Hg, Pb, Ni, Cr, Cu, and others in both freshwater sources and industrial effluent is a serious health and environmental hazard (Oyaro et al., 2007). The efficacy of biopolymers in wastewater treatment has been established. When grafting functionalization or chelating agents are used to chemically modify biopolymeric materials, they become more effective. Biopolymer-based nanocomposites are being studied and employed in water treatment

Figure 3. Biopolymer used in wastewater treatment.

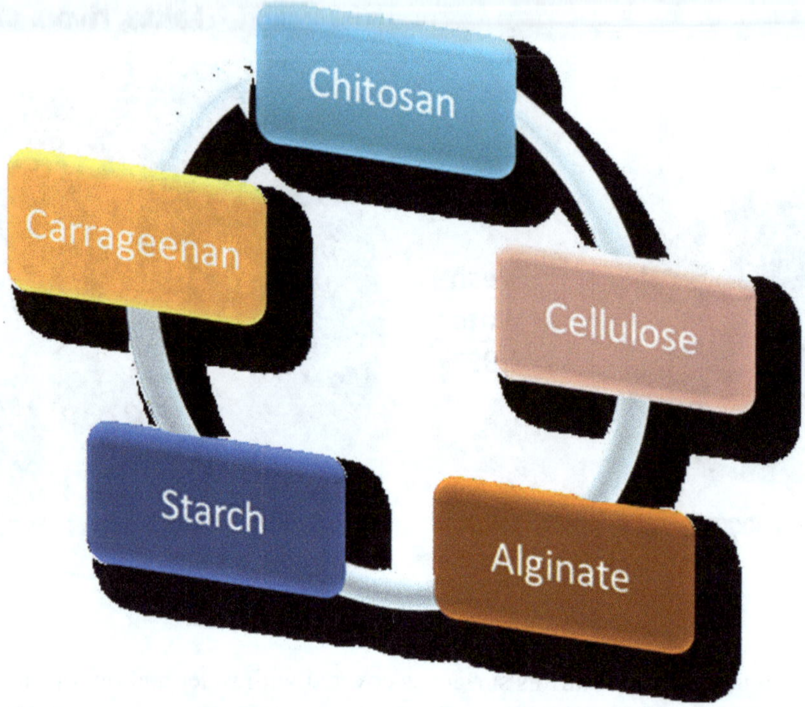

systems in a variety of ways. Biopolymer-based nanocomposites can be used for desalination, filtration, and wastewater treatment, to name a few. Biopolymer nanocomposites combine cutting-edge purification and wastewater treatment technologies with core knowledge research and development. These innovative materials contain a one-of-a-kind small separation between organic and inorganic phases that can be easily adjusted to work under certain conditions. Because of the growing concern about heavy metal contamination, new, more precisely tailored sorbent materials are needed for effective detoxification. In this field, biopolymers such as cellulose, chitosan, alginate, starch, and carrageenan have shown to be very effective in wastewater treatment. This chapter will give an overview of current advances in the field of biopolymer-based nanocomposites and their use in wastewater treatment.

There has been a surge in interest in creating biopolymer-based alternative sorbents in recent years. Synthesis of functionalized bio-based particulate matter/nanoparticles for pollutant removal in stimulated aqueous systems. Polymer analog reactions increased the efficiency of the newly generated value-added product. The synthesis approach proved successful in terms of zero waste generation. The potential of particulate matter for the desired contaminants was investigated due to the dynamic character of an aquatic system (Berber, 2020; Yeamin et al., 2021; Mohamed et al., 2018). In recent years, there has been a lot of interest in the development of various nanocomposites for wastewater applications. Nanotechnology provides a platform for generating custom-made smart multi-purpose materials with a wide range of uses in industry and commerce, as the need for environmentally friendly materials grows. A biodegradable and biocompatible polymer, whether natural or synthetic, is essential for supplying matrices for nano-filler reinforcement to generate a composite with the highest potential performance (Ali et al., 2007).

Figure 4. Water treatment processes

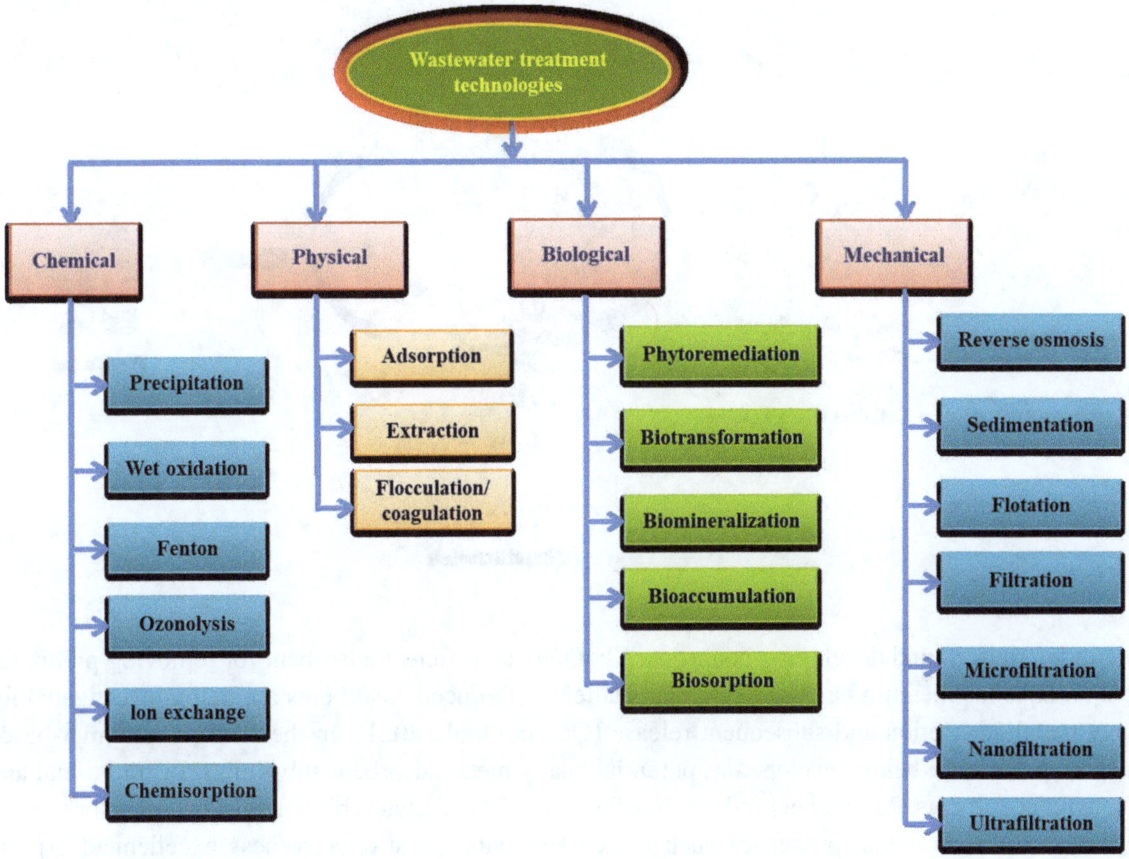

Nanotechnology is engineering at the atomic or molecular level. It is a group of enabling technologies that involves the manipulation of matter at the nanoscale to create new materials, structures, and devices (Allhoff, 2007). Chemical and physical properties of materials, such as color, magnetism, and the ability to conduct electricity, can change at this very small scale, allowing for whole new ways based on molecular self-assembly to be developed from traditional mechanisms (Nadeem Baig et al., 2021). Membrane separation (Ezugbe & Rathilal, 2020), chemical-physical, and biological treatment techniques (Ashkanani et al., 2019), coagulation/ flocculation (Al-Qodah et al., 2020), ion exchange (Wu et al., 2021), chemical precipitation (Wawrzkiewicz et al., 2017), reverse osmosis (Gao et al., 2021), photocatalysis (Almomani et al., 2019), advanced oxidation methods (Al-Bsoul et al., 2020), solvent extraction (Yadav et al., 2020) are some of the physicochemical and biological treatment technologies used in wastewater purification procedures.

Water Treatment Processes Using Biopolymer-Based Nanocomposites

Contamination by heavy metals of water has become a global environmental hazard, demanding the treatment and disposal of tainted industrial wastewater. There are a variety of biopolymer-based nanocomposites with heavy metal removal capabilities. Heavy metal pollution is a major issue that affects

Figure 5. Water treatment processes using biopolymer-based nanocomposites

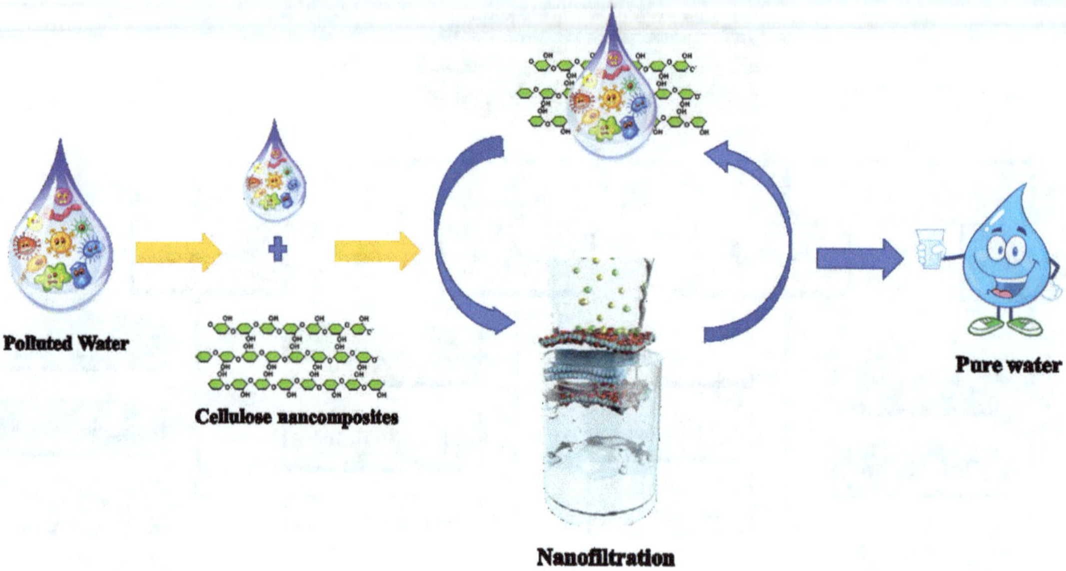

both industrialized and developing countries. Choosing an efficient adsorbent for removing pollutants from an aqueous medium has always been a challenge. Reduced adsorbents are being investigated for heavy metal adsorption and subsequent release (Qasem et. al., 2021). In the field, biopolymer-based nanocomposites are being developed as potential heavy metal adsorbent substitutes for traditional and expensive adsorbents. Researchers and policymakers have been drawn to biopolymer-based nanocomposites because of their unique properties, such as their abundance, cost-effectiveness, excellent adsorption capacity, biocompatibility, biodegradability, and simplicity of structural modification (Ahmad et al., 2015; Rendón-Villalobos et al., 2016). The use of nanoparticles as an integrated material in cellulose, alginate, and chitosan nano-based biopolymer composites for heavy metal removal from wastewater is one of the most widely used nanocomposites products ever developed. In the presence of other heavy metals, biopolymer-based nanocomposites have been shown to have high selectivity for heavy metals (Zia et al., 2020). There is now a lot of potential for producing various types of adsorbents using biopolymer-based nanocomposites, as well as assessing their practicality for wastewater treatment. Because of their acute and long-term toxicity, the presence of hazardous metal ions in the environment is a matter of concern (Ghaedi and Mosallanejad, 2013). Pesticide manufacture, mining, paper production, fossil fuel refining, textile production, and biomedicine all produce hazardous metal ions, which are found in water. As a result, industrial effluents include high levels of metal ions, posing a considerable environmental risk (Dixit et al., 2015; Bolan et al., 2014). Hg, Pb, and Cd are resistant to chemicals and environmental conditions, as well as microbial attacks. Because of these characteristics, standard biological treatments have a hard time removing harmful metal ions from effluents. Cr, Cu, Pb, Cd, Hg, and Ca, for example, are widely used in the textile and chemical sectors (Agarwal and Singh, 2017; Barakat, 2011). Toxicity, mutagenicity, and carcinogenicity are all well-known characteristics. It is necessary to find effective technologies for removing hazardous metals from wastewater.

Figure 6. Chemical structure of chitin and chitosan.

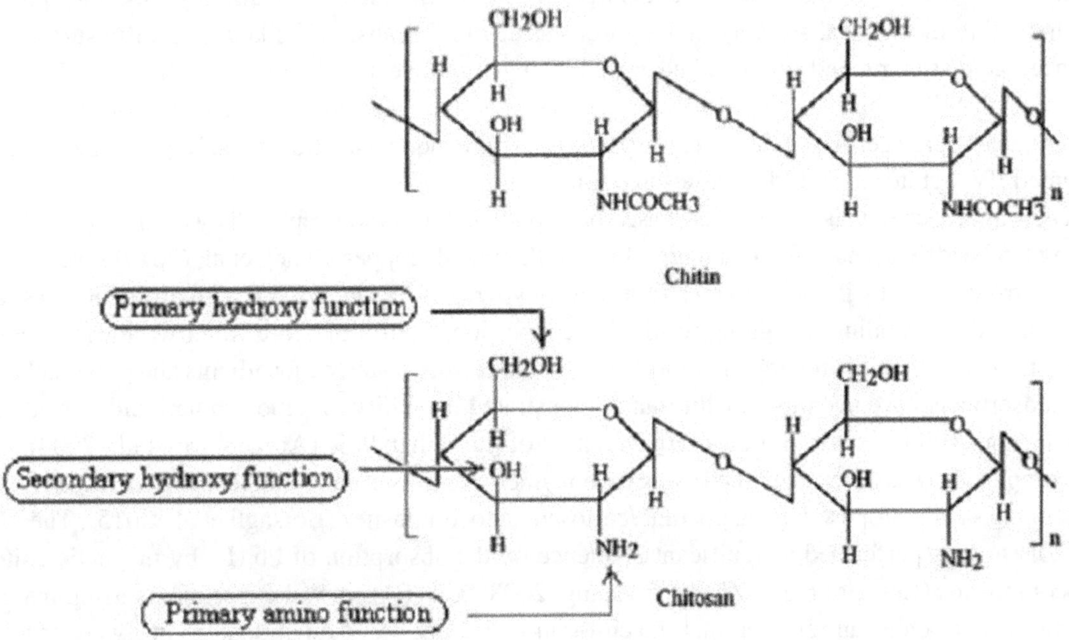

Adsorption of heavy metals on nanomaterials seems to be a promising approach for their removal from contaminated water (Fu F. and Wang Q. 2011). Many adsorbents are reported for the removal of these metal ions such as natural and modified biopolymer, cellulose, alginate, starch, chitosan, and carrageenan.

Chitosan

The development of chitosan biopolymer-based products has received a great deal of interest in current history. Chitosan is a crustacean exoskeleton-derived alkaline deacetylated chitin derivative. It is non-toxic, hydrophilic, and biocompatible (Salehi, E. t al., 2016; Kasiri, M. B. 2018). Chitosan is capable of adsorbing a range of metal ions due to its amino group's ability to behave as chelation sites (Ngah, W. S. W., et al., 2011). Chitosan has three major reactive groups that can be used to modify it chemically. They are C2-NH_2, C3-OH, and C6-OH, as shown in Fig. 6. Researchers are particularly interested in their biodegradability and antibacterial properties, which can be exploited in wastewater treatment. These are the most basic materials that can be employed as adsorbents for a variety of hazardous metal ions (Olivera, S. et al., 2016). A chitosan derivative with potential for optical properties in water treatment applications must be created to verify that chemical procedures used in converting biopolymer to usable material are safe (Bhatnagar and Sillanpää, 2009). The removal of heavy metals from wastewater streams has been extensively documented using chitosan-based adsorbents and chitosan derivatives (Wu et al. 2010). Physical and chemical approaches could be used to modify chitosan-based adsorbents, allowing them to be used in more applications. The physical structure of chitosan, which has a larger surface area, higher porosity, and smaller size, can speed up the adsorption process and reduce the time it takes to reach equilibrium (Zhang et al., 2016).

Su et al. (2016) developed nanoscale zero-valent iron/chitosan composite foams (ICCFs) for notorious arsenic which is exhibited excellent adsorption activity of arsenite remarkably superior to the two with including mechanical stability and a porous structure, because of its larger specific surface area, enhanced acid strength and improved adsorption capacity. Recently, diverse nanostructures have been coupled with chitin/chitosan-based bio(nano)composites (i.e., titanium dioxide-TiO_2, iron oxide-Fe_3O_4, etc.) (Anaya-Esparza et al., 2020; Li et al., 2010). To remove aqueous nitrate ions, Rajeswari et al. (2016) presented PVA-chitosan and PEG-chitosan composites.

A new approach to making chitosan-based bead photocatalysts was reported by Pinus et al. (2018) by coating CS beads with nano-TiO_2 and crosslinking them with copper. Zhang et al. (2019) investigated a nanoelectrode made by in-situ polymerization of polypyrrole (PPy)/CS/CNT with different mass ratios of PPy and CS, revealing intriguing results for copper ion absorption from water/wastewater. For the biosorption of pesticides and organic/inorganic pollutants from water, nanosorbents and polysaccharide-based adsorbents have been used. Chitosan demonstrated its ability to remove a herbicide (oxadiazon) from aqueous solutions with removal effectiveness of more than 90% (Arvand, M. et al., 2009). Cd^{2+}/Cr^{6+} complexation has been examined using carboxymethyl chitosan nanocomposites, whereas Pb^{2+} complexation has been proposed using goethite/chitosan nanocomposites (Borsagli et al., 2015). The –NH2 and –OH groups performed a significant influence on the absorption of Pb (II) by magnetic chitosan nanocomposites (Liu, Hu, Fang, Zhang & Zhang, 2008). Cu (II) removal experiments using magnetic nanoparticles of chitosan revealed that the chitosan –NH2 and –OH functional groups were crucial in adsorbing Cu (II) (Yuwei & Jianlong, 2011). Electrostatic attraction between oppositely charged ions was identified as the main mechanism in another research of nano chitosan–chromium (VI) interactions (Sivakami et al., 2013). By using an ion-exchange mechanism, a chitin/chitosan nano-hydroxyapatite composite could remove Cu (II) (Gandhi et al., 2011).

Cellulose

Cellulose is the most abundant biomass source and has tremendous potential. Because of its tremendous strength and rigidity, cellulose has gotten a lot of attention. Cellulose is a high-molecular-weight, renewable, biodegradable biopolymer composed of 1,4-anhydro-D-glucopyranose monomers linked together in a chain (Moon et al., 2011). Because of their improved mechanical, thermal, and biodegradation capabilities, cellulose nanocomposites can be employed as reinforcement in composite materials (Ray et al., 2015). Because of the improved mechanical, thermal, and biodegradation qualities of composites, cellulose nanofibers can be employed as reinforcement. Due to the hydrophilicity of cellulose nanocomposites, increasing their surface roughness is required for the production of composites with improved characteristics (Trache et al., 2017). The majority of cellulose is made up of crystallites, with amorphous regions with low degree of order interspersed. These cellulose crystallites have attracted a lot of attention in the last two decades due to their remarkable mechanical characteristics and nanoscale dimensions, as well as their high surface area-to-volume ratios and potential for surface functionalization (Shatkin et al., 2013).

These characteristics make them ideal for detecting hazardous metal ions, for example wood, plants (wheat, maize, potato, cotton, and corn), algae, tunicates, and bacteria are some of the sources of cellulose. To ensure that chemical techniques used in converting biopolymer to usable material are safe, a cellulose derivative with potential for water treatment applications must be developed (Carpenter et al., 2015). Cellulosic nanocomposites are structured in micro fibrils in plants and are sequentially combined

Figure 7. Structure of cellulose

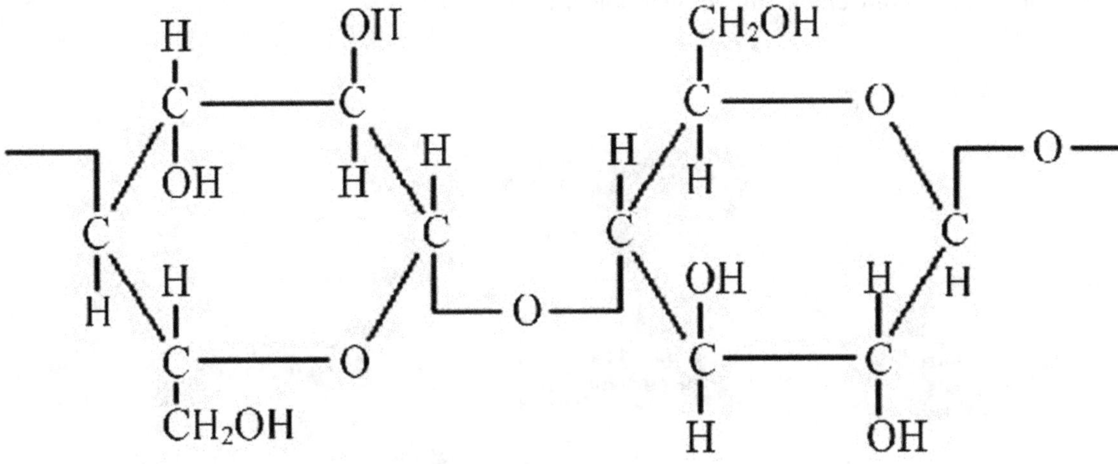

with other chemical components such as hemicelluloses, lignin, and pectin. For polymeric materials, it has a lot of strength and reinforcing potential. The chemical treatment of acid hydrolysis and the physical/mechanical treatment of the pressure homogenizer can produce cellulosic nanocomposites and micro fibrils. Because cellulose nanocomposites are renewable and biodegradable, they are being used as reinforcement for the sorption of organic contaminants (O'Connell et al., 2008). Varghese (2019) demonstrated the dye and heavy metal adsorption capacity of several lignocellulosic adsorbents, both before and after physical and chemical treatment.

Cellulose nanocomposite materials adsorbing various toxic metal ions (Jamshaid et al., 2017). Nano-sized cellulose and chitosan–based polymers are being considered as a possible wastewater treatment solution. Unlike standard submicron and micron–sized cellulose and chitosan materials, which have severe limits in terms of adsorption behaviour, nano–sized counterparts can offer appealing qualities such as smaller size, increased porosities leading to bigger surface area, and the absence of internal diffusion and quantum size effects (Lam E. et al., 2012). As an adsorbent for the removal of Cr, amino-functionalized magnetic cellulose nanocomposite was used (VI) (Sun X. et al., 2014). A study of Pb^{2+} biosorption using cellulosic nanofibers found that 94.2% of Pb^{2+} was absorbed (Kardam et al., 2012). A nanocellulosic fiber–modified carbon paste electrode was used in another investigation to determine Pb^{2+} and Cd^{2+} (Rajawat et al., 2013). Ultrahigh effective and reversible removal of Cu (II) ions from water using multi-wall perforated nanocellulose-based polyethylenimine aerogels (Mo L. et. al., 2019). The focus is on developing sensor materials for detecting heavy metal ions that are quick, simple, and easy to use by non-experts. The chemical interactions of dangerous metal ions such as mercury with nanomaterials, as well as the effect of metal ion concentration and medium pH on their interaction (Johari, 2016).

Dyes

Azo dyes are the most often used type of dye in business. As a result, industrial effluents include high levels of dyes, posing a significant environmental risk. Azo dyes are chemically stable, resistant to environmental conditions, and microbially resistant (Yan et al., 2020). Standard biological treatments are unable to remove the colours from effluents due to these properties. For example, methyl orange dye has a wide range of uses in the textile, automotive, and chemical industries. The properties of toxicity, mutagenicity, and carcinogenicity are all well-known.

Figure 8. Types of dyes

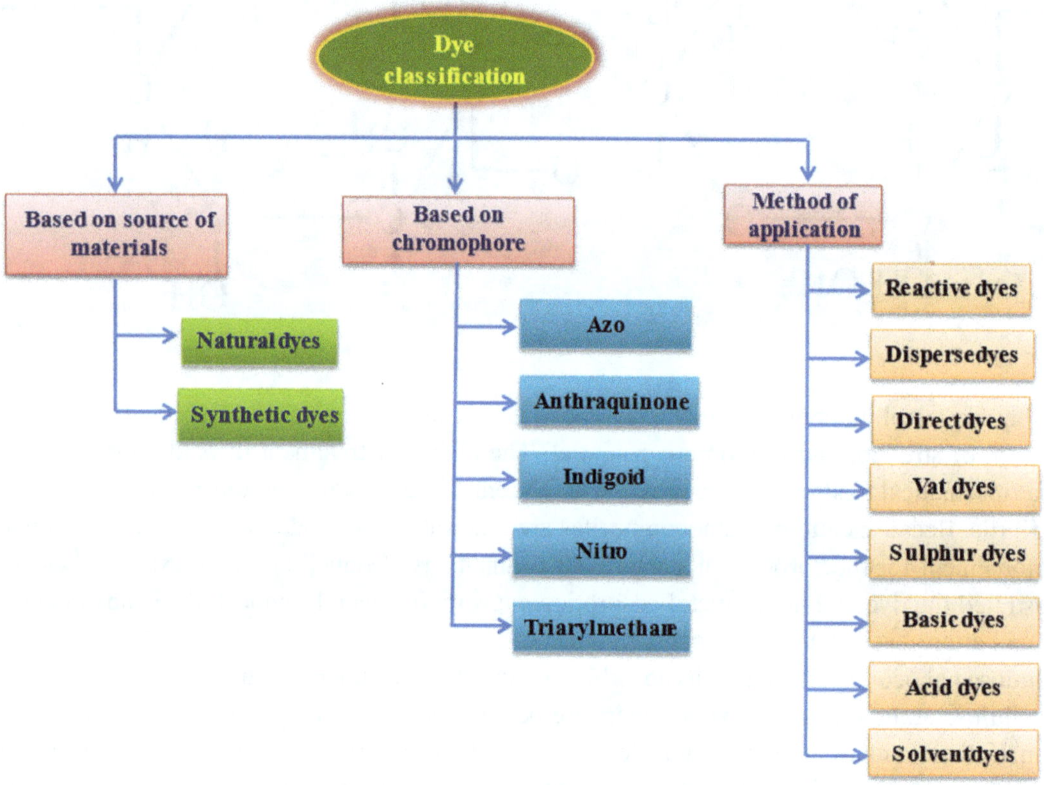

CONCLUSION

Biopolymer-based nanocomposites have gained increasing attention due to the high strength and stiffness, biodegradability and renewability and their production and application in development of polymer composites. The use of cellulose nanofibers in composites development is a relatively recent field of study. Because of the improved mechanical, thermal, and biodegradation properties of composites, biopolymer macro- and nanofibers can be employed as reinforcement in composite materials. Because biopolymers are hydrophilic, increasing their surface roughness is required for the production of composites with improved characteristics. The purification of wastewater by heavy metals utilising various ways has been described in this article. Heavy metal contamination is a serious problem that affects both developed and developing countries. Reduced adsorbents are being investigated for heavy metal adsorption and subsequent release. Biopolymers are preferred because of their natural source and abundance. Novel functional nanocomposites materials with integrated nanoparticles are at the heart of nano-based water treatment. One of the most cost-effective nanocomposites products ever devised is recently created nano-based polymer composites using nanoparticles as an integrated material and its applications in heavy metal removal from wastewater. There are a variety of polymer-based nanocomposites having heavy metal removal properties built in. The hybrid composite surpasses typical heavy metal removal technologies when paired with metal nanoparticles.

ACKNOWLEDGMENT

Author is grateful to University Grants Commission, New Delhi, India and Department of Chemistry, Dr. Harisingh Gour Central University, Sagar for financial support.

REFERENCES

Agarwal, M., & Singh, K. (2017). Heavy metal removal from wastewater using various adsorbents: A review. *Journal of Water Reuse and Desalination*, *7*(4), 387–419.

Ahmad, A., Mohd-Setapar, S. H., Chuong, C. S., Khatoon, A., Wani, W. A., Kumar, R., & Rafatullah, M. (2015). Recent advances in new generation dye removal technologies: Novel search for approaches to reprocess wastewater. *RSC Advances*, *5*(39), 30801–30818. doi:10.1039/C4RA16959J

Ahmad, M., Ahmed, S., Swami, B. L., & Ikram, S. (2015). Adsorption of heavy metal ions: Role of chitosan and cellulose for water treatment. *Langmuir*, *79*, 109–155.

Ahuja, S. (2015). Overview of Global Water Challenges and Solutions. *ACS Symposium Series*, *1206*, 1–25. doi:10.1021/bk-2015-1206.ch001

Al-Bsoul, A., Al-Shannag, M., Tawalbeh, M., Al-Taani, A. A., Lafi, W. K., Al-Othman, A., & Alsheyab, M. (2020). Optimal conditions for olive mill wastewater treatment using ultrasound and advanced oxidation processes. *The Science of the Total Environment*, *700*, 134576.

Al-Qodah, Z., Tawalbeh, M., Al-Shannag, M., Al-Anber, Z., & Bani-Melhem, K. (2020). Combined electrocoagulation processes as a novel approach for enhanced pollutants removal: A state-of-the-art review. *The Science of the Total Environment*, *744*, 140806.

Ali, I., & Gupta, V. K. (2007). Advances in water treatment by adsorption technology. *Nature Protocols*, *1*(6), 2661–2667. doi:10.1038/nprot.2006.370 PMID:17406522

Allhoff, F. (2007). On the autonomy and justification of nanoethics. *NanoEthics*, *1*(3), 185–210. doi:10.100711569-007-0018-3

Almomani, F., Al Ketife, A., Judd, S., Shurair, M., Bhosale, R. R., Znad, H., & Tawalbeh, M. (2019). Impact of CO2 concentration and ambient conditions on microalgal growth and nutrient removal from wastewater by a photobioreactor. *The Science of the Total Environment*, *662*, 662–671.

Anaya-Esparza, L. M., Ruvalcaba-Gómez, J. M., Maytorena-Verdugo, C. I., González-Silva, N., Romero-Toledo, R., Aguilera-Aguirre, S., ... Montalvo-González, E. (2020). Chitosan-TiO2: A versatile hybrid composite. *Materials (Basel)*, *13*(4), 811.

Arvand, M., Latify, L., Tajmehri, H., Yagubov, A. I., Nuriyev, A. N., Pourhabib, A., ... Abolhassani, M. R. (2009). Comparative study for the removal of oxadiazon from aqueous solutions by adsorption on chitosan and activated carbon. *Analytical Letters*, *42*(6), 856–869.

Ashkanani, A., Almomani, F., Khraisheh, M., Bhosale, R., Tawalbeh, M., & AlJaml, K. (2019). Bio-carrier and operating temperature effect on ammonia removal from secondary wastewater effluents using moving bed biofilm reactor (MBBR). *The Science of the Total Environment*, *693*, 133425.

Baig, N., Kammakakam, I., & Falath, W. (2021). Nanomaterials: A review of synthesis methods, properties, recent progress, and challenges. *Mater. Adv.*, *2*(6), 1821–1871. doi:10.1039/D0MA00807A

Barakat, M. A. (2011). New trends in removing heavy metals from industrial wastewater. *Arabian Journal of Chemistry*, *4*(4), 361–377.

Bayramoglu, G., Kunduzcu, G., & Arica, M. Y. (2020). Preparation and characterization of strong cation exchange terpolymer resin as effective adsorbent for removal of disperse dyes. *Polymer Engineering and Science*, *60*(1), 192–201. doi:10.1002/pen.25272

Berber, M. R. (2020). Current advances of polymer composites for water treatment and desalination. *Journal of Chemistry*.

Bharagava, R. N., & Chandra, R. (2010). Biodegradation of the major color containing compounds in distillery wastewater by an aerobic bacterial culture and characterization of their metabolites. *Biodegradation*, *21*(5), 703–711. doi:10.100710532-010-9336-1 PMID:20146090

Bharagava, R. N., Saxena, G., Mulla, S. I., & Patel, D. K. (2018). Characterization and identification of recalcitrant organic pollutants (ROPs) in tannery wastewater and its phytotoxicity evaluation for environmental safety. *Archives of Environmental Contamination and Toxicology*, *75*(2), 259–272. doi:10.100700244-017-0490-x PMID:29243159

Bhatnagar, A., & Sillanpää, M. (2009). Applications of chitin-and chitosan-derivatives for the detoxification of water and wastewater—A short review. *Advances in Colloid and Interface Science*, *152*(1-2), 26–38.

Bolan, N., Kunhikrishnan, A., Thangarajan, R., Kumpiene, J., Park, J., Makino, T., ... Scheckel, K. (2014). Remediation of heavy metal (loid) s contaminated soils–to mobilize or to immobilize? *Journal of Hazardous Materials*, *266*, 141–166.

Borsagli, F. G. M., Mansur, A. A., Chagas, P., Oliveira, L. C., & Mansur, H. S. (2015). O-carboxymethyl functionalization of chitosan: Complexation and adsorption of Cd (II) and Cr (VI) as heavy metal pollutant ions. *Reactive & Functional Polymers*, *97*, 37–47.

Briggs, A. M., Cross, M. J., Hoy, D. G., Blyth, F. H., Woolf, A. D., & March, L. (2016). Musculoskeletal Health Conditions Represent a Global Threat to Healthy Aging: A Report for the 2015World Health Organization World Report on Ageing and Health. *The Gerontologist*, *56*(Suppl 2), 243–255. doi:10.1093/geront/gnw002 PMID:26994264

Carpenter, A. W., de Lannoy, C. F., & Wiesner, M. R. (2015). Cellulose nanomaterials in water treatment technologies. *Environmental Science & Technology*, *49*(9), 5277–5528.

Crini, G. (2005). Recent developments in polysaccharide-based materials used as adsorbents in wastewater treatment. *Progress in Polymer Science*, *30*(1), 38–70. doi:10.1016/j.progpolymsci.2004.11.002

Dixit, R., Malaviya, D., Pandiyan, K., Singh, U. B., Sahu, A., Shukla, R., ... Paul, D. (2015). Bioremediation of heavy metals from soil and aquatic environment: An overview of principles and criteria of fundamental processes. *Sustainability*, *7*(2), 2189–2212.

Ezugbe, E. O., & Rathilal, S. (2020). Membrane technologies in wastewater treatment: A review. *Membranes*, •••, 10.

Fu, F., & Wang, Q. (2011). Removal of heavy metal ions from wastewaters: A review. *Journal of Environmental Management*, *92*(3), 407–418.

Gandhi, M. R., Kousalya, G., & Meenakshi, S. (2011). Removal of copper (II) using chitin/chitosan nano-hydroxyapatite composite. *International Journal of Biological Macromolecules*, *48*(1), 119–124.

Gao, B., Iftekhar, S., Srivastava, V., Doshi, B., & Sillanpää, M. (2018). Insights into the generation of reactive oxygen species (ROS) over polythiophene/ZnIn 2 S 4 based on different modification processing. *Catalysis Science & Technology*, *8*(8), 2186–2194. doi:10.1039/C8CY00303C

Gao, L., Liu, G., Zamyadi, A., Wang, Q., & Li, M. (2021). Life-cycle cost analysis of a hybrid algae-based biological desalination–low pressure reverse osmosis system. *Water Research*, *195*, 116957.

Ghaedi, M., & Mosallanejad, N. (2013). Removal of heavy metal ions from polluted waters by using of low cost adsorbents. *Journal of Chemical Health Risks*, *3*, 7–21.

Houa, Y., Yan, S., Huang, G., Yang, Q., Huang, S., & Cai, J. (2020). Fabrication of N-doped carbons from waste bamboo shoot shell with high removal efficiency of organic dyes from water. *Bioresource Technology*, *303*, 122939. doi:10.1016/j.biortech.2020.122939 PMID:32045864

Jamshaid, A., Hamid, A., Muhammad, N., Naseer, A., Ghauri, M., Iqbal, J., ... Shah, N. S. (2017). Cellulose-based Materials for the Removal of Heavy Metals from Wastewater–An Overview. *ChemBioEng Reviews*, *4*(4), 240–256.

Jassby, D., Cath, T. Y., & Buisson, H. (2018). The role of nanotechnology in industrial water treatment. *Nature Nanotechnology*, *13*(8), 670–672. doi:10.103841565-018-0234-8 PMID:30082807

Johari, K., Saman, N., Song, S. T., Chin, C. S., Kong, H., & Mat, H. (2016). Adsorption enhancement of elemental mercury by various surface modified coconut husk as eco-friendly low-cost adsorbents. *International Biodeterioration & Biodegradation*, *109*, 45–52.

Kardam, A., Raj, K. R., Arora, J. K., & Srivastava, S. (2012). Artificial neural network modeling for biosorption of Pb (II) ions on nanocellulose fibers. *BioNanoScience*, *2*(3), 153–160.

Kasiri, M. B. (2018) Application of chitosan derivatives as promising adsorbents for treatment of textile wastewater. *The Impact and Prospects of Green Chemistry for Textile Technology*, 417-469.

Khan, I., Saeed, K., & Khan, I. (2019). Nanoparticles: Properties, applications and toxicities. *Arabian Journal of Chemistry*, *12*(7), 908–931. doi:10.1016/j.arabjc.2017.05.011

Kishor, R., Bharagava, R. N., & Saxena, G. (2018). Industrial wastewaters: the major sources of dye contamination in the environment, ecotoxicological effects, and bioremediation approaches. In *Recent advances in environmental management* (pp. 1–25). CRC Press. doi:10.1201/9781351011259-1

Kumar, S., Ahlawat, W., Bhanjana, G., Heydarifard, S., Nazhad, M. M., & Dilbaghi, N. (2014). Nanotechnology-based water treatment strategies. *Journal of Nanoscience and Nanotechnology*, *14*(2), 1838–1858. doi:10.1166/jnn.2014.9050 PMID:24749460

Lam, E., Male, K. B., Chong, J. H., Leung, A. C., & Luong, J. H. (2012). Applications of functionalized and nanoparticle-modified nanocrystalline cellulose. *Trends in Biotechnology*, *30*(5), 283–290.

Li, W., Xiao, L., & Qin, C. (2010). The characterization and thermal investigation of chitosan-Fe3O4 nanoparticles synthesized via a novel one-step modifying process. *Journal of Macromolecular Science. Part A*, *48*(1), 57–64.

Liu, X., Hu, Q., Fang, Z., Zhang, X., & Zhang, B. (2008). Magnetic chitosan nanocomposites: A useful recyclable tool for heavy metal ion removal. *Langmuir*, *25*(1), 3–8.

Lytle, D. A., Sorg, T., Wang, L., & Chen, A. (2014). The accumulation of radioactive contaminants in drinking water distribution systems. *Water Research*, *50*, 396–407. doi:10.1016/j.watres.2013.10.050 PMID:24275108

Mo, L., Pang, H., Tan, Y., Zhang, S., & Li, J. (2019). 3D multi-wall perforated nanocellulose-based polyethylenimine aerogels for ultrahigh efficient and reversible removal of Cu (II) ions from water. *Chemical Engineering Journal*, *378*, 122157.

Mohamed, R. R., Abu Elella, M. H., Sabaa, M. W., & Saad, G. R. (2018). Synthesis of an efficient adsorbent hydrogel based on biodegradable polymers for removing crystal violet dye from aqueous solution. *Cellulose (London, England)*, *25*(11), 6513–6529. doi:10.100710570-018-2014-x

Moon, R. J., Martini, A., Nairn, J., Simonsen, J., & Youngblood, J. (2011). Cellulose nanomaterials review: Structure, properties and nanocomposites. *Chemical Society Reviews*, *40*(7), 3941–3994.

Nasrollahzadeh, M., Sajjadi, M., Iravani, S., & Varma, R. S. (2021). Starch, cellulose, pectin, gum, alginate, chitin and chitosan derived (nano) materials for sustainable water treatment: A review. *Carbohydrate Polymers*, *251*, 116986. doi:10.1016/j.carbpol.2020.116986 PMID:33142558

Ngah, W. S. W., Teong, L. C., & Hanafiah, M. A. K. M. (2011). Adsorption of dyes and heavy metal ions by chitosan composites: A review. *Carbohydrate Polymers*, *83*(4), 1446–1456.

O'Connell, D. W., Birkinshaw, C., & O'Dwyer, T. F. (2008). Heavy metal adsorbents prepared from the modification of cellulose: A review. *Bioresource Technology*, *99*, 6709–6724.

Olivera, S., Muralidhara, H. B., Venkatesh, K., Guna, V. K., Gopalakrishna, K., & Kumar, Y. (2016). Potential applications of cellulose and chitosan nanoparticles/composites in wastewater treatment: A review. *Carbohydrate Polymers*, *153*, 600–618.

Oyaro, N., Ogendi, J., Murago, E. N., & Gitonga, E. (2007). *The contents of Pb, Cu, Zn and Cd in meat in Nairobi*. Academic Press.

Pincus, L. N., Melnikov, F., Yamani, J. S., & Zimmerman, J. B. (2018). Multifunctional photoactive and selective adsorbent for arsenite and arsenate: Evaluation of nano titanium dioxide-enabled chitosan cross-linked with copper. *Journal of Hazardous Materials*, *358*, 145–154.

Qasem, N. A., Mohammed, R. H., & Lawal, D. U. (2021). Removal of heavy metal ions from wastewater: A comprehensive and critical review. *Npj Clean Water*, *4*(1), 1–15.

Rajasekhar, B., Nambi, I. M., & Govindarajan, S. K. (2018). Human health risk assessment of ground water contaminated with petroleum PAHs using Monte Carlo simulations: A case study of an Indian metropolitan city. *Journal of Environmental Management*, *205*, 183–191. doi:10.1016/j.jenvman.2017.09.078 PMID:28985597

Rajawat, D. S., Kardam, A., Srivastava, S., & Satsangee, S. P. (2013). Nanocellulosic fibermodified carbon paste electrode for ultra trace determination of Cd (II) and Pb (II) in aqueous solution. *Environmental Science and Pollution Research International*, *20*(5), 3068–3076.

Rajeswari, A., Amalraj, A., & Pius, A. (2016). Adsorption studies for the removal of nitrate using chitosan/PEG and chitosan/PVA polymer composites. *Journal of Water Process Engineering*, *9*, 123–134.

Ray, P. Z., & Shipley, H. J. (2015). Inorganic nano-adsorbents for the removal of heavy metals and arsenic: A review. *RSC Advances*, *5*(38), 29885–29907.

Rendón-Villalobos, R., Ortíz-Sánchez, A., Tovar-Sánchez, E., & Flores-Huicochea, E. (2016). The role of biopolymers in obtaining environmentally friendly materials. *Composites from renewable and sustainable materials*, 151-159.

Ritter, K. S., Sibley, P., Hall, K., Keen, P., Mattu, G., & Linton, B. L. (2002). Sources, pathways, and relative risks of contaminants in surface water and groundwater: A perspective prepared for the Walkerton inquiry. *Journal of Toxicology and Environmental Health. Part A.*, *65*(1), 1–142. doi:10.1080/152873902753338572 PMID:11809004

Rivas, B. L., Urbano, B. F., & Sánchez, J. (2018). Water-soluble and insoluble polymers, nanoparticles, nanocomposites and hybrids with ability to remove Hazardous inorganic pollutants in water. *Frontiers in Chemistry*, *6*, 320. doi:10.3389/fchem.2018.00320 PMID:30109224

Rodriguez-Proteau, R., & Grant, R. L. (2006). Toxicity evaluation and human health risk assessment of surface and ground water contaminated by recycled hazardous waste materials. Handb. Environ. Chem. Vol. 5 Water Pollut., 5 F, 133–189.

Salehi, E., Daraei, P., & Shamsabadi, A. A. (2016). A review on chitosan-based adsorptive membranes. *Carbohydrate Polymers*, *152*, 419–432.

Saxena, G., Purchase, D., Mulla, S. I., Saratale, G. D., & Bharagava, R. N. (2019). Phytoremediation of heavy metal-contaminated sites: Eco-environmental concerns, field studies, sustainability issues, and future prospects. *Reviews of Environmental Contamination and Toxicology*, *249*, 71–131. PMID:30806802

Shatkin, J. A., Wegner, T. H., & Neih, W. (2013). Incorporating life-cycle thinking into risk assessment for nanoscale materials: Case study of nanocellulose. *2013 TAPPI International Conference on Nanotechnology for Renewable Materials*.

Sivakami, M., Gomathi, T., Venkatesan, J., Jeong, H.-S., Kim, S.-K., & Sudha, P. (2013). Preparation and characterization of nano chitosan for treatment wastewaters. *International Journal of Biological Macromolecules*, *57*, 204–212.

Su, F. C., Zhou, H. J., Zhang, Y. X., & Wang, G. Z. (2016). Three-dimensional honeycomb-like structured zero-valent iron/chitosan composite foams for effective removal of inorganic arsenic in water. *Journal of Colloid and Interface Science*, *478*, 421–429.

Sun, X., Yang, L., Li, Q., Zhao, J., Li, X., Wang, X., & Liu, H. (2014). Amino-functionalized magnetic cellulose nanocomposite as adsorbent for removal of Cr (VI): Synthesis and adsorption studies. *Chemical Engineering Journal*, *241*, 175–183.

Trache, D., Hussin, M. H., Haafiz, M. M., & Thakur, V. K. (2017). Recent progress in cellulose nanocrystals: Sources and production. *Nanoscale*, *9*(5), 1763–1786.

Varghese, A. G., Paul, S. A., & Latha, M. S. (2019). Remediation of heavy metals and dyes from wastewater using cellulose-based adsorbents. *Environmental Chemistry Letters*, *17*(2), 867–877.

Vert, M., Doi, Y., Hellwich, K. H., Hess, M., Hodge, P., Kubisa, P., Rinaudo, M., & Schué, F. (2012). Terminology for biorelated polymers and applications (IUPAC Recommendations 2012). *Pure and Applied Chemistry*, *84*(2), 377–410. doi:10.1351/PAC-REC-10-12-04

Wawrzkiewicz, M., Wiśniewska, M., Wołowicz, A., Gun'ko, V. M., & Zarko, V. I. (2017). Mixed silica-alumina oxide as sorbent for dyes and metal ions removal from aqueous solutions and wastewaters. *Microporous and Mesoporous Materials*, *250*, 128–147. doi:10.1016/j.micromeso.2017.05.016

Wu, F. C., Tseng, R. L., & Juang, R. S. (2010). A review and experimental verification of using chitosan and its derivatives as adsorbents for selected heavy metals. *Journal of Environmental Management*, *91*, 798–806.

Wu, J., Wang, T., Wang, J., Zhang, Y., & Pan, W. P. (2021). A novel modified method for the efficient removal of Pb and Cd from wastewater by biochar: Enhanced the ion exchange and precipitation capacity. *The Science of the Total Environment*, *754*, 142150. doi:10.1016/j.scitotenv.2020.142150 PMID:32920404

Yadav, M., Gupta, R., Arora, G., Yadav, P., Srivastava, A., & Sharma, R. K. (2020). Current status of heavy metal contaminants and their removal/recovery techniques. In Contaminants in Our Water: Identification and Remediation Methods (pp. 41-64). American Chemical Society.

Yan, L., Liu, B., Li, W., Zhao, T., Wang, Y., & Zhao, Q. (2020). Multiscale cellulose based self-assembly of hierarchical structure for photocatalytic degradation of organic pollutant. *Cellulose (London, England)*, *27*(9), 5241–5253.

Yeamin, M. B., Islam, M. M., Chowdhury, A. N., & Awual, M. R. (2021). Efficient encapsulation of toxic dyes from wastewater using several biodegradable natural polymers and their composites. *Journal of Cleaner Production*, *291*, 125920. doi:10.1016/j.jclepro.2021.125920

Yuwei, C., & Jianlong, W. (2011). Preparation and characterization of magnetic chitosan nanoparticles and its application for Cu (II) removal. *Chemical Engineering Journal*, *168*(1), 286–292.

Zhang, L., Zeng, Y. X., & Cheng, Z. J. (2016). Removal of heavy metal ions using chitosan and modified chitosan: A review. *Journal of Molecular Liquids*, *214*, 175–191.

Zhang, Y., Wu, B., Xu, H., Liu, H., Wang, M., He, Y., & Pan, B. (2016). Nanomaterials-enabled water and wastewater treatment. *NanoImpact*, *3*, 22–39. doi:10.1016/j.impact.2016.09.004

Zhang, Y. J., Xue, J. Q., Li, F., & Dai, J. Z. (2019). Preparation of polypyrrole/chitosan/carbon nanotube composite nano-electrode and application to capacitive deionization process for removing Cu2+. *Chemical Engineering and Processing-Process Intensification*, *139*, 121–129.

Zia, Z., Hartland, A., & Mucalo, M. R. (2020). Use of low-cost biopolymers and biopolymeric composite systems for heavy metal removal from water. *International Journal of Environmental Science and Technology*, *17*(10), 4389–4406.

Chapter 8
Photodegradation of Thymol Blue Under Visible Light by the Novel Photocatalyst Titania/PAni/GO

Azad Kumar
https://orcid.org/0000-0001-5162-7768
M.L.K. P.G. College, India

ABSTRACT

The TiO2/PAni and TiO2/PAni/GO nanocomposites were prepared by one-step in situ oxidative polymerization of aniline hydrochloride using ammonium persulphate as an oxidant in the presence of powder of TiO2 nanoparticles cooled in an ice bath. The obtained nanocomposites were characterized by XRD, TEM, SEM, BET, FTIR, and DRS. The obtained results showed that TiO2 nanoparticles have been encapsulated by PAni. The FTIR characterization confirms that the TiO2/GO molecules are well combined with polyaniline structure. The maximum photodegradation of Thymol blue was found in TiO2/PAni/GO at 25 ppm concentration of dye, 1600 mg/L amount of photocatalyst, pH 7, and 120 min irradiation of visible light. Hence, the photocatalytic activity of Titania has been increased by the coating of PAni and Graphene oxide.

1. INTRODUCTION

Titania can be used to destroy a wide range of organic pollutants present in water and aqueous effluents. The TiO_2 absorbs only 5% UV radiation of the solar light spectrum due to the large band gap energy and recombination of electron (e-) - hole (h+) pairs (Li et al 2013, Moghaddam & Nasirian 2014, Su 2017). Consequently, hundreds of TiO_2 variants and other oxide/non-oxides have been developed and tested in propose to conquer the recombination process and reduction of band gap energy (Clifford et al 2017). It is believed that availability of visible light absorbing photocatalysts would largely solve the technological problems of photo reactor considerations. For harnessing of sunlight the band gap energy

DOI: 10.4018/978-1-6684-4553-2.ch008

should be reduced (Nasirian et al 2014). The coating of conducting polymers on TiO_2 surface is improve the visible light photoactivity and electron transfer performance; e.g., Polyaniline/TiO_2 (Su & Gan 2012), polypyrrole/ TiO_2 (Arora 2015) and polythiophene/TiO_2 (Chowdhury & Balasubramanian 2014). Many published reports focussed on the preparation and photo catalytic studies on nanocomposites consisting of polyaniline and TiO_2 (PAni/TiO_2) (Al-Hussaini et al 2016, Patil et al 2015, Lai et al 2010). morphology of TiO_2/PAni and TiO_2/PAni/GO nanocomposites. The prepared nanocomposites were used as photocatalyst for the degradation of Thymol blue dye at different parameters i.e. concentration, dose, time and pH. The TiO_2/PAni/GO nanocomposite shows the highest activity amongst all the synthesized materials.

2. METHODS AND MATERIALS

2.1. Synthesis of TiO_2/PAni and TiO_2/PAni/GO nanocomposite

10 mL of CCl_4 and 4.0 mL of TTIP were placed in a beaker to which 1 mL of aniline were added. The entire system was stirred constantly on an ice bath. To the above dispersion of aniline, the solution of oxidant (0.5 M APS in 100 mL of 1M HCl) was added drop-wise, which simultaneous initiated the polymerization of aniline and the synthesis of TiO_2. The reaction mixture soon turned into greenish black slurry, which was filtered and washed with water and acetone to remove the excess APS and PAni oligomers. Same method was used for the synthesis of TiO_2/PAni/GO only added 60 mg of graphene oxide with aniline (Prasad et al 2006).

3. RESULTS AND DISCUSSION

3.1. XRD

The XRD patterns of TiO_2, PAni, TiO_2/PAni and TiO_2/PAni/GO nanocomposite are showing in Fig.1. In Fig. 1(a), a series of characteristic peaks: $2\theta = 25.32°$ (101), 37.86° (004), 48.06° (200), 55.09° (211) and 62.75° (204) are observed, which correspond to the tetragonal anatase phase of TiO2 (JCPDS file No: 86-1157). The XRD pattern of PAni in Fig. 1(b) resembles the Emeraldine Salt crystalline form of PAni. The XRD patterns of the TiO_2/PAni and TiO_2/PAni nanocomposites in Fig. 1(c and d) show characteristic peaks of PAni and TiO_2 (Wang et al 2016).

3.2. SEM and TEM analysis

The SEM images of TiO_2, PAni, TiO_2/PAni and TiO_2/PAni/GO are indicating that the particle morphology is in spherical shape and in nanodiamension Shown in Fig.2. The TiO_2 molecule is agglomerate with PAni to form chips like structures which are partially spherical and disc shape. The TEM images of TiO_2, PAni, TiO_2/PAni and TiO_2/PAni/GO are shown in Fig 2. It was morphologically spherical with the average particle size of 10 nm.

Figure 1. XRD Pattern of (a) TiO_2 (b) PAni (c) TiO_2/PAni (d) TiO_2/PAni/GO

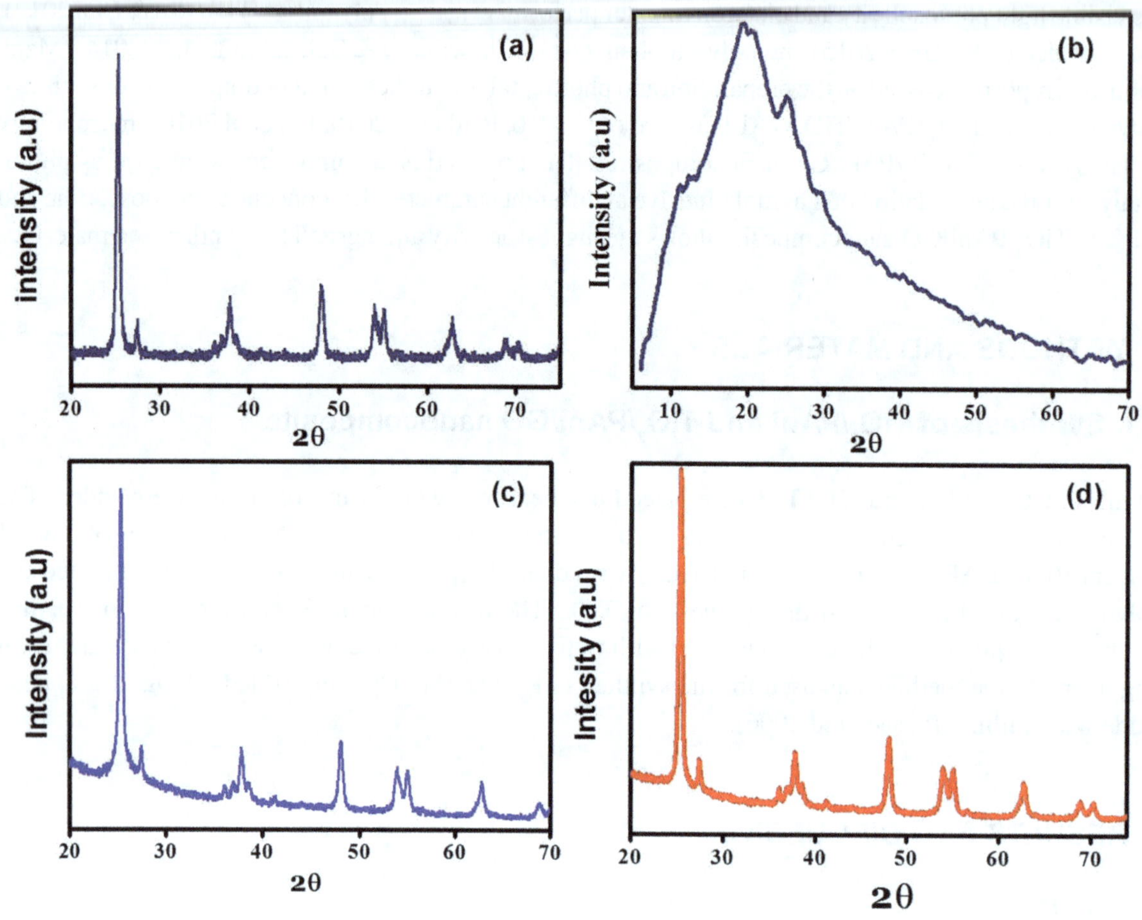

3.3. Bruner Earner Taylor (BET)

The specific surface areas (from BET and Surface area, pore volume and the pore radius of the TiO_2, TiO_2/PAni and TiO_2/PAni/GO nanocomposite are showing in Table 1. The surface area of TiO_2, TiO_2/PAni and TiO_2/PAni/GO nanocomposites were observed 37. 52, 76.68 and 96.24 m^2/g respectively. There is an increase in pore volume (Vp) of TiO_2, TiO_2/PAni and TiO_2/PAni/GO nanocomposite and pore radius is decreased. From these results, it may be concluded that the high surface area of the TiO_2/PAni/GO nanocomposite may favour rapid electron transport and high ion diffusion, allowing improved photochemical performance. Moreover, the BET surface areas increased remarkably in the TiO_2/PAni/GO nanocomposite, which suggests that TiO_2 is well intercalated in PAni matrix and may also provide direct conduction pathway for electrons. The formation of TiO_2 with PAni by co-deposition oxidation synthesis resulted in the generation of well dispersed TiO_2 in PAni Matrix giving one TiO_2/PAni system with a unique set of properties.

Figure 2a. SEM photographs of (a) TiO_2 (b) PAni (c) PAni/TiO_2 (d) PAni/TiO_2/GO

Figure 2b. TEM photographs of (a) TiO_2 (b) PAni (c) PAni/TiO_2 (d) PAni/TiO_2/GO

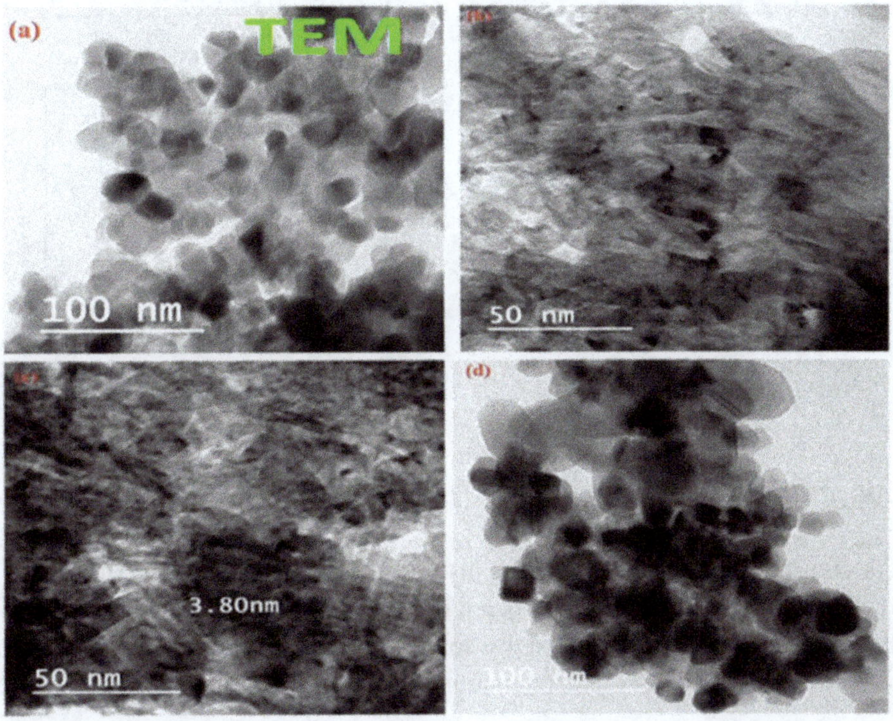

Table 1 showing the surface area, pore volume, pore radius of the TiO_2, TiO_2/PAni and PAni/TiO_2/GO

Sample	Surface area (m^2/g)	Pore volume (cm^3/g)	Pore radius (nm)
TiO_2	37. 52	3.132	1.21
TiO_2/PAni	76.68	6.5124	1.64
TiO_2/PAni/GO	96.24	9.5124	1.84

3.4. Band Gap Energy

The band gap of samples was calculated by tauc plot. Fig.3 yields an *Eg* value of 3.2 eV for TiO_2 and 3.0 for PAni/TiO_2. The slight decrease in band gap energy in case of PAni/TiO_2, is due to formation of sub-band level between valence band and conduction band caused coating of PAni on TiO_2 host (Zhang et al 2006).

Figure 3. Band gap energy of (a) TiO_2 (b) pure PAni (c) TiO_2/PAni (d) TiO_2/PAni/GO

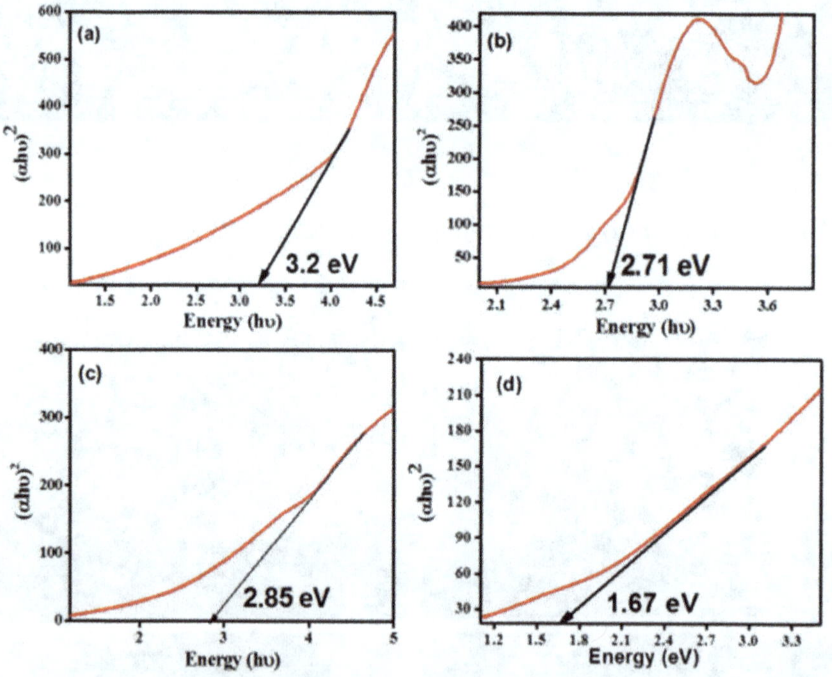

3.5. Photodegradation

The photo-catalytic degradation of Thymol Blue in the presence of TiO_2, TiO_2/PAni and TiO_2/PAni/GO has been studied. The solution of dye was prepared in 5:1 (V/V) ratio of water and alcohol. The known amount of photocatalyst was dispersed in the dye solution. The reaction mixture was illuminated under visible light, while kept continuously under agitation, for the different time intervals. The residual concentration of dye in the reaction mixture was measured spectrophotometrically. The results obtained for the degradation of Thymol Blue is shown in Fig 4. Photocatalytic degradation efficiencies (η) are obtained by using following equation (Albelda et al 2017).

$$= \frac{TB_O - TB_F}{TB_O}$$

where TB_0 is the initial absorbance and TB_F is the final sampled absorbance for a given time.

3.5.1. Effect of Concentration of Dye

Effect of dye concentration Keeping the catalyst loading concentration constant at 800 mg/litre of the dye solution, the effect of varying concentration of the dye was studied on its rate of degradation (25, 50,75, 100 and 125 ppm) as given in Fig.4. The rate of photodegradation was decrease with increasing concentration of TB. This is because as the number of dye molecules increase, the amount of light (quantum of photons) penetrating into the dye solution to reach the catalyst surface is reduced owing to the hindrance in the path of light (Ameen et al 2011).

3.5.2. Effect of Photocatalyst Amount

The effect of photocatalyst amount is showing in Fig.4. The photodegradation of Thymol blue was increased with increase the amount of photocatalyst. It is observed that TiO_2/PAni/GO is the more effective photocatalyst than TiO_2 and TiO_2/PAni. When the photocatalyst amount increases, the number of active site increase for the reaction of dyes. The amount of photocatalyst increases two times the rate of photodegradation increase about 30% and 60 %, in presence of TiO_2, TiO_2/PAni and TiO_2/PAni/GO respectively (Tjong et al 2006).

3.5.3. Effect of Irradiation Time

The effect of irradiation time of visible light was investigated. TiO_2/PAni/GO seems to be more effective as photo-catalyst for the degradation of Thymol Blue (TB). The prominent degradation of Thymol Blue was found in 180 min (Fig.4) study in the presence of TiO_2/PAni/GO in comparison to the prepared TiO_2 and TiO_2/PAni. This is due to the coating of polyaniline of Titania surface which provide the electron from the HOMO to LUMO. The electrons of HOMO get excited into LUMO which is further jump into the conduction band of Titania (Arora et al 2014).

3.5.4. Effect of pH

The photodegradation was carried out under varying pH conditions from (3 to 11), by adjusting with H_2SO_4 and NaOH, with TiO_2 kept at constant amounts of photocatalyst of 800 mg/L and 25 ppm concentration of dye solutions (Fig. 4). The photodegradation was found highest rates at neutral ranges of pH. While at lower pH it was found to decrease. In the basic condition, the photodegradation rate was found slow and very poor degradation. Hence highly acidic and basic condition is not favourable for the

Figure 4. Photodegradation of Thymol blue at various condition with (a) TiO_2 (b) TiO_2/PAni (c) TiO_2/PAni/GO

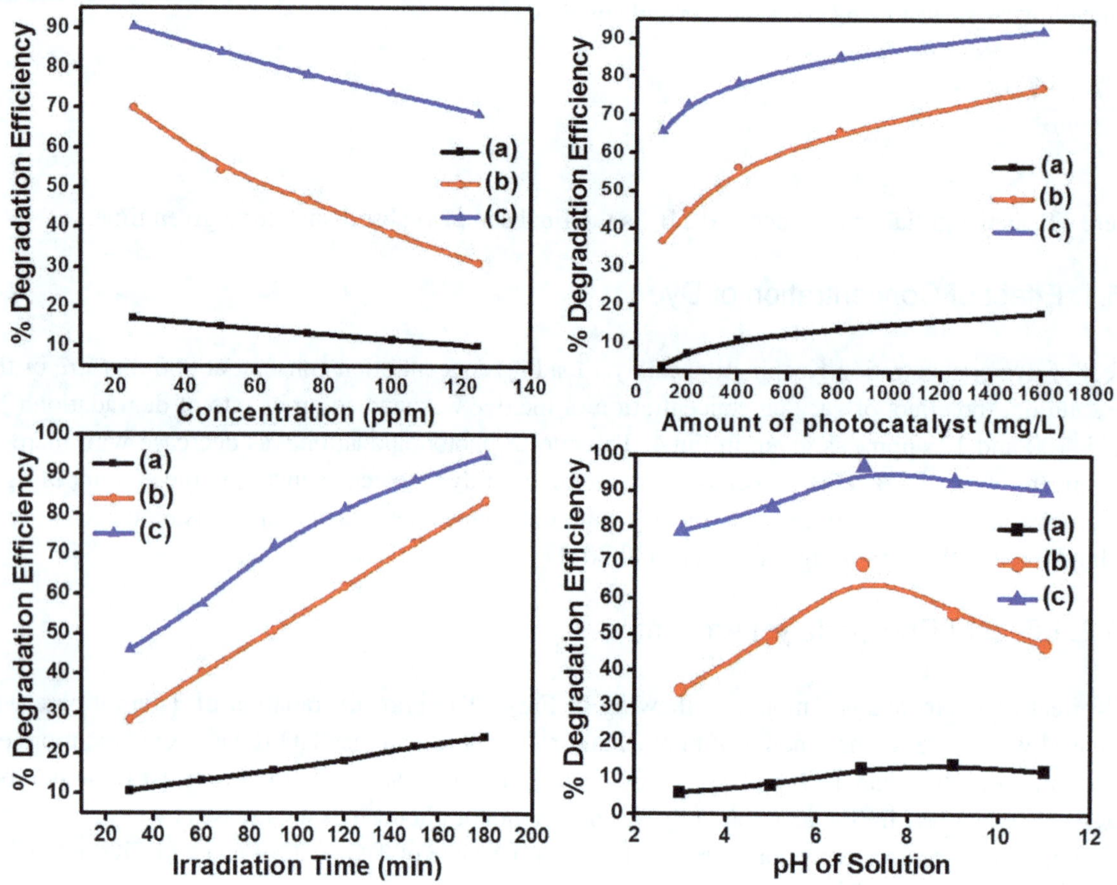

degradation of TB. This implies that neutral conditions are favourable towards the formation of the reactive intermediates that is hydroxyl radicals is significantly enhanced, which further help in enhancing the reaction rate. On the other hand, in basic and acidic conditions, the formation of reactive intermediates is relatively less favourable and hence less spontaneous (Sultana et al 2012, Zięba et al 2010).

3.6. Recyclability of Photocatalyst

The photocatalyst recyclability has been studied. The photocatalyst and Thymol blue mixture was agitated, illuminated with visible light and after desired time, the mixture was centrifuge to remove the photocatalyst. The removed photocatalyst washed three times with distilled water and finally kept in oven for 24 h at 60 °C temperature and further it is reuse for the degradation of Thymol Blue.

The photodegradation of Thymol blue by the recyclized Photocatalyst showed in Fig. 5. The result shows that the recyclized photocatalyst efficiency is decreased due to the loss of some active sites and decrease of collection efficiency of photon (Epling & Lin 2002).

Figure 5. Recyclability of Photocatalyst for the degradation of Thymol Blue.

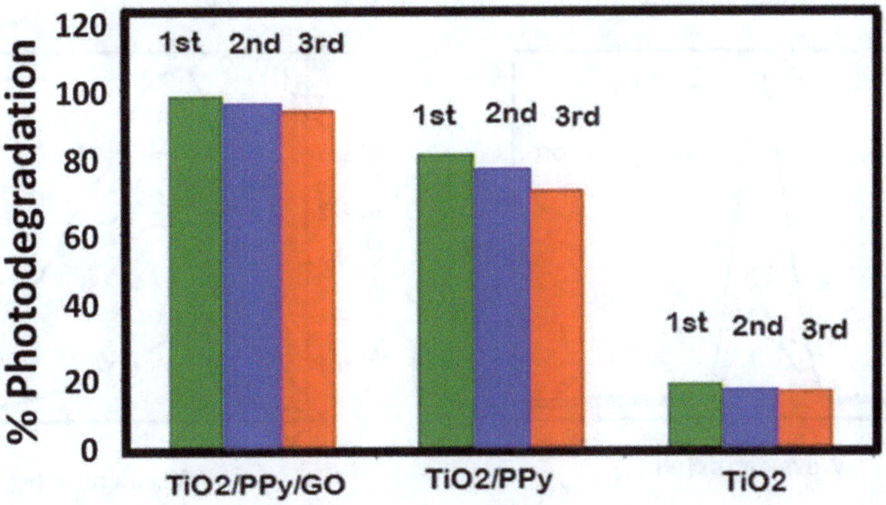

3.7. Lowering of Electron-Hole Recombination

Photoluminescence spectra have been used to examine the mobility of the charge carriers to the surface as well as the recombination process involved by the electron-hole pairs in semiconductor particles. PL emission results from the radiative recombination of excited electrons and holes. In other words, it is a critical necessity of a good photocatalyst to have minimum electron-hole recombination. To study the recombination of charge carriers, PL studies of synthesized materials have been undertaken. PL emission intensity is directly related to recombination of excited electrons and holes. Fig. 6 shows the photoluminescence spectra of synthesized photocatalysts. It means TiO_2 and TiO_2/PAni with strong PL intensity has high recombination of charge carriers where as TiO_2/PAni/GO has weak intensity. The weak PL intensity of TiO_2/PAni/GO may arise due to the coating of polypyrrole on Titania lattice, so that decrease in band gap of TiO_2/PAni/GO was found which resulting the decolourisation of photo excited electrons. This delays the electrons- holes recombination process and hence is utilized in the redox reaction leading to improved photocatalytic activity (Kordouli et al 2015).

3.8. Hydroxyl Radical Formation

To determine whether reactive oxygen species involved in the photocatalytic degradation of dyes is hydroxyl radical or not, terephthalic acid photoluminescence probing technique was used. In this, alkaline solution of terephthalic acid, having TiO_2, TiO_2/PAni and TiO_2/PAni/GO nanocomposites was irradiated with visible light. After 30 min of irradiation, sample was withdrawn from the reaction mixture and was centrifuged to separate photocatalyst particles. The photoluminescence spectrum of sample was recorded between 335 and 600 nm at an excitation wavelength of 325 nm and variation in intensity of peak at 425 nm was monitored using Perkin Elmer LS 55 Fluorescence Spectrometer.

As hydroxyl radical performs the key role for the decomposition of the organic pollutants, it is necessary to investigate the amount of hydroxyl radicals produced by each photocatalyst. Thus, there is a technique to establish the formation of hydroxyl radicals using terephthalic acid (TA) as a probe

Figure 6. (I) Photolumiscence Spectra of TiO_2, TiO_2/PAni and TiO_2/PAni/GO, (II) PL spectra of photocatalyst with terephthalic acid (0.001M) TiO_2, TiO_2/PAni and TiO_2/PAni/GO

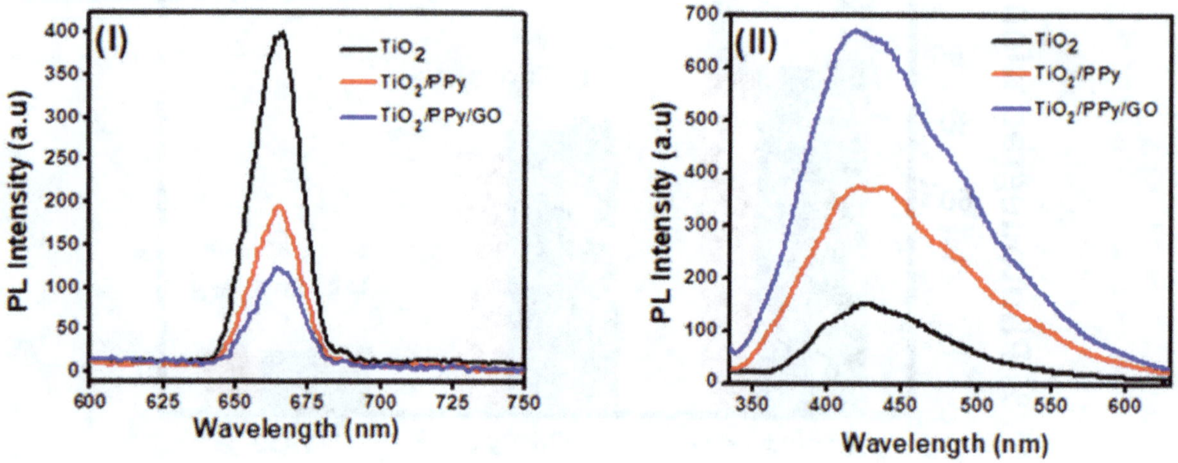

molecule. In this method, TA was directly attacked by OH radical forming 2- hydroxyl terephthalic acid (TAOH) which gives a fluorescence signal at 426 nm. Fig. 6 depicts the fluorescent signal of all the photocatalysts after reacting with TA solution. The fluorescent intensity is linearly related to the number of hydroxyl radicals formed by the photocatalysts. It means higher is the generation of hydroxyl radical, yield of TAOH will be more and hence more intense will be the fluorescence peak. Thus, TiO_2/PAni/GO with highest intensity confirms the generation more number of hydroxyl radicals compared to other photocatalysts. The fluorescence intensity follows the trend (i,e. TiO_2, < TiO_2/PAni < TiO_2/PAni/GO) of photocatalytic performance of all the photocatalyst (Wang et al 2008).

3.9. Kinetic Study of Photocatalytic Degradation

For kinetic study of photocatalytic degradation, a control experiment was first carried out under two conditions, vis (i) dye + Visible light (no catalyst) (ii) catalyst+ dye in dark without any irradiation (Fig. 7). It can be seen that in under dark conditions, the amount of catalyst adsorbed becomes constant after 20 min, where adsorption equilibrium is achieved. For the kinetic study of bleaching of Thymol Blue, the initial concentration of the dyes was varied and the experiments were first conducted in dark for 20 min and then immediately followed by irradiation (Fig. 7). The amount of catalyst was kept constant (0.2 g) throughout the experiment.

Applying the Langmuir Hinshelwood model for determining the oxidation rate of the photocatalysis of dye:

$$Rate(r) = -\frac{dC}{dt} = k\theta = \theta = \frac{kK_A C}{1+K_A C} \quad (1)$$

Where k is the rate constant (mg/L min^{-1}), C is the concentration of dye, K_A is the adsorption constant of the dye (L/mg), and t is the illumination time (min).

Figure 7. % Adsorption of Thymol Blue dye under dark condition in presence of (a) TiO_2, (b) TiO_2/PAni and (c) TiO_2/PAni/GO

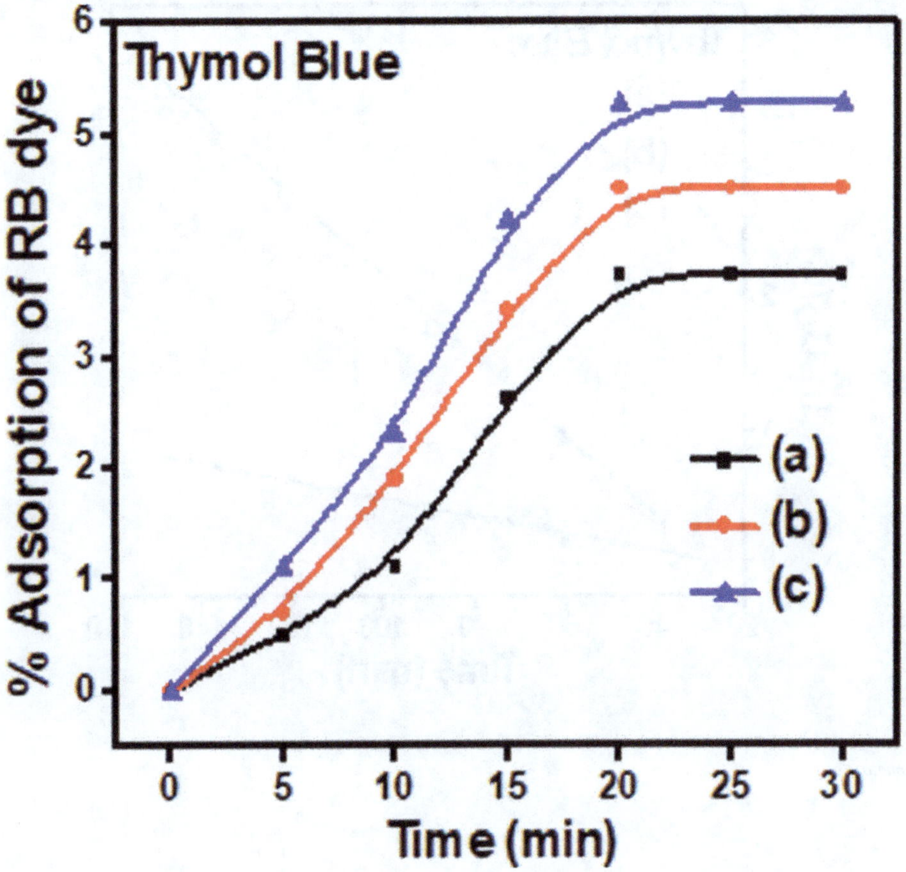

During the course of reaction, the initial pH, amount of catalyst, and photo intensity were kept same. In addition to it, the formation of intermediates may interfere in the rate determination; hence the calculation was done at the beginning of irradiation. The rate expression can be written as:

$$r_o = \frac{kK_A C_o}{1 + K_A C_o} \tag{2}$$

Where r_o is the initial rate of degradation of Thymol Blue and C_o is the initial concentration (almost equal to Ceq). When the initial concentration $C_{initial}$ is very small, Co will also be small and Eq. (2) can be simplified as a first-order equation 4:

$$-\frac{dC}{dt} = kK_A C_o = \frac{\ln C_o}{C} = kK_A t \tag{3}$$

Figure 8. Linear first order reaction of Langmuir Hinshelwood kinetics of Thymol Blue dye vs. time (a) TiO_2 (b) TiO_2/PAni (c) TiO_2/PAni/GO

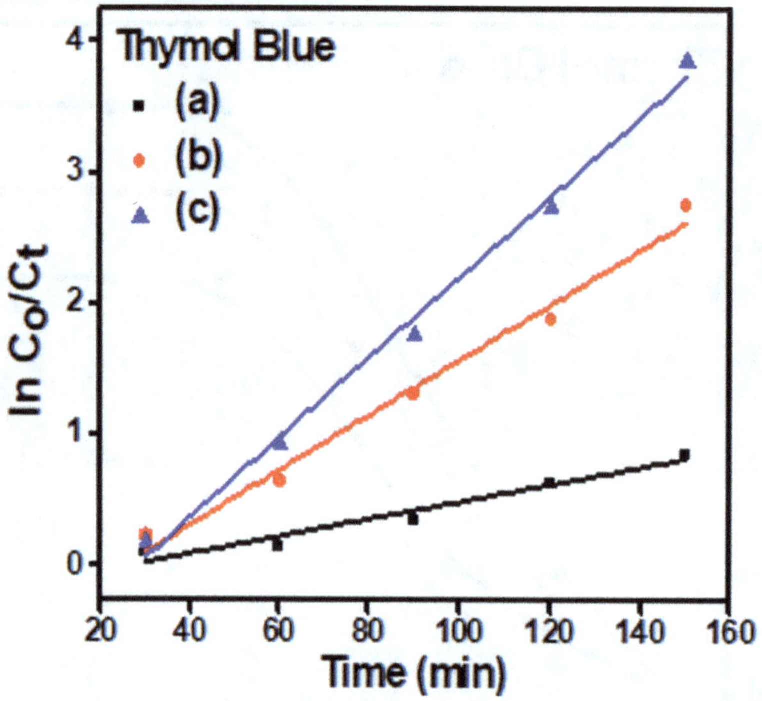

$$C = C_o e^{-k_f photot} \qquad (4)$$

Where

$k_{f,\,Photo=}\ k\, k_A$

The value of $k_{f,photo}$ can be determined from the plot of ln C_t/C_o vs. t (Fig.8).

The slope of the straight line obtained will be the value of first order rate constant (Guettaı & Ait Amar 2005). The Value of apparent rate constant were determining at definite concentrations of dye solution for photocatalysis reaction in presence of TiO_2, TiO_2/PAni and TiO_2/PAni/GO showing in Fig.8.

This is confirmed that photocatalytic degradation of Thymol Blue follows first order kinetic in presence of TiO_2, TiO_2/PAni and TiO_2/PAni/GO.

CONCLUSION

In this research, the different techniques were used for the characterisation of the nano composite such as XRD, SEM, TEM, FTIR and UV spectrophotometer. The SEM study conðrms that spherical morphology of the nanocomposite. The TEM analysis confirms that the size of nanocomposite (10 nm). The FTIR characterisation confirms that the TiO_2/GO molecules are well combined with polyaniline structure.

The maximum photodegradation of Thymol blue was found in TiO_2/PAni/GO at 25 ppm concentration of dye, 1600 mg/L amount of photocatalyst, pH 7 and 120 min irradiation of visible light. Hence, the photocatalytic activity of Titania has been increased by the coating of PAni and Graphene oxide.

REFERENCES

Al-Hussaini, A. S., Eltabie, K. R., & Rashad, M. E. E. (2016). One-pot modern fabrication and characterization of TiO2@terpoly(aniline, anthranilic acid and o-phenylenediamine) core-shell nanocomposites via polycondensation. *Polymer*, *101*, 328–337. https://doi.org/10.1016/j.polymer.2016.08.104

Albelda, J. A. V., Uzunoglu, A., Santos, G. N. C., & Stanciu, L. A. (2017). Graphene-titanium dioxide nanocomposite based hypoxanthine sensor for assessment of meat freshness. *Biosensors & Bioelectronics*, *89*(1), 518–524. https://doi.org/10.1016/j.bios.2016.03.041

Ameen, S., Akhtar, M. S., Kim, Y. S., & Shin, H. S. (2011). Nanocomposites of poly(1-naphthylamine)/SiO2 and poly(1-naphthylamine)/TiO2: Comparative photocatalytic activity evaluation towards methylene blue dye. *Applied Catalysis B: Environmental*, *103*(1–2), 136–142. https://doi.org/10.1016/j.apcatb.2011.01.019

Arora, R., Mandal, U. K., Sharma, P., & Srivastav, A. (2014). Effect of fabrication technique on microstructure and electrical conductivity of polyaniline-TiO2 -PVA composite material. *Procedia Materials Science*, *6*, 238–243. https://doi.org/10.1016/j.mspro.2014.07.029

Arora, R., Mandal, U. K., Sharma, P., & Srivastav, A. (2015). Article. *Materials Today: Proceedings*, *2*, 2215–2225.

Chowdhury, S., & Balasubramanian, R. (2014). Article. *Applied Catalysis B: Environmental*, *160–161*, 307–324.

Clifford, D. M., Castano, C. E., & Rojas, J. V. (2017). Supported transition metal nanomaterials: Nanocomposites synthesized by ionizing radiation. *Radiation Physics and Chemistry*, *132*, 52–64. https://doi.org/10.1016/j.radphyschem.2016.12.001

Epling, G. A., & Lin, C. (2002). Photoassisted bleaching of dyes utilizing TiO2 and visible light. *Chemosphere*, *46*(4), 561–570. https://doi.org/10.1016/s0045-6535(01)00173-4

Guettaı, N., & Ait Amar, H. A. (2005). Photocatalytic oxidation of methyl orange in presence of titanium dioxide in aqueous suspension. Part II: Kinetics study. *Desalination*, *185*(1–3), 439–448. https://doi.org/10.1016/j.desal.2005.04.049

Kordouli, E., Bourikas, K., Lycourghiotis, A., & Kordulis, C. (2015). The mechanism of azo-dyes adsorption on the titanium dioxide surface and their photocatalytic degradation over samples with various anatase/rutile ratios. *Catalysis Today*, *252*, 128–135. https://doi.org/10.1016/j.cattod.2014.09.010

Lai, C., Li, G. R., Dou, Y. Y., & Gao, X. P. (2010). Mesoporous polyaniline or polypyrrole/anatase TiO2 nanocomposite as anode materials for lithium-ion batteries. *Electrochimica Acta*, *55*(15), 4567–4572. https://doi.org/10.1016/j.electacta.2010.03.010

Li, Y., Yu, Y., Wu, L., & Zhi, J. (2013). Processable polyaniline/titania nanocomposites with good photocatalytic and conductivity properties prepared via peroxo-titanium complex catalyzed emulsion polymerization approach. *Applied Surface Science, 273*, 135–143. https://doi.org/10.1016/j.apsusc.2013.01.213

Moghaddam, H. M., & Nasirian, S. (2014). Article. *Applied Surface Science, 317*, 117–124.

Nasirian, S., & Milani Moghaddam, H. M. (2014). Hydrogen gas sensing based on polyaniline/anatase titania nanocomposite. *International Journal of Hydrogen Energy, 39*(1), 630–642. https://doi.org/10.1016/j.ijhydene.2013.09.152

Patil, P. T., Anwane, R. S., & Kondawar, S. B. (2015). Development of Electrospun Polyaniline/ZnO Composite Nanofibers for LPG Sensing. *Procedia Materials Science, 10*, 195–204. https://doi.org/10.1016/j.mspro.2015.06.041

Prasad, G. K., Takei, T., Yonesaki, Y., Kumada, N., & Kinomura, N. (2006). Hybrid nanocomposite based on NbWO6 nanosheets and polyaniline. *Materials Letters, 60*(29–30), 3727–3730. https://doi.org/10.1016/j.matlet.2006.03.097

Su, C. (2017). Environmental implications and applications of engineered nanoscale magnetite and its hybrid nanocomposites: A review of recent literature. *Journal of Hazardous Materials, 322*(A), 48–84. https://doi.org/10.1016/j.jhazmat.2016.06.060

Su, L., & Gan, Y. X. (2012). Experimental study on synthesizing TiO2 nanotube/polyaniline (PANI) nanocomposites and their thermoelectric and photosensitive property characterization. *Composites. Part B, Engineering, 43*(2), 170–182. https://doi.org/10.1016/j.compositesb.2011.07.015

Sultana, S., Rafiuddin, Z., Khan, M., & Umar, K. (2012). Synthesis and characterization of copper ferrite nanoparticles doped polyaniline. *Journal of Alloys and Compounds, 535*, 44–49. https://doi.org/10.1016/j.jallcom.2012.04.081

Tjong, S. C. (2006). Structural and mechanical properties of polymer nanocomposites. *Materials Science and Engineering R Reports, 53*(3–4), 73–197. https://doi.org/10.1016/j.mser.2006.06.001

Wang, D., Wang, Y., Li, X., Luo, Q., An, J., & Yue, J. (2008). Sunlight photocatalytic activity of polypyrrole–TiO_2 nanocomposites prepared by "in situ" method. *Catalysis Communications, 9*(6), 1162–1166. doi:10.1016/j.catcom.2007.10.027

Wang, H., Lin, J., & Shen, Z. X. (2016). Journal of Science. *Advanced Materials and Devices, 1*, 225–255.

Zhang, L., Liu, P., & Su, Z. (2006). Preparation of PANI–TiO2 nanocomposites and their solid-phase photocatalytic degradation. *Polymer Degradation & Stability, 91*(9), 2213–2219. https://doi.org/10.1016/j.polymdegradstab.2006.01.002

Zięba, A., Drelinkiewicz, A., Konyushenko, E. N., & Stejskal, J. (2010). Activity and stability of polyaniline-sulfate-based solid acid catalysts for the transesterification of triglycerides and esterification of fatty acids with methanol. *Applied Catalysis A, 383*(1–2), 169–181. https://doi.org/10.1016/j.apcata.2010.05.042

Chapter 9
Carbon Nanotubes in Water Treatment:
Progress and Challenges

Wilfrida Nyanduko Nyairo
Maseno University, Kenya

Victor Odhiambo Shikuku
https://orcid.org/0000-0002-2295-293X
Kaimosi Friends University, Kenya

Yacouba Sanou
University Joseph Ki-Zerbo, Burkina Faso

ABSTRACT

Access to safe drinking water is one of the most pressing challenges in the 21st century. New and better technologies for the treatment of wastewater are critically needed. Carbon nanotubes are emerging as effective and environmentally friendly alternative adsorbents for water purification due to their porous structure, relatively large specific surface areas, and strong hydrophobicity. Nevertheless, carbon nanotubes also suffer the inherent challenges of nanomaterials with potential health risks. This chapter presents a detailed review of the progress made in the utilization of carbon nanotubes and their composites in the sequestration of organic and inorganic pollutants from water. The factors affecting performance, the adsorption capacities, and mechanisms are concisely discussed. Additionally, the associated health risks of carbon nanotubes are highlighted, and risk assessment strategies are recommended. Overall, carbon nanotubes are shown to be suitable candidates for water treatment regimes.

INTRODUCTION

Many technologies have been developed for water treatment, however, and a majority require high capital investment for implementation especially in developing countries. Adsorption is described as a simple,

DOI: 10.4018/978-1-6684-4553-2.ch009

techno-economical and efficient method for the removal of organic and inorganic contaminants from water (Tome et al., 2021; Shikuku and Mishra, 2021). Among the various adsorbents, those derived from lignocellulosic biomass such as activated carbons (ACs), are the widely used type of adsorbents in water treatment, because of their local availability, chemical inertness, broad-spectrum removal capability toward pollutants, and thermal stability (Sanou et al., 2016; Sanou et al., 2019; Liu et al., 2012). However, the use of ACs in water treatment is limited by several bottlenecks, such as slow adsorption kinetics and difficulty for regeneration (Liu et al., 2012). To overcome these constraints, activated carbon fibers (ACFs) were developed as the second generation of carbonaceous adsorbents and their pores are directly opening on the surface of carbon matrix, which shortens the diffusion distance of pollutants to the adsorption sites. Consequently, adsorption kinetics of ACFs is higher than those of ACs. On the other hand, carbon nanotubes (CNTs) are considered like miniaturized ACFs with one dimensional structure and all adsorption sites are located on their inner and outer layer surface. Theoretically, CNTs may be a promising third generation of carbonaceous adsorbents with a tunable surface chemistry. These new types of adsorbents have been used for adsorption of metal ions (Li et al., 2002), anionic contaminants (Li et al., 2001; Peng et al., 2005), organic compounds (Cho et al., 2011), removal of biological contaminants (Venkata et al., 2009), and for the softening of hard water (Maryam and Toraj, 2011). A great number of organic compounds and heavy metals have been studied as the target pollutants on CNTs in water treatment with various physical structures and surface chemistry. The influence of operating conditions on solution chemistry, including solution pH, ionic strength, and co-existing matter, must be investigated for application in industrial scale to clean real wastewaters. In addition, the study of kinetic and isotherm models is important in order to explain the adsorption mechanisms. The use of CNTs for water treatment is not without difficulties. When using CNT powder as an adsorbent on an industrial scale, combining the CNTs with hard water with ultrasonic agitation is neither economical nor technically feasible. Furthermore, without centrifugation process, it is impossible to entirely recover the spent and contaminant-laden CNTs powder from treated water following the adsorption process. Adsorbent recovery by filtration is also difficult since the CNTs can quickly clog the filter (Maryam and Toraj, 2011). Furthermore, CNTs have been reported to be potentially toxic and their occurrence in water is contentious from a safety standpoint (Das et al., 2018). This chapter discusses the preparation and application of CNTs and their composites for the removal of dyes, heavy metals, and emerging contaminants from water. Catalytic properties of CNTs have also been highlighted. The associated toxicity and risks of CNTs are also presented.

Preparation of Carbon Nanotubes (CNTs)

The main methods employed in the synthesis of CNTs are chemical vapour deposition, arc discharge and laser ablation (Prasek et al., 2011). Chemical vapour deposition technique is considered the most feasible and economic of the three methods because it requires low power input, and it produces CNTs of relatively high purity (Prasek et al., 2011; Anzar et al., 2020).

Generally, the carbon sources used in the fabrication of CNTs include methane, ethylene, benzene, xylene, carbon monoxide, among others (Endo et al., 1993; Yamamoto, Inoue, Matsumura, 2017; Lv el al., 2021). Chemical vapour deposition involves pumping the carbon source into a quartz tube that acts as a furnace in the presence of an inert gas. The carbon is vaporized at a temperature of about 500-900þC in argon gas to form carbon atoms which join through van der Waals forces to form CNTs. First described by Iijima (1991), arc discharge technique utilizes pure graphite electrodes placed in a quartz chamber at a pressure of 500 torr in the presence of helium. A high temperature (>3000°C) ionizes carbon atoms which then move to the cathode where they are deposited and with proper cooling, uniform CNTs are formed. This method mostly produces multi-walled carbon nanotubes (MWCNTs), however, incorporation of metal catalysts such as iron, cobalt, nickel, and yttrium, at either the anode or cathode produces single walled carbon nanotubes (SWCNTs). The size and purity of the resulting CNTs depend on the metal concentration, type and pressure of inert gas, and temperature used (Arora and Sharma, 2014). Laser ablation is similar to the arc discharge process except that it uses a laser beam at a temperature of 1200°C and to produce SWCNTs, metal catalysts are added to the graphite target (Yuge et al 2012). The quality of the CNTs produced depend on the amount and type of catalyst used, type of inert gas, the laser power and wavelength, temperature and pressure used (Arepalli, 2004). These synthesis processes have undergone several improvements with time, in order to attain scale-up production of size-controlled CNTs for commercial purposes.

Modification of CNTs

As-produced CNTs tend to aggregate due to their bundle-like nature with high van der Waals attractions. Aggregation reduces the availability of the outer space of as-produced CNTs for adsorption of pollutants and prevents wettability of the CNTs. To overcome this challenge, ultra-sonication is employed when working CNTs into aqueous media (Zhang et al. 2012). Modification of CNTs through processes such oxidation also reduces aggregation by altering the CNTs surface chemistry and improve their dispersibility in water and other solvents (Datsyuk et al 2008). Various methods of oxidation of CNTs including wet chemical methods (Andrade et al., 2013), photo-induced oxidation (Savage et al., 2003), plasma treatment (Ren et al., 2011) have been reported. The most common oxidation technique applied is the wet chemical treatment of CNTs, in which acids or oxidizing reagents such as nitric acid, sulfuric acid, and potassium permanganate among others are used. The oxidation removes metal impurities left on the CNTs during their synthesis and also introduces oxygen functionalities on the CNTs surfaces (Datsyuk et al 2008; Wepasnick et al., 2011). These functional groups including –COOH, OH and –C=O, act as sites upon which pollutants attach themselves (Ihsanulah et al., 2016), thus enhancing the adsorption capacity of the CNTs for the target water pollutants. The surface characteristics of the CNTs have also been modified to form CNT-based materials by metal/metal oxide impregnation (Mallakpour and Khadem, 2016), doping with heteroatoms (Yi et al., 2020) or grafting functional molecules (Guo et al., 2019) which play a selective and synergic role of trapping of pollutants of interest. Magnetic material can also be loaded onto CNTs for easy separation of adsorption materials from aqueous media using an external magnet (Bhatia et al., 2019).

APPLICATON CNTs IN THE REMOVAL OF ORGANIC POLLUTANTS

Organic pollutants include pesticides, organic dyes, pharmaceuticals, phenols, aromatic compounds among others. These pollutants find their way into the environment when effluents industries such as textile, pharmaceutical, paper and leather tanning, are not properly treated before discharge. CNTs and CNTs-based materials have been applied as adsorbents, catalytic support, electrodes or membranes for the removal of these organic pollutants from wastewater. Adsorptive and catalytic applications of CNT materials for removal of organic pollutants are discussed herein.

Due to their porous structure, relatively large specific surface area and strong hydrophobicity CNTs have shown a great potential as adsorbents for organic pollutants such as 1, 2-dicholorobenzene (Peng et al 2003), polycyclic aromatic hydrocarbons (PAHs) (Akinpelu et al., 2018), dyes such as methylene blue, methyl orange, rhodamine B etc. (Li et., 2020; Saxena et al., 2020; Zhao et al., 2013; Pete et al., 2021), phenols (Lawal et al., 2019), and isonicotinic acid (Bhatia et al., 2019) among others.

The organic pollutants can be trapped in the pores or adsorbed onto the CNTs surfaces through different interaction such as hydrophobic interactions, π-π interactions, van der Waals forces, electrostatic attraction, and hydrogen bonding; in most cases these interactions act simultaneously (Pyrzynska, 2008). The sorption process of some organic compounds especially organic dyes as also been reported to be dependent on the morphology of the adsorbate. Liu et al. (2008) observed that the most easily adsorbed dyes are the planar polynuclear dyes (RhB), followed by non- planar conjugated molecules (Orange G) and the least, non- planar non-conjugated (bromothymol blue). Langmuir and Freudlinch equations are the most common models used to describe the adsorption equilibrium data. The Langmuir equation describes monolayer sorption on homogenous surfaces while Freudlinch equation is fitted on surfaces considered heterogeneous (Lawal et al., 2019; Saxena et al., 2020). The change in adsorption with time is usually fitted in the pseudo-first-order and pseudo-second-order kinetic models. The pseudo-first-order reaction implies that the reaction rate depends only on the concentration of one the reactants while the pseudo-second-order rate is determined by the concentration of two reactants simultaneously. Thus, the pseudo-second-order rate indicates that both the concentrations of the adsorbent and adsorbates are involved in the adsorption process (Bhatia et al., 2019). Among organic contaminants, dyes and chemicals of emerging concern are hereunder discussed.

ADSORPTION OF DYES

Zhao et al. (2013) investigated the sorption of methyl orange from water using MWCNTs at varying adsorption conditions. The study observed that the adsorption capacity of the MWCNTs increased with increase in stirring speed, methyl orange concentration and temperature but it decreased with increase in the mass of adsorbent. Pete et al. (2021) compared the adsorption performance of a binary composite of MWCNTs with polyaniline (PANI-MWCNTs) to that of MWCNTs in the removal of methyl orange. The research established that the composite had a greater removal efficiency for methyl orange than when PANI or MWCNTs were applied individually. The maximum adsorption capacity of PANI-MWCNTs for methyl orange was reported as 149.25 mgg^{-1} which was well described by Langmuir isotherm model while the kinetic studies fitted well on the pseudo-second order model. Furthermore, the PANI-MWCNTs adsorbent was found to have excellent reusability and it could be regenerated using 1 M HCl. In another study, the removal of methylene blue from water using MWCNTs functionalized with tyrosine (CNT-TYR) showed a maximum adsorption capacity of 440 mgg^{-1} (Saxena et al., 2020). The adsorption mechanism was reported to be as a result of different interactions namely; electrostatic, π-π interactions and hydrogen bonding. The reaction fitted well with pseudo-second order (PSO) kinetic model and Langmuir adsorption model. Lawal and group (2019), also investigated the removal of crystal violet (CRV) and phenol (PHE) using CNTs and CNTs functionalized with eutectic solvent (CNT-DES). The group found out that adsorption of capacity of CNTs was 128.6 mgg^{-1} and 312 mgg^{-1} and CNT-DES was 298 mgg^{-1} and 394 mgg^{-1} for CRV and PHE, respectively. The adsorption reaction was best described by pseudo second order kinetics and Freundlich isotherms, while the reaction mechanism was hydrophobic interactions

Table 1. Application CNT-based adsorbents for various organic pollutants from wastewater

Adsorbent	Dye	Q_{max} (mgg^{-1})	References
SWCNTs	Reactive red 120	426.4	Walker et al., 2004
PANI-MWCNTs	Methyl orange	149.25	Pete et al., 2021
MWCNTs	Crystal violet	312	Lawal et al., 2019
MWCNT-DES	Crystal violet	394	Lawal et al., 2019
MWCNT-TYR	Methylene blue	440	Saxena et al., 2020
MWCNTs	Maxilon blue	260.7	Alkaim et al., 2015
MWCNTs	Congo red	352.1	Zare et al., 2015b
MWCNTs	Malachite green	142.8	Shirmardi et al., 2013
MWCNTs	Acid red 18	166.6	Shirmardi et al., 2012

through π-π interactions. Table 1 is a summary of the adsorption capacities of CNTs for different dyes of varying structural frameworks under varying experimental conditions.

Adsorption of Chemicals of Emerging Concern (CEC)

With increasing robustness of analytical tools and methods, the detection of previously unreported and unregulated chemicals in water resources has been on the rise. These include pharmaceutically active ingredients (PAIs), and personal care products (PCPs). These have been shown to be toxic even at low concentrations and antibiotics in the environment have been linked with the development of drug-resistant bacterial strains, raising serious environmental concerns. Such unregulated chemicals are generally referred to us chemicals of emerging concern (CEC) or emerging contaminants (ECs).

Carbon nanotubes and its composites have been shown to a suitable adsorbent for the removal of various classes of emerging contaminants. Cho et al. (2011) described the adsorptive properties of single-walled carbon nanotubes (SWCNTs) and multi-walled carbon nanotubes (MWCNTs) for the uptake of phenol and ibuprofen and triclosan from water. The adsorption capacity of SWNCT was 223 mg/g and 558 mg/g for ibuprofen and triclosan, respectively. SWCNTs exhibited higher adsorption capacity than SWCNTs owing to the large surface area. A similar observation was reported for the adsorption of perfluorooctane sulfonate (PFOS) onto SWCNTs and MWCNTs (Chen et al., 2011b). Elsewhere, Carabineiro et al. (2011) reported that MWCNT has an adsorption capacity of 150 mg/g for ciprofloxacin. Besides facilitating separation of adsorbent from solution by use of an external magnet, Bhatia et al. (2019) observed that there was synergistic effect between Fe_3O_4 and MWCNTs when the binary composite MWCNTs/Fe_3O_4 was applied in the removal of isonicotinic acid. The results showed that the maximum adsorption capacity for isonicotinic acid by MWCNTs and MWCNTs/Fe_3O_4 was 671 and 1234 mgg^{-1}, respectively. Moreover, both adsorbents exhibited good reusability, thus indicating that both adsorbents had great potential for removing isonicotinic acid from wastewater. Table 2 lists several other emerging contaminants evaluated against CNTs.

Table 2. Application CNT-based adsorbents for removal of emerging contaminants

Adsorbent	CEC	Q_{manx} (mgg^{-1})	References
MWCNTs	Perchlorate	3.55	Fang and Chen, 2012
MWCNTs	Tetrabromobisphenol A	33.7	Ji et al., 2012
MWCNTs	Roxarsone	13.5	Hu et al., 2012
MWCNTs	Diuron	28.37	Deng et al., 2012a
MWCNTs	Isonicotinic acid	671	Bhatia et al., 2019
Magnetic MWCNTs	Isonicotinic acid	1234	Bhatia et al., 2019

APPLICATION CNTs AND ITS DERIVATIVES IN THE REMOVAL OF HEAVY METALS

Inorganic pollutants such as heavy metal ions, released into the environment, particularly due to industrialization have resulted to serious water pollution. In trace levels some heavy metal ions are essential for to human lives. However, when high levels are ingested, they cause harmful effects on life.

There are many reports on the application of CNTs and its derivatives in the removal of heavy metal ions from aqueous solution (Li et al., 2003; Ihsanulah et al., 2016). The treatment of wastewater with CNTs for removal of heavy metal ions is attributed to their superior properties including large specific surface area (SSA), porosity, fast adsorption kinetics and relatively high adsorption capacity (Ren et al., 2011). Most studies have described the driving force in the sorption of heavy metal ions as mainly; electrostatic interaction, ion exchange, surface complexation and physisorption (Gusain, Kumar & Ray, 2020). Similarly, to determine the maximum adsorption capacities the most commonly used isotherm models are those of Langmuir and Freudlinch. Noteworthy, is the fact that adsorption performance of the acid treated CNTs is much higher than that of pristine CNTs (Vukovic et al., 2011; Nyairo et al., 2018). This is attributed to the negative charge on CNTs surfaces that interacts with the heavy metal cations through electrostatic attraction. Furthermore, studies have also indicated that the modification of CNTs with other nanoparticles/molecules enhances their adsorptive performances as indicated in Table 2. For instance, MWCNTs modified by 2-vinylpyridine showed a better adsorption capacity for Pb(II) that as-produced MWCNTs (Ren et al., 2011). A study carried out by Gusain et al. (2019) shows that the composite of molybdenum sulfide/thiol functionalized MWCNTs (MoS_2/SH-MWCNTs) had a significantly better adsorption capacity for Pb(II) than o-MWCNTs. The performance was attributed to the additional functional groups (S, O, and C) present on the composite that promoted lead-sulphur complexation. At the same time, the zeta potential of the composite was comparatively higher than that of o-MWCNTs implying that there were higher metal ion interactions with the higher negatively charged surface of the composite. Table 3 presents the performance of CNTs-based adsorbents for the removal of heavy metals from water.

CATALYTIC PROPERTIES OF CARBON NANOTUBES

Recently, CNTs-based materials have been employed as catalysts/catalytic support in advanced oxidation process (AOPs) for removal of organic pollutants from wastewater through ozonation, fenton-like

Table 3. The sorption capacities of raw, oxidized and CNTs modified with other molecules in aqueous solutions.

Adsorbent	Metal ions	q_{max} (mgg^{-1})	Adsorption conditions	Isotherm models	Kinetic models	References
MWCNTs	Pb(II)	15.9	pH=6	Langmuir	-	Ren et al. (2011)
MWCNTs/2-vinylpyridine	Pb(II)	37.0	-	Langmuir	-	Ren et al. (2011)
DTC/MWCNT	Cd(II) Cu(II) Zn(II)	167.2 98.1 11.2	pH=6, T=298K, t =150 min	Langmuir	PSO	Li et al. (2015)
p-MWCNTs	Cu(II)	4.27	pH=5, T=298K	Langmuir	PSO	Nyairo et al. (2018)
o-MWCNTs	Cu(II)	16.95	pH=5, T=298K	Langmuir	PSO	Nyairo et al. (2018)
MWCNTs/Ppy	Cu(II)	24.39	pH=5, T=298K	Langmuir	PSO	Nyairo et al. (2018)
o-MWCNTs	Pb(II) Cd(II)	27.027 24.4	pH=6, T=298K	Langmuir	PSO	Gusain et al. (2019)
MoS_2/SH-MWCNTs	Pb(II) Cd(II)	90 66.6	pH=6, T=298K	Freudlinch	PSO	Gusain et al. (2019)
Key: DTC- Dithiocarbamate, Ppy-polypyrrole, MoS_2/SH –Molybdenum sulfide thiol, T-temperature, t-time						

systems, and persulfates among others. In this process, the CNTs-based materials are used to catalyze different oxidants which produce more reactive oxidative species (ROS) that destroy the organic pollutants (Liu et al., 2011).

Catalytic ozonation by CNT-based materials has successfully been employed in removing organic pollutants such as indigo (Qu et al., 2015), sulphamethoxazole (Goncalves, Orfa and Pereira, 2012), oxalic acid (Liu et al., 2011) from wastewater. Generally, the catalytic ozonation activity of the MWCNTs/O_3 is attributed to oxygen functionalities on the CNTs' surface. According to Zhang et al. (2015), the suggested mechanism follows two pathways, namely; production of $HO^\bullet$ when ozone attacks the active sites on the surface of CNTs and conversion of ozone to $O_2^{\bullet-}$ radicals by CNT initiators. Liu et al. (2011) investigated the effect of operating conditions (such as initial solution pH and initial concentration of oxalic acid) in water on the catalytic performance of MWCNTs. The study established that the degradation efficiency of oxalic acid was pH dependent, and this was attributed to the surface characteristics of the MWCNTs. In another study, Goncalves et al. (2012) compared the catalytic activity of MWCNTs to that of activated carbon in the ozonation of sulphamethoxazole (SMX). The results from the study indicated that MWCNTs had a higher performance than activated carbon, which was attributed to the surface chemistry of the MWCNTs. Furthermore, the lowest acute toxicity was achieved when MWCNTs/O_3 was used compared to activated carbon.

CNT-Fenton like systems which involves the reaction of peroxides (mostly H_2O_2) with iron ions to form radicals that oxidize the target compounds have also been used in the abatement of organic pollutants such as methyl orange (Yu et al., 2012), SMX (Nawaz et al., 2020), methylene blue (Fayazi, 2021) in wastewater. Nawaz et al. (2020) investigated the photo-Fenton effect of different contents (10, 15, 20, 25, 50 wt%) of MWCNTs in a composite of nickel ferrite and MWCNTs (NiFe-CNT). The study applied the highly photo active reagent under UV-A light (in the presence and absence of H_2O_2) to determine the degradation of SMX in aqueous solution. It was observed that by increasing MWCNTs content, the photocatalytic activity of the composite was enhanced with a maximum being achieved at 25 wt%. The findings also showed that under UV-A, the composite containing 25 wt% MWCNTs, completely degraded 5 mg/L SMX within 2 h.

TOXICITY AND RISKS OF CNTs

CNTs are classified in two main categories, single wall carbon nanotubes (SWCNTs) and multiwall carbon nanotubes (MWCNTs), according to the number of layers. They are considered to have carcinogenicity and can cause lung tumors. However, the carcinogenicity of CNTs may attenuate if the fiber length is shorter. The available data provide initial information on the potential reproductive and developmental toxicity of CNTs. They can be altered to resist biodegradation, increased cellular uptake, reactivity and toxicity to terrestrial, aquatic and aerial flora and fauna. In other part, CNTs have similar pathological effects to asbestos causing other societal perceptions intending to ban CNTs.

CNTs can induce proliferation and differentiation of neurons and osteoblasts and serve as drug and vaccine delivery vehicles for cancer treatment (Gorityala et al., 2010). Some investigations have reported toxic effects following the exposure of several cell types to both SWCNTs and MWCNTs. It has been demonstrated that a number of factors such as impurities, surface modification, structure, and exposure routes that may influence the toxicity of CNTs. The presence of metal impurities could lead to conflicting data about the biocompatibility, toxicity, and risk assessment of CNTs and may limit their further industrial applications. Contamination of CNTs by catalyst residues is unavoidable during large-scale production by chemical vapor deposition techniques (Pumera, 2007).

As described above, research regarding CNTs hazard has been driven primarily by the potential for exposure via respiratory and dermal routes, particularly during their manufacture. In addition, the longevity at the exposure site and potential for distribution within the body subsequent to exposure has also been addressed within a limited number of conducted studies. Some investigations also assess the biocompatibility of CNTs used in biomedical applications, such as drug delivery. Different exposures including pulmonary, oral, dermal and cardiovascular, absorption distribution metabolism excretion and intraperitoneal injection of CNTs have been discovered as consequences of CNTs for humans (Johnston et al., 2010). Previous studies reported that the CNTs exposure to the respiratory system could lead to asthma, bronchitis, emphysema and lung cancer (Lam et al., 2006). It is important to note that some factories are dustier possibly due to the lack of industrial hygiene standards (Lam et al., 2006). Working with pulverized CNTs or mixtures that contain fine CNT particles could pose a risk of inhalation. Many experimental studies conducted on inhalation exposure have contributed to the assessment of the effects of CNTs on respiratory tract and identification of exposure limits (Das et al., 2018).

The circulation of CNTs within the blood may affect blood components (specifically cells) or allow for their distribution to a number of targets that are potentially detrimentally affected by exposure (see later). Entry into the blood could of course occur in a medical application via direct intravenous injection, but might also occur if their translocation into the circulation from other organs (including the lungs, skin or gastrointestinal tract) is realized (Johnston et al., 2010). Due to the cost and technical difficulty of tracing CNT translocation following inhalation or instillation, very few studies of this nature have been studied to date. Intraperitoneal exposure of the mesothelial lining of the abdominal cavity to CNTs has been used as a surrogate for the mesothelial lining of the pleural cavity surrounding the lungs. Direct injection into the pleural cavity is technically difficult and so the intraperitoneal model provides a convenient model for investigating mesothelial responses (Johnston et al., 2010). The limited number of available studies suggests that there is a hazard associated with the exposure of skin to CNTs, with the response being primarily inflammatory in nature. In future studies it is necessary to consider the systemic availability of CNTs following dermal exposure. The recurrent appearance of granulomas following subcutaneous exposure is interesting as it suggests that this pathological hallmark is not limited to the respiratory system or mesothelium, but that such a response may occur in many different tissue types, independent of the route of delivery. Furthermore, the results of many studies suggest that granuloma formation might be related to the agglomeration status of the nanotubes, with greater agglomeration being more likely to induce more extensive pathology. In addition, as demonstrated in the lung, inflammatory and oxidative driven responses appear to be replicated within the skin, thus reinforcing the importance of their development when assessing CNTs toxicity.

CONCLUSION

Carbon nanotubes (CNTs) have the potential to support point of use-based treatment approach for removal of water hardness, chemical, and biological contaminants from water. Generally, CNTs exhibit higher adsorption capacities in the removal of heavy metals, dyes and emerging contaminants relative to other adsorbents. This is attributable to their fibrous shape with high aspect ratio, large surface area and well developed mesopores. The relatively high cost of CNTs stands as a major constraint towards application of CNTs o industrial scale for water purification. Additionally, the release of contaminant-laden unrecovered CNTs into the environment and concomitant human exposure to CNTs remains contentious due to the associated health risks. Adsorbent recovery strategies need to be further explored including the use of life cycle assessment (LCA) tool in adsorption studies both at laboratory and pilot scale experiments.

REFERENCES

Akinpelu, A. A., Ali, M. E., Johan, M. R., Saidu, R., Qurban, M. A., & Saleh, T. A. (2019). Polycyclic aromatic hydrocarbons extraction and removal from wastewater by carbon nanotubes: A review of the current technologies, challenges and prospects. *Process Safety and Environmental Protection*, *122*, 68–82. doi:10.1016/j.psep.2018.11.006

Alkaim, A. F., Sadik, Z., Mahdi, D. K., Alshrefi, S. M., Al-Sammarraie, A. M., Alamgir, F. M., Singh, P. M., & Aljeboree, A. M. (2015). Preparation, structure and adsorption properties of synthesized multiwall carbon nanotubes for highly effective removal of maxilon blue dye. *Korean Journal of Chemical Engineering*, *32*(12), 2456–2462. doi:10.100711814-015-0078-y

Andrade, N. F., Martinez, D. S. T., Paula, A. J., Silveira, J. V., Alves, O. L., & Fiho, A. G. S. (2013). Temperature effects on the nitric acid oxidation of industrial grade multi-walled carbon nanotubes. *Journal of Nanoparticle Research*, *15*(7), 1767. doi:10.100711051-013-1761-8

Anzar, N., Hasan, R., Tyagi, M., Yadav, N., & Narang, J. (2020). Carbon nanotube - A review on synthesis, properties and plethora of applications in the biomedical science. *Sensors International*, *1*, 100003. doi:10.1016/j.sintl.2020.100003

Arepalli, S. (2004). Laser ablation process for single-walled carbon nanotube production. *Journal of Nanoscience and Nanotechnology*, *4*(4), 317–325. doi:10.1166/jnn.2004.072 PMID:15296222

Arora, N., & Sharma, N. (2014). Arc discharge synthesis of carbon nanotubes. A comprehensive review. *Diamond and Related Materials*, *50*, 135–150. doi:10.1016/j.diamond.2014.10.001

Bhatia, D., Datta, D., Joshi, A., Gupta, S., & Gote, Y. (2019). Adsorption of isonicotinic acid from aqueous solution using multi-walled carbon nanotubes/Fe_3O_4. *Journal of Molecular Liquids*, *276*, 163–169. doi:10.1016/j.molliq.2018.11.127

Carabineiro, S. A. C., Thavorn-Amornsri, T., Pereira, M. F. R., & Figueiredo, J. L. (2011). Adsorption of ciprofloxacin on surface-modified carbon materials. *Water Research*, *45*(15), 4583–4591. doi:10.1016/j.watres.2011.06.008 PMID:21733541

Chen, X., Xia, X., Wang, X., Qiao, J., & Chen, H. (2011). A comparative study on sorption of perfluorooctane sulfonate (PFOS) by chars, ash and carbon nanotubes. *Chemosphere*, *83*(10), 1313–1319. doi:10.1016/j.chemosphere.2011.04.018 PMID:21531440

Cho, H.-H., Huang, H., & Schwab, K. (2011). Effects of solution chemistry on the adsorption of ibuprofen and triclosan onto carbon nanotubes. *Langmuir*, *27*(21), 12960–12967. doi:10.1021/la202459g PMID:21913654

Das, R., Leo, B. F., & Murphy, F. (2018). The Toxic Truth About Carbon Nanotubes in Water Purification: A Perspective View. *Nanoscale Research Letters*, *13*(1), 183. doi:10.118611671-018-2589-z PMID:29915874

Deng, J., Shao, Y., Gao, N., Deng, Y., Tan, C., Zhou, S., & Hu, X. (2012). Multiwalled carbon nanotubes as adsorbents for removal of herbicide diuron from aqueous solution. *Chemical Engineering Journal*, *193–194*, 339–347. doi:10.1016/j.cej.2012.04.051

Endo, M., Takeuchi, K., Igarashi, S., Kobori, K., Shiraishi, M., & Kroto, H. W. (1993). The production and structure of pyrolytic carbon nanotubes (PCMTs). *Journal of Physics and Chemistry of Solids*, *54*(12), 1841–1848. doi:10.1016/0022-3697(93)90297-5

Fayazi, M. (2021). Preparation and characterization of carbon nanotubes/pyrite nanocomposite for degradation of methylene blue by heterogeneous Fenton reaction. *Journal of the Taiwan Institute of Chemical Engineers*, *120*, 229–235. doi:10.1016/j.jtice.2021.03.033

Goncalves, A. G., Orfao, J. J. M., & Pereira, M. F. R. (2012). Catalytic ozonation of sulphamethoxazole in the presence of carbon materials: Catalytic performance and reaction pathways. *Journal of Hazardous Materials*, *239-240*, 167–174. doi:10.1016/j.jhazmat.2012.08.057 PMID:23009796

Gorityala, B., Ma, J., Wang, X., Chen, P., & Liu, X. (2010). Carbohydrate Functionalized Carbon Nanotubes and Their Applications. *Chemical Society Reviews*, *39*(8), 2925–2934. doi:10.1039/b919525b PMID:20585681

Guo, M., Wang, J., Strong, P. J., Jiang, P., Ok, Y. S., & Wang, H. (2019). Carbon nanotube grafted chitosan and its adsorption capacity for phenol in aqueous solution. *The Science of the Total Environment*, *682*, 340–347. doi:10.1016/j.scitotenv.2019.05.148 PMID:31125747

Gusain, R., Kumar, N., Fosso-Kankeu, E., & Ray, S. S. (2019). Efficient removal of Pb(II) and Cd(II) for industrial mine water by a hierarchical MoS_2/SH-MWCNT nanocomposite. *ACS Omega*, *4*(9), 13922–13935. doi:10.1021/acsomega.9b01603 PMID:31497710

Gusain, R., Kumar, N., & Ray, S. S. (2020). Recent advances in carbon nanomaterials-based adsorbents for water purification. *Coordination Chemistry Reviews*, *405*, 213111. doi:10.1016/j.ccr.2019.213111

Hu, J., Tong, Z., Hu, Z., Chen, G., & Chen, T. (2012). Adsorption of roxarsone from aqueous solution by multi-walled carbon nanotubes. *Journal of Colloid and Interface Science*, *377*(1), 355–361. doi:10.1016/j.jcis.2012.03.064 PMID:22513167

Ihsanullah, A. A., Al-Amer, A. M., Laoui, T., Al-Marri, M. J., Nasser, M. S., Khraisheh, M., & Atieh, M. A. (2016). Heavy metal removal from aqueous solution by advanced carbon nanotubes: A critical review of adsorption applications. *Separation and Purification Technology*, *157*, 141–161. doi:10.1016/j.seppur.2015.11.039

Iijama, S. (1991). Helical microtubules of graphitic carbon. *Nature*, *354*(6348), 56–58. doi:10.1038/354056a0

Ji, L., Zhou, L., Bai, X., Shao, Y., Zhao, G., Qu, Y., Wang, C., & Li, Y. (2012). Facile synthesis of multiwall carbon nanotubes/iron oxides for removal of tetrabromobisphenol A and Pb(II). *Journal of Materials Chemistry*, *22*(31), 15853–15862. doi:10.1039/c2jm32896h

Johnston, H. J., Hutchison, G. R., Christensen, F. M., Peters, S., Hankin, S., Aschberger, K., & Stone, V. (2010). A critical review of the biological mechanisms underlying the in vivo and in vitro toxicity of carbon nanotubes: The contribution of physico-chemical characteristics. *Nanotoxicology*, *4*(2), 207–246. doi:10.3109/17435390903569639 PMID:20795897

Lam, C. W., James, J. T., McCluskey, R., Arepalli, S., & Hunter, R. L. (2006). A review of carbon nanotube toxicity and assessment of potential occupational and environmental health risks. *Critical Reviews in Toxicology*, *36*(3), 189–217. doi:10.1080/10408440600570233 PMID:16686422

Li, Q., Yu, J., Zhou, F., & Jiang, X. (2015). Synthesis and characterization of dithiocarbamate carbon nanotubes for removal of heavy metal ions from aqueous solutions. *Colloids and Surfaces. A, Physicochemical and Engineering Aspects*, *482*, 306–314. doi:10.1016/j.colsurfa.2015.06.034

Li, Y.-H., Ding, J., Luan, Z., Di, Z., Zhu, Y., Xu, C., Wu, D., & Wei, B. (2003). Competitive adsorption of Pb(II)· Cu(II), Cd^{2+} ions from aqueous solutions by multiwalled carbon nanotubes. *Carbon*, *41*(14), 2787–2792. doi:10.1016/S0008-6223(03)00392-0

Li, Y. H., Wang, S. G., Cao, A. Y., Zhao, D., Zhang, X. F., Xu, C. L., Luan, Z. K., Ruan, D. B., Liang, J., Wu, D. H., & Wei, B. Q. (2001). Adsorption of fluoride from water by amorphous alumina supported on carbon nanotubes. *Chemical Physics Letters*, *350*(5-6), 412–416. doi:10.1016/S0009-2614(01)01351-3

Li, Y. H., Wang, S. G., Wei, J. Q., Zhang, X. F., Xu, C. L., Luan, Z. K., Wu, D. H., & Wei, B. Q. (2002). Lead adsorption on carbon nanotubes. *Chemical Physics Letters*, *357*(3-4), 263–266. doi:10.1016/S0009-2614(02)00502-X

Li, Z., Sellaoui, L., Franco, D., Netto, M. S., Georgin, J., Dotto, G. L., & Li, Q. (2020). Adsorption of hazardous dyes in functionalized multiwalled carbon nanotubes in single and binary systems. Experimental study and physicochemical interpretation of adsorption mechanism. *Chemical Engineering Journal*, *389*, 124467. doi:10.1016/j.cej.2020.124467

Liu, C.H., Li, J.J., Zhang, H.L., Li, B. R., & Guo, Y. (2008). Structure dependent interaction between organic dyes and carbon nanotubes. *Colloid Surface, A, 313*, 9-12.

Liu, X., Zhang, S., & Pan, B. (2012). Potential of Carbon Nanotubes in Water Treatment. INTECH. doi:10.5772/51332

Liu, Z. Q., Ma, J., Cui, Y. H., Zhao, L., & Zhang, B. P. (2011). Factors affecting the catalytic activity of multiwalled carbon nanotubes for ozonation of oxalic acid. *Separation and Purification Technology*, *78*(2), 147–153. doi:10.1016/j.seppur.2011.01.034

Lv, X., Zhang, Y., Wang, Y., Zhang, G., & Liu, J. (2021). Formation of carbon nanofibres/nanotubes by chemical vapor deposition using Al_2O_3/KOH. *Diamond and Related Materials*, *113*, 108265. doi:10.1016/j.diamond.2021.108265

Mallakpour, S., & Khadem, E. (2016). Carbon nanotube-metal oxide nanocomposites: Fabrication, properties and applications. *Chemical Engineering Journal*, *302*, 344–367. doi:10.1016/j.cej.2016.05.038

Maryam, A. T., & Toraj, M. (2011). Permanent hard water softening using carbon nanotube sheets. *Desalination*, *268*(1-3), 208–213. doi:10.1016/j.desal.2010.10.028

Nawaz, M., Shahzad, A., Tahir, K., Kim, J., Moztahida, M., & Lee, D. S. (2020). Photo-Fenton reaction for the degradation of sulfamethoxazole using a multi-walled carbon nanotubes- $NiFe_2O_4$ composite. *Chemical Engineering Journal*, *382*, 123053. doi:10.1016/j.cej.2019.123053

Nyairo, W. N., Eker, Y. R., Kowenje, C., Akin, I., Bingol, H., Tor, A., & Ongeri, D. M. (2018). Efficient adsorption of lead(II) and copper(II) from aqueous solution phase using oxidized multiwalled carbon nanotubes/polypyrrole composite. *Separation Science and Technology*, *58*(1), 1–13.

Peng, X., Li, Y., Luan, Z., Di, Z., Wang, H., Tian, B., & Jia, Z. (2003). Adsorption of 1, 2- dichlorobenzene from water to carbon nanotubes. *Chemical Physics Letters*, *376*(1-2), 154–158. doi:10.1016/S0009-2614(03)00960-6

Peng, X. J., Luan, Z. K., Ding, J., Di, Z. H., Li, Y. H., & Tian, B. H. (2005). Ceria nanoparticles supported on carbon nanotubes for the removal of arsenate from water. *Materials Letters*, *59*(4), 399–403. doi:10.1016/j.matlet.2004.05.090

Pete, S., Kattil, R. A., & Thomas, L. (2021). Polyaniline-multiwalled carbon nanotubes (PANI-MWCNTs) composite revisited: An efficient and reusable material for methyl orange removal. *Diamond and Related Materials*, *117*, 108455. doi:10.1016/j.diamond.2021.108455

Prasek, J., Drbohlavova, J., Chomoucka, J., Hubalek, J., Jasek, O., Adam, V., & Kizek, R. (2011). Methods for carbon nanotubes synthesis-Review. *Journal of Materials Chemistry*, *21*(40), 15872–15884. doi:10.1039/c1jm12254a

Pumera, M. (2007). Carbon Nanotubes Contain Residual Metal Catalyst Nanoparticles even after Washing with Nitric Acid at Elevated Temperature because These Metal Nanoparticles Are Sheathed by Several Graphene Sheets. *Langmuir*, *23*(11), 6453–6458. doi:10.1021/la070088v PMID:17455966

Pyrzynska, K. (2008). Carbon nanotubes as a new solid phase extraction material for removal and enrichment of organic pollutants in water. *Separation and Purification Reviews*, *37*(4), 372–389. doi:10.1080/15422110802178843

Qu, R., Xu, B., Meng, L., Wang, L., & Wang, Z. (2015). Ozonation of indigo enhanced by carboxylated carbon nanotubes: Performance optimization, degradation products, reaction mechanism and toxicity evaluation. *Water Research*, *68*, 316–327. doi:10.1016/j.watres.2014.10.017 PMID:25462739

Ren, X. M., Shao, D. D., Zhao, G. X., Sheng, G. D., Hu, J., Yang, S. T., & Wang, X. K. (2011). Plasma induced multiwalled carbon nanotube grafted with 2-vinylpyridine for preconcentration of Pb(II) from aqueous solutions. *Plasma Processes and Polymers*, *8*(7), 589–598. doi:10.1002/ppap.201000192

Sanou, Y., Nguyen, T. T. P., Pare, S., & Nguyen, V. P. (2019). The removal As (V) from aqueous solutions using Ferromagnetic Activated Carbon: Equilibrium and kinetic studies. *Revue des Sciences de l'Eau*, *32*(2), 179–192. doi:10.7202/1065206ar

Sanou, Y., Pare, S., Baba, G., Segbeaya, N. K., & Bonzi-Coulibaly, L. Y. (2016). Removal of COD in Wastewaters by Activated Charcoal from Rice Husk. *Revue des Sciences de l'Eau*, *29*(3), 265–277. doi:10.7202/1038927ar

Savage, T., Bhattachanya, S., Sanadan, B., Gaillard, J., Tritt, T., Sun, Y.-P., & Rao, A. M. (2003). Photo-induced oxidation of nanotubes. *Journal of Physics Condensed Matter*, *15*(35), 5915–5921. doi:10.1088/0953-8984/15/35/301

Saxena, M., Sharma, N., & Saxena, R. (2020). Highly efficient and rapid removal of toxic dye: Adsorption kinetics, isotherm, and mechanism studies of functionalized multiwalled carbon nanotubes. *Surfaces and Interfaces*, *21*, 100639. doi:10.1016/j.surfin.2020.100639

Shikuku, V. O., & Mishra, T. (2021). Adsorption isotherm modeling for methylene blue removal onto magnetic kaolinite clay: A comparison of two-parameter isotherms. *Applied Water Science*, *11*(6), 103. doi:10.100713201-021-01440-2

Shirmardi, M., Mahvi, A. H., Hashemzadeh, B., Naeimabadi, A., Hassani, G., & Niri, M. V. (2013). The adsorption of malachite green (MG) as a cationic dye onto functionalized multi walled carbon nanotubes. *Korean Journal of Chemical Engineering*, *30*(8), 1603–1608. doi:10.100711814-013-0080-1

Shirmardi, M., Mesdaghinia, A., Mahvi, A. H., Nasseri, S., & Nabizadeh, R. (2012). Kinetics and equilibrium studies on adsorption of Acid Red 18 (azo-dye) using multiwall carbon nanotubes (MWCNTs) from aqueous solution. *E-Journal of Chemistry*, *9*(4), 9. doi:10.1155/2012/541909

Tome, S., Dzoujo, H., Shikuku, V., & Otieno, S. (2021). Synthesis, characterization and application of acid and alkaline activated volcanic ash-based geopolymers for adsorptive remotion of cationic and anionic dyes from water. *Ceramics International*, *47*(15), 20965–20973. Advance online publication. doi:10.1016/j.ceramint.2021.04.097

Upadhyayula, V. K. K., Deng, S., Mitchell, M. C., & Smith, G. B. (2009). Application of carbon nanotube technology for removal of contaminants in drinking water: A review. *The Science of the Total Environment*, *408*(1), 1–13. doi:10.1016/j.scitotenv.2009.09.027 PMID:19819525

Vukovic, G. D., Marinkovic, A. D., Skapin, S. D., Ristic, M. D., Aleksic, R., Peric-Grujic, A. A., & Uskokovic, P. S. (2011). Removal of lead from water by amino modified multiwalled carbon nanotubes. *Chemical Engineering Journal*, *173*(3), 855–865. doi:10.1016/j.cej.2011.08.036

Walker, D. J., Clemente, R., & Bernal, M. P. (2004). Contrasting effects of manure and compost on soil pH, heavy metal availability and growth of *Chenopodiumalbum L.* in a soil contaminated by pyritic mine waste. *Chemosphere*, *57*(3), 215–224. doi:10.1016/j.chemosphere.2004.05.020 PMID:15312738

Wepasnick, K. A., Smith, B. A., Schrote, K. E., Wilso, H. K., Diegemann, S. R., & Fairbrothe, D. H. (2011). Surface and structural characterization of multi-walled carbon nanotubes following different oxidative treatments. *Carbon*, *49*(1), 24–36. doi:10.1016/j.carbon.2010.08.034

Yamamoto, Y., Inoue, S., & Matsumura, Y. (2017). Thermal decomposition products of various carbon sources in chemical vapor deposition synthesis of carbon nanotube. *Diamond and Related Materials*, *75*, 1–5. doi:10.1016/j.diamond.2016.11.017

Yi, L., Zuo, L., Wei, C., Fu, H., Qu, X., Zheng, S., Xu, S., Guo, Y., Li, H., & Zhu, D. (2020). Enhanced adsorption of bisphenol A, tylosin, and tetracycline from aqueous solution to nitrogen- doped multiwall carbon nanotubes via cation-π and π-π electron-donor-acceptor (EDA) interaction. *The Science of the Total Environment*, *719*, 137389. doi:10.1016/j.scitotenv.2020.137389 PMID:32120097

Yu, F., Chen, J., Chen, L., Huai, J., Gong, W., Yuan, Z., & Ma, J. (2012). Magnetic carbon nanotubes synthesis by Fenton's reagent method and their potential application for removal of azo dye from aqueous solution. *Journal of Colloid and Interface Science*, *378*(1), 175–183. doi:10.1016/j.jcis.2012.04.024 PMID:22564767

Yuge, R., Toyama, K., Ichihashi, T., Ohkara, T., Aoki, Y., & Manako, T. (2012). Characterization and field mission properties of multiwalled carbon nanotubes with fine crystallinity prepared by CO_2 laser ablation. *Applied Surface Science*, *258*(18), 6958–6962. doi:10.1016/j.apsusc.2012.03.143

Zare, K., Sadegh, H., Shahryari-ghoshekandi, R., Maazinejad, B., Ali, V., Tyagi, I., Agarwal, S., & Gupta, V. K. (2015). Enhanced removal of toxic Congo red dye using multi walled carbon nanotubes: Kinetic, equilibrium studies and its comparison with other adsorbents. *Journal of Molecular Liquids*, *212*, 266–271. doi:10.1016/j.molliq.2015.09.027

Zhang, S., Quan, X., Zheng, J. F., & Wang, D. (2017). Probing the interphase "HO• zone" originated by carbon nanotubes during catalytic ozonation. *Water Research*, *122*, 86–95. doi:10.1016/j.watres.2017.05.063 PMID:28595124

Zhang, S. J., Shao, T., Kose, H. S., & Karanfil, T. (2012). Adsorption kinetics of aromatic compounds on carbon nanotubes and activated carbons. *Environmental Toxicology and Chemistry*, *31*(1), 79–85. doi:10.1002/etc.724 PMID:22021047

Chapter 10
Microencapsulation in Textiles: An Overview

Poonam Kumari
https://orcid.org/0000-0002-2952-6706
Chaudhary Charan Singh Haryana Agricultural University, India

ABSTRACT

Microencapsulation is a well-established process of enveloping or surrounding one substance into another substance that provides capsules having range from one micron to many hundred microns in size. One among the highly efficient methods is microencapsulation. The encapsulation efficiency of microcapsules, micro particles, or microspheres depend on various factors like solubility of polymer in solvent, concentration of polymer, solubility of organic solvent in water, rate of solvent removal, etc. Substances are often encapsulated in such a way that the core material is confined within capsule shells (coating material) for a particular interval of time. Different types of techniques are used for preparation of microcapsules. These techniques are utilized in different fields like pharmaceutical, agriculture, textile, food, printing, and defence. This text covers a review on microencapsulation and materials involved, microencapsulation techniques, and use of microencapsulation in textiles.

INTRODUCTION

Environment concerns and demands for environment friendly processing of textiles has led to the event of many new, cleaner, and greener technologies (Gunay, 2013). Science has come up with many technologies for the eco processing of textiles, which incorporate, Enzymatic Finishing, Plasma Technology, Finishing by Natural products and Microencapsulation (Shrimali and Dedhia, 2015). Fragrances play a crucial role in revitalizing the mind and also the body. Addition of fragrances to textiles has been there for several years within the sort of fabric conditioners for the aim of washing and while tumble drying. These products were designed to impart fresh aroma to the textiles treated with it, but couldn't retain the fragrances for an extended time. This became possible only with the appearance of the microencapsulation technology in textiles. It's an efficient method wont to control the discharge properties of active ingredients that lengthens the effect of fragrances. Microencapsulation is the technique where small solid particles,

DOI: 10.4018/978-1-6684-4553-2.ch010

liquid droplets, or minute bubbles of gas are coated, and when certain quantity or pressure is exerted or the surface is scratched, it releases the compounds inside the capsules. Currently microencapsulation is employed in textiles for anti-bacterial treatments, UV protection, for moisturizing and skin treatments, vital sign regulation, repellence, Industrial chemicals, Agro chemicals, Food additives, Adhesives and for perfume or fragrance releases. The applying of the technique for fragrance releases however is of high demand. (Umer et al., 2011). The core substances enclosed within the capsule are released instantly on the material, when a mechanical movement like abrasion, deformation and friction occurs and ultimately the active agents are released on the skin. There are various methods and techniques of microencapsulation which involves tiny solid, liquid or gas particles which are between 50 nm and 2 mm and are covered with a natural or synthetic polymeric membrane. The smaller the capsules, greater are the covering of the merchandise, and consequently longer the fragrances would last. The material that's encapsulated is named the fill, internal phase or core material. The outer wall that encloses the fill is termed the shell or the coating. Various materials are utilized in coating like modified polysaccharides, gums, lipids, protein materials, and ample of synthetic polymers. Utilizing a specific sort of core material and the covering relies upon the technique utilized for micro-encapsulation of aromas.

Encapsulation is basic to shield fragrances from oxidation in view of warmth, light, dampness and openness to different components in the course of their life. It likewise forestalls unpredictable mixes present in fragrances from getting dissipated. There are numerous physiochemical, compound, and mechanical strategies. Ensuing are a few strategies of micro-encapsulation of fragrances in textiles (Fibre2Fashion, n.d.).

TECHNIQUES OF MICROENCAPSULATION

Various techniques are existing for encapsulation of core material. Mainly the techniques are dived into three types listed in Table 1.

1. Physio-chemical
2. Physio -mechanical and
3. Chemical methods

Physio-Chemical

Phase Separation and Coacervation

Phase coacervation, including simple and complex coacervation, is one among the oldest and widely used microencapsulation techniques. This technique is based on the separation of macromolecular solution into two immiscible liquid phases corresponding to the coacervate and the dilute equilibrium phases. The simple coacervation method requires the use of one colloidal solute, whereas within the complex coacervation two oppositely charged colloid polymers are used. The microencapsulation process is administered in three or four consecutive steps under stirring, i.e., (i) dispersion of the active principle in a solution containing the surface-active hydrocolloid; (ii) precipitation of the hydrocolloid onto the dispersed droplets by decreasing the solubility of the hydrocolloid, with the utilization of a non-solvent,

Table 1. Different techniques used for microencapsulation

Physio-chemical methods	Physio –mechanical methods	Chemical methods
• Phase separation and Coacervation • Solvent evaporation • Supercritical CO2 assisted microencapsulation	• Spray drying and congealing • Air-suspension coating • Pan coating/fluid bed • centrifugal extrusion	• Interfacial polymerization • In situ polymerization • Matrix polymerization • Poly condensation

a pH or/and temperature change, or the addition of an electrolyte solution; (iii) addition of a second hydrocolloid to induce the polymer-polymer complex in the case of complex coacervation; and (iv) stabilization and hardening of the microcapsule shell by crosslinking agent addition, such formaldehyde, glutaraldehyde, tripolyphosphate or genipin. Essential oils are one among the foremost active principles, for textile applications, encapsulated via either simple coacervation with gum acacia (Sharma and Goel, 2018) or ethyl cellulose (Türkoğlu et al., 2017) or via complex coacervation with chitosan/gum Arabic (Sharkawy et al., 2017; Poonam et al., 2015) chitosan/ carboxymethyl cellulose (Roy et al., 2017) gelatin/ gum Arabic or gelatin/carboxy cellulose (Wijesirigunawardana and Perera, 2018).

Solvent Evaporation

This technique is based on the evaporation of a volatile solvent (such as chloroform or dichloromethane). Firstly, the solvent, active ingredient, and coating polymer are mixed then this mixture is dispersed in a continuous medium in which the polymer and active ingredient are not miscible, and the solvent is not miscible. This step is generally carried out by mechanical agitation, but can also be obtained by extrusion or by using static mixers, which allow the rapid and continuous production of an emulsion with a narrow size distribution. The formation of microparticle is obtained by removing the solvent, causing the solubility of the polymer to decrease and then precipitate. The solvent can be extracted by diluting the medium, adding a co-solvent. The microparticles obtained are then recovered by filtration, rinsed and dried. This process easily achieves excellent yields, close to 100%, rapid production of large quantities of suitable quality capsules and predictable release of the active ingredient. Some parameters such as low pressure contributing to extraction the high viscosity of the solvent or the high viscosity of the dispersed phase thus favour the quality the encapsulation of active drug ingredients. However, it is limited by using a volatile solvent (which must be recycled) and a polymer that is not soluble in the continuous phase. The particles obtained are microspheres containing 30 to 40% by weight of the active ingredient. The particle size depends on the formulation and solubility of the polymer, emulsion parameters, evaporation conditions, and physico-chemical parameters of the products used (Zimniewska et al., 2019).

Supercritical CO_2 Assisted Microencapsulation

Compressed carbon dioxide in the liquid or supercritical state is attractive as a solvent in microencapsulation processes. - Carbon dioxide is non-toxic, non-flammable, and inexpensive. - The high volatility of carbon dioxide allows it to be easily separated from polymeric materials by lowering pressure. - The supercritical fluid state is reached when the temperature and pressure of a substance are above its critical temperature and pressure. For carbon dioxide, the critical temperature is 31°C and the critical pressure is 74 bar. Phase diagram of CO_2. -Generally there are three steps in the impregnation: First, the polymer materials are exposed to supercritical CO2 for a while; then the solution of additives in CO_2 is introduced and the solute is transferred from CO_2 to polymer - Last, CO_2 is released and the solute is trapped in the polymer material. - When suspensions of polymer particles in water are exposed to supercritical CO_2 with the presence of additives in water, the transport of the additive into polymer particles can also be enhanced. After releasing CO_2, additives can be trapped in colloidal polymer particles (Essam. 2012).

Physio-Mechanical Methods

A low-cost commercial process in which microencapsulation is done by spray-drying and it is mostly used for the encapsulation of fragrances, flavors and oils. In which core particles are dispersed in a polymer solution and sprayed into hot chamber. The shell material solidifies onto the core particles as the solvent evaporates such that the microcapsules obtained are of poly-nuclear or matrix type. Very often the encapsulated particles are aggregated and the use of large amounts of core material can lead to uncoated particles. However, higher loadings of core particles of up to 50–60% have been reported. Water-soluble polymers are mainly used as shell materials because a solvent-borne system produces unpleasant odors and environmental problems (Ghosh, 2006).

- Spray drying and congealing
- Fluid bed coating
- Pan coating

Spray Drying and Congealing

It is a low-cost process in which microencapsulation is done by spray-drying. When an active material is dissolved or suspended during a melt or polymer solution and becomes trapped within the dried particle. The main advantage is that the power to handle labile materials thanks to the short contact time within the dryer, additionally the operation is economical. In modern spray dryers the viscosity of the solutions to be sprayed are often as high as 300mPa. Spray drying and spray congealing processes are similar therein both involve dispersing the core material during a liquefied coating substance and spraying or introducing the core - coating mixture into some condition, whereby, relatively rapid solidification (and formation) of the coating is affected. The principal difference between the two methods is that how coating solidification is accomplished. Coating solidification within the case of spray drying is affected by rapid evaporation of a solvent during which the coating material is dissolved. Coating solidification in spray congealing methods, however, is accomplished by thermally congealing a molten coating material or by solidifying a dissolved coating by introducing the coating - core material mixture into a non-solvent. Removal of the non-solvent or solvent from the coated product is then accomplished by sorption, extraction, or evaporation techniques (Garg et al., 2018).

Spray drying serves as a microencapsulation technique when an active material is dissolved or suspended in a melt or polymer solution and becomes trapped in the dried particle. The main advantages is the ability to handle labile materials because of the short contact time in the dryer, in addition, the operation is economical. In modern spray dryers the viscosity of the solutions to be sprayed can be as high as 300mPa. Spray drying and spray congealing processes are similar in that both involve dispersing the core material in a liquefied coating substance and spraying or introducing the core - coating mixture into some environmental condition, whereby, relatively rapid solidification (and formation) of the coating is affected. The principal difference between the two methods is the means by which coating solidification is accomplished. Coating solidification in the case of spray drying is effected by rapid evaporation of a solvent in which the coating material is dissolved. Coating solidification in spray congealing methods, however, is accomplished by thermally congealing a molten coating material or by solidifying a dissolved coating by introducing the coating - core material mixture into a non-solvent. Removal of the non-solvent or solvent from the coated product is then accomplished by sorption, extraction, or evaporation techniques.

Air-Suspension Coating

This procedure gives a better control and flexibility. The particles are coated while suspended in an upward-moving air stream. They are supported by a perforated plate having different patterns of holes inside and outside a cylindrical insert. Just sufficient air is permitted to rise through the outer annular space to fluidize the settling particles. Most of the rising air (usually heated) flows inside the cylinder, causing the particles to rise rapidly. At the top, as the air stream diverges and slows, they settle back onto the outer bed and move downward to repeat the cycle. The particles pass through the inner cylinder many times in a few minutes' methods. The air suspension process offers a wide variety of coating materials candidates for microencapsulation. The process has the capability of applying coatings in the form of solvent solutions, aqueous solution, emulsions, dispersions, or hot melt in equipment ranging in capacities from one pound to 990 pounds. Core materials comprised of micron or submicron particles can be effectively encapsulated by air suspension techniques, but agglomeration of the particles to some larger size is normally achieved (Bansode et al., 2010)

Pan Coating/Fluid Bed

Fluid-bed coating is a microencapsulation technique used extensively to encapsulate pharmaceuticals into coated particles or tablets (Teunou and Poncelet, 2002). It is a variation of the pan coating method. The pan coating process, widely used in the pharmaceutical industry, is among the oldest industrial procedures for forming small, coated particles or tablets. The particles are tumbled in a pan or other device while the coating material is applied slowly. The pipe of the blower stretches into pot for an evenly heating distribution while the coating pan is rotating. Solid particles greater than 600 microns in size are generally considered essential for effective coating (Thoke, 2012).

Chemical Methods

- Interfacial polymerization
- In situ polymerization
- Poly condensation
- Matrix polymerization

Interfacial Polymerization

According to this technique the monomer (alkyl acrylates) is added dropwise to the stirred aqueous polymerisation medium containing the fabric to be encapsulated (core material) and an appropriate emulsifier. The polymerisation begins and initially produced polymer molecules precipitate within the aqueous medium to make primary nuclei because the polymerisation proceeds, these nuclei grow gradually and simultaneously entrap the core material to make the ultimate microcapsules. Generally lipophilic materials (insoluble or scarcely soluble in water) are more suitable for encapsulation by this system. Insulin loaded poly (alkyl cyanoacrylate) nanocapsules are synthesised by using this system. Additionally, to the entrapment of drug during microcapsule formation, drug loading also can be accomplished by incubation of cyanoacrylate nanocapsules (empty nanocapsules) with the dissolved or finely dispersed drug (Salaün et al., 2008).

In Situ Polymerization

In situ polymerization is one among the chemical microencapsulation processes often used for technical applications, including textiles. The process takes place in oil-in-water emulsions; the result's nicely smooth, spherical, reservoir-type microcapsules with transparent polymeric pressure-sensitive microcapsule walls. Typical wall materials for in situ polymerization are aminoplast resins, such as melamine–formaldehyde, urea-formaldehyde, urea–melamine-formaldehyde or resorcinol-modified melamine–formaldehyde polymers. In situ processes can start either directly from amine and aldehyde monomers or from the precondensates. Typically, all materials for the formation of microcapsule wall originate from the continual aqueous phase of the oil-in-water emulsion system and thus need to be water-soluble. To achieve better process control and improved mechanical properties of microcapsules, modifying agents/protective colloids are added, like styrene-maleic acid anhydride copolymers, polyacrylic acid or acrylamidopropylsulfonate and methacrylic acid/acrylic acid copolymers (Jadupati et al., 2012; Dubey, 2009).

Interfacial Polycondensation

In interfacial polycondensation, the two reactants in a polycondensation meet at an interface and react rapidly. The basis of this method is the classical Schotten-Baumann reaction between an acid chloride and a compound containing an active hydrogen atom, such as an amine or alcohol, polyesters, polyurea, polyurethane. Under the right conditions, thin flexible walls form rapidly at the interface. A solution of the pesticide and a diacid chloride are emulsified in water and an aqueous solution containing an amine and a polyfunctional isocyanate is added. Base is present to neutralize the acid formed during the reaction. Condensed polymer walls form instantaneously at the interface of the emulsion droplets.

Matrix Polymerization

In a number of processes, a core material is imbedded in a polymeric matrix during formation of the particles. A simple method of this type is spray-drying, in which the particle is formed by evaporation of the solvent from the matrix material. However, the solidification of the matrix also can be caused by a chemical change ("Micro-encapsulation", 2022).

MICROENCAPSULES IN TEXTILES

Fragrance Finishes

Fragrance finishes are directly applied on to fibers and fabrics numerous times, but the aroma doesn't last for quite two wash cycles. Microencapsulation of fragrances may be a technique which when used on the material gives extended effect. This system is usually utilized in aromatherapy during which microcapsules may contain volatile oil flavours like lavender, rosemary, pine etc. This is basically done to treat insomnia, headache, and to stop bad odour.

Phase-Change Materials

Phase-change materials perform the function of adjusting the aggregation from solid to liquid within certain range of temperature. Microcapsules of phase-change materials reduce the effect of utmost variations in temperatures. This facilitates the thermoregulation of clothing and therefore the constant temperature is provided. These sorts of microcapsules are applied to different materials, vests, parkas, snowsuits, blankets, mattresses, duvets etc. (Nelson, 2002).

Fire Retardants

Microcapsules with fire retardant core were developed to beat the matter of reduced softness which is caused by the direct application of fireside retardant materials. They're applied to fabrics utilized in military applications like tentage (Nelson, 2002).

Polychromic and Thermo-chromic Microcapsules (Color-changing Technology)

The colour changing systems changes color response to temperature, which is termed as thermo chromatic and therefore the other changes color response to UV light, this is often referred to as photo chromatic. In textiles, polychromic and thermo-chromic microcapsules are often found in product labelling, medical and security applications. There are microencapsulated thermo-chromatic dyes that change colour at specific temperature - in response of human contact (Nelson, 2002).

Antimicrobials

Bacteria often cause microbiological decay of materials which successively causes loss of varied useful properties of materials. These problems are often prevented by the utilization of anti-microbial finishes which will be applied with the assistance of microencapsulation. This finish is particularly for textiles for medical and technical use. (Shrimali and Dedhia, 2015).

Counterfeiting

Imitation of high added value textiles, branded and designer goods are often addressed by the utilization of microencapsulation. Microcapsules applied to label contain a color former or an activator. By the utilization of UV Microencapsulation for textile finishing or a solvent, microcapsules break open, the content is released, colour is developed and during this way detection is achieved (Nelson, 2002)

CONCLUSION

In today's world of developing technologies, the technique of microencapsulation is applied in most the fields. It's become a prominently effective technique which reinforces the property imparted to the material and assures its durability. It's fascinating that our clothing is now ready to actively moisturize, heal and even can release fragrances to scale back anxiety. The growing health awareness among consumers is further propelling researchers to undertake and test all possible ingredients to deliver expected performance. New materials are being explored and a serious shift is towards the utilization of organic compounds both in sheath and core. There's little question that this technology features a promising future, however, one aspect that seems critical is that the intended delivery of the encapsulated core on particular external stimulus. There's a requirement to optimize the methods of manufacturing microcapsules and extend the time period of treated materials to realize large scale industrial production for every specific application. The samples of application of this system discussed during this paper are just a couple of of very interesting ones. A huge use of this system is often witnessed in functional finish fabrics, medical and healthcare textiles, aromatherapy, cosmetic textiles, and lots more.

REFERENCES

Bansode, S. S., Banarjee, S. K., Gaikwad, D. D., Jadhav, S. L., & Thorat, R. M. (2010). Microencapsulation: A Review. *International Journal of Pharmaceutical Sciences Review and Research, 1*(2).

Dubey, R. (2009). Microencapsulation technology and applications. *Defence Science Journal, 59*(1), 82.

Essam, J. (2012 April 12). *Microencapsulation methods* [PowerPoint slides]. https://www.slideshare.net/JehanEssam/microencapsulation-methods-12259395

Fibre2Fashion. (n.d.). *Microencapsulation of fragrances in textiles*. Fibre2Fashion. Retrieved from https://www.fibre2fashion.com/industry-article/6962/microencapsulation-of-fragrances-in-textiles

Garg, A., Chhipa, K., & Kumar, L. (2018). Microencapsulation techniques in pharmaceutical formulation. *European Journal Pharmaceutical and Medical Research, 5*(3), 199–206.

Ghosh, S. K. (2006). Functional coatings and microencapsulation: a general perspective. *Functional Coatings: By Polymer Microencapsulation*, 1-28.

Gunay, M. (Ed.). (2013). *Eco-friendly textile dyeing and finishing*. BoD–Books on Demand. doi:10.5772/3436

Jadupati, M., Tanmay, D., & Souvik, G. (2012). Microencapsulation: An indispensable technology for drug delivery system. *Int. Res. J. Pharm, 3*(4), 8–13.

Micro-encapsulation. (2022, April 28). In *Wikipedia*. https://en.wikipedia.org/wiki/Micro-encapsulation

Nelson, G. (2002). Application of microencapsulation in textiles. *International Journal of Pharmaceutics, 242*(1-2), 55–62.

Poonam, K., Rose, N. M., & Singh, S. S. J. (2015). Microencapsulation of lime essential oil for fragrant textiles. *Annals of Agri Bio Research, 20*(1), 152–157.

Roy, J. C., Ferri, A., Salaün, F., Giraud, S., Chen, G., & Jinping, G. (2017, October). Chitosan-carboxymethylcellulose based microcapsules formulation for controlled release of active ingredients from cosmeto textile. *IOP Conference Series. Materials Science and Engineering, 254*(7), 072020. doi:10.1088/1757-899X/254/7/072020

Salaün, F., Devaux, E., Bourbigot, S., & Rumeau, P. (2008). Preparation of multinuclear microparticles using a polymerization in emulsion process. *Journal of Applied Polymer Science, 107*(4), 2444–2452.

Sharkawy, A., Fernandes, I. P., Barreiro, M. F., Rodrigues, A. E., & Shoeib, T. (2017). Aroma-loaded microcapsules with antibacterial activity for eco-friendly textile application: Synthesis, characterization, release, and green grafting. *Industrial & Engineering Chemistry Research, 56*(19), 5516–5526. doi:10.1021/acs.iecr.7b00741

Sharma, R., & Goel, A. (2018). Development of insect repellent finish by a simple coacervation microencapsulation technique. *International Journal of Clothing Science and Technology, 30*(2), 152–158. doi:10.1108/IJCST-02-2017-0022

Shrimali, K., & Dedhia, E. M. (2015). Microencapsulating for Textile Finishing. *IOSR Journal of Polymer and Textile Engineering, 2*(2). https://www.iosrjournals.org/iosr-jpte/papers/Vol2-issue2/A0220104.pdf

Teunou, E., & Poncelet, D. (2002). Batch and continuous fluid bed coating–review and state of the art. *Journal of Food Engineering, 53*(4), 325–340.

Thoke, S. B. (2012 April 11). *A Seminar on Microencapsulation Techniques and Application* [PowerPoint slides]. Department of Pharmaceuticals, S.N.D College of Pharmacy. https://www.slideshare.net/thokesagar/sagar-thoke

Türkoğlu, G. C., Sariişik, A. M., Erkan, G., Kayalar, H., Kontart, O., & Öztuna, S. (2017). *Determination of antioxidant capacity of capsule loaded textiles*. Academic Press.

Umer, H., Nigam, H., Tamboli, A. M., & Nainar, M. S. M. (2011). Microencapsulation: Process, techniques and applications. *International Journal of Research in Pharmaceutical and Biomedical Sciences*, 2(2), 474–481.

Wijesirigunawardana, P. B., & Perera, B. G. K. (2018). Development of a cotton smart textile with medicinal properties using lime oil microcapsules. *Acta Chimica Slovenica, 65*(1), 150–159. doi:10.17344/acsi.2017.3727 PMID:29562116

Zimniewska, M., Pawlaczyk, M., Krucinska, I., Frydrych, I., Mikolajczak, P., Schmidt-Przewozna, K., Komisarczyk, A., Herczynska, L., & Romanowska, B. (2019). The influence of natural functional clothing on some biophysical parameters of the skin. *Textile Research Journal, 89*(8), 1381–1393. doi:10.1177/0040517518770680

Chapter 11
Potential Applications of Carbon Nanotubes for Environmental Protection

Ratnesh Das
Dr. Harisingh Gour Central University, India

Atish Roy
Dr. Harisingh Gour Central University, India

Pratibha Mishra
Dr. Harisingh Gour Central University, India

Indu Kumari
Chandigarh Group of Colleges, India

Arunesh K. Mishra
Dr. Harisingh Gour Central University, India

Sushil Kashaw
Dr. Harisingh Gour Central University, India

Anil K. Bahe
Dr. Harisingh Gour Central University, India

ABSTRACT

Carbon nanotubes (CNTs) are a unique carbon material with physical, chemical, mechanical, optical, structural, and electrical characteristics researched and tested for a wide range of uses. The safeguards of environmental health have been identified as one of the most critical sustainability goals in recent decades. When it concerns identifying atmospheric toxins, carbon nanotube-based detectors offer great sensibility and precision, along with carbon nanotubes displaying the ability for adsorption to remove impurities with great rates and excellent amelioration competency. Carbon nanotubes have made essential contributions to a responsible future in wastewater treatment, air pollution management, biotechnologies, nano sensors, and sorbents. Carbon nanotubes are also utilized as a reinforcing material in green nanocomposites, which are essential for achieving desired characteristics and are ecologically benign. The utilisation of carbon nanotubes as hybrid filters, nano sensors, sorbents, and other materials is covered in this chapter, as well as its advantages for the environment.

DOI: 10.4018/978-1-6684-4553-2.ch011

INTRODUCTION

In recent years, innovative nanometer-scale elements have been discovered, developed, and, in some cases, massively processed and distributed. These new nanomaterials are made up of inorganic or organic ingredients, and that they have just recently been studied in terms of their environmental impact. Carbon nanotubes are used in all of them (CNTs). Carbon nanotubes (CNTs) are carbon allotropes that are tubular and composed of graphite. Their diameter and chirality determine their values. Carbon nanotubes offer a broad radius-length ratio, muscular mechanical strength and long service life. Carbon nanotubes may be utilized as metallic, semi-conductive, or insulating materials, and they are proving to be quite valuable right now. The employment of high mechanical characteristics, exceptional electrical features, high chemical and thermal stability, and large, particular surface areas is expanding in prospective methods, such as wastewater treatment, air pollution, nanosensors, compost filters, absorbers and antimicrobial agents, etc.

Figure 1. Single wall nanotube (SWCNT) schematic designs and multi wall carbon nanotube schematics (MWCNT) (1).

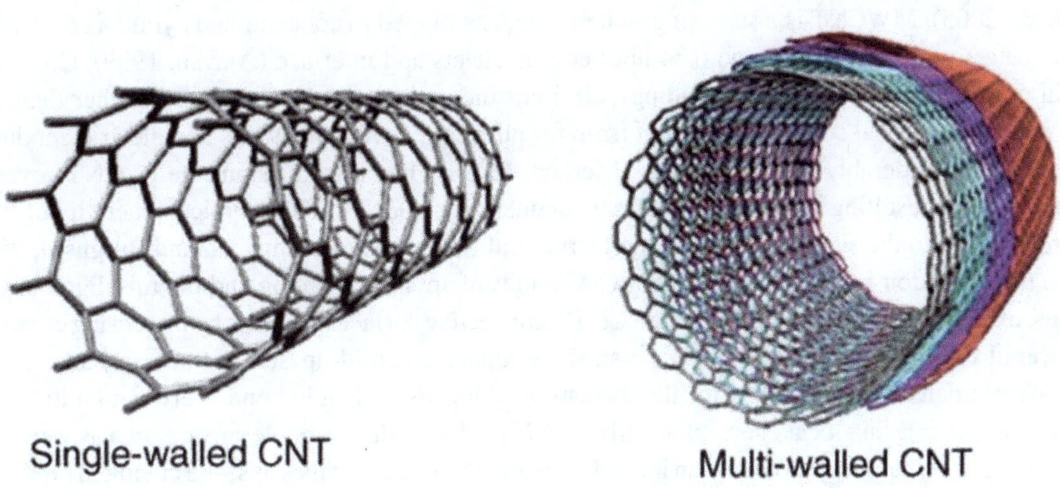

Radushkevich and Lukyanovich (1952) published explicit pictures of carbon tubes with diameters of 50 nanometers in the Soviet Journal of Physical Chemistry in 1952. In a 1976 publication, Oberlin, Endo and Koyama showed hollow, nanometer-wide carbon fibers utilizing a steam-growth technique. In addition, John Abrahamson gave evidence of carbon nanotubes during the 14th Carbon Biennial Conference in 1979 at Penn State University. According to a conference report, carbon nanotubes are carbon fibers produced on carbon anodes during arc release (Abrahamson et al., 1999). The conclusions of chemical and structural characterization of carbon nanoparticles synthesised by thermocatalytical disproportionation of carbon monoxide were published by Soviet researchers in 1981. The authors claimed that their "carbon multi-layer tubular crystals" were based on TEM images and XRD patterns to roll graphene layers into cylinders (Izvestiya Akademii Nauk SSSR, 1982). Hyperion Catalysis' In 1987, Howard G. Tenen received a US patent for "cylindrical discrete carbon fibroids" with a "constant diameter between approximately 3.5 and 70 nanometers, 102-fold in size and several layers of decreed carbon atoms and a separate inner centre, essentially continuous in the external area". In 1991, according to a significant number of scholarly and general literature, the Sumio Iijima company of Nippon Electric company was credited for the development of hollow, nanometer-sized pipes formed of graphical carbon. In 2006, a journal, carbon, published by Marc Monthioux and Vladimir Kuznetsov, identified the carbon nanotube's genesis (Monthioux and Kuznetsov, 2006). Carbon nanotubes contain a sp^2 bond comparable to graphite, with each atom linked to three neighbors. Therefore, the tubes can be seen as roll-up graphite sheets (graphene is a single layer of graphite). This binding mechanism, which is more robust than the sp^3 link in diamonds, gives the molecules unique strength. Nanotubes may merge at high pressure and exchange some sp^2 connections to produce sp^3 connections, which allow solid, end-to-end wires to form through high-pressure nanotubes. Carbon nanotubes (CNTs) have been the focus of considerable study and development since their discovery by Iijima (1991), because of their unique electrical, chemical, mechanical, and structural properties. CNTs, like graphene sheets, are produced in hexagonal networks from sp^2 carbon atoms (Merkoçi, 2006). These could be classified as single-walled carbon nanotubes (SWCNT) or multi-walled carbon nanotubes (MWCNTs). SWCNT is made by rolling a sheet of graphene into a cylinder and cutting it with hemispheric ends caused by the pentagon in its hexagonal carbon network. SWCNTs are around 0.4 to 3 nm in diameter and are many μm long (Balasubramanian and Burghard, 2005). MWCNT is a stack of graphene sheets wrapped in concentrated cylinders of 2 and 25 nm diameters with a gap of around 0.34 nm between sheets and interface (Ajayan, 1999). CNTs have several unique physical features, including 100 times the strength of steel's tensile, higher than most other materials thermal conductivity, apart from for pure copper-like diamonds and electric conductivity but have considerably higher currents (Merkoçi, 2006). The transmission rate of CNTs over the sidewall is high, resulting in the high electrical conductivity of CNTs. The chemical reactivity of CNTs is primarily due to the sidewall curvature and structural flaws (Balasubramanian and Burghard, 2008). Due to the distortion from the typically flat C–C bonds of sp^2 hybridization and the misalignment of p orbitals, exoedral chemical reactivity on the CNT's convective surface increases by increasing curvature. The overall chemical reactivity of CNTs is strongly reliant on build-up defects such as vacancies and Stone–Wales defects, which allow for the formation of localised double bonds between faulty carbon atoms, hence increasing local chemical reactivity. CNTs absorb the available surface area as well as the nanotube's pore type. The nanotubes can indeed be opened by removing the caps, and chemical processing of CNTs can substantially alter the adsorption characteristics by altering the surface and porous structure. In general, CNTs had greater condensing pressures and lower adsorption temperatures than graphite, which resulted in superior adsorption (Masenelli-Varlot et al., 2002). CNTs have two different areas: the sides of the CNT and tube ends and contribute to their electrochemical behavior. CNT purification, a result of chemical or thermal transformation, CNTs are partially oxidised, oxidised functional groups at tubular ends can be formed and defects eliminated (Gong et al., 2005).

The purified CNTs are cyclically voltammetric (CV) in a redox process, and the peaks attributable to the oxygenated functional groups in the CNTs are a redox process (Luo et al., 2001; Barisci et al., 2000). The CNT's sidewall's electrochemical behavior resembles the base level of highly orientated pyrolytic graphite, and their tube ends electrochemical conduct resembles the edge plane of highly oriented pyrolytic graphite (Banks and Compton, 2006). Carbon nanotubes may be utilized as energy storage, as a gaseous adsorption membrane, and applications to absorb hydrogen and separate gases for the environment. In this section, we explain the key principles underlying the function of carbon nanotubes, in particular for the treatment of wastewater, air pollution, creation of green nanocomposites, composite filters, antimi-

Table 1. Species that can be discovered using CNTs electrocatalysis (17).

Species	Media	CNTs
Catechol	PBS (pH 7.0)	MWNT
Dopamine	PBS (pH 6.9)	SW/MWNT
Epinephrine	B-R buffer (pH 6.9)	SWNT
AA	PBS (pH 7.0)	MWNT
DOPAC	PBS (pH 7.0)	MWNT
Nitric oxide	PBS (pH 7.0)	MWNT
Cysteine	0.2 M H2SO4	CNTPE
Glutathione	PBS (pH 7.0)	MWNT
Homocysteine	PBS (pH 7.0)	MWNT
NADH	PBS (pH 7.4)	MWNT
H_2O_2	PBS (pH 7.4)	MWNT
Uric acid	PBS (pH 7.0)	SWNT
Oxygen	0.1 M NaOH	MWNT
Glucose	0,1 M NaOH	MWNT
Adenine	Acetate buffer (pH 5.0)	CNTPE
Guanine	Acetate buffer (pH 5.0)	CNTPE
Insulin	PBS (pH 7.4)	MWNT

crobials, etc. We also address the possible applications of CNTs for environmental monitoring sensors and adsorbents and CNT-based green nanocomposite creation to protect the environment.

Nanotubes in the realm of nanotechnology are relatively young yet well researched materials. Due to their extraordinary mechanical, electrical, and chemical features, carbon nanotubes are promising to bring new and better technology in the environment, as sensors and sorbent material, to detect and treat current pollutants and avoid further environmental and engineering contamination. Nanotubes, existing as tiny particles, can potentially offer ecological risks on their own (Masciangioli and Zhang, 2003). Nanotechnology offers numerous problems and potential which scientists have just recently begun to investigate. Control of the environment authorities, academics and the public are worried that trace elements are increasingly mobilized and released into the environment. Heavy metals e.g., mercury (Hg), lead (Pb), cadmium (Cd)) and organic compounds are the substances of considerable significance for soil remediation, sediment, and groundwater (e.g., Cyclohexa-1,3,5-triene, chlorinated solvents and toluene). However, numerous types of trace pollutants were adsorbed by nanotubes (Li et al., 2002; 2003a). The physical and chemical composition of nanotube surfaces connected to sorbent capacity and interference with the gas mixture of other constituents lack basic comprehension in the water. Many items might be lighter, stronger, cleaner, cheaper, and more precise with nanotechnology. Nanotechnology can enhance environmental protection considerably. This chapter is, therefore, an attempt to describe the ecological protection status of carbon nanotubes.

WASTEWATER TREATMENT OF CARBON NANOTUBES

Water resources can become contaminated by a variety of processes, including metallurgy, mining, tanning, chemical production, fossil fuel refinery, battery manufacturing, and the creation of plastics, which often uses metal compounds, particularly as heat stabilisers. Additionally, more pollutants are emerging while traditional pollutants' problems have yet to be fully overcome. Clean water for all and environmental conservation are hampered by the removal of toxic metals (Muyibi et al., 2008; Bansal and Goyal, 2005). Water contaminated with metals can lead to a number of illnesses, including anaemia, cancer, kidney disease, and damage to the nervous system (Friberg et al., 1998; Calderón et al., 2001; Li et al., 2002). When carbon nanotubes (CNTs), a unique alloform of the carbon family, were first found in 1991 (Iijima, 1991), they displayed extraordinary physical, chemical, mechanical, electrical, and mechanical features (Ajayan, 1999; Terrones, 2003; Dai and Mau, 2001). This is the consequence of promising future applications in various fields, including applications in nanoelectronics (Collins et al., 1997), microelectronic devices (Javey et al., 2003), field emissions (de Heer et al., 1995; Wang et al., 1998), catalyst support (Planeix et al., 1994; Che et al., 1998), chemical sensors (Kong et al., 2000; Collins et al., 2000), and strengthening applications for composite materials (Dalton et al., 2003). CNTs are good candidates for kinetic adsorption because of their large, specific region, great temperature stability, and simple, mass manufacture (Wang et al., 2002). Comparing CNTs to activated carbon, the former has a substantially higher capacity for adsorption (AC). This is due to the surface area's expansion, which encourages CNT and dioxin contact. (Long and Yang, 2001). These CNTs have the potential to be heavy-weight, making them useful for removing radioactive nuclides (Li et al., 2003; Wang et al., 2005; Lin et al., 2006; Chen et al., 2007) as well as the heavier metal ions zinc (Zn) (Lu and Chiu, 2006; Lu et al., 2006), cadmium (Cd) (Li et al., 2003a; Vuković et al., 2010; Gao et al., 2008), lead (Pb) (Li et al., 2006 Peng et al., 2005), nickel (Ni) (Kandah and Meunier, 2007; Lu et al., 2006; Chen et al., 2006), copper (Li et al., 2003b, 2003c), fluoride (Li et al., 2001 Li et al., 2003d), and nickel (Ni).

CNTs have a great deal of promise and are also highly interested in gas adsorption. Studies on the adsorption of various gases, including as ammonia (Tuzen and Soylak, 2007; Bauschlicher and Ricca, 2004), nitrogen and methane (Bienfait et al., 2004; Talapatra and Migone, 2002), hydrogen (Dillon et al., 1997 Chen et al., 1999; Liu et al.,1999; Lee and Lee, 2000), ozone (Yim and Liu, 2004), carbon monoxide and carbon dioxide (Varghese et al., 2001), 1,2-dichlorobenzene (Lin et al., 2002 Peng et al., 2003), and dioxin, have been detailed in the CNTs (Long and Yang, 2001; Fagan et al., 2007). The use of CNTs for the adsorption of single or binary/tertiary wastewater heavy metal separation is also proposed as an important topic for discussion. The most promising choices for heavy metal separation and removal from wastewater treatment systems are CNTs. The CNTs have significant adsorption capacity due to their porosity, surface area, and broad spectrum of functional surface groups. The attraction, precipitation, and chemical interaction between metal ions and CNT surface function groups is what makes methods for integrating metal ions with CNTs so challenging (Rao et al., 2007). CNTs are essential for the removal of certain organic salts, toxic colours, and heavy metals from water. However, when CNTs fail to operate, a more exceptional adsorption ability towards non-polar molecules, such as polycyclic aromatic hydrocarbons, is observed (Yang et al., 2006; Wang et al., 2008). Lu et al. (2008) Ni^{2+}.'s adsorption and desorption in CNTs were demonstrated by regeneration study. However, after a few cycles, the amount of granular activated carbon (GAC) significantly decreased (Lu et al., 2008). It is more difficult to desorb Ni2+ because of the GAC's porous nature because the ions must travel from the inside. CNTs are currently used in technology as nano filters to reduce wastewater particles and serve as a sorbent for organic and inorganic contaminants (Srivastava et al., 2004; Jin et al., 2007; Tahaikt et al., 2007).

Similar to sorbents, different functionalities can be added to the pores to manage a specific CNT filter selectivity (Fornasiero et al., 2008). In spite of their hydrophobic characteristics, CNTs have demonstrated outstanding efficacy in water transportation. Due to the hydrophobic property of CNT pores, which denotes weak interactions with water molecules, quick, almost immature water circulation molecular dynamic simulations are possible (Noy et al., 2007). Hummer et al. (2001) claims that nanoscale confinement, which results in a narrowing of the interaction energy distribution and a reduction in the contact with water, is the source of the frictionless water flow. Additionally, recent CNT-based filtration studies have demonstrated the ability of Nanofilters CNT for the removal of pathogens from the surface of CNTs, such as bacteria, viruses, and protozoa, employing a deep-filtration process (Bohonak and Zydney, 2005; Brady-Estévez et al., 2008). Brady-Estévez et al. (2008) have reported the effective method for getting rid of E. coli bacteria using SWCNT filters. Using a microporous membrane made of the SWCNT's internal nanotube bundles, E. coli cells could be completely captured and held (PVDF). Also suggested was a modification that would immobilise SWCNTs on a microporous ceramic filter. In a different study, Mostafavi et al. (2009) created an adjustable nano-scale porosity CNT-based filter with maximal efficacy in MS2 virus eradication at a pressure of 8–11 bar using a spray pyrolysis approach. The usage of CNTs in wastewater treatment goes beyond filtration and sorbent; multiple studies found that CNTs had superb antibacterial properties (Mostafavi et al., 2009). This property makes carbon nanotubes an innovative and efficient alternative to chemical disinfectants for the treatment of micro-biological diseases. Utilizing CNTs in water disinfection treatment avoids the formation of dangerous disinfection byproducts (DBPs) such trihalomethanes, haloacetic acids, and aldehydes because they are not strong oxidants and are often inert in water (Savage and Diallo, 2005; Kang et al., 2007; Li et al., 2008; Nepal et al., 2008; Cortes et al., 2009).

CARBON NANOTUBES IN AIR POLLUTION

Researchers wanted to examine the possible uses of CNTs as sensing components for detecting and monitoring the concentration of hazardous gas discharged into the environment through their exceptional electrical, electrochemical, and optical characteristics (Wei et al., 2006 Van Hieu et al., 2008 Bondavalli et al., 2009; Di Francia et al., 2009; Lu et al., 2009; Penza et al., 2009a, 2009b; Zhang and Zhang, 2009). The CNTs feature unique and adjustable electronic elements, which substantially impact their metallic or semiconducting structure, including their size and chirality. CNT gas sensor has several benefits over conventional semi-conductor metal oxide gas sensors, such as low power consumption, low operating temperatures and high sensitivity (Endo et al., 2007). A 180-micron thick glass insulator is a thin film range of CNTs acting as a cathode. The single CNTs produce a strong electric field close to the ultra-fine tips and enhance the total area to accelerate the gas ionization process (Bogue, 2004). Due to direct gas contact, the detection by this gas sensor is dependent on the resistance or conducting change in CNT.

In numerous research studies, CNT-based gaseous sensors were utilized for the detection (Ueda et al., 2008a, 2008b) of nitrogen oxides (NO_2), (Kong et al., 2000; Cantalini et al., 2003; Valentini et al., 2004; Cho et al., 2006; Moon et al., 2001), ammonia (NH_3) (Nguyen et al., 2006; Quang et al., 2006; Nguyen et al., 2007), and room temperature of sulfide dioxide (SO_2) (Suehiro et al., 2005). Although the findings demonstrated a high and quick gas sensor response, the time-consuming recovery is challenging. Several techniques were suggested to improve gas desorption from the sensor. For example, the heating and flow rate of purging gases might increase the gas degradation from the sensor by utilizing UV lighting. Attempts have been undertaken to enhance CNT-based gas sensors' greater sensitivity and selectivity by polymerization (Wei et al., 2006; Lu et al., 2009).

CNTs in conveyor polymers like polyaniline and polypyrrole lead to increased sensitivity for specific gases or vapors. A polymer coating research (Qi et al., 2003) has shown that a polyethylene-coated CNT-based gas sensor offering a solid NO_2-detection affinity at concentrations less than one ppb without NH_3 interference due to the poor binding relationship and the adhesive coefficient of NH_3 on electron-rich

CNTs. In contrast to the PEI-coated sensor, a nafion (polymeric sulfonic acid ionomer) CNT-based gas sensor has increased selective capacity for CNT sensing by inhibiting the NO_2 of CNT. Nafion is also available. The production of a small wireless gas sensor based on a composite chemical resistor CNT/poly(methylmethacrylate) (PMMA) was also shown (Abraham et al., 2004) to enhance sensitivity to exposures such as dichloromethane, chloroform, or acetone vapor to volatile organic chemicals. Although CNT-based gas sensors compete with conventional metal oxide sensors, non-stop advancements to enhance conventional gas sensors have been implemented. In addition to the high working temperature limitations, the integration of CNTs in metal oxide sensors has been observed (Wei et al., 2004; Espinosa et al., 2007; Duc Hoa et al., 2007; Van Duy et al., 2008; Van Hieu et al., 2008; Wang et al., 2008). In addition to this, considerable improvements were seen in sensitivity and responsiveness time for detecting at room temperature of NO_2 and NH_3 using CNT/metal oxide sensors. The improved performance has been attributed to the efficient access of the gas to nanopassages and the change of conduct following CNT gas adsorption (Duc Hoa et al., 2007 Gong et al., 2008).

GREEN NANOCOMPOSITE DESIGN OF CARBON NANOTUBES

The production of waste is equal to the economic growth of the globe. Wastes, in particular synthetic polymer waste, create adverse environmental consequences. To solve that problem, it has been suggested by the Europe to develop a waste management strategy based on two complementary strategies: waste avoidance via improved product design and increased waste recycling with a concentration on life cycle assessment (Baillie, 2004). Within the wave of next-generation materials, the prospect for waste management has led to the problem of synthesizing the green nanocomposites by employing biodegradable polymers (Mojumdar and Raki, 2005; Wang et al., 2005). The green nanocomposite movement using renewable natural resources covers the idea of life cycle assessment that promotes waste recycling and reuse. Biodegradable polymers have a high potential economic value and, because of their degradability, have garnered a significant deal of attention from researchers to replenish non-renewable pyrotechnic polymers. However, most biodegradable polymers have lower mechanical characteristics and low-temperature distortion, limiting their usage in broad applications. As a result, CNTs can serve as nano reinforcements for biodegradable polymers to improve mechanical parts, increase durability, and increase thermal stability in a range of composite materials. The biodegradable nanocomposite quality is typically based on the alignment with CNT, the adhesion of CNT-biodegradable polymers and the distribution of CNT in a biodegradable matrix of polymers (Grossiord et al., 2005; Ray et al., 2006; Vaudreuil et al., 2006).

Song and Qiu (2009) confirmed that the thermal strength of poly (butylene succinate) (PBSU) has been increased by around 10°C after inclusion into CNTs. Sitharaman et al. (2008) have shown that ultra-short CNTs may substantially strengthen poly propylene fumarate (PPF) that is generally restricted to load-bearing applications because of their mechanical characteristics. Besides, CNT/PPF nanocomposites have compatibility in vivo that promise use as a bone tissue engineering prototype scaffold. The green nanocomposites give the capacity to recycle the inserted CNTs owing to the degradability of the biodegradable polymer. The biodegradation of the polymer can be accomplished under specified pH and temperature settings by both microbial degradation and enzyme degradation. The recovered CNTs can function as strengtheners for the production of newly developed composites following deterioration.

The method of reuse and recycling of CNTs can at the same time minimize waste disposal and make material treatment economical.

Carbon Nanotubes in Biotechnology

In recent years, the fast rise in biotechnology, where live organisms are used to produce goods or processes for specific usage, has resulted in an increased need for new, environmentally friendly technologies. The development of biotechnology allows CNTs to be involved particular in biological fuel cells (biofuel cells).

Biofuel cells, as specified by (Palmore and Whitesides, 1994), are fuel cells that use biocatalytic activity in power production. They may generally be categorized as microbial fuel cells (MFC) or enzyme biofuel cell (EFC). The MFC uses microbial catabolism for electrical power generation. Even as material resource containing intricate organic waste and renewable biomass in wastewater they have been regarded as future alternatives in wastewater treatment (Logan et al., 2006; Watanabe, 2008). But, mainly due to low performance and lack of technological maturity, the technology is not practicably practicable. Due to their high conductivity and wide area, much study was focused on using and modifying CNTs as electrodes to enhance MFC power generation (Morozan et al., 2007; Qiao et al., 2007). By coating CNTs on carbon cloth, Tsai et al. (2007) have produced a novel electrode design to develop a highly conductive MFC particular area electrode. The inclusion of CNTs improved the power density by 250% compared to an electrode not covered with a CNT. Sharma et al. (2008) tested the CNT electron performance against a graphite-flavored electrode; the CNT electrode was shown to increase the power density of the electrode by about six times over that of the graphite electrode. MFC biocompatibility of CNTs with microorganisms may be produced to optimize power density further.

Although CNTs have been asserted to be antibacterial, the cytotoxicity comportment diminishes after being modified and functionalized, as previously mentioned (Sayes et al., 2006). Staphylococcus aureus biocompatibility with Morozan et al. (2007) obligations in CNT modified cell culture media resulted in rapid microorganism multiplication. If MFC anodic design is applied, this suggests the possibility of decreasing energy loss. Instead of microbial catalytic activity, EFC employs enzymes or protein catalysis to convert chemical energy into electric energy. EFC's short-life, low stability, and low power density of enzymes continue to limit its use as a power source for low-power sensors, communications devices, and medical implants (Kim et al., 2006; Minteer et al., 2007). Nevertheless, CNTs are a significant advance for EFC as biologic electrodes. Besides the mediation of load transfer, CNTs showed considerable effectiveness in EFC by providing a robust enzyme immobilization platform (Fischback et al., 2006; Asuri et al., 2007; Li et al., 2008 Zhao et al., 2009). CNTs enable the covalent attachment of enzyme molecules to their surface and encourage large loads of enzymes by allowing cross-linked enzymes to be assembled. The stability of the Enzyme in EFC is significant because it gives outstanding operational resilience, which is expected to pave the way to increase the power density and extend the life of EFC. For example, the stable activity of CNT-coated glucose Oxidase allowed for more than 16 hours of continuous operation by an EFC (Fischback et al., 2006). Furthermore, CNTs can improve enzyme stability in the nano-scale environment. The covalent connecting of enzymes to CNTs have been shown to have a high level of peace because of the covalent connection (Govardhan, 1999; Sheldon, 2007). In another sense, CNT curvature increases the distance between enzyme molecules, reducing harmful interactions between enzymes and improving the enzymes (Asuri et al., 2006).

CONCLUSION

CNTs applications are increasingly important from a research perspective from their prospective usage for future energy conversion and storage technology, environmental monitoring and wastewater treatment, and green nanocomposite design. This review looks at the characteristics of CNTs in electrical and energy appliances as the electrode material or conductive filter. In addition to the growth as adsorbents to remove contaminants in wastewater treatment, the potential development of CNTs as a sensing material for environmental monitoring in nanosensors is highlighted. The concluding section describes the benefits of green nanocomposites conferred by CNTs and their effect on the environment. CNTs with their exceptional structural, electrical and mechanical characteristics revealed beneficial potential in energy and environmental applications. Developing integrated energy conversion technology developed by CNT with an effective energy storage system has demonstrated successful advances in tackling energy challenges for clean, sustainable energy sources in the future. In the field of environmental pollution analysis, industrial emissions monitoring and control as well as wastewater treatment, CNTs are a fantastic option for developing sensor and adsorbent products. In addition, the creation of green nanocomposites solves the environmental problems of synthetic polymers and finds intriguing potentials in several sectors. In short, CNTs provide new possibilities for energy and ecological growth with beneficial potential. Intensive research is needed to enhance further the performance and the practical uses of integrated CNT devices towards marketing and protection of the environment.

REFERENCES

Abraham, J. K., Philip, B., Witchurch, A., Varadan, V. K., & Reddy, C. C. (2004). A compact wireless gas sensor using a carbon nanotube/PMMA thin film chemiresistor. *Smart Materials and Structures*, *13*(5), 1045–1049. https://doi.org/10.1088/0964-1726/13/5/010

Abrahamson, J., Wiles, P. G., & Rhoades, B. L. (1999). Structure of carbon fibers found on carbon arc anodes. *Carbon*, *37*(11), 1873–1874. doi:10.1016/S0008-6223(99)00199-2

Ajayan, P. M. (1999). Nanotubes from carbon. *Chemical Reviews*, *99*(7), 1787–1800. https://doi.org/10.1021/cr970102g

Asuri, P., Bale, S. S., Pangule, R. C., Shah, D. A., Kane, R. S., & Dordick, J. S. (2007). Structure, function, and stability of enzymes covalently attached to single-walled carbon nanotubes. *Langmuir*, *23*(24), 12318–12321. https://doi.org/10.1021/la702091c

Asuri, P., Karajanagi, S. S., Yang, H., Yim, T. J., Kane, R. S., & Dordick, J. S. (2006). Increasing protein stability through control of the nanoscale environment. *Langmuir*, *22*(13), 5833–5836. https://doi.org/10.1021/la0528450

Baillie, C. (2004). Green composites. In Polymer Composites and the Environment. CRC Press.

Balasubramanian, K., & Burghard, M. (2005). Chemically functionalized carbon nanotubes. *Small*, *1*(2), 180–192. doi:10.1002mll.200400118 PMID:17193428

Balasubramanian, K., & Burghard, M. (2008). Electrochemically functionalized carbon nanotubes for device applications. *Journal of Materials Chemistry, 18*(26), 3071–3083. doi:10.1039/b718262g

Banks, C. E., & Compton, R. G. (2006). New electrodes for old: From carbon nanotubes to edge plane pyrolytic graphite. *Analyst (London), 131*(1), 15–21. doi:10.1039/B512688F PMID:16425467

Bansal, R. C., & Goyal, M. (2005). *Activated carbon adsorption*. Taylor & Francis Group.

Barisci, J. N., Wallace, G. G., & Baughman, R. H. (2000). Electrochemical studies of single-wall carbon nanotubes in aqueous solutions. *Journal of Electroanalytical Chemistry (Lausanne, Switzerland), 488*(2), 92–98. doi:10.1016/S0022-0728(00)00179-0

Bauschlicher, C. W., & Ricca, A. (2004). Binding of NH3 to graphite and to a (9,0) carbon nanotube. *Physical Review. Part B, 70*(11), 115409–115413. doi:10.1103/PhysRevB.70.115409

Bienfait, M., Zeppenfeld, P., Dupont-Pavlovsky, N. D., Muris, M., Johnson, M. R., Wilson, T., DePies, M., & Vilches, O. E. (2004). Thermodynamics and structure of hydrogen, methane, argon, oxygen, and carbon dioxide adsorbed on single-wall carbon nanotube bundles. *Physical Review. Part B, 70*(3), 035410–035419. doi:10.1103/PhysRevB.70.035410

Bogue, R. W. (2004). Nanotechnology: What are the prospects for sensors? *Sensor Review, 24*(3), 253–260. https://doi.org/10.1108/02602280410545362

Bohonak, D. M., & Zydney, A. L. (2005). Compaction and permeability effects with virus filtration membranes. *Journal of Membrane Science, 254*(1–2), 71–79. https://doi.org/10.1016/j.memsci.2004.12.035

Bondavalli, P., Legagneux, P., & Pribat, D. (2009). Carbon nanotubes based transistors as gas sensors: State of the art and critical review. *Sensors and Actuators. Part B, 140*(1), 304–318. doi:10.1016/j.snb.2009.04.025

Brady-Estévez, A. S., Kang, S., & Elimelech, M. (2008). A single-walled-carbon-nanotube filter for removal of viral and bacterial pathogens. *Small, 4*(4), 481–484. https://doi.org/10.1002/smll.200700863

Calderón, J., Navarro, M. E., Jimenez-Capdeville, M. E., Santos-Diaz, M. A., Golden, A., Rodriguez-Leyva, I., Borja-Aburto, V., & Díaz-Barriga, F. (2001). Exposure to arsenic and lead and neuropsychological development in Mexican children. *Environmental Research, 85*(2), 69–76. https://doi.org/10.1006/enrs.2000.4106

Cantalini, C., Valentini, L., Lozzi, L., Armentano, I., Kenny, J. M., & Santucci, S. (2003). NO2 gas sensitivity of carbon nanotubes obtained by plasma enhanced chemical vapor deposition. *Sensors and Actuators. Part B, 93*(1–3), 333–337. doi:10.1016/S0925-4005(03)00224-7

Che, G., Lakshmi, B. B., Fisher, E. R., & Martin, C. R. (1998). Carbon nanotubule membranes for electrochemical energy storage and production. *Nature, 393*(6683), 346–349. https://doi.org/10.1038/30694

Chen, C., Li, X., Zhao, D., Tan, X., & Wang, X. (2007). Adsorption kinetic, thermodynamic and desorption studies of Th(IV) on oxidized multi-wall carbon nanotubes. *Colloids and Surfaces. A, Physicochemical and Engineering Aspects, 302*(1–3), 449–454. https://doi.org/10.1016/j.colsurfa.2007.03.007

Chen, C., & Wang, X. (2006). Adsorption of Ni(II) from aqueous solution using oxidized multiwall carbon nanotubes. *Industrial & Engineering Chemistry Research*, *45*(26), 9144–9149. https://doi.org/10.1021/ie060791z

Chen, P., Wu, X., Lin, J., & Tan, K. L. (1999). High H2 uptake by alkali-doped carbon nanotubes under ambient pressure and moderate temperatures. *Science*, *285*(5424), 91–93. https://doi.org/10.1126/science.285.5424.91

Cho, W. S., Moon, S. I., Paek, K. K., Lee, Y. H., Park, J. H., & Ju, B. K. (2006). Patterned multiwall carbon nanotube films as materials of NO2 gas sensors. *Sensors and Actuators. Part B, 119*(1), 180–185. doi:10.1016/j.snb.2005.12.004

Collins, P. G., Bradley, K., Ishigami, M., & Zettl, A. (2000). Extreme oxygen sensitivity of electronic properties of carbon nanotubes. *Science*, *287*(5459), 1801–1804. https://doi.org/10.1126/science.287.5459.1801

Collins, P. G., Zettl, A., Bando, H., Thess, A., & Smalley, R. E. (1997). Nanotube nanodevice. *Science*, *278*(5335), 100–102. https://doi.org/10.1126/science.278.5335.100

Cortes, P., Deng, S., & Smith, G. B. (2009). The adsorption properties of bacillus atrophaeus spores on singlewall carbon nanotubes. *Journal of Sensors*, *2009*, 1–6. https://doi.org/10.1155/2009/131628

Dai, L., & Mau, A. W. H. (2001). Controlled synthesis and modification of carbon nanotubes and C60: Carbon nanostructures for advanced polymeric composite materials. *Advanced Materials*, *13*(12–13), 899–913. https://doi.org/10.1002/1521-4095(200107)13:12/13<899::AID-ADMA899>3.0.CO;2-G

Dalton, A. B., Collins, S., Muñoz, E., Razal, J. M., Ebron, V. H., Ferraris, J. P., Coleman, J. N., Kim, B. G., & Baughman, R. H. (2003). Super-tough carbonnanotube fibers. *Nature*, *423*(6941), 703. https://doi.org/10.1038/423703a

de Heer, W. A. D., Châtelain, A., & Ugarte, D. (1995). A carbon nanotube fieldemission electron source. *Science*, *270*(5239), 1179–1180. https://doi.org/10.1126/science.270.5239.1179

Di Francia, G., Alfano, B., & La Ferrara, V. (2009). Conductometric gas nanosensors. *Journal of Sensors*, *2009*, 1–18. https://doi.org/10.1155/2009/659275

Dillon, A. C., Jones, K. M., Bekkedahl, T. A., Kiang, C. H., Bethune, D. S., & Heben, M. J. (1997). storage of hydrogen in single walled carbon nanotubes. *Nature*, *386*(6623), 377–379. https://doi.org/10.1038/386377a0

Duc Hoa, N., Van Quy, N., Suk Cho, Y., & Kim, D. (2007). Nanocomposite of SWNTs and SnO2 fabricated by soldering process for ammonia gas sensor application. *Physica Status Solidi*, *204*(6), 1820–1824. https://doi.org/10.1002/pssa.200675318

Endo, M., Strano, M. S., & Ajayan, P. M. (2007). Potential applications of carbon nanotubes. *Topics in Applied Physics*, 13–62. doi:10.1007/978-3-540-72865-8_2

Espinosa, E. H., Ionescu, R., Chambon, B., Bedis, G., Sotter, E., Bittencourt, C., Felten, A., Pireaux, J. J., Correig, X., & Llobet, E. (2007). Hybrid metal oxide and multiwall carbon nanotube films for low temperature gas sensing. *Sensors and Actuators. Part B, 127*(1), 137–142. doi:10.1016/j.snb.2007.07.108

Fagan, S. B., Santos, E. J. G., Souza Filho, A. G., Mendes Filho, J., & Fazzio, A. (2007). Ab initio study of 2,3,7,8-tetrachlorinated dibenzo-p-dioxin adsorption on single wall carbon nanotubes. *Chemical Physics Letters*, *437*(1–3), 79–82. https://doi.org/10.1016/j.cplett.2007.01.071

Fischback, M., Youn, J., Zhao, X., Wang, P., Park, H., Chang, H., Kim, J., & Ha, S. (2006). Miniature biofuel cells with improved stability under continuous operation. *Electroanalysis*, *18*(19–20), 2016–2022. https://doi.org/10.1002/elan.200603626

Fornasiero, F., Park, H. G., Holt, J. K., Stadermann, M., Grigoropoulos, C. P., Noy, A., & Bakajin, O. (2008). Ion exclusion by sub-2-nm carbon nanotube pores. *Proceedings of the National Academy of Sciences of the United States of America*, *105*(45), 17250–17255. https://doi.org/10.1073/pnas.0710437105

Friberg, L., Nordberg, G. F., & Vouk, V. B. (1979). *Handbook on the toxicology of metals*. Elsevier/North-Holland.

Gao, Z., Bandosz, T. J., Zhao, Z., Han, M., Liang, C., & Qiu, J. (2008). Investigation of the role of surface chemistry and accessibility of cadmium adsorption sites on open-surface carbonaceous materials. *Langmuir*, *24*(20), 11701–11710. https://doi.org/10.1021/la703638h

Gong, J., Sun, J., & Chen, Q. (2008). Micromachined solgel carbon nanotube/SnO2 nanocomposite hydrogen sensor. *Sensors and Actuators. Part B*, *130*(2), 829–835. doi:10.1016/j.snb.2007.10.051

Gong, K., Yan, Y., Zhang, M., Su, L., Xiong, S., & Mao, L. (2005). Electrochemistry and electroanalytical applications of carbon nanotubes: A review. *Analytical Sciences*, *21*(12), 1383–1393. doi:10.2116/analsci.21.1383 PMID:16379375

Govardhan, C. P. (1999). Crosslinking of enzymes for improved stability and performance. *Current Opinion in Biotechnology*, *10*(4), 331–335. https://doi.org/10.1016/S0958-1669(99)80060-3

Grossiord, N., Loos, J., & Koning, C. E. (2005). Strategies for dispersing carbon nanotubes in highly viscous polymers. *Journal of Materials Chemistry*, *15*(24), 2349. https://doi.org/10.1039/b501805f

Hummer, G., Rasaiah, J. C., & Noworyta, J. P. (2001). Water conduction through the hydrophobic channel of a carbon nanotube. *Nature*, *414*(6860), 188–190. https://doi.org/10.1038/35102535

Iijima, S. (1991). Helical microtubules of graphitic carbon. *Nature*, *354*(6348), 56–58. doi:10.1038/354056a0

Izvestiya Akademii Nauk, S. S. S. R. (1982). Article. *Metals*, *3*, 12–17.

Javey, A., Guo, J., Wang, Q., Lundstrom, M., & Dai, H. J. (2003). Ballistic carbon nanotube field-effect transistors. *Nature*, *424*(6949), 654–657. https://doi.org/10.1038/nature01797

Jin, S., Fallgren, P. H., Morris, J. M., & Chen, Q. (2007). Removal of bacteria and viruses from waters using layered double hydroxide nanocomposites. *Science and Technology of Advanced Materials*, *8*(1–2), 67–70. https://doi.org/10.1016/j.stam.2006.09.003

Kandah, M. I., & Meunier, J. L. (2007). Removal of nickel ions from water by multi-walled carbon nanotubes. *Journal of Hazardous Materials*, *146*(1–2), 283–288. https://doi.org/10.1016/j.jhazmat.2006.12.019

Kang, S., Pinault, M., Pfefferle, L. D., & Elimelech, M. (2007). Single-walled carbon nanotubes exhibit strong antimicrobial activity. *Langmuir*, *23*(17), 8670–8673. https://doi.org/10.1021/la701067r

Kim, J., Jia, H., & Wang, P. (2006). Challenges in biocatalysis for enzyme-based biofuel cells. *Biotechnology Advances*, *24*(3), 296–308. https://doi.org/10.1016/j.biotechadv.2005.11.006

Kong, J., Franklin, N. R., Zhou, C., Chapline, M. G., Peng, S., Cho, K., & Dai, H. (2000). Nanotube molecular wires as chemical sensors. *Science*, *287*(5453), 622–625. https://doi.org/10.1126/science.287.5453.622

Lee, S. M., & Lee, Y. H. (2000). Hydrogen storage in single-walled carbon nanotubes. *Applied Physics Letters*, *76*(20), 2877–2879. https://doi.org/10.1063/1.126503

Li, Q., Mahendra, S., Lyon, D. Y., Brunet, L., Liga, M. V., Li, D., & Alvarez, P. J. J. (2008). Antimicrobial nanomaterials for water disinfection and microbial control: Potential applications and implications. *Water Research*, *42*(18), 4591–4602. https://doi.org/10.1016/j.watres.2008.08.015

Li, X., Zhou, H., Yu, P., Su, L., Ohsaka, T., & Mao, L. (2008b). A Miniature glucose/O2 biofuel cell with single-walled carbon nanotubes-modified carbon fiber microelectrodes as the substrate. *Electrochemistry Communications*, *10*(6), 851–854. https://doi.org/10.1016/j.elecom.2008.03.019

Li, Y. H., Ding, J., Luan, Z., Di, Z., Zhu, Y., Xu, C., Wu, D., & Wei, B. (2003c). Competitive adsorption of Pb2+, Cu2+ and CD2+ ions from aqueous solutions by multiwalled carbon nanotubes. *Carbon*, *41*(14), 2787–2792. https://doi.org/10.1016/S0008-6223(03)00392-0

Li, Y. H., Luan, Z., Xiao, X., Zhou, X., Xu, C., Wu, D., & Wei, B. (2003b). Removal $Cu2^{+}$ ions from aqueous solutions by carbon nanotubes. *Adsorption Science and Technology*, *21*, 475–485.

Li, Y. H., Wang, S., Cao, A., Zhao, D., Zhang, X., Xu, C., Luan, Z., Ruan, D., Liang, J., Wu, D., & Wei, B. (2001). Adsorption of fluoride fromwater by amorphous alumina supported on carbon nanotubes. *Chemical Physics Letters*, *350*(5–6), 412–416. https://doi.org/10.1016/S0009-2614(01)01351-3

Li, Y. H., Wang, S., Wei, J., Zhang, X., Xu, C., Luan, Z., Wu, D., & Wei, B. (2002). Lead adsorption on carbon nanotubes. *Chemical Physics Letters*, *357*(3-4), 263–266. doi:10.1016/S0009-2614(02)00502-X

Li, Y. H., Wang, S., Zhang, X., Wei, J., Xu, C., Luan, Z., & Wu, D. (2003d). Adsorption of fluoride from water by aligned carbon nanotubes. *Materials Research Bulletin*, *38*(3), 469–476. https://doi.org/10.1016/S0025-5408(02)01063-2

Li, Y. H., Wang, S. W., Luan, Z. L., Ding, J. D., Xu, C. X., & Wu, D. (2003a). Adsorption of cadmium (II). *Carbon*, *41*(5), 1057–1062. https://doi.org/10.1016/S0008-6223(02)00440-2

Li, Y. H., Zhu, Y., Zhao, Y., Wu, D., & Luan, Z. (2006). Different morphologies of carbon nanotubes effect on the lead removal from aqueous solution. *Diamond and Related Materials*, *15*(1), 90–94. https://doi.org/10.1016/j.diamond.2005.07.004

Lin, H. F., Ravikrishna, R., & Valsaraj, K. T. (2002). Reusable adsorbents for dilute solution separation. 6. Batch and continuous reactors for the adsorption and degradation of 1,2- dichlorobenzene from dilute wastewater streams using titania as a photocatalyst. *Separation and Purification Technology*, *28*(2), 87–102. https://doi.org/10.1016/S1383-5866(02)00017-5

Lin, K., Xu, Y., He, G., & Wang, X. (2006). The kinetic and thermodynamic analysis of Li ion in multi-walled carbon nanotubes. *Materials Chemistry and Physics*, *99*(2–3), 190–196. https://doi.org/10.1016/j.matchemphys.2005.09.035

Liu, C., Fan, Y. Y., Liu, M., Cong, H. T., Cheng, H. M., & Dresselhaus, M. S. (1999). Hydrogen storage in single-walled carbon nanotube at room temperature. *Science*, *286*(5442), 1127–1129. https://doi.org/10.1126/science.286.5442.1127

Logan, B. E., Hamelers, B., Rozendal, R., Schröder, U., Keller, J., Freguia, S., Aelterman, P., Verstraete, W., & Rabaey, K. (2006). Microbial fuel cells: Methodology and technology. *Environmental Science & Technology*, *40*(17), 5181–5192. https://doi.org/10.1021/es0605016

Long, R. Q., & Yang, R. T. (2001). Carbon nanotubes as superior sorbent for dioxin removal. *Journal of the American Chemical Society*, *123*(9), 2058–2059. https://doi.org/10.1021/ja0038301

Lu, C., & Chiu, H. (2006). Adsorption of zinc (II). *Chemical Engineering Science*, *61*(4), 1138–1145. https://doi.org/10.1016/j.ces.2005.08.007

Lu, C., Chiu, H., & Liu, C. (2006). Removal of zinc (II). *Industrial & Engineering Chemistry Research*, *45*(8), 2850–2855. https://doi.org/10.1021/ie051206h

Lu, C., & Liu, C. (2006). Removal of nickel (II). *Journal of Chemical Technology and Biotechnology (Oxford, Oxfordshire)*, *81*(12), 1932–1940. https://doi.org/10.1002/jctb.1626

Lu, C., Liu, C., & Rao, G. P. (2008). Comparisons of sorbent cost for the removal of Ni^{2+} from aqueous solution by carbon nanotubes and granular activated carbon. *Journal of Hazardous Materials*, *151*(1), 239–246. https://doi.org/10.1016/j.jhazmat.2007.05.078

Lu, J., Kumar, B., Castro, M., & Feller, J. F. (2009). Vapour sensing with conductive polymer nanocomposites (CPC): Polycarbonate-carbon nanotubes transducers with hierarchical structure processed by spray layer by layer. *Sensors and Actuators. Part B*, *140*(2), 451–460. doi:10.1016/j.snb.2009.05.006

Luo, H., Shi, Z., Li, N., Gu, Z., & Zhuang, Q. (2001). Investigation of the electrochemical and electro catalytic behaviour of single-wall carbon nanotube film on a glassy carbon electrode. *Analytical Chemistry*, *73*(5), 915–920. doi:10.1021/ac0009671 PMID:11289436

Masciangioli, T., & Zhang, W. X. (2003). Environmental technologies at the nanoscale. *Environmental Science & Technology*, *37*(5), 102A–108A. doi:10.1021/es0323998 PMID:12666906

Masenelli-Varlot, K., McRae, E., & Dupont-Pavlovsky, N. (2002). Comparative adsorption of simple molecules on carbon nanotubes. *Applied Surface Science*, *196*(1–4), 209–215. doi:10.1016/S0169-4332(02)00059-4

Merkoçi, A. (2006). Carbon nanotubes in analytical sciences. *Mikrochimica Acta*, *152*(3–4), 157–174. doi:10.100700604-005-0439-z

Minteer, S. D., Liaw, B. Y., & Cooney, M. J. (2007). Enzyme-based biofuel cells. *Current Opinion in Biotechnology*, *18*(3), 228–234. https://doi.org/10.1016/j.copbio.2007.03.007

Mojumdar, S. C., & Raki, L. (2005). Preparation and properties of calcium silicate hydrate- poly(vinyl alcohol) nanocomposite materials. *Journal of Thermal Analysis and Calorimetry*, *82*(1), 89–95. doi:10.1007/s10973-005-0846-8

Monthioux, M., & Kuznetsov, V. L. (2006). Who should be given the credit for the discovery of carbon nanotubes? *Carbon, 44*(9), 1621–1623. doi:10.1016/j.carbon.2006.03.019

Moon, S. I., Paek, K. K., Lee, Y. H., Park, H. K., Kim, J. K., Kim, S. W., & Ju, B. K. (2008). Biasheating recovery of MWCNT gas sensor. *Materials Letters*, *62*(16), 2422–2425. https://doi.org/10.1016/j.matlet.2007.12.027

Morozan, A., Stamatin, L., Nastase, F., Dumitru, A., Vulpe, S., Nastase, C., Stamatin, I., & Scott, K. (2007). The biocompatibility microorganisms-carbon nanostructures for applications in microbial fuel cells. *Physica Status Solidi*, *204*(6), 1797–1803. https://doi.org/10.1002/pssa.200675344

Mostafavi, S. T., Mehrnia, M. R., & Rashidi, A. M. (2009). Preparation of nanofilter from carbon nanotubes for application in virus removal from water. *Desalination*, *238*(1–3), 271–280. https://doi.org/10.1016/j.desal.2008.02.018

Muyibi, S. A., Ambali, A. R., & Eissa, G. S. (2008). Development-induced water pollution in Malaysia: Policy implication from an econometric analysis. *Water Policy*, *10*(2), 193–206. doi:10.2166/wp.2008.039

Nepal, D., Balasubramanian, S., Simonian, A. L., & Davis, V. A. (2008). Strong antimicrobial coatings: Single-walled carbon nanotubes armored with biopolymers. *Nano Letters*, *8*(7), 1896–1901. https://doi.org/10.1021/nl080522t

Nguyen, H. Q., & Huh, J. S. (2006). Behavior of singlewalled carbon nanotube-based gas sensors at various temperatures of treatment and operation. *Sensors and Actuators. Part B, 117*(2), 426–430. doi:10.1016/j.snb.2005.11.056

Nguyen, L. H., Phi, T. V., Phan, P. Q., Vu, H. N., Nguyen-Duc, C., & Fossard, F. (2007). Synthesis of multi-walled carbon nanotubes for NH3 gas detection. *Physica E, Low-Dimensional Systems and Nanostructures*, *37*(1–2), 54–57. https://doi.org/10.1016/j.physe.2006.12.006

Noy, A., Park, H. G., Fornasiero, F., Holt, J. K., Grigoropoulos, C. P., & Bakajin, O. (2007). Nanofluidics in carbon nanotubes. *Nano Today*, *2*(6), 22–29. https://doi.org/10.1016/S1748-0132(07)70170-6

Oberlin, A., Endo, M., & Koyama, T. (1976). Filamentous growth of carbon through benzene decomposition. *Journal of Crystal Growth*, *32*(3), 335–349. doi:10.1016/0022-0248(76)90115-9

Palmore, G. T. R., & Whitesides, G. M. (1994). *Enzymatic conversion of biomass for fuels cells, 566*. American Chemical Society.

Peng, X., Li, Y., Luan, Z., Di, Z., Wang, H., Tian, B., & Jia, Z. (2003). Adsorption of 1,2-dichlorobenzene from water to carbon nanotubes. *Chemical Physics Letters*, *376*(1–2), 154.

Peng, X., Luan, Z., Di, Z., Zhang, Z., & Zhu, C. (2005). Carbon nanotubes–iron oxides magnetic composites as adsorbent for removal of Pb(II) and Cu(II) from water. *Carbon*, *43*(4), 880–883. https://doi.org/10.1016/j.carbon.2004.11.009

Penza, M., Rossi, R., Alvisi, M., Cassano, G., & Serra, E. (2009a). Functional characterization of carbon nanotube networked films functionalized with tuned loading of Au nanoclusters for gas sensing applications. *Sensors and Actuators. Part B, 140*(1), 176–184. doi:10.1016/j.snb.2009.04.008

Penza, M., Rossi, R., Alvisi, M., Signore, M. A., Cassano, G., Dimaio, D., Pentassuglia, R., Piscopiello, E., Serra, E., & Falconieri, M. (2009b). Characterization of metal-modified and vertically- aligned carbon nanotube films for functionally enhanced gas sensor applications. *Thin Solid Films, 517*(22), 6211–6216. https://doi.org/10.1016/j.tsf.2009.04.009

Planeix, J. M., Coustel, N., Coq, B., Brotons, V., Kumbhar, P. S., Dutartre, R., Geneste, P., Bernier, P., & Ajayan, P. M. (1994). Application of carbon nanotubes as supports in heterogeneous catalysis. *Journal of the American Chemical Society, 116*(17), 7935–7936. https://doi.org/10.1021/ja00096a076

Qi, P., Vermesh, O., Grecu, M., Javey, A., Wang, Q., Dai, H., Peng, S., & Cho, K. J. (2003). Toward large arrays of multiplex functionalized carbon nanotube sensors for highly sensitive and selective molecular detection. *Nano Letters, 3*(3), 347–351. https://doi.org/10.1021/nl034010k

Qiao, Y., Li, C. M., Bao, S. J., & Bao, Q. L. (2007). Carbon nanotube/polyaniline composite as anode material for microbial fuel cells. *Journal of Power Sources, 170*(1), 79–84. https://doi.org/10.1016/j.jpowsour.2007.03.048

Quang, N. H., Van Trinh, M., Lee, B. H., & Huh, J. S. (2006). Effect of NH3 gas on the electrical properties of single-walled carbon nanotube bundles. *Sensors and Actuators. Part B, 113*(1), 341–346. doi:10.1016/j.snb.2005.03.089

Radushkevich, L. V., & Lukyanovich, V. M. (1952). Soviet. *Journal of Physical Chemistry, 26*, 88–95.

Rao, G. P., Lu, C., & Su, F. (2007). Sorption of divalent metal ions from aqueous solution by carbon nanotubes: A review. *Separation and Purification Technology, 58*(1), 224–231. https://doi.org/10.1016/j.seppur.2006.12.006

Ray, S. S., Vaudreuil, S., Maazouz, A., & Bousmina, M. (2006). Dispersion of multi-walled carbon nanotubes in biodegradable poly(butylene succinate) matrix. *Journal of Nanoscience and Nanotechnology, 6*(7), 2191–2195. https://doi.org/10.1166/jnn.2006.368

Savage, N., & Diallo, M. S. (2005). Nanomaterials and water purification: Opportunities and challenges. *Journal of Nanoparticle Research, 7*(4–5), 331–342. https://doi.org/10.1007/s11051-005-7523-5

Sayes, C. M., Liang, F., Hudson, J. L., Mendez, J., Guo, W., Beach, J. M., Moore, V. C., Doyle, C. D., West, J. L., Billups, W. E., Ausman, K. D., & Colvin, V. L. (2006). Functionalization density dependence of single-walled carbon nanotubes cytotoxicity in vitro. *Toxicology Letters, 161*(2), 135–142. https://doi.org/10.1016/j.toxlet.2005.08.011

Sharma, T., Mohanareddy, A. L. M., Chandra, T. S., & Ramaprabhu, S. (2008). Development of carbon nanotubes and nanofluids based microbial fuel cell. *International Journal of Hydrogen Energy, 33*(22), 6749–6754. https://doi.org/10.1016/j.ijhydene.2008.05.112

Sheldon, R. A. (2007). Enzyme immobilization: The quest for optimum performance. *Advanced Synthesis & Catalysis, 349*(8–9), 1289–1307. https://doi.org/10.1002/adsc.200700082

Sitharaman, B., Shi, X., Walboomers, X. F., Liao, H., Cuijpers, V., Wilson, L. J., Mikos, A. G., & Jansen, J. A. (2008). In vivo biocompatibility of ultrashort singlewalled carbon nanotube/biodegradable polymer nanocomposites for bone tissue engineering. *Bone, 43*(2), 362–370. https://doi.org/10.1016/j.bone.2008.04.013

Song, L., & Qiu, Z. (2009). Crystallization behavior and thermal property of biodegradable poly(butylene succinate)/functional multi-walled carbon nanotubes nanocomposite. *Polymer Degradation & Stability, 94*(4), 632–637. https://doi.org/10.1016/j.polymdegradstab.2009.01.009

Srivastava, A., Srivastava, O. N., Talapatra, S., Vajtai, R., & Ajayan, P. M. (2004). Carbon nanotubes filters. *Nature Materials, 3*(9), 610–614. https://doi.org/10.1038/nmat1192

Suehiro, J., Zhou, G., & Hara, M. (2005). Detection of partial discharge in SF6 gas using a carbon nanotube-based gas sensor. *Sensors and Actuators. Part B, 105*(2), 164–169. doi:10.1016/S0925-4005(04)00415-0

Tahaikt, M., El Habbani, R., Ait Haddou, A., Achary, I., Amor, Z., Taky, M., Alami, A., Boughriba, A., Hafsi, M., & Elmidaoui, A. (2007). Fluoride removal from groundwater by nanofiltration. *Desalination, 212*(1–3), 46–53. https://doi.org/10.1016/j.desal.2006.10.003

Talapatra, S., & Migone, A. D. (2002). Adsorption of methane on bundles of closed-ended single-wall carbon nanotubes. *Physical Review. Part B, 65*(4), 045416-045421. doi:10.1103/PhysRevB.65.045416

Terrones, M. (2003). Science and technology of the twenty-first century: Synthesis, properties and applications of carbon nanotubes. *Annual Review of Materials Research, 33*(1), 419–501. https://doi.org/10.1146/annurev.matsci.33.012802.100255

Tsai, H. Y., Wu, C. C., Lee, C. Y., & Shih, E. P. (2009). Microbial fuel cell performance of multiwall carbon nanotubes on carbon cloth as electrodes. *Journal of Power Sources, 194*(1), 199–205. https://doi.org/10.1016/j.jpowsour.2009.05.018

Tuzen, M., & Soylak, M. (2007). Multiwalled carbon nanotubes for speciation of chromium in environmental samples. *Journal of Hazardous Materials, 147*(1–2), 219–225. https://doi.org/10.1016/j.jhazmat.2006.12.069

Ueda, T., Bhuiyan, M. M. H., Norimatsu, H., Katsuki, S., Ikegami, T., & Mitsugi, F. (2008a). Development of carbon nanotube-based gas sensors for NOx gas detection working at low temperature. *Physica E, Low-Dimensional Systems and Nanostructures, 40*(7), 2272–2277. https://doi.org/10.1016/j.physe.2007.12.006

Ueda, T., Katsuki, S., Takahashi, K., Narges, H. A., Ikegami, T., & Mitsugi, F. (2008b). Fabrication and characterization of carbon nanotube based high sensitive gas sensors operable at room temperature. *Diamond and Related Materials, 17*(7–10), 1586–1589. https://doi.org/10.1016/j.diamond.2008.03.009

Valentini, L., Cantalini, C., Armentano, I., Kenny, J. M., Lozzi, L., & Santucci, S. (2004). Highly sensitive and selective sensors based on carbon nanotubes thin films for molecular detection. *Diamond and Related Materials, 13*(4–8), 1301–1305. https://doi.org/10.1016/j.diamond.2003.11.011

Van Duy, N., Van Hieu, N., Huy, P. T., Chien, N. D., Thamilselvan, M., & Yi, J. (2008). Mixed SnO2/TiO2 included with carbon nanotubes for gas-sensing application. *Physica E, Low-Dimensional Systems and Nanostructures, 41*(2), 258–263. https://doi.org/10.1016/j.physe.2008.07.007

Van Hieu, N., Thuy, L. T. B., & Chien, N. D. (2008). Highly sensitive thin film NH_3 gas sensor operating at room temperature based on SnO_2/MWCNTs composite. *Sensors and Actuators. Part B, 129*(2), 888–895. doi:10.1016/j.snb.2007.09.088

Varghese, O. K., Kichambre, P. D., Gong, D., Ong, K. G., Dickey, E. C., & Grimes, C. A. (2001). Gas sensing characteristics of multi-wall carbon nanotubes. *Sensors and Actuators. Part B, 81*(1), 32–41. doi:10.1016/S0925-4005(01)00923-6

Vaudreuil, S., Labzour, A., Sinha-Ray, S., El Mabrouk, K. E., & Bousmina, M. (2007). Dispersion characteristics and properties of poly(methylmethacrylate)/multiwalled carbon nanotubes nanocomposites. *Journal of Nanoscience and Nanotechnology*, *7*(7), 2349–2355. https://doi.org/10.1166/jnn.2007.419

Vuković, G. D., Marinković, A. D., Čolić, M., Ristić, M. Đ., Aleksić, R., Perić-Grujić, A. A. P., & Uskoković, P. S. (2010). Removal of cadmium from aqueous solutions by oxidized and ethylenediamine-functionalized multi-walled carbon nanotubes. *Chemical Engineering Journal*, *157*(1), 238–248. https://doi.org/10.1016/j.cej.2009.11.026

Wang, H., Xu, P., Zhong, W., Shen, L., & Du, Q. (2005). Transparent poly(methylmethacrylate)/silica/zirconia nanocomposites with excellent thermal stabilities. *Polymer Degradation & Stability*, *87*(2), 319–327. https://doi.org/10.1016/j.polymdegradstab.2004.08.015

Wang, Q. H., Setlur, A. A., Lauerhaas, J. M., Dai, J. Y., Seelig, E. W., & Chang, R. P. H. (1998). A nanotube-based field-emission flat panel display. *Applied Physics Letters*, *72*(22), 2912–2913. https://doi.org/10.1063/1.121493

Wang, X., Chen, C., Hu, W., Ding, A., Xu, D., & Zhou, X. (2005). Sorption of 234Am(III) to multiwall carbon nanotubes. *Environmental Science & Technology*, *39*(8), 2856–2860. https://doi.org/10.1021/es048287d

Wang, X., Lu, J., & Xing, B. (2008). Sorption of organic contaminants by carbon nanotubes: Influence of adsorbed organic matter. *Environmental Science & Technology*, *42*(9), 3207–3212. https://doi.org/10.1021/es702971g

Wang, Y., Wei, F., Luo, G., Yu, H., & Gu, G. (2002). The large-scale production of carbon nanotubes in a nano-agglomerate fluidized-bed reactor. *Chemical Physics Letters*, *364*(5–6), 568–572. https://doi.org/10.1016/S0009-2614(02)01384-2

Watanabe, K. (2008). Recent developments in microbial fuel cell technologies for sustainable bioenergy. *Journal of Bioscience and Bioengineering*, *106*(6), 528–536. https://doi.org/10.1263/jbb.106.528

Wei, B. Y., Hsu, M. C., Su, P. G., Lin, H. M., Wu, R. J., & Lai, H. J. (2004). A novel SnO2 gas sensor doped with carbon nanotubes operating at room temperature. *Sensors and Actuators. Part B, 101*(1–2), 81–89. doi:10.1016/j.snb.2004.02.028

Wei, C., Dai, L., Roy, A., & Tolle, T. B. (2006). Multifunctional chemical vapor sensors of aligned carbon nanotube and polymer composites. *Journal of the American Chemical Society*, *128*(5), 1412–1413. https://doi.org/10.1021/ja0570335

Yang, K., Zhu, L., & Xing, B. (2006b). Adsorption of polycyclic aromatic hydrocarbons by carbon nanomaterials. *Environmental Science & Technology*, *40*(6), 1855–1861. https://doi.org/10.1021/es052208w

Yim, W. L., & Liu, Z. F. (2004). A reexamination of the chemisorptions and desorption of ozone on the exterior of a (5,5) single-walled carbon nanotube. *Chemical Physics Letters*, *398*(4–6), 297–303. https://doi.org/10.1016/j.cplett.2004.09.082

Zhang, W. D., & Zhang, W. H. (2009). Carbon nanotubes as active components for gas sensors. *Journal of Sensors*, *2009*, 1–16. https://doi.org/10.1155/2009/160698

Zhao, H. Y., Zhou, H. M., Zhang, J. X., Zheng, W., & Zheng, Y. F. (2009). Carbon nanotube-hydroxyapatite nanocomposite: A novel platform for glucose/O2 biofuel cell. *Biosensors & Bioelectronics*, *25*(2), 463–468. https://doi.org/10.1016/j.bios.2009.08.005

ADDITIONAL READING

Ebbesen, T. W., Lezec, H. J., Hiura, H., Bennett, J. W., Ghaemi, H. F., & Thio, T. (1996). Electrical conductivity of individual carbon nanotubes. *Nature*, *382*(6586), 54–56. doi:10.1038/382054a0

Ruoff, R. S., & Lorents, D. C. (1995). Mechanical and thermal properties of carbon nanotubes. *Carbon*, *33*(7), 925–930. doi:10.1016/0008-6223(95)00021-5

Treacy, M. M. J., Ebbesen, T. W., & Gibson, J. M. (1996). Exceptionally high Young's modulus observed for individual carbon nanotubes. *Nature*, *381*(6584), 678–680. doi:10.1038/381678a0

Chapter 12
Synthesis of Plant-Mediated Metallic Nanoparticles for Wastewater Treatment

Manish Kumar Dwivedi
Indira Gandhi National Tribal University, India

ABSTRACT

Nanotechnology is broadly used in the different fields of science such as biomedicine, pharmaceuticals, electronics, diagnostic instruments, and environmental detection. Nanoparticles have great potential to purify wastewater and decontaminate wastewater. Nanoparticles can eliminate inorganic/organic pollutants, heavy metals, and chemical dye from contaminated water. Nanoparticles are synthesized with various methods such as physical, chemical, and biosynthesized. Plant extract is used for the synthesis of metallic nanoparticles because plant extract contains different types of primary and secondary metabolites. These metabolites act as stabilizing and reducing agents in the synthesis of novel metallic nanoparticles. The size and shape of nanoparticles have unique properties; thus, they are widely used for removing pollutants from water. The chapter discussed green synthesized metallic nanoparticles and their application in the treatment of wastewater.

INTRODUCTION

The nanotechnology field mainly covers chemistry, biology, physics, and material science. It develops new therapeutic nanoparticles with significant applications in different areas of science like energy, biomass, nutrition, and medicine (Chandran et al., 2006). The size and shape of nanoparticles are challenging work in the field of biomedical science and also developed novel nanoparticles in the pharma industry to cure different types of viral and bacterial diseases (Song and Kim, 2009). The synthesis of nanoparticles with the different methods but the biosynthesis methods/green synthesis methods have more compensation over the chemical synthesis methods due to the eco-friendly procedures. Today, the researchers focused on drugs from medicinal plants and also rich biodiversity regions, thus plant parts' easy availability has been extremely explored for the nanoparticle's synthesis (Monda et al., 2011). Medicinal plants contain

DOI: 10.4018/978-1-6684-4553-2.ch012

several types of novel secondary metabolites like terpenoids, diterpenoids, alkaloids, flavonoids, phenolic acid/phenol, and other secondary metabolites. These secondary metabolites are responsible for the formation of nanoparticles/metallic nanoparticles into bulk (Figure 1) (Aromal and Philip, 2012). The plant metabolites are involved in the redox reaction to synthesize eco-friendly metallic nanoparticles and also reported several studies, biosynthesized nanoparticles effectively controlled apoptosis-related changes, genotoxicity, and oxidative stress (Kim et al., 2007). Furthermore, biosynthesized nanoparticles have extensive application in the plant sciences and agriculture industry.

Water is an essential thing for the survival and development of living beings. Water is covered about 70% earth's surface due to the reflecting blue colour. Water is further catharized in saline water cover about 97.5%, while freshwater cover about 2.5% in the different form of glacier, snow, and ice. Freshwater is only 0.3% easily available and is an elementary quality for a flourishing society as well as a thriving economy (Kurniawan et al., 2012). The rapid development of urban regions and industrialization are causing contamination of the environment and deterioration of the quality of water (Singh et al., 2020; Kamali et al., 2019), which has become a severe problem for humans and other living organisms. Today, it is a big issue to protect the drinking water from contamination such as organic dyes, metal anions, cations, and other pollutants (Karthikeyan et al., 2020; Karthikeyan and Meenakshi, 2019). Drinking water is a vital element for all living organisms and also a demand in agriculture for large production. The freshwater is high consumption by households, industry, and another sector is about 8%, 22%, and 70% respectively (Naseem and Durrani, 2020). Modern industries such as food, leather, and cosmetic industries most of them are toxins and harmful to the life of humans and all living organisms (Sirajudheen et al., 2020). The pollutants (anionic) like nitrate, phosphate ions in water have become a serious issue and cause several carcinogenic and mutagenic diseases/disorders like lung cancer, liver inflammation, dermatitis, and chronic ulcers and also affected environment (Karthikeyan et al., 2019). Thus, it is important to remove/eliminate these ions from water for humans and all other living organisms. Nanomaterials/nanoparticles is commonly developed at atomic and molecular level to craft new structure/system have optical, electronic, magnetic, and conductive properties. Therefore, nanoparticles have been found to be effective in removal of different types of pollutants from wastewater (Kumar et al., 2014). In this chapter, I have discussed green synthesized nanoparticles apply for the elimination of pollutants from wastewater to become potable drinking water.

BACKGROUND

Worldwide people face diverse challenges of water supply, around 1.2 billion people do not have access to drinking water (Herschy, 2012). About 2.6 billion population struggles to fulfil basic sanitation and polluted water communicated different types of diseases thus million people, mostly children, have lost their lives and (Kumar et al., 2014). Mostly developing countries several parts are affected because their wastewater organization is commonly non-existent. The wastewater is mainly divided into two types municipal and industrial wastewater based on the source, which contains urine, industrial, food waste, and agricultural wastewater sources (Figure 2) (Abou El-Nour et al., 2010). Municipal wastewater contains several types of contaminates like inorganic/organic soluble compounds, microorganisms, and different types of heavy metals (Figure 3). These contaminants are changed to clean water properties (Abou El-Nour et al., 2010). Nowadays, wastewater treatment is essential due to the toxic effects of microorganisms, agriculture, animals, hazards on humans, and the environment. The treatment of wastewater

Figure 1. Synthesis mechanism of metallic nanoparticles using plant extracts.

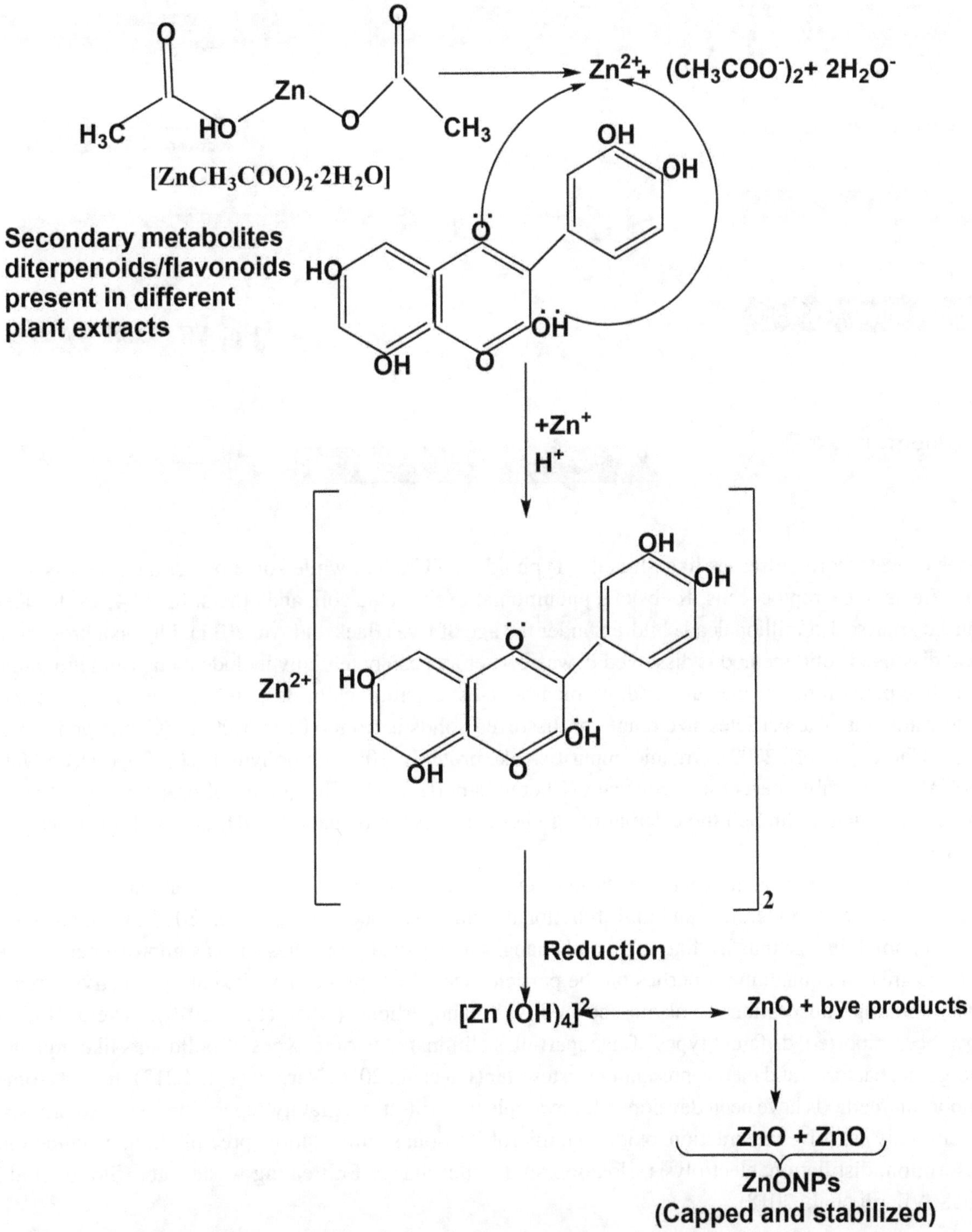

can involve biological, chemical, and physical procedures (dye, volatile, dissolved, total solid) (Bitton, 2005; Borgohain and Mahamuni, 2002). In wastewater, several types of bacteria can cause various

Figure 2. Represented different sources of wastewater.

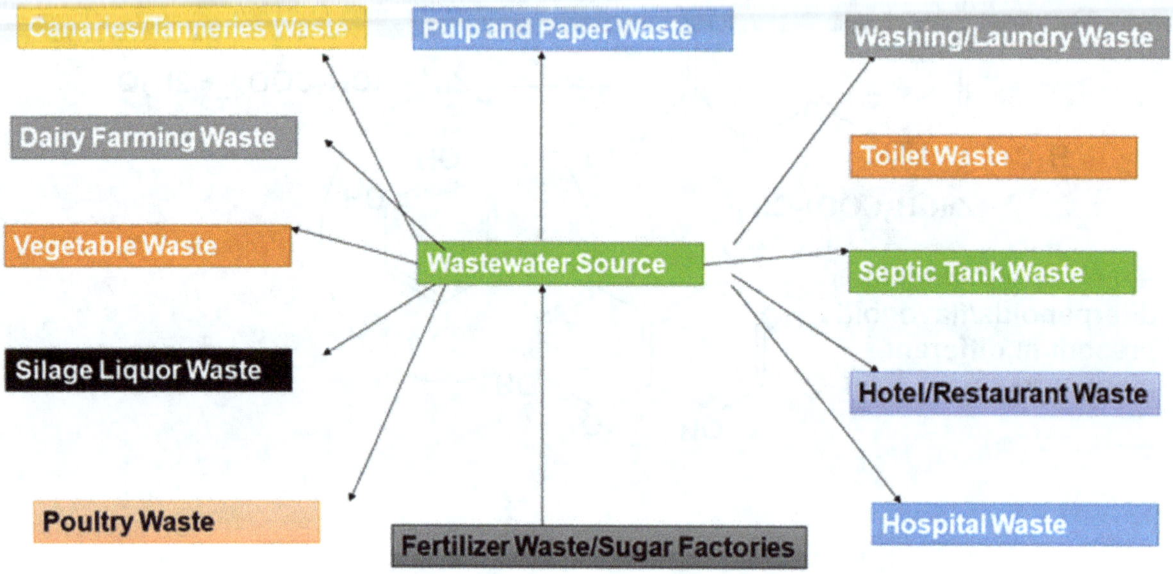

types of waterborne diseases like Shigella, Typhoid, and Cholera while some bacteria cause less serious effects like streptococcus, Klebsiella pneumonia, Escherichia coli, and others. In 2004, wastewater caused approx. 1.5 million dead children under the age of five (Baek and An, 2011). Physical procedure total dissolved solid method is dissolved in wastewater and can potentially include metals and inorganic salts like potassium, magnesium, sodium bicarbonates, calcium, chlorides, and also organic materials in low amount. The particles size range for dissolved solids is from 0.01 to 1.00μm (Chang and Zeng, 2004; Cheremisinoff, 2002). Organic impurities like proteins (10%), carbohydrates (40%), oils, and fats (50%) are generally present in wastewater (Cheremisinoff, 2002). The quality of organic impurities in water is calculated through the calculation of chemical oxygen demand (COD) and biological oxygen demand (BOD).

Nanotechnology is an inexpensive wastewater treatment compared to a huge organization because it is a very effective, integrated, and multifunctional field of science (Vilardi et al., 2017). Nanoparticles are very small in size thus leading to several changes in physical properties such as improvement of the surface area and quantum properties on the particle size. Other properties like catalytic activity, reactivity, and high adsorption are also associated with nanoparticles (Khan et al., 2019). Several studies have been reported, different types of nanoparticles eliminated various types of pollutants like organic, inorganic, bacteria, and metals present in wastewater (Du et al., 2020; Varjani et al., 2017). In past, some important methods have been developed for example ultrafiltration, gravity separation, reverse osmosis, electrodialysis, solvent extraction, oxidation, microfiltration, sedimentation, precipitation, coagulation, adsorption, distillation, electrolysis, flotation, and ion exchange for treating wastewater (Stoller et al., 2018; Vilardi et al., 2018).

Figure 3. Represented various types of pollutants present in wastewater.

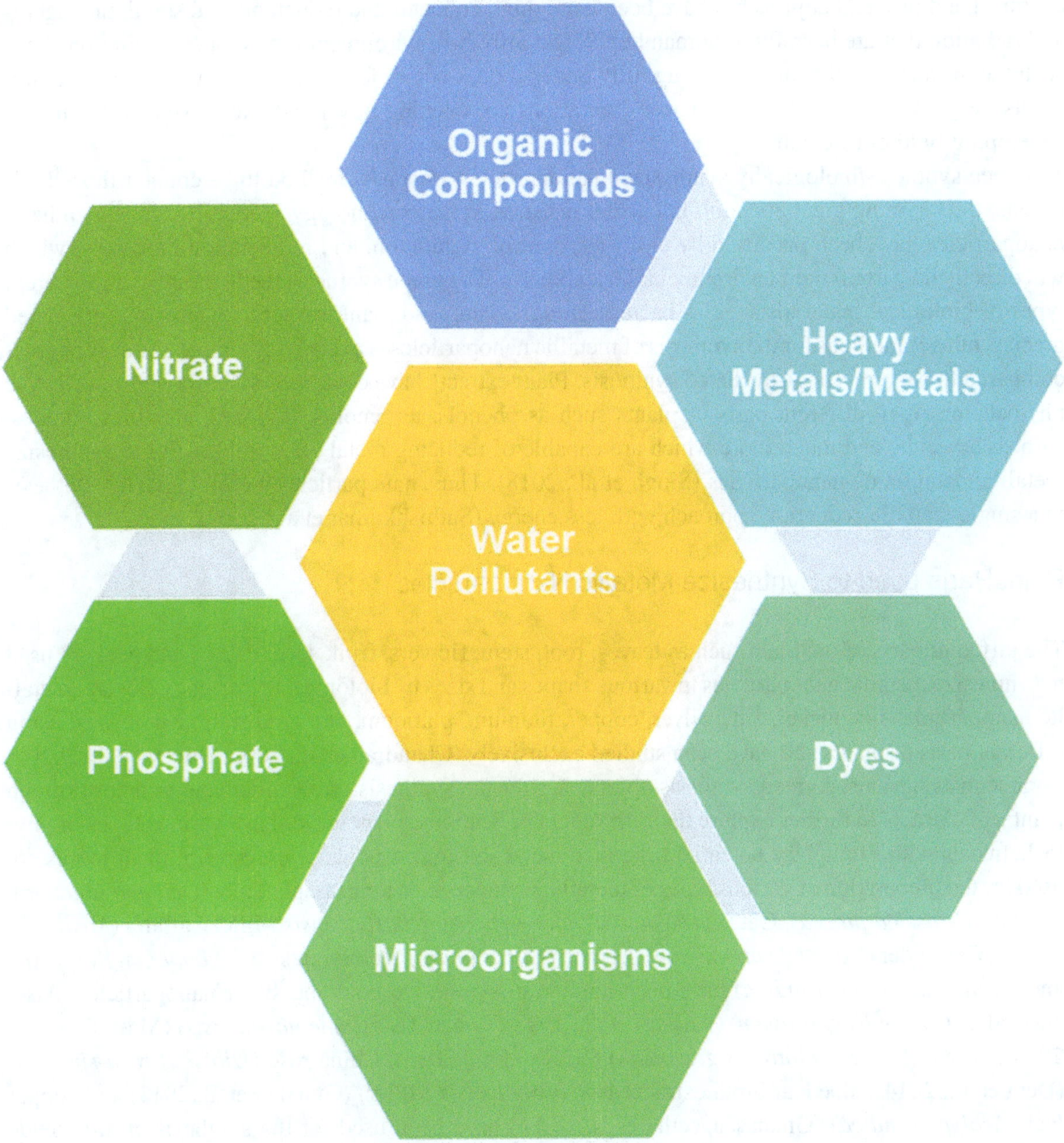

Synthesis of Metallic Nanoparticles

Metal oxide nanoparticles are considered the maximum promising as having remarkable biological applications because of their large surface area, which is of interest to scientists due to their increasing vast application in science (Ahmed et al., 2016). Metallic nanoparticles are synthesized using several methods like chemical (electrodeposition, chemical vapor deposition, Langmuir Blodgett method, catalytic route, hydrolysis, wet chemical method, etc.), physical (spray pyrolysis, ball milling, ultra-thin films, thermal evaporate, plasma arcing, etc.), biological, and enzymatic (Joerger et al., 2000; Gan et al., 2012). Both chemical and physical approaches have been using high concentrated reductants and stabilizing agents and radiation that are harmful to human health and also the environment. Some other limitations have such the mechanism of action not being fully understood, toxicity features, expansive analysis requirements, recycle/reuse/regeneration. Those limitations are opening novel and excessive chances in this developing field of research.

Green synthesis/biologically synthesized nanoparticles are a new method to be counter those limitations, thus gaining great attention in current research in nanotechnology. Generally, green synthesis nanoparticles have been produced by clean-up, control, regulation, and remediation processes, which will directly help uplift their environmental friendliness. The green synthesis methods are using different types of biological materials such as bacteria, fungi, algae, and plant extracts. Plant extract-mediated green synthesis methods for the synthesis of metallic nanoparticles is a relatively simple and easy process compared to bacteria/fungi-mediated synthesis. Plant extracts have contained several types of effective phytochemicals in different parts of plants such as phenols, terpenoids, flavones, alkaloids, amides, carboxylic acids, and much more which are capable of reducing metal salts and helpful to synthesize metallic/metal oxide nanoparticles (Singh et al., 2018). Thus, nanoparticles are biologically synthesis in a single-step bio-reduction approach with less energy (Sathishkumar et al., 2009).

Plant Parts used to Synthesize Metallic Nanoparticles

The different parts of the plant such as leaves, root, stem, flowers, fruit, seed, callus, and peel are used to synthesize metallic nanoparticles in various shapes and sizes by biological approaches. Several metallic nanoparticles like nickel, zinc, silver, copper, titanium, platinum, and gold were synthesized using different parts of plants and have been studied exclusively (Chandran et al., 2006; Dubey et al., 2010).

Scientists have used green synthesis approaches for the synthesis of metallic nanoparticles through plant leaf extracts to further explore their different applications. Plant leaves have various types of phytochemicals with exemplary potential to reduce metal salt into nanoparticles. Several plant leaves are used for the preparation of various types of metallic nanoparticles for example Tulsi (*Ocimum sanctum*), Mustard (*Brassica juncea*), Oat (*Avena sativa*), aloe vera (*Aloe barbadensis* Miller), alfalfa (*Medicago sativa*), Coriander (*Coriandrum sativum*), Lemon (*Citrus limon*), Lemongrass (*Cymbopogon flexuosus*) and Neem (*Azadirachta indica*) have been applied to synthesize gold and silver nanoparticles. Also, coriander (*Coriandrum sativum*) (Anastas et al., 1998), Green Tea (*Camellia sinensis*) (Maensiri et al., 2008), crown flower (*Calotropis gigantean*) (Vidya et al., 2013), China rose (*Hibiscus rosa-sinensis*) (Devi et al., 2014), aloe leaf broth extract (*Aloe barbadensis* Miller) (Gunalan et al.,2012) and copper leaf (*Acalypha indica*) (Gnanasangeetha et al., 2014) have been used for the synthesis of zinc oxide nanoparticles with a great diversity of plant leaf extracts. This type of study has explored the *ex vivo* synthesis of nanoparticles for example observed *in vivo* synthesis of nanoparticles like cobalt, copper, nickel, and zinc in sunflower (*Helianthus annuus*), alfalfa (*Medicago sativa*), and mustard (*Brassica juncea*) (Marchiol et al., 2012). These metallic nanoparticles have been applied to treat microbial activities and display good antimicrobial activity.

Plant parts stem is also used for the synthesis of metallic nanoparticles. *Callicarpa maingayi* was prepared methanolic extract using stem and it is used for the synthesis of silver nanoparticles. The plant parts stem extract displays, various types of functional groups, like carboxyl, amine, phenolic, aldehyde, amide I, and other compounds. The stem extract of the plant contains an aldehyde group which can reduce silver ions to silver nanoparticles. The other secondary metabolites such as polypeptides, amide I are responsible for the capping of ionic substances into silver nanoparticles (Shameli et al., 2021). Vanaja et al. has also reported the synthesis of silver nanoparticles using *Cissus quadrangularis* stem extracts (Vanaja et al., 2013). Garibo et al. reported the synthesis of silver nanoparticles using *Lysiloma acapulcensis* stem to distill water extract and applied it against microorganisms (Garibo et al., 2020).

Seed extracts of medicinal plants contained high flavonoids and other phytochemicals like lignin, saponin, and vitamins. A seed extract has good reducing agents and also acts as a better surfactant for the synthesis of gold nanoparticles (Mittal et al., 2013). The other example of *Macrotyloma uniflorum* aqueous seed extract increased the reduction rate of silver ions for the synthesis of silver nanoparticles. Caffeic acid is present in the plant extract, which is enhanced the reduction reaction few minutes (Kuppusamy et al., 2014). The seed extract of *Tectona grandis* has been used for the synthesis of silver nanoparticles and applied against microbial activity (Rautela et al., 2019).

Astragalus tribuloides root extract has been used as a key drug in traditional medicine in China, Russia, and Bulgaria. The root extract has rich sources of flavonoids, saponins, polysaccharides, and phenolics (Li et al., 2019; Li et al., 2014). It is applied for the treatment of throat, eye diseases, tumors, liver diseases, and chest pain (Chaudhary et al., 2008). *A. tribuloides* root extract is used for the synthesis of silver nanoparticles and is active against microbial activity (Sharifi-Rad et al., 2020). Dangi et al. also reported the synthesis of silver nanoparticles using aqueous root extract of *Berberis asiatica* and treated human pathogens such as *Staphylococcus aureus* positive bacteria and negative bacteria *Escherichia coli, K. pneumonia,* and *Salmonella typhimurium.* Silver nanoparticles showed good antibacterial activity (Doan et al., 2020). Bekele et al. (2020) reported the synthesis of Titanium Oxide nanoparticles at different volume compositions (1:2, 1:1, and 2:1) of titanium tetrabutoxide using root extract of *Kniphofia foliosa* respectively. TiO2 nanoparticles have been applied against human pathogen bacteria strains of *Streptococcus pyogenes, S. aureus, K. pneumonia*, and *E. coli* and showed potential activity against the human pathogen (Bekele et al., 2020).

The extract of rose petals has been used for the environmentally friendly synthesis of gold nanoparticles. Petals of Rose extract contained proteins and sugars and these functional compounds reduced tetrachloroaurate salt in bulk gold nanoparticles (Noruzi et al., 2011). *Clitoria ternatea, Nyctanthes arbortristis,* and *Catharanthus roseus* different groups of flowers are used for the synthesis of metal oxide nanoparticles. Synthesized nanoparticles have been efficiently controlling human pathogenic bacteria (Das et al., 2011). Vankar and Bajpai also reported the synthesis of gold nanoparticles using an aqueous extract of *Mirabilis jalapa* flowers that acts as a reducing agent (Vankar and Bajpai, 2010). *Abelmoschus esculentus* (L.) Moench is also known as okra and an economically important malvaceous vegetable crop which a rich source of vitamins, minerals, and nutrients. The extracts of the flowers of *A. esculentus* were used for the synthesis of silver nanoparticles. Silver nanoparticles have been applied for the treatment of Gram-positive pathogens *B. subtilis, S. aureus, S. epidermidis,* and *S. pyogenes* and the Gram-negative pathogens *K. pneumonia, E. coli, Pseudomonas aeruginosa, Proteus vulgaris, Salmonella typhimurium,* and *Shigella sonnei* and found promising antibacterial activity (Devanesan et al., 2021).

The fruit extracts have bioactive phytomolecules that are known to possess a significant therapeutic potential and are also used for the synthesis of metallic nanoparticles. *Tribulus terrestris* fruit extracts are used for the synthesis of silver nanoparticles with different concentrations (Gopinath et al., 2012). The extract has contained various types of active phytochemicals that are responsible for the reduction reaction and synthesize nanoparticles. Silver nanoparticles were prepared through *T. terrestris* fruits extract applied against multidrug-resistant human pathogens and found admirable antimicrobial activity

(Gopinath et al., 2012). Amarnath et al. (2012) also reported synthesizing palladium nanoparticles using polyphenol from grapes and treating efficiently against bacterial diseases (Amarnath et al., 2012). Several plant extracts of fruits have been used for the synthesis of metallic nanoparticles for the example *Cornus mas* L. (copper oxide nanoparticles), *Citrullus lanatus* (gold nanoparticles), *Vitis vinifera* (zinc oxide nanoparticles), *Terminalia arjuna* (silver nanoparticles), and *Punica granatum* L (silver nanoparticles) (Timoszyk et al., 2018; Shende et al., 2015; Patra et al., 2015; Devanesan et al., 2018).

Wastewater Treatment Using Nanoparticles

Metallic nanoparticles/metal oxide nanoparticles are huge applications in different fields of science like personal care products (cosmetics), medicine, drug delivery, and another field of science (physics, chemistry, and material sciences). Metallic nanoparticles can be prepared through different methods like chemical synthesis and green synthesis methods. Today mainly plants are used for the green synthesis of metallic nanoparticles compared to microbes and enzymes because they have unique properties like sustainability, renewable suppliers, ability to light energy transformed into chemical energy, antioxidant, etc. thus plants are considered the main factory for the green synthesis of metallic nanoparticles/metal oxide nanoparticles. Metallic nanoparticles have inimitable opto-electrical features due to recognized surface plasmon resonance characteristics and also few metals like Ag, Cu, and Au have wide absorption bands. Metallic nanoparticles' size and shape are also of significant importance (Dreaden et al., 2012). Nanoparticles have unique characteristics, including a high removal capacity and heavy metals selectivity. The treatment of wastewater through metallic nanoparticles depends on different factors like size, shape, stability, and aggregation. Here discussed green synthesized metal oxide/metallic nanoparticles used for the removal/elimination of pollutant from wastewater.

Zinc Nanoparticles

Zinc oxide has high chemical stability and excellent photocatalytic activity thus it is a good agent for removing water pollutants. Zinc oxide has an exciton binding energy of 60meV at room temperature and also has a wide bandgap of 3.37 eV (Naseem and Durrani, 2020). Different types of zinc oxide nanoparticles developed but hollow spheres are better for wastewater treatment because of their high surface area, light-harvesting efficiencies, low density, good surface permeability, and highly enhanced photocatalytic activity. Several studies have been reported, zinc oxide nanoparticles removed bacterial contamination from wastewater (Gondal et al., 2011; Ray et al., 2008). Zinc oxide nanoparticles are tested in various concentrations against bacterial and found potential antibacterial activity due to reactive oxygen species. The smaller size of zinc nanoparticles is also able to kill bacteria present in wastewater. Zinc oxide nanoparticles are quickly aggregated with wastewater compared to pure water (Tso et al., 2010). Esmailzadeh et al. (2016) reported zinc oxide nanoparticles and mixing low-density polyethylene (nanocomposite) tested against *B. subtilis* and *Enterobacter aerogenes* (Esmailzadeh et al. 2016). The results displayed nanoparticles more effective against gram-positive bacteria. Adams et al. (2006) also testified zinc oxide nanoparticles are antibacterial against *B. subtilis* and *Escherichia coli* (Adams et al., 2006). EI Saeed et al. (2015) also reported different type concentrations of zinc oxide nanoparticles (0.1% to 2.0%) displayed antibacterial activity. The finding showed gram-positive bacteria were more sensitive to gram-negative bacteria under 2% zinc oxide nanoparticles (El Saeed et al., 2015). Zinc oxide nanoparticles have been synthesized using Cassava starch and Aloe vera. Both metal oxide nanoparticles are able to able to removed copper from wastewater. It is displayed at a low Cu2+ ion concentration with the same removal efficiency (Primo et al., 2020).

Copper Nanoparticles

Copper oxide is a semiconductor and many useful chemical and physical characteristics (Ren et al., 2009). Thus, copper oxide nanoparticles have various types of applications like catalysis/hydro-catalysis. The use of Copper oxide nanoparticles with smaller size distributions for these applications would further encourage the chemical reactivity of nanoparticles as the particle size reduces the surface-to-volume ratio and thus the number of reactive sites (Dagher et al., 2014). Copper oxide nanoparticles were synthesized through the green synthesis method and tested against *S. aureus* and *E. coli* (Wang et al., 2002). The finding showed copper oxide nanoparticles are more active against *E. coli*, gram-positive bacteria (Borgohain and Mahamuni, 2002). Heinlaan et al., (2008) also tested copper oxide nanoparticles and bulk-copper oxide against *Thamnocephalus platyurus, Vibrio fischeri*, and *Daphnia magna* (Heinlaan et al., 2008). The finding displayed copper oxide nanoparticles are more antibacterial than bulk copper oxide (Heinlaan et al., 2008). Green synthesized cubic copper oxide nanoparticles confirm their efficacy in the photodegradation of wastewater and antimicrobial activity (Nwanya et al. 2019). Nwanya et al. (2019) reported copper oxide nanoparticles able to stop the growth of *E. coli* and *S. aureus*, it is better suited for *Pseudomonas aeruginosa* and *Bacillus licheniformis* (Nwanya et al. 2019).

Silver Nanoparticles

Silver oxide nanoparticles have potential antibacterial activity and are used for many commercial products (Martinez and Silley, 2010). Silver oxide nanoparticles were synthesized using *Lippia citriodora* plant powder and tested against *S. aureus*, finding potential antibacterial activity (Li et al., 2019). Shah et al. (2019) reported that green synthesized silver nanoparticles using fresh leaves extract of *Paeonia emodi* and performed antibacterial activity against two gram-positive and two gram-negative bacteria (Shah et al., 2019). The results displayed a strong growth inhibitor of gram-positive bacteria. Silver nanoparticles are also able to reduce 97.78% methylene blue in 3hours (Shah et al., 2019). Li et al. (2019) and Manikandan et al. (2017) both were green synthesized silver nanoparticles using different types of plant extracts and *ficus benghalensis* prop root extract respectively (Li et al., 2019; Manikandan et al., 2017). Synthesized silver oxide nanoparticles displayed strong growth inhibiter against *S. aureus, Lactobacilli sp.*, and *S. mutans.* Abdulrazzak et al. (2018) were synthesized silver nanoparticles as a combination of Ag2O, AgO, and Ag from hydrogen peroxide and silver nitrate. The synthesized nanoparticles displayed an inhibitory effect on *Acinetobacter baumannii* bacteria (Abdulrazzak et al., 2018). Therefore, silver oxide nanoparticles are able to eliminate microorganism form wastewater.

Titanium Nanoparticles

Titanium oxide (TiO2) is the most widely studied metal oxide due to its reasonable price, and physical characteristics like photostability, photocatalytic activity, and chemical stability (Guesh et al., 2016). TiO2 has a large bandgap energy of 3.2ev and ultraviolet excitation due to these properties' particles separated based on charge. TiO2 nanoparticles can kill a wide array of microorganisms (bacteria, viruses, algae, fungi, and protozoa) due to their photocatalytic properties (Foster et al., 2011). Several researchers synthesized TiO2 nanoparticles and used them for the removal of contamination in water. TiO2 nanoparticles were able to remove Pb2+, Ni, Cd, Cu, Cu2+, Fe2+, Zn, and Zn2+ from tap water and contaminated water (Engates et al., 2011; Youssef et al., 2014). Mousa et al., (2022) have also prepared green synthesized TiO2 nanoparticles using three different plant aqueous extracts Seder, Pomegranate (Pom), and Beta vulgaris. The biologically synthesized titanium oxide nanoparticles have greater photocatalytic activity compared to chemically synthesized titanium oxide nanoparticles. The finding indicated TiO2 nanoparticles can able to remove organic pollutants from wastewater (Mousa et al., 2022). The biologically synthesized TiO2 nanoparticles using *Syzygium cumini* aqueous extract. TiO2 nanoparticles displayed good photocatalytic activity and were applied for the removal of lead from industrial wastewater treatment (Sethy et al., 2020).

Iron Nanoparticles

Iron bases nanoparticles are widely used for the removal of heavy metals due to properties like an improved membrane, strength, high surface area, and small particle size (Lei et al., 2014). Iron oxide Fe3O4 (Magnetic), γ-Fe2O4 (magnetic maghemite), and α-Fe2O3 (nonmagnetic) are used as nano adsorbents, and these are also easily separate and recovered from the system. Iron oxide nanoparticles have been successfully used for the removal of various types of heavy metals from wastewater as sorbent material (Ngomsik et al., 2012). Bibi et al. (2019) reported synthesized Iron oxide nanoparticles (Fe2O3 NPs) using pomegranate (*Punica granatum*) seeds extract. The seed extracts have contained magnolol, benzoic acid, 3-deoxyflavonoids, methyl gallate, pinocembrin, catechin, gallic acid, kaempferol-3-O-sophoroside, ferulic acid, and vanillic acid, confirm through the LCMS/MS which can work as reducing agents at the time of synthesis of iron oxide nanoparticles. Biosynthesized Iron nanoparticles displayed excellent photocatalytic activity and were able to degrade dyes in wastewater (Bibi et al., 2019). Green synthesized Iron oxide (α-Fe2O3) nanoparticles were synthesized using papaya (*Carica papaya*) leaf extract. Iron oxide nanoparticles were degraded remazol yellow RR dye due to photocatalytic activity. Nanoparticles were also displayed potential antibacterial activity against *Klebsiella* spp., *E. Coli, Pseudomonas* spp., *S. aureus* bacterial strains (Bhuiyan et al., 2020). Biosynthesized Iron oxide nanoparticles were using different leaf extracts such as *A. indica, Magnolia champaca, Murraya Koenigii,* and Mangifera indica. The synthesized nanoparticles were able to remove phosphates and ammonia nitrogen from wastewater. Among the various green synthesized iron nanoparticles, *A. indica* displayed 98.08% of phosphate, 84.32% ammonia nitrogen removed from wastewater (Devatha et al., 2016).

FUTURE PROSPECTS

Metal oxide nanoparticles have played a vital role in different fields of science especially in biological science and nanomedicine. Metallic nanoparticles have also been used for the removal of contamination agents from water. Thus, it is required to synthesize nanoparticles on a commercial scale. The plant parts that have been used to synthesized nanoparticles should be low cost, environmentally friendly, and sustainable. For future research, it is significant to produce monodispersed nanoparticles. Currently, nanoparticles' mechanism of action is not clear so future research should focus on the mechanism of action and synthesis of nanoparticles. The toxicity of metallic nanoparticles is also an important challenge thus in the future reduce toxicity and expand the use of nanoparticles in therapeutic applications. Novel approaches are being technologically advanced to overcome such challenges by the use of new metal nanoparticles by changes in nanotechnology, therefore, their effects on anthropological health factors should be taken into account before their widespread use.

CONCLUSION

Green synthesis of metallic nanoparticles using plant parts extracts is hugely studied in the last few years. Plant parts have contained metabolites (primary and secondary) which are induce the production of metal oxide nanoparticles in an environmentally friendly manner. The plant's extract and purified metabolites is a novel substrate for the large-scale production of metallic nanoparticles with eco-friendly. Biosynthesized metallic nanoparticles have to be used in different areas like medicine, therapeutics, commercial products, and energy. Today, several research published to eliminate pollutants from wastewater using green synthesized nanoparticles. Nanoparticles have physical and chemical properties to kill microorganisms and remove pollutants from wastewater. Metallic nanoparticles are able to absorption of heavy metals and other organic waste. The nanoparticle's success has been credited due to their chemical and physical properties but nanoparticle uses are still limited in wastewater treatment. In this chapter, I have discussed in detail an overview of four metallic nanoparticles zinc, copper, silver, and titanium nanoparticles synthesized through the green synthesis method. Zinc nanoparticles are mostly used for various types of wastewater treatment compared to the other three nanoparticles. Silver nanoparticles are mostly used for the treatment of antimicrobial activities and the other three nanoparticles are used in various applications like adsorption, photocatalytic and antibacterial activities. Overall, these green synthesized metal oxide nanoparticles are useful for the treatment of wastewater with eco-friendly.

REFERENCES

Abdulrazzak, F. (2018). Preparation and Characterization of Silver Oxide Nanoparticles AgNPs and Evaluation the Ratios of Oxides. *Journal of Engineering and Applied Sciences (Asian Research Publishing Network), 14*. Advance online publication. doi:10.36478/jeasci.2019.9177.9184

Abou El-Nour, K. M. M., Eftaiha, A., Al-Warthan, A., & Ammar, R. A. A. (2010). Synthesis and applications of silver nanoparticles. *Arab. J. Chem., 3*(3), 135–140. https://www.sciencedirect.com/science/article/pii/S1878535210000377

Adams, L. K., Lyon, D. Y., & Alvarez, P. J. J. (2006). Comparative eco-toxicity of nanoscale TiO2, SiO2, and ZnO water suspensions. *Water Research, 40*(19), 3527–3532. doi:10.1016/j.watres.2006.08.004 PMID:17011015

Ahmed, S., Ahmad, M., Swami, B. L., & Ikram, S. (2016). A review on plants extract mediated synthesis of silver nanoparticles for antimicrobial applications: A green expertise. *Journal of Advanced Research, 7*(1), 17–28. doi:10.1016/j.jare.2015.02.007 PMID:26843966

Amarnath, K., Kumar, J., Reddy, T., Mahesh, V., Ayyappan, S. R., & Nellore, J. (2012). Synthesis and characterization of chitosan and grape polyphenols stabilized palladium nanoparticles and their antibacterial activity. *Colloid Surf. B, 92*, 254–261. doi:10.1016/j.colsurfb.2011.11.049 PMID:22225943

Anastas, P. T., & Warner, J. C. (1998). *12 principles of green chemistry. Green chemistry: theory and practice*. Oxford University Press.

Aromal, S. A., & Philip, D. (2012). Green synthesis of gold nanoparticles using Trigonella foenum-graecum and its size dependent catalytic activity. *Spectrochimica Acta. Part A: Molecular Spectroscopy*, *97*, 1–5. doi:10.1016/j.saa.2012.05.083 PMID:22743607

Baek, Y. W., & An, Y. J. (2011). Microbial toxicity of metal oxide nanoparticles (CuO, NiO, ZnO, and Sb2O3) to Escherichia coli, Bacillus subtilis, and Streptococcus aureus. *The Science of the Total Environment*, *409*(8), 1603–1608. doi:10.1016/j.scitotenv.2011.01.014 PMID:21310463

Bekele, E. T., Gonfa, B. A., Zelekew, O. A., Belay, H. H., & Sabir, F. K. (2020). Synthesis of Titanium Oxide Nanoparticles Using Root Extract of Kniphofia foliosa as a Template, Characterization, and Its Application on Drug Resistance Bacteria. *Journal of Nanomaterials*. . doi:10.1155/2020/2817037

Bhuiyan, M., Miah, M. Y., Paul, S. C., Aka, T. D., Saha, O., Rahaman, M. M., Sharif, M., Habiba, O., & Ashaduzzaman, M. (2020). Green synthesis of iron oxide nanoparticle using Carica papaya leaf extract: application for photocatalytic degradation of remazol yellow RR dye and antibacterial activity. *Heliyon*, *6*(8). doi:10.1016/j.heliyon.2020.e04603

Bibi, I., Nazar, N., Ata, S., Sultan, M., Ali, A., Abbas, A., Jilani, K., Kamal, S., Sarim, F. M., Khan, M. I., Jalal, F., & Iqbal, M. (2019). Green synthesis of iron oxide nanoparticles using pomegranate seeds extract and photocatalytic activity evaluation for the degradation of textile dye. *Journal of Materials Research and Technology*, *8*(6), 6115–6124. doi:10.1016/j.jmrt.2019.10.006

Bitton, G. (2005). *Wastewater Microbiology*. Wiley-Liss. doi:10.1002/0471717967

Borgohain, K., & Mahamuni, S. (2002). Formation of single-phase CuO quantum particles. *Journal of Materials Research*, *17*(5), 1220–1223. doi:10.1557/JMR.2002.0180

Chandran, S. P., Chaudhary, M., Pasricha, R., Ahmad, A., & Sastry, M. (2006). Synthesis of gold nanotriangles and silver nanoparticles using Aloe vera plant extract. *Biotechnology Progress*, *22*(2), 577–583. doi:10.1021/bp0501423 PMID:16599579

Chang, Y., & Zeng, H. C. (2004). Controlled synthesis and self-assembly of single-crystalline CuO Nanorods and Nanoribbons. *Crystal Growth & Design*, *4*(2), 397–402. doi:10.1021/cg034127m

Chaudhary, L. B., Rana, T. S., & Anand, K. K. (2008). Current status of the systematics of Astragalus L. (Fabaceae) with special reference to the Himalayan species in India. *Taiwania*, *53*, 338–355.

Cheremisinoff, N. P. (2002). *Handbook of Water and Wastewater Treatment Technologies*. Butterworth-Heinemann.

Dagher, S., Haik, Y., Ayesh, A.I., & Tit, N. (2014). Synthesis and optical properties of colloidal CuO nanoparticles. *J. Lumin.*, *151*, 149–154. .02.015 doi:10.1016/j.jlumin.2014

Das, R. K., Gogoi, N., & Bora, U. (2011). Green synthesis of gold nanoparticles using Nyctanthes arbortristis flower extract. Bioprocess. *Biosystems Engineering*, *34*, 615–619. PMID:21229266

Devanesan, S., AlSalh, M. S., Balaji, R. V., Ranjitsingh, A. J. A., Ahamed, A., Alfuraydi, A. A., AlQahtani, F. Y., Aleanizy, F. S., & Othman, A. H. (2018). Antimicrobial and cytotoxicity e_ects of synthesized silver nanoparticles from Punica granatum peel extract. *Nanoscale Research Letters*, *13*(1), 315. doi:10.118611671-018-2731-y PMID:30288618

Devanesan, S., Jayamala, M., AlSalhi, M. S., Umamaheshwari, S., & Ranjitsingh, A. J. A. (2021, May). Antimicrobial and anticancer properties of Carica papaya leaves derived di-methyl flubendazole mediated silver nanoparticles. *Journal of Infection and Public Health*, *14*(5), 577–587. doi:10.1016/j.jiph.2021.02.004 PMID:33848887

Devatha, C. P., Thalla, A., & Katte, S. (2016). Green synthesis of iron nanoparticles using different leaf extracts for treatment of domestic wastewater. *Journal of Cleaner Production*, *139*, 1425–1435. Advance online publication. doi:10.1016/j.jclepro.2016.09.019

Devi, H. S., & Singh, T. D. (2014). Synthesis of copper oxide nanoparticles by a novel method and its application in the degradation of methyl orange. *Adv Electron Electr Eng.*, *4*, 83–88.

Doan, V. D., Luc, V. S., Nguyen, T. L. H., Nguyen, T. D., & Nguyen, T. D. (2020). Utilizing waste corn-cob in biosynthesis of noble metallic nanoparticles for antibacterial effect and catalytic degradation of contaminants. *Environmental Science and Pollution Research International*, *27*(6), 6148–6162. doi:10.100711356-019-07320-2 PMID:31863387

Dreaden, E. C., Alkilany, A. M., Huang, X., Murphy, C. J., & El-Sayed, M. A. (2012). The golden age: Gold nanoparticles for biomedicine. *Chemical Society Reviews*, *41*(7), 2740–2779. doi:10.1039/C1CS15237H PMID:22109657

Du, C., Song, Y., Shi, S., Jiang, B., Yang, J., & Xiao, S. (2020). Preparation and characterization of a novel Fe3O4-graphene-biochar composite for crystal violet adsorption. *The Science of the Total Environment*, *711*, 134662. doi:10.1016/j.scitotenv.2019.134662 PMID:31831251

Dubey, S. P., Lahtinen, M., Särkkä, H., & Sillanpää, M. (2010). Bioprospecting of Sorbus aucuparia leaf extract in development of silver and gold nanocolloids. *Colloid Surf. B*, *80*(1), 26–33. doi:10.1016/j.colsurfb.2010.05.024 PMID:20620889

El Saeed, A. M., El-Fattah, M. A., & Azzam, A. M. (2015). Synthesis of ZnO nanoparticles and studying its influence on the antimicrobial, anticorrosion and mechanical behavior of polyurethane composite for surface coating. *Dyes and Pigments*, *121*, 282–289. doi:10.1016/j.dyepig.2015.05.037

Engates, K., & Shipley, H. (2011). Adsorption of Pb, cd, cu, Zn, and Ni to titanium dioxide nanoparticles: Effect of particle size, solid concentration, and exhaustion. *Environmental Science and Pollution Research International*, *18*(3), 386–395. doi:10.100711356-010-0382-3 PMID:20694836

Esmailzadeh, H., Sangpour, P., Shahraz, F., Hejazi, J., & Khaksar, R. (2016). Effect of nanocomposite packaging containing ZnO on growth of Bacillus subtilis and Enterobacter aerogenes. *Materials Science and Engineering C*, *58*, 1058–1063. doi:10.1016/j.msec.2015.09.078 PMID:26478403

Foster, H. A., Ditta, I. B., Varghese, S., & Steele, A. (2011). Photocatalytic disinfection using titanium dioxide: Spectrum and mechanism of antimicrobial activity. *Applied Microbiology and Biotechnology*, *90*(6), 1847–1868. doi:10.100700253-011-3213-7 PMID:21523480

Gan, P. P., Ng, S. H., Huang, Y., & Li, S. F. Y. (2012). Green synthesis of gold nanoparticles using palm oil mill effluent (POME): A low-cost and eco-friendly viable approach. *Bioresource Technology, 113*, 132–135. doi:10.1016/j.biortech.2012.01.015 PMID:22297042

Garibo, D., Borbón-Nuñez, H. A., de León, J. N. D., García Mendoza, E., Estrada, I., Toledano-Magaña, Y., Tiznado, H., Ovalle-Marroquin, M., Soto-Ramos, A. G., Blanco, A., Rodríguez, J. A., Romo, O. A., Chávez-Almazán, L. A., & Susarrey-Arce, A. (2020). Green synthesis of silver nanoparticles using Lysiloma acapulcensis exhibit high-antimicrobial activity. *Scientific Reports, 10*(1), 12805. doi:10.103841598-020-69606-7 PMID:32732959

Gnanasangeetha, D., & Saralathambavani, D. (2014). Biogenic production of zinc oxide nanoparticles using Acalypha indica. *Journal of Chemical, Biological and Physical Sciences, 4*, 238–246.

Gondal, M. A., Dastageer, M. A., Khalil, A., Hayat, K., & Yamani, Z. H. (2011). Nanostructured ZnO synthesis and its application for effective disinfection of Escherichia coli microorganism in water. *Journal of Nanoparticle Research, 13*(8), 3423–3430. doi:10.100711051-011-0264-8

Gopinath, V., MubarakAli, D., Priyadarshini, S., Priyadharsshini, N. M., Thajuddin, N., & Velusamy, P. (2012). Biosynthesis of silver nanoparticles from Tribulus terrestris and its antimicrobial activity: A novel biological approach. *Colloid Surf. B, 96*, 69–74. doi:10.1016/j.colsurfb.2012.03.023 PMID:22521683

Guesh, K., Mayoral, Á., Márquez-Álvarez, C., Chebude, Y., & Díaz, I. (2016). Enhanced photocatalytic activity of TiO2 supported on zeolites tested in real wastewaters from the textile industry of Ethiopia. *Microporous and Mesoporous Materials, 225*, 88–97. doi:10.1016/j.micromeso.2015.12.001

Gunalan, S., Sivaraj, R., & Rajendran, V. (2012). Green synthesized ZnO nanoparticles against bacterial and fungal pathogens. *Progress in Natural Science, 22*(6), 693–700. doi:10.1016/j.pnsc.2012.11.015

Heinlaan, M., Ivask, A., Blinova, I., Dubourguier, H.-C., & Kahru, A. (2008). Toxicity of nanosized and bulk ZnO, CuO and TiO2 to bacteria Vibrio fischeri and crustaceans Daphnia magna and Thamnocephalus platyurus. *Chemosphere, 71*(7), 1308–1316. doi:10.1016/j.chemosphere.2007.11.047 PMID:18194809

Herschy, R. W. (2012). Earth Sci. Ser.: Vol. 876–883. *Water quality for drinking: WHO guidelines, Encycl.* doi:10.1007/978-1-4020-4410-6_184

Joerger, R., Klaus, T., & Granqvist, C. G. (2000). Biologically produced silver–carbon composite materials for optically functional thin-film coatings. *Advanced Materials, 12*(6), 407–409. doi:10.1002/(SICI)1521-4095(200003)12:6<407::AID-ADMA407>3.0.CO;2-O

Kamali, M., Persson, K. M., Costa, M. E., & Capela, I. (2019). Sustainability criteria for assessing nanotechnology applicability in industrial wastewater treatment: Current status and future outlook. *Environment International, 125*, 261–276. doi:10.1016/j.envint.2019.01.055 PMID:30731376

Karthikeyan, P., Banu, H., & Meenakshi, S. (2019). Synthesis and characterization of metal-loaded chitosan-alginate biopolymeric hybrid beads for the efficient removal of phosphate and nitrate ions from aqueous solution. *International Journal of Biological Macromolecules, 130*, 407–418. doi:10.1016/j.ijbiomac.2019.02.059 PMID:30802518

Karthikeyan, P., Elanchezhiyan, S. S. D., Preethi, J., Meenakshi, S., & Park, C. M. (2020). Mechanistic performance of polyaniline-substituted hexagonal boron nitride composite as a highly efficient adsorbent for the removal of phosphate, nitrate, and hexavalent chromium ions from an aqueous environment. *Applied Surface Science*, *511*, 145543. doi:10.1016/j.apsusc.2020.145543

Karthikeyan, P., & Meenakshi, S. (2019). In-situ fabrication of cerium incorporated chitosan-β- cyclodextrin microspheres as an effective adsorbent for toxic anions removal. *Environmental Nanotechnology, Monitoring & Management*, *12*, 100272. doi:10.1016/j.enmm.2019.100272

Kim, J. S., Kuk, E., Yu, K. N., Jong-Ho, K., Park, S. J., Lee, H. J., & Kim, S. H. (2007). Antimicrobial effects of silver nanoparticles. *Nanomedicine (London)*, *3*(1), 95–101. doi:10.1016/j.nano.2006.12.001 PMID:17379174

Kumar, S., Ahlawat, W., Bhanjana, G., Heydarifard, S., Nazhad, M. M., & Dilbaghi, N. (2014). Nanotechnology-Based Water Treatment Strategies. *Journal of Nanoscience and Nanotechnology*, *14*(2), 1838–1858. doi:10.1166/jnn.2014.9050 PMID:24749460

Kuppusamy, P., Yusoff, M. M., Maniam, G. P., & Govindan, N. (2016). Biosynthesis of metallic nanoparticles using plant derivatives and their new avenues in pharmacological applications – An updated report. *Saudi Pharmaceutical Journal*, *24*(4), 473–484. doi:10.1016/j.jsps.2014.11.013 PMID:27330378

Kurniawan, T. A., Sillanpää, M. E., & Sillanpää, M. (2012). Nanoadsorbents for Remediation of Aquatic Environment: Local and Practical Solutions for Global Water Pollution Problems. *Critical Reviews in Environmental Science and Technology*, *42*(12), 1233–1295. doi:10.1080/10643389.2011.556553

Lei, Y., Chen, F., Luo, Y., & Zhang, L. (2014). Three-dimensional magnetic graphene oxide foam/Fe3O4 nanocomposite as an efficient absorbent for Cr (VI) removal. *Journal of Materials Science*, *49*(12), 4236–4245. doi:10.100710853-014-8118-2

Li, R., Chen, Z., Ren, N., Wang, Y., Wang, Y., & Yu, F. (2019). Biosynthesis of silver oxide nanoparticles and their photocatalytic and antimicrobial activity evaluation for wound healing applications in nursing care. J. Photochem. Photobiol. *Boletín Biológico*, *199*, 111593. doi:10.1016/j.jphotobiol.2019.111593 PMID:31505420

Li, X., Qu, L., Dong, Y., Han, L., Liu, E., Fang, S., Zhang, Y., & Wang, T. (2014). A review of recent research progress on the Astragalus genus. *Molecules (Basel, Switzerland)*, *19*(11), 18850–18880. doi:10.3390/molecules191118850 PMID:25407722

Li, Y., Guo, S., Zhu, Y., Yan, H., Qian, D. W., Wang, H. Q., Yu, J. Q., & Duan, J. A. (2019). Comparative analysis of twenty-five compounds in different parts of Astragalus membranaceus var. mongholicus and Astragalus membranaceus by UPLC-MS/MS. *Journal of Pharmaceutical Analysis*, *9*(6), 392–399. doi:10.1016/j.jpha.2019.06.002 PMID:31890338

Maensiri, S., Laokul, P., & Klinkaewnarong, J. (2008). Indium oxide (in 2O3) nanoparticles using aloe vera plant extract: Synthesis and optical properties. *Journal of Optoelectronics and Advanced Materials*, *10*, 161–165.

Manikandan, V. (2017). Green synthesis of silver oxide nanoparticles and its antibacterial activity against dental pathogens. *Biotech.*, *3*(7). . doi:10.1007/s13205-017-0670-4

Marchiol, L. (2012). Synthesis of metal nanoparticles in living plants. *Italian Journal of Agronomy, 7*(3), 274–282. doi:10.4081/ija.2012.e37

Martinez, M., & Silley, P. (2010). *Antimicrobial Drug Resistance.* In F. Cunningham, J. Elliott, & P. Lees (Eds.), *Comparative and Veterinary Pharmacology* (pp. 227–264). Springer Berlin Heidelberg.

Mittal, A. K., Chisti, Y., & Banerjee, U. C. (2013). Synthesis of metallic nanoparticles using plant extracts. *Biotechnology Advances, 31*(2), 346–356. doi:10.1016/j.biotechadv.2013.01.003 PMID:23318667

Monda, S., Roy, N., Laskar, R. A., Sk, I., Basu, S., Mandal, D., & Begum, N. A. (2011). Biogenic synthesis of Ag, Au and bimetallic Au/Ag alloy nanoparticles using aqueous extract of mahogany (Swietenia mahogani JACQ.) leaves. *Colloid Surf. B, 82*(2), 497–504. doi:10.1016/j.colsurfb.2010.10.007 PMID:21030220

Mousaa, S. A., Shalan, A. E., Hassan, H. H., Ebnawale, A. A., & Khairya S. A. (2022). Enhanced the photocatalytic degradation of titanium dioxide nanoparticles synthesized by different plant extracts for wastewater treatment. *Journal of Molecular Structure, 1250*(3).

Naseem, T., & Durrani, T. (2020). The role of some important metal oxide nanoparticles for wastewater and antibacterial applications: A review. *Environmental Chemistry and Ecotoxicology, 3*, 59–75. doi:10.1016/j.enceco.2020.12.001

Ngomsik, A. F., Bee, A., Talbot, D., & Cote, G. (2012). Magnetic solid-liquid extraction of Eu (III), La (III), Ni (II) and Co(II) with maghemite nanoparticles. *Separation and Purification Technology, 86*, 1–8. doi:10.1016/j.seppur.2011.10.013

Noruzi, M., Zare, D., Khoshnevisan, K., & Davoodi, D. (2011). Rapid green synthesis of gold nanoparticles using Rosa hybrida petal extract at room temperature. Spectrachem. *Acta Part A, 79*(5), 1461–1465. doi:10.1016/j.saa.2011.05.001 PMID:21616704

Nwanya, A. C., Razanamahandry, L. C., Bashir, A. K. H., Ikpo, C. O., Nwanya, S. C., Botha, S., Ntwampe, S. K. O., Ezema, F. I., Iwuoha, E. I., & Maaza, M. (2019). Industrial textile effluent treatment and antibacterial effectiveness of Zea mays L. Dry husk mediated bio-synthesized copper oxide nanoparticles. *Journal of Hazardous Materials, 375*, 281–289. doi:10.1016/j.jhazmat.2019.05.004 PMID:31078988

Patra, J. K., & Baek, K. H. (2015). Novel green synthesis of gold nanoparticles using Citrullus lanatus rind and investigation of proteasome inhibitory activity, antibacterial, and antioxidant potential. *International Journal of Nanomedicine, 10*, 7253–7264. PMID:26664116

Primo, J. O., Bittencourt, C., Acosta, S., Sierra-Castillo, A., Colomer, J. F., Jaerger, S., Teixeira, V. C., & Anaissi, F. J. (2020). Synthesis of Zinc Oxide Nanoparticles by Ecofriendly Routes: Adsorbent for Copper Removal From Wastewater. *Frontiers in Chemistry, 8*, 571790. doi:10.3389/fchem.2020.571790 PMID:33330360

Rautela, A., & Rani, J., & Debnath (Das), M. (2019). Green synthesis of silver nanoparticles from Tectona grandis seeds extract: Characterization and mechanism of antimicrobial action on different microorganisms. *Journal of Analytical Science and Technology, 10*, 5. doi:10.118640543-018-0163-z

Ray, B., Jones, N., Manna, A. C., & Ranjit, K. T. (2008). Antibacterial activity of ZnO nanoparticle suspensions on a broad spectrum of microorganisms. *FEMS Microbiology Letters*, *279*(1), 71–76. doi:10.1111/j.1574-6968.2007.01012.x PMID:18081843

Ren, G., Hu, D., Cheng, E. W. C., Vargas-Reus, M. A., Reip, P., & Allaker, R. P. (2009). Characterization of copper oxide nanoparticles for antimicrobial applications. *International Journal of Antimicrobial Agents*, *33*(6), 587–590. doi:10.1016/j.ijantimicag.2008.12.004 PMID:19195845

Sathishkumar, M., Sneha, K., Won, S. W., Cho, C. W., Kim, S., & Yun, Y. S. (2009). Cinnamon zeylanicum bark extract and powder mediated green synthesis of nano-crystalline silver particles and its bactericidal activity. *Colloid Surf. B*, *73*(2), 332–338. doi:10.1016/j.colsurfb.2009.06.005 PMID:19576733

Sethy, N. K., Arif, Z., Mishra, P. K., & Kumar, P. (2020). Green synthesis of TiO2 nanoparticles from Syzygium cumini extract for photo-catalytic removal of lead (Pb) in explosive industrial wastewater. *Green Process Synth.*, *9*(1), 171–181. doi:10.1515/gps-2020-0018

Shah, A., Haq, S., Rehman, W., Waseem, M., Shoukat, S., & Rehman, M. (2019). Photocatalytic and antibacterial activities of paeonia emodi mediated silver oxide nanoparticles. *Mater. Res. Express, 6*(4). . doi:10.1088/2053-1591/aafd42

Shameli, K., Ahmad, M., Al-Mulla, E. A. J., Ibrahim, N. A., Shabanzadeh, P., Rustaiyan, A., & Abdollahi, Y. (2012). Green biosynthesis of silver nanoparticles using Callicarpa maingayi stem bark extraction. *Molecules (Basel, Switzerland)*, *17*(7), 8506–8517. doi:10.3390/molecules17078506 PMID:22801364

Sharifi-Rad, M., Pawel, P., Francesco, E., & Álvarez-Suarez, J. M. (2020). Green Synthesis of Silver Nanoparticles Using Astragalus tribuloides Delile. Root Extract: Characterization, Antioxidant, Antibacterial, and Anti-Inflammatory Activities. *Nanomaterials, 10*(12). . doi:10.3390/nano10122383

Shende, S., Ingle, A. P., Gade, A., & Rai, M. (2015). Green synthesis of copper nanoparticles by Citrus medica Linn. (Idilimbu) juice and its antimicrobial activity. *World Journal of Microbiology & Biotechnology*, *31*(6), 865–873. doi:10.100711274-015-1840-3 PMID:25761857

Singh, I., & Mishra, P. K. (2020). Nano-membrane filtration a novel application of nanotechnology for waste water treatment. *Materials Today: Proceedings*, *29*, 327–332. doi:10.1016/j.matpr.2020.07.284

Singh, J., Dutta, T., Kim, K. H., Rawat, M., Samddar, P., & Kumar, P. (2018). Green' synthesis of metals and their oxide nanoparticles: Applications for environmental remediation. *Journal of Nanobiotechnology*, *16*(1), 84. doi:10.118612951-018-0408-4 PMID:30373622

Sirajudheen, P., Karthikeyan, P., Vigneshwaran, S., & Meenakshi, S. (2020). Synthesis and characterization of La (III) supported carboxymethylcellulose-clay composite for toxic dyes removal: Evaluation of adsorption kinetics, isotherms, and thermodynamics. *International Journal of Biological Macromolecules*, *161*, 1117–1126. doi:10.1016/j.ijbiomac.2020.06.103 PMID:32553962

Song, J. Y., & Kim, B. S. (2009). Biological synthesis of bimetallic Au/Ag nanoparticles using Persimmon (Diospyros kaki) leaf extract. *Korean Journal of Chemical Engineering*, *25*(4), 808–811. doi:10.100711814-008-0133-z

Stoller, M., Sacco, O., Vilardi, G., Pulido, J. M. O., & Di Palma, L. (2018). Technical–economic evaluation of chromium recovery from tannery wastewater streams by means of membrane processes. Desalin. *Water Treat*, *127*, 57–63. doi:10.5004/dwt.2018.22533

Timoszyk, A. (2018). A review of the biological synthesis of gold nanoparticles using fruit extracts: Scientific potential and application. *Bulletin of Materials Science*, *41*(6), 154. doi:10.100712034-018-1673-4

Tso, C., Zhung, C., Shih, Y., Tseng, Y.-M., Wu, S., & Doong, R. (2010). Stability of metal oxide nanoparticles in aqueous solutions. *Water Science and Technology*, *61*(1), 127–133. doi:10.2166/wst.2010.787 PMID:20057098

Vanaja, M., Rajeshkumar, S., Paulkumar, K., Gnanajobitha, G., Malarkodi, C., & Annadurai, G. (2013). Phytosynthesis and characterization of silver nanoparticles using stem extract of Coleus aromaticus. *Int. J. Mater. Biomat. Appl.*, *3*, 1–4.

Vankar, P. S., & Bajpai, D. (2010). Preparation of gold nanoparticles from Mirabilis jalapa flowers. *Indian Journal of Biochemistry & Biophysics*, *47*, 157–160. PMID:20653286

Varjani, S. J., Gnansounou, E., & Pandey, A. (2017). Comprehensive review on toxicity of persistent organic pollutants from petroleum refinery waste and their degradation by microorganisms. *Chemosphere*, *188*, 280–291. doi:10.1016/j.chemosphere.2017.09.005 PMID:28888116

Vidya, C., Hiremath, S., & Chandraprabha, M. N. (2013). Green synthesis of ZnO nanoparticles by Calotropis gigantea. *Int J Curr Eng Technol.*, *1*, 118–120.

Vilardi, G., Ochando-Pulido, J. M., Stoller, M., Verdone, N., & Di Palma, L. (2018). Fenton oxidation and chromium recovery from tannery wastewater by means of iron-based coated biomass as heterogeneous catalyst in fixed-bed columns. *Chemical Engineering Journal*, *351*, 1–11. doi:10.1016/j.cej.2018.06.095

Vilardi, G., Stoller, M., Verdone, N., & Di Palma, L. (2017). Production of nano zero-valent iron particles by means of a spinning disk reactor. *Chemical Engineering Transactions*, *57*, 751–756. doi:10.3303/CET1757126

Wang, H., Xu, J.-Z., Zhu, J.-J., & Chen, H.-Y. (2002). Preparation of CuO nanoparticles by microwave irradiation. *Journal of Crystal Growth*, *244*(1), 88–94. doi:10.1016/S0022-0248(02)01571-3

Youssef, A. M., & Malhat, F. M. (2014). Selective removal of heavy metals from drinking water using titanium dioxide nanowire. *Macromolecular Symposia*, *337*(1), 96–101. doi:10.1002/masy.201450311

Chapter 13
Emerging Nanocomposites and Their Impact on Effective Dye Degradation

P. Kanchana
PSGR Krishnammal College for Women, India

S. Shanmuga Sundari
PSGR Krishnammal College for Women, India

S. Karthika
PSGR Krishnammal College for Women, India

ABSTRACT

Nanocomposites are a class of nanomaterials where there is more than one phase in nano dimension. Simply, they are multiphase solid materials in nano dimension. Nanocomposites are found to be attractive because of their large aspect ratio. There are different types of nanocomposites like metal, organic, metal-organic polymer, carbon nano tubes, and nano fibres. Different methodologies were adopted to synthesize the nanocomposite which includes sol-gel, wet chemical method, thermal decomposition, Pechini method, insitu polymerisation, solution blending methods, etc. The nanocomposites are found to be used mainly in photodegradation, drug delivery, sensors, biomedical applications, artificial implants, and batteries. Significant research has been carried out to analyse the dye degradation property of the nanocomposites. This chapter particularly concentrates to explain the synthesis of nanocomposite and their catalytic activity towards dye degradation.

INTRODUCTION

Increasing population and industrialization remains a threat to the world due to the environmental pollution particularly to the water bodies. The abandoned release of industrial wastes into the water assets deteriorates the quality of water. The contagions like pigments, drugs, chemicals, pesticides and other

DOI: 10.4018/978-1-6684-4553-2.ch013

organic contaminations are the main cause for pollution. (Prabhuraj et al., 2021). In this modern scenario the dyes are playing inevitable role in various industries such as printing, papermaking, textile, food processing, pharmaceutical, and cosmetics industries. The dyes concentrations in the textile wastewaters can reach values as high as 300 mg L^{-1} (Nidheesh et al., 2018) near the exit of the factories. The untreated water cause irritation to the eyes, skin, digestive region, and respiratory system when affected. Therefore, it is necessary to apply efficient treatment methods such as physical separation (membrane filtration), and chemical treatments, to reduce dye contaminants, organic, and inorganic pollutants before they enter in to the ecosystem (Ahmad et al., 2021). However, the operation cost and maintenance costs of these methods are considered as the basic and determining factors for large-scale applications (Adeleye et al., 2016). Many different techniques have been applied for industrial effluent water treatment such as coagulation, flocculation, membrane filtration, adsorption, precipitation, biological oxidation, photochemical decolorization, chlorination and ozonation. Although each of these processes has their own merits and demerits. Nowadays, eradication of environmental pollution due to dye pollutants, by photocatalytic transformation of the organic dyes into nontoxic molecules, becomes one of the widely researched topics (Kajbafvala et al., 2012; Keshavarz). Metal nano particles have attracted a great interest in scientific research and industrial applications owing to large surface to-volume ratios and quantum-size effects. Metal nanoparticles generally exhibit small sizes, well defined and regular shapes and a narrow size distribution curve. Industrial catalysts usually work on the surface of metals, the metal nanoparticles which possess much larger surface area per unit volume or weight of metal than the bulk metal, have been considered as promising materials for catalysts. Photocatalysis is one of the environmentally friendly methods, in which depending upon absorption of photons by the catalyst (NCs) degrades greatly reactive radical dyes (Zang & Tang, 2015). In recent years, nanomaterials in their different forms, shapes, and sizes have been discovered to be efficient in the removal of dye contaminants through photocatalytic activities (Sharma et al., 2019; Singh et al., 2018). This is attributed to their unique physicochemical properties such as their structures, high mechanical strength, high width-to-height ratio, high thermal and electrical conductivities, slight advantage metal/semi-metallic weight and behaviour, and high surface area (Pugazhendhi et al., 2019; Salem & Fouda, 2021). Various types of nanomaterials, such as copper, zinc, iron and titanium, are employed in various treatments of dyes, including precipitation, decolorization, adsorption, and photo-degradation, as well as in the treatment of textile waste dyes (Aminuzzaman et al., 2017; Bhuyan et al., 2015; Thandapani et al., 2018). The main textual features obtained when two metal oxides semiconductors are coupled include high thermal stability and high surface area, which accelerate their reaction by enabling more active sites on their surfaces. Also, this coupling induces mass and electron transfer without photo-corrosion of the nanocomposites and improves their efficacy. Therefore, nanocomposites have a broad scope of bio-applications due to their physical, chemical, and low toxicity properties (Elfeky et al., 2020; Fouda et al., 2019).

Photocatalytic activity in metal oxide nanocomposites is mainly due to their morphology, band gap and electronic structure of the catalyst. The photo degradation kinetics of the dyes usually follows Langmuir–Hinshelwood mechanism

$$r = \frac{dC}{dt} = -k_n C^n$$

Where n is the order of the reaction and k_n is rate constant.(Choo, 2018)

The photocatalytic decoloration efficiency was calculated using the below formula

$$Decoloration\% = \frac{C_o - C_{to}}{C_o} * 100$$

where C_0 is the initial concentration of dye solution and C_{to} is final the concentration of dye solution with respective time.

In the photocatalytic mechanism, when light falls on metal oxide semiconductors the electrons (e^-) get excited from valence band (VB) to conduction band (CB) and leaving the holes in VB. Generally, the photoinduced electrons and holes (h^+) have tendency to recombine as soon as possible and only limited electrons and holes can reach the electrolyte solution for photocatalytic reaction. Presence of impurities providing the way to transfer the photoinduced electrons to the surface of photocatalyst, consequently, react with the oxygen in the solution to form oxygen radical (O^{2-}), which can oxidize the dye in solution for degradation (Li et al., 2013; Zhang et al., 2015). This hydroxyl radical produces efficient oxidants which produce enormous heat enormous heat, the organic dyes are degraded (El-Bindary et al., 2019; Mekasuwandumrong et al., 2010).

The photocatalytic behaviour also depends on the surface property like the hydrophilic-hydrophobic balance (Jung et al., 2010). Many metal oxide nano materials like CuO, CuO_2, ZnO, AgO, TiO_2, WO_3, CeO_2, Fe_2O_3, Graphene oxide (GO), Al_2O_3, SiO_2 enhances the catalytic behaviour of materials (Dhivya and Yadav, 2022; Alhebshi et al., 2020; Aly and Abd-Elhamid, 2018; Gan et al., 2016; Nabi et al., 2021; Neves et al., 2017; Salem & Kazemi, 2020; Wang et al., 2020; Yang et al., 2020). In this CuO, AgO, ZnO, and Fe_2O_3 are investigated due to their high efficiency, photochemical stability, non-toxic nature and low cost. Certain green synthesised nano particles play a major role in dye degradation. They possess an advantage of single-step reaction, ease of synthesis and low incubation period (Menon et al., 2021). As the amount of catalyst is increased, the collision between the particles increases, leading to the formation of aggregates which results in decreased surface area. The high catalyst loading also induces opacity to the solution leading to decrease in the photo flux and hence reducing the rate of degradation. The analysis for the effect of pH on dye degradation is a very difficult task. The pH of the solution alters the electrical double layer of the metal oxide and dye solution interface. This affects the adsorption/desorption and separation of the photo-generated electron-hole pairs from the surface of the semiconductor particles. Factors such as pH, zero-point charge of pH, size of the photocatalyst, adsorption-desorption equilibrium, efficiency and reusability rate play a major role in dye degradation using nanocomposites. Additionally, the geometry of vessel has a strong influence on the rate of degradation.

SILVER BASED NANOCOMPOSITES

The important silver-based nanocomposites are Ag-AgO_2 -$BiNbO_4$, Ag-ZnO, Ag-ZnO carbon nanospheres, silver doped ZrO_2, biomass derived polymer with dopamine Ag aggregates, Ag-CdS-PrTiO_2, Ag-Cu nanocatalyst, Ag-Ni, AgTiO_2, Cu_2O-AgO-Ag nanoparticles, Ag-RGO-g-C_3N_4 ternary OD and 2D nanocomposite, MoO_3 and Ag synthesized TiO_2, NiO-ZnO-Ag nanocomposite, Ag loaded ZnO graphene hybrid nanocomposite (Aydoghmish et al., 2019; Hasan Khan Neon & Islam, 2019, 2019; Liu et

Figure 1. (a) Schematic illustration of charges transfer and radical generation over Ag-ZnO composite for dye degradation, (b) Photocatalytic mechanism of Ag deposited Nanocomposite photocatalyst degradation.
Sources: (a) Liu et al. (2019) and (b) Singhal et al. (2018)

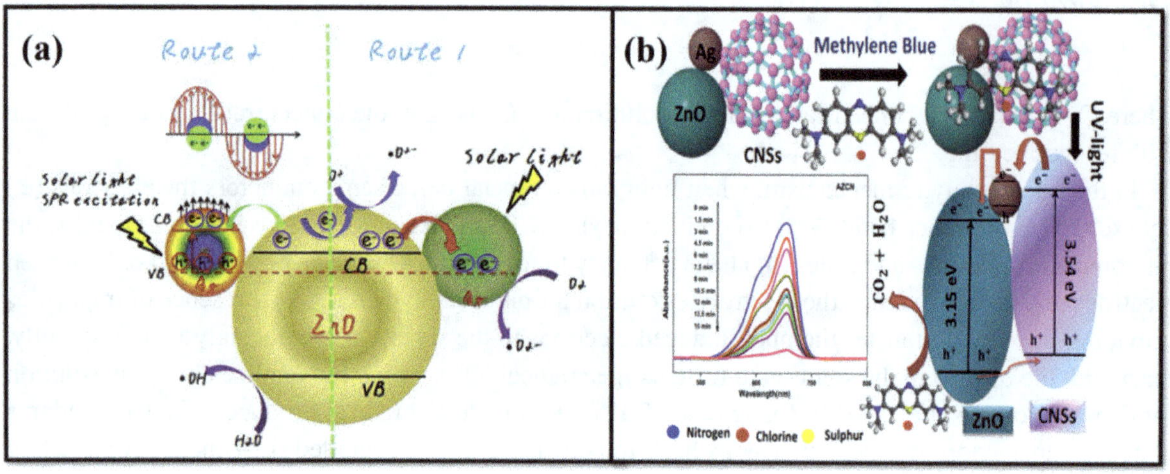

al., 2019; Pragathiswaran et al., 2021; Stanley et al., 2021; Rashid Al-Mamun et al., 2021; Singh et al., 2021; Singhal et al., 2018; Wan et al., 2019; Xu et al., 2017).

It is seen that traditional technologies like adsorption, sedimentation, filtration, osmosis degradation is carried out for effluent treatments (Guo et al., 2014; Gupta & Suhas, 2009; Ihsanullah et al., 2016; Singhal et al., 2018). Ag-ZnO composites are prepared by hydrothermal and reduction and various other routes (Andrade et al., 2017; Chen et al., 2017; Jadhav & Biswas, 2018; Liu et al., 2019; Liu et al., 2018; Mendoza-Mendoza et al., 2018; Xu et al., 2017; Zhang et al., 2017). Methods like membrane separation (Hu et al., 2020), adsorption (Li et al., 2019), oxidation methods like Fenton and Ozone oxidation (Hao et al., 2020; Szpyrkowicz et al., 2001) are already employed for waste water treatments. Coupling of semiconductor nanomaterials causes the photocatalyst to absorb more solar energy since their band gaps are larger (Borges et al., 2016; Douafer et al., 2019; Singh et al., 2021).

Microwave assisted one-pot synthesis of Zinc acetate with sodium peroxide and silver nitrate resulted the product as investigated by Liu et al. (2019). Increase in silver content enhances the photocatalytic activity of the nanocomposite towards Rhodamine B dye degradation when compared with ZnO when irradiated with solar light. Here Ag nanoparticle acts as store house of electrons and prevents the combination of hole and excited electrons and is shown in figure 1 (a) (Liu et al., 2019).

Benzene is pyrolyzed at a temperature of 1000 °C in a quartz substrate. Composite of ZnO carbon nanospheres when sonicated with silver nitrate in ethanol medium gave precipitates which was annealed in Argon. Ag deposited Nanocomposite degrades Methylene blue dye effectively say about 95% by UV irradiation when compared to with ZnO-CNS nanocomposite as shown in figure 1 (b) (Singhal et al., 2018). Similarly, in Ag-ZnO nanocomposites, the absorption is shifted from UV to visible region when Ag is added to ZnO. Photocatalytic degradation of Methylene blue, Rhodamine B and Methyl orange dyes were performed under sunlight. The electron hole union was hindered by Ag addition. Hydroxyl free radical and oxide anion radical are found as active species towards Methylene blue dye degradation (Stanley et al., 2021).

Figure 2. (a) Schematic illustration and preparation of BHP-PDA-Ag, (b) Mechanistic pathway for catalytic-degradation of dyes in aqueous solution by BHPs-PDA-Ag catalyst in the presence of $NaBH_4$ reducing agent and without $NaBH_4$.
Sources: (a) Raza et al. (2021) and (b) Raza et al. (2021)

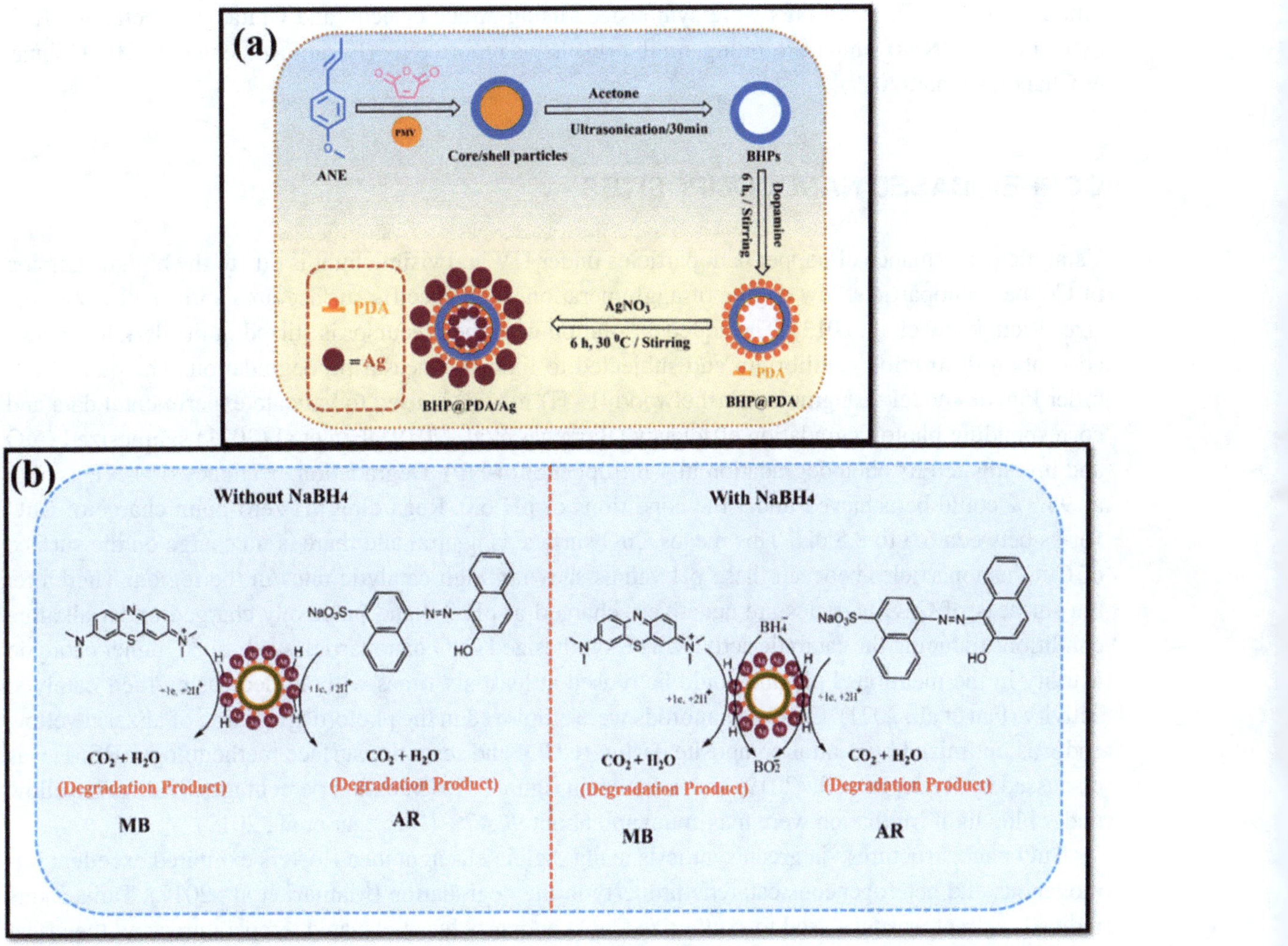

Biomass derived catalysts have significant applications due to their cost effectiveness, simplicity, higher efficacy, and sustainability (Crini & Lichtfouse, 2019; Crini et al., 2019; Ong et al., 2018). Polymeric material with dopamine and silver (BHP-PDA-Ag) was prepared by biomass of Trans-anethole (ANE) and Maleic anhydride (MAH) and the schematic and mechanism is shown in figure 2 (a) and (b).

Silver-CdS-Pr-TiO_2 shell nanomaterials were synthesized by sol-gel technique and seems to have 98% dye degradation efficiency against methyl orange dye in 30 mins time duration (Singh et al., 2021). whereas Bimetallic Ag-Cu nanomaterials due to the synergetic influence of Ag-Cu in the ratio of 25:75 gave good degradation efficiency for methylene blue and methyl orange dyes. The mechanism behind this degradation is oxidative and reductive degradation (Tantawy et al., 2021).

Green synthesis derived Ag-TiO_2 from activated carbon of tea for dyes (Wu et al., 2021) and using Beta vulgaris outer skin extract are discussed (Jayapriya & Arulmozhi, 2021). Ag-TiO_2 from activated carbon of tea due to high porosity of material it gives nearly 95% efficiency in methylene blue dye removal. Photocatalyst has good recyclability and cost effectiveness. The catalytic efficiency of the synthesized Ag-TiO_2 nanocomposite using Beta vulgaris outer skin extract exhibit higher degradation efficiency of 92%, 84%, and 88% for Methylene blue (MB), Congo red (CR) and Methyl orange (MO) in 9 min, 20 min and 10 min. Pure Ag NPs were synthesized using Areca catechu and Crataegus pentagyna fruit extract (CP-AgNPs) which exhibiting similar results as reported by (Ebrahimzadeh et al., 2020; Vinay & Chandrasekhar, 2019).

COPPER BASED NANOCOMPOSITES

Catalytic performance of copper nanoparticles under UV and visible light is due to the high dispersion of Copper nanoparticle, low degree of agglomeration, high specific surface area and small crystallite size (Wongkaew et al., 2013). The aqueous solution with the precursor is stirred in the dark to achieve adsorption-desorption equilibrium and subjected to light for measuring degradation. The quasi first-order kinetic model (Langmuir–Hinshelwood (L–H) model) is used to validate experimental data and corresponding photo-degradation efficiency (Barzegar et al., 2019). Fan et al (2021) synthesized CuO and investigated its photodegradation in 4-nitrophenol (4-NP). Degradation efficiency of 4-NP as high as 95.3% could be achieved under the conditions of pH 6.0. Regarding pH zero-point charge of CuO varies between 6.5 to 8.5 pH. This means CuO surface is neutral and there is no charge on the surface of CuO nanoparticles between these pH values showing high catalytic rates in the region. Therefore, the surfaces of CuO particles are negatively charged at pH 9.0 and positively charged under alkaline conditions reducing the catalytic activity. The synthesized CuO nanoparticles exhibited higher catalytic activity in the mentioned pH and could be reused at least six times without decreasing their catalytic activity (Fan et al., 2021). Cu-CuO nanorods were employed in the photo-degradation of alizarin yellow and was optimized by central composite design (CCD) and response surface methodology (RSM) was discussed by Barzegar et al. (2019) and is shown in figure 3. Degradation percentages of alizarin yellow under blue light irradiation were maximum and about 96.47% (Barzegar et al., 2019).

CuO nano structures via green synthesis using Aglaia elaeagnoidea flowers exhibited excellent homogenous and heterogeneous catalytic property in the degradation (Manjari et al., 2017). Tamarindus indica L extract was also used to synthesize CuO nanoparticles at elevated temperatures and the effect of heat treatment on the size of nanoparticles were discussed. The microspheres made of agglomerated nanosheets were used to carry out the photocatalytic degradation of rhodamine B (RhB) and xylenol orange (XO) dyes exhibits photodegraded 96% of RhB and 87% of XO in 150 and 90 min of light illumination, respectively(Gudipati et al., 2021).

Heterostructured Cu-based catalysts, such as Cu_2O-TiO_2, Co_3O_4-CuO, and Cu-Cu2O, were found to be superior when compared to their unitary components. The CuO-Cu mesh has proven to be an excellent photoactive catalyst with 100% degradation for MB which maintains at 98.8% efficiency after two successive photocatalytic cycles. The mechanism of photocatalytic activity is shown in figure 4 (Alhebshi et al., 2020). In addition, CuO-Cu_2O nanocomposite on Cu foil shows good degradation under different LED colors and resulted that blue one has the highest photonic efficiency (Behjati et al., 2020; Xu et al., 2021).

Synthesis of hybrid composites using the organic-inorganic hybrid strategy for synthesis of catalyst such as CuO-PVDF (polyvinylidene fluoride) (Gao et al., 2015) and Sepiolite-Cu_2O-Cu (P. Wang et al.,

Figure 3. Schematic mechanism of photocatalytic degradation of alizarin yellow.
Source: Barzegar et al. (2019)

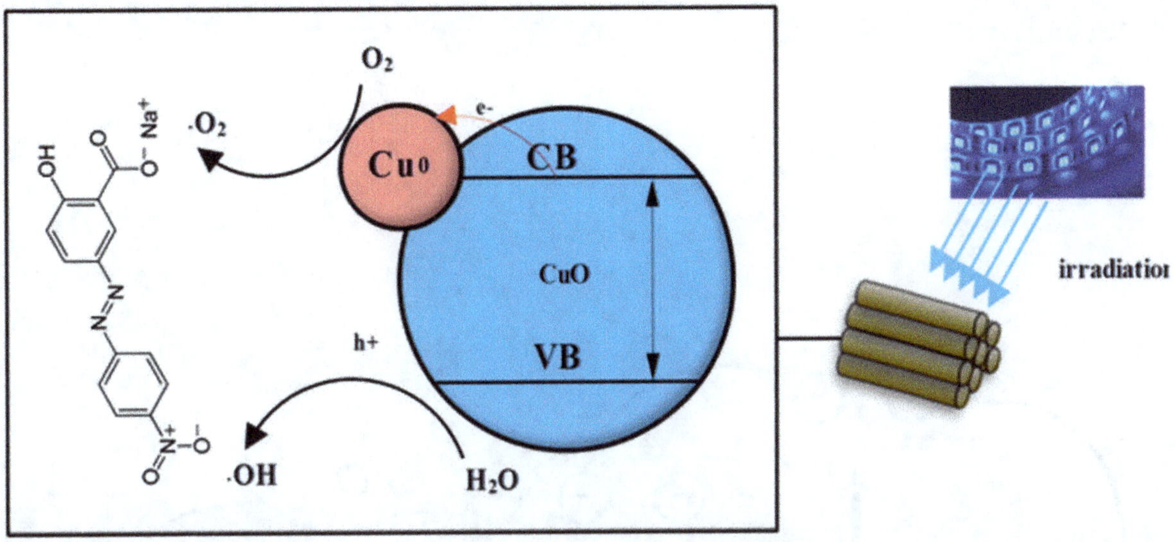

2019) shows 95.8% and 95.1% efficiency in visible light which plays major role in treating wastewater. When considering pure chitosan with CS-CuO nanocomposite, the CS-CuO is found to be more degradation efficiency of 97% with congo red and eriochrome black dye (Srivastava & Choubey, 2021). La doped CuO nanoparticles in the degradation of (~98% in 90 minutes) MB dye under visible light, thus delivering high Photo-Fenton activity and stability for lanthanum doped copper oxide nanoparticles (Rodney et al., 2018).

Direct Blue 71 (DB 71) degradation with Cu doped ZnO showed excellent effect at pH 6.8 and has a reusability of 3 cycles with an efficiency of 97% under visible light(Fouda et al., 2020; Sharma et al., 2020; Thennarasu & Sivasamy, 2016). The g-C_3N_4-ZnO-Cu_2O (Rajendran et al., 2021) and Cu-Cu_2O-ZrO_2 (Zhao et al., 2021) composite photocatalyst revealed a supreme photocatalytic efficiency in Rhodamine B dye of 91.4% (Rajendran et al., 2021). These are found to have good efficiency than CuO-TiO_2 nanocomposite which showed 85-90% degradation of methylene blue (Simamora et al., 2012; Udayabhanu et al., 2020). When we consider the effect of CuO – transition metal nano composites George et al. (2022) stated that when compared to CuO doped with Ni, the Zn or Fe doped samples exhibited good degradation of methylene blue dye. A degradation efficiency of around 63% and 62% was witnessed on doping CuO with Zn or Fe respectively.

IRON BASED NANOCOMPOSITES

Iron based nanocomposites such as Fe-Fe_2O_3, Fe_2O_3-graphene-CuO, Fe_3O_4- Ag nano alloys, *NRGO-CoWO$_4$*-Fe_2O_3 nanocomposite, Fe_2O_3 -Mn_2O_3 nanocomposite, Fe_2O_3 nanoparticle doped with In_2O_3 have provided to be good photocatalysts for degradation of dyes and preventing pollution of water bodies which is discussed below (Nuengmatcha et al., 2019; Yang et al., 2020).

Figure 4. Schematic diagram of proposed photocatalytic activity.
Source: Alhebshi et al. (2020)

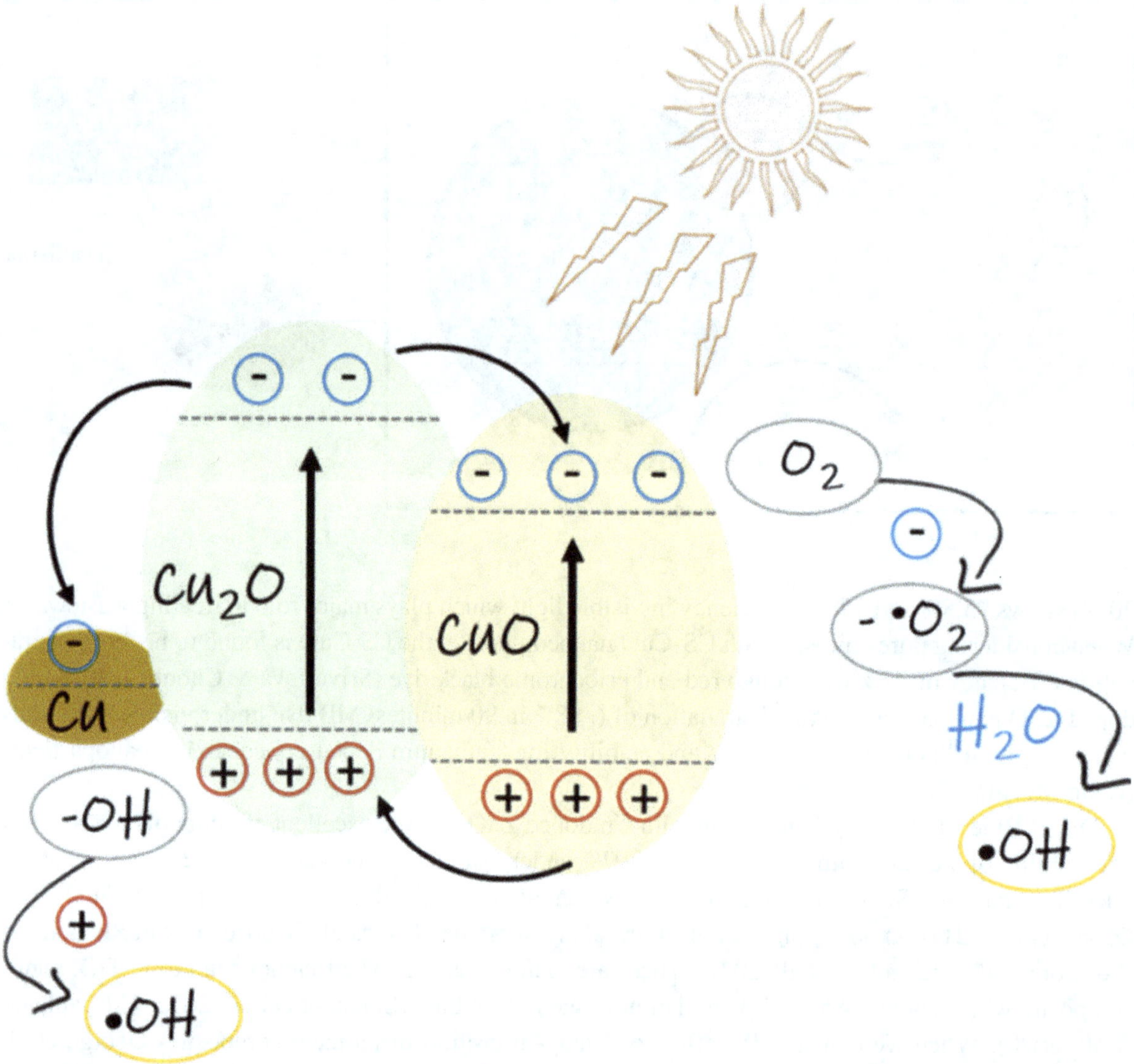

Fe-Fe_2O_3 core shell nanoparticle assisted by $NaHSO_3$ is utilized for degradation of orange II dye. The dye degradation occurs at solid-liquid interface. Generally, Zerovalent iron is used to remove arsenic and organic substances. Iron degrades by oxidation of Fe^0 to Fe^{II}, the presence of Fe^0 with Fe_2O_3 enhances molecular oxygen which is activated by phosphate ion. The phosphate ion prevents the corrosion of iron materials. The degradation of dye is accelerated by fast electron transfer of Fe-Fe_2O_3 and its mechanism is shown in figure 5. The catalyst was applied in acidic and alkaline conditions and it showed 90% efficiency against Orange II in 5 mins (Yang et al., 2020).

Numerous Iron nanoparticles are reported in the literature (Atla et al., 2018; Atrak et al., 2019; Kubiak et al., 2020; Sayadi et al., 2021; Wang et al., 2019). Fe_2O_3-graphene-CuO (FGC) possess a band gap of 1.82 eV which is prepared by solvothermal method and is employed for Methylene blue dye

Figure 5. Mechanism of Fe-Fe_2O_3 system on orange II degradation
Source: Yang et al. (2020).

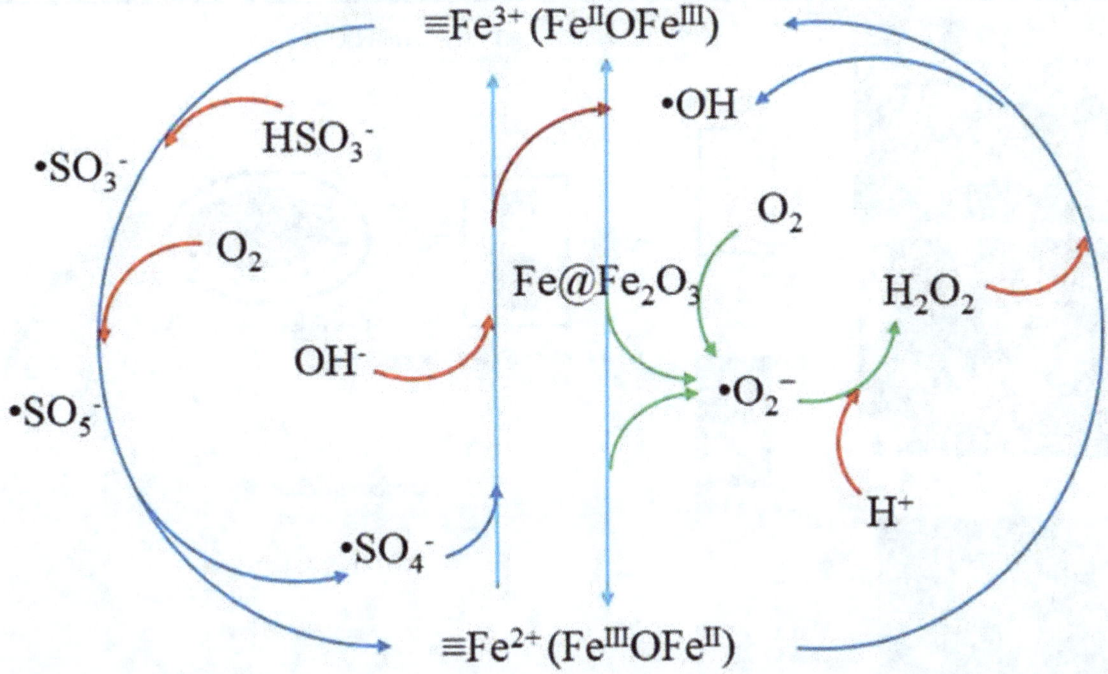

degradation. The degradation was carried out under visible light in a closed UV chamber at normal temperatures which resulted in 90% efficiency at 40°C. The maximum efficiency of dye removal was 73.83% at pH 5.8 (Nuengmatcha et al., 2019). Nano Fe_3O_4 modified granular activated carbon was used as catalyst for degradation of Azo dye Reactive Red 2 in anaerobic condition using microorganisms like Halomonas, Pseudomonas and Bacillus bacteria in reactors. Direct interspecies electron transfer called DIET occurs in pseudomonas and methanosarcina species in presence of catalyst nano Fe_3O_4 modified granular activated carbon (Wan et al., 2021). Alpha-Fe_2O_3-Cu_2O nanomaterials reported by Norouzi & Nezamzadeh-Ejhieh (2020) was prepared by solvothermal method. The band gap of Fe_2O_3 which was 2.28 eV gives a bathochromic shift to 1.95 eV by Cu_2O introduction. Methylene blue dye was degraded using this catalyst which followed first order kinetics with a rate constant of 0.025 min^{-1} It is understood that 1 mg of dye molecules can be degraded in 27 minutes.

Fe_3O_4- Ag nanoalloy synthesized by green routes using Justicia spicigera plant by sonication technique. The catalytic material is employed against Red Cango dye degradation. It is a good magnetic material (Ruíz-Baltazar, 2020). Iron oxide prepared using Fan palm, Dombeya wallichi and Pyrus comminis extracts and depicted in figure 6. Maximum absorption was seen at pH 3 for a dosage of 0.01g/50 mL adsorbent at 65° C for an initial concentration of 200 mg/L with an average of 80 mins contact time towards reactive blue dye (Noreen et al., 2020).

Fe_2O_3 - Mn_2O_3 nanocomposite is used for methylene blue degradation. Catalyst has α- Fe_2O_3 and β-Mn_2O_3 and possess a band gap of 1.11 eV, such low value show that they are efficient light absorbers in visible and UV- light. It showed 80-95% efficiency of dye degradation in 75 s. The active species are hydroxyl free radical and holes which are responsible for degradation (Ghaffari et al., 2020). α- Fe_2O_3-

Figure 6. Iron oxide prepared using Fanpalm, Dombeya wallichi and Pyrus comminis extracts
Source: Noreen et al. (2020).

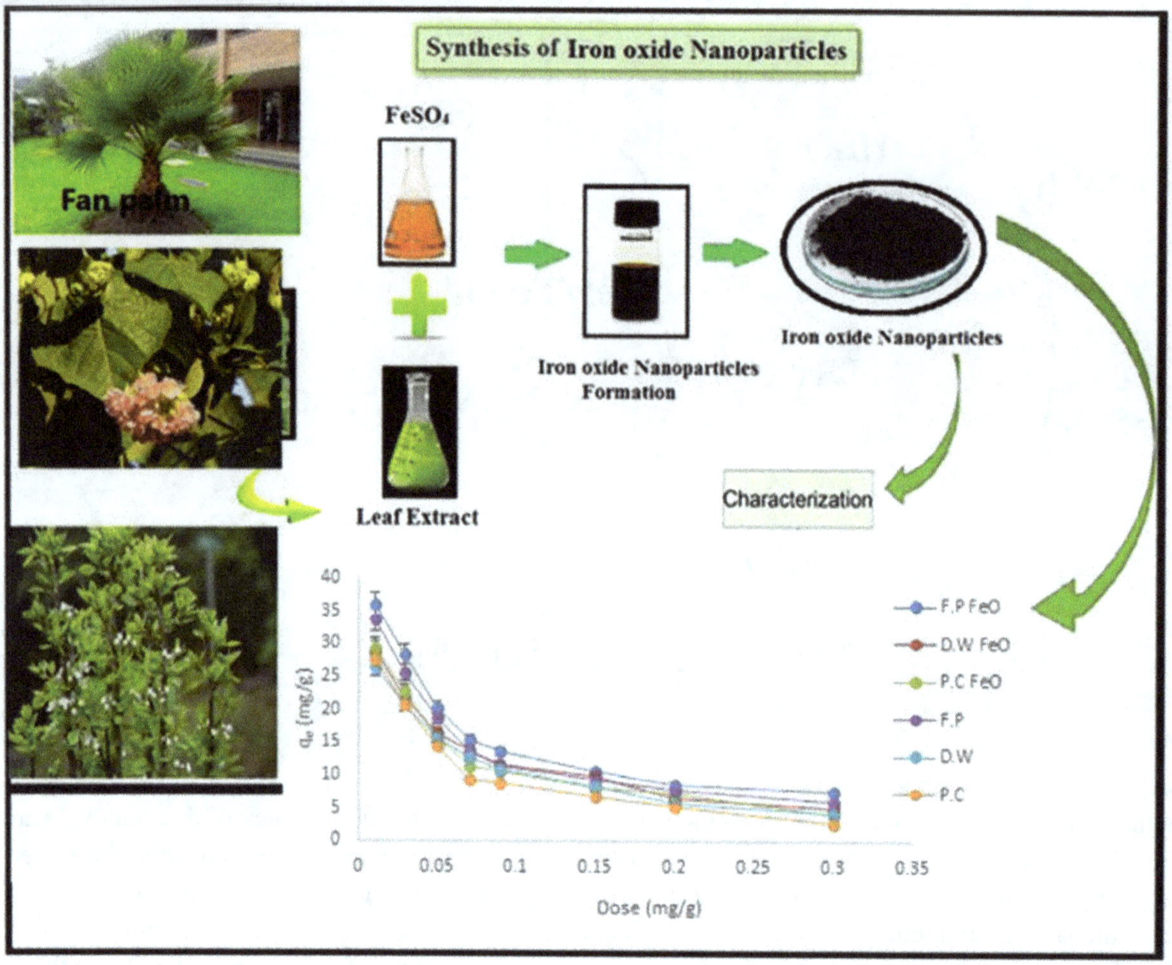

WO_3 nanocomposite is used for Reactive Blue (RB19) dye degradation under UV and LED lights. The mechanism behind the dye removal is it stops the recombination of holes and electrons. The removal efficiency is good for 6 continuous cycles and aids in good reusability (Delnavaz et al., 2021).

Fe_2O_3 nanoparticle doped with In_2O_3 was reported by Guo et al. (2020). The degradation percentage of Rhodamine was 94% with 10% In_2O_3. The Rhodamine B (100 mg/L) removal was efficient at pH 4 in presence of H_2O_2. The active hydroxyl free radical attack the pi system of Rhodamine B at the same time the -C_2H_5 group is attacked by hydroxyl free radical during which denitrification and decarboxylation happens. Finally evolution of CO_2 and H_2O occurs (Guo et al., 2020). Iron oxide and SiO was used to get the composite material by facile impregnation process. About 5 nm sized iron nanoparticle got dispersed in SiO_2. The dye molecules namely Janus green B, methylene blue, tartrazine and cango red were degraded using this composite material. The dye degradation % for Tartrazine is 98.5%. The degradation efficiency of dyes was found to be 82% after the fifth cycle (Vu et al., 2019).

ZINC BASED NANOCOMPOSITES

Among the metal oxides semiconductor, Zinc Oxide (ZnO), is an important n-type semiconducting materials has been extensively used in wastewater treatment because of its low toxicity, earth abundance and high electron mobility (Kumar et al., 2013; Li & Haneda, 2003). It has materials with exceptional features like high electron mobility, tunable band positions, high catalytic activity, high photo sensitivity, excellent chemical and thermal stability, non-toxicity and cost effectiveness. ZnO is a versatile semiconductor with a wide direct band gap (3.37eV) and a large exciton binding energy (60 meV) at room temperature has a wide range of applications in solar cells, optical and electrical fields, gas and chemical sensing, biomedical applications and also in catalysis and photocatalysis (Basnet et al., 2018; Dhatshanamurthi et al., 2017). Mostly the ZnO nanoparticles are prepared from Zinc Acetate Dihydrate ($(CH_3COO)_2Zn.2H_2O$, 99%) by solvothermal, hydrothermal method sol-gel method, solution combustion, ultra-sonic spray pyrolysis and sonochemical method, laser ablation, inert gas condensation, and vapour phase process etc (Dhivya and Yadav, 2022; Ahmed Quraishi et al., 2020; Fouda et al., 2020; Mohaghegh et al., 2014; Pavithra & Jessie Raj, 2021; Prabhuraj et al., 2021; Saad et al., 2020; Velumani et al., 2020; Vidya et al., 2017; Zyoud et al., 2016). ZnO exist in hexagonal wurtzite structure and the shape of the nanoparticles are decided by the impurities, surfactants, catalysts and other preparation conditions like temperature and pH. Mainly the zinc oxide nano particles are applied to degrade azo dyes like methylene blue (MB), Rhodamine (Rh), Orange M2R, malachite green (MG), Methyl Orange (MO), Eosin Y (EY) and Congo Red (CR) (Dhivya and Yadav, 2022; Fouda et al., 2020; Prabhuraj et al., 2021; Zyoud et al., 2016). There are three different reaction mechanisms which contribute to dye degradation, (1) Attack by hydroxyl radical, (2) Direct oxidation by the positive hole on the valence band, and (3) Direct reduction by the electron in the conduction band

$$ZnO + hv(light) \rightarrow ZnO(e^- + h^+)$$

$$ZnO(e^-) + O_2 \rightarrow O^{2-}$$

$$ZnO(h^+) + OH^- \rightarrow OH$$

All the produced radicals react with dye and results the efficient decolouration along with by-products. The photocatalytic reactions mostly take place on the surface of ZnO catalysts. Hence, the ZnO surface area is a key factor in the kinetic and efficiency of photocatalytic reaction. It was known that supported photocatalysts with high adsorption ability can (i) attract the pollutant substances near the reactive surface of the catalyst particles, (ii) create a lot of active sites for adsorption of intermediates, (iii) extend lifetime and reusability of the photocatalyst and(iv)decrease recombination rate of photogenerated electron–hole pairs. So, it can promote the degradation rate and enhance photocatalytic activity of ZnO (Kahouli et al., 2015; Ong et al., 2014).

However, there are hurdles to attain the high photocatalytic efficiency due to i) wide band gap (3.37 eV) could limit the visible light absorption and ii) predominant photoexcited charge recombination rate. The pure form of metal oxides efficiency is low due to its high recombination rate and low quantum efficiency affect the hands-on applications of metal oxides. So, the researchers used some strategy to enhance the efficiency which are doping with some other additives or composite with additional semiconductor materials (Hisatomi et al., 2015). Photo degradation of the pure ZnO is around 63% whereas the addition of dopants like graphene oxide increases its efficiency to 96% (Dhivya and Yadav, 2022). The efficiency of the degradation varies and depend on the source of light used. The band gap of ZnO is 3.3 eV and bandgap of this material is easily can tuned by doping with transition metal ions (Mn, Fe, Ag, Au, Pd Pb, Ti) and rare earth ions (Ce, Nd, Eu and etc.) (Achouri et al., 2016; Basnet et al., 2019; Basnet; Kumaresan et al., 2017; Saleh & Djaja, 2014; Ullah & Dutta, 2008). The t_2g orbital level of Mn is very adjacent to the valence band of ZnO and so it can overlap straightforward with the d-electrons of Mn which reduces the bandgap of ZnO (Yang et al., 2010). Depending upon the concentration of Manganese (Mn) in ZnO decreases the band gap and it varies from 2.95 eV to 2.755 eV. Decrease in the band gap considerably enriches the photocatalytic activities by suppressing the electron-hole recombination. Dhivya and Yadav (2022) reported the addition of H_2O_2 on Mn doped Zn hinder the recombination of electron- hole and improved the oxidation and reduction mechanism.

The visible light absorption could be extended by doping with metal ions or introduction of light sensitizers like quantum dots and carbon dots on the surface of ZnO and also which could suppress the charge recombination rate (Phang & Tan, 2019). Graphene Oxide (GO) is considered as one of the best impurities to Zn for photodegradation as it shows 96.9% efficiency even after 5 cycles in rhodamine die under UV light irradiation (Ong et al., 2014; Prabhuraj et al., 2021). The high adsorption activity and low recombination rate of hybrid metal oxides-graphene composite shows enhanced photocatalytic activity(Liu et al., 2010; Mishra et al., 2020). Zn-GO nanocomposite shows around 24% of catalytic enhancement and reduce the bandgap to 2.97 eV from 3.3 eV, due to the wide-ranging background absorption of GO (Bao et al., 2011; Gu et al., 2017). Carbon dots (CQD) Zn nano composite is yet another important favourable material to suppress the charge recombination rate (Phang & Tan, 2019). *H. Bozetine et al.* reported the synthesis of ZnO-carbon quantum dots (ZnO-CQDs) nanocomposites by hydrothermal method and studied their photocatalytic activity by degradation of Rhodamine B (RhB) under simulated solar light. The degradation efficiency was 94% degradation; it is higher than compared with pure ZnO (Bozetine et al., 2016). The degradation rate was reached 96% for ZnO-C-dots nanocomposite when the photocatalytic reaction time reaches 30 min. Meanwhile only 63% photocatalytic degradation efficiency was obtained for pure ZnO (Velumani et al., 2020).

One of the main advantages of CNTs in the interaction with ZnO is the charge transfer process through the channels of the tube which preventing the agglomeration of ZnO nanoparticles therefore, CNTs-ZnO nanostructure could be the optimum choice for the degradation process. ZnO-CNT nanocomposites were successfully synthesized by the laser ablation process under particular conditions of time change in laser ablation duration of zinc target immersed in dispersed functionalized CNT solution producing ZnO-CNT nanocomposite with different amounts of ZnO nanoparticles by

Mostafa et al. (2021) observed that the efficiency of the prepared Zn-CNT nanocomposite with different amounts of nanoparticles in catalytic degradation was mainly depends on the amount of deposited ZnO as a factor of ablation time. Also, they concluded that it is not necessary as the concentration of ZnO on the surface of CNT increases, the efficiency of catalytic degradation process increases but it increases to a critical value then the efficiency would be decreased which might be related to the change in the mechanical properties of the CNT when the amount of decoration was extreme leading to a drop in the catalytic degradation efficiency.

Among various metal dopants, lead (Pb) substitution on the ZnO lattice plays a major role in enlightening the photocatalytic and antibacterial activity (Panchal et al., 2020). The decrease of band hole vitality incredibly affects the quantity of photo-generated electrons and gaps in photocatalytic response (Kong et al., 2009). Towards this effort, to remove anionic textile dyes from an aqueous solution, lead ions (Pb) have been chosen as a dopant to improve the Azo dye degradation and antibacterial efficiency through altering the electronic structure, optical and morphological property of zinc oxide (Kahouli et al., 2015). Pb-ZnO nanocomposites ultrasonicated for 21600 s has a huge surface area that has the capability to absorb significant amounts of hydroxyl groups. 5 mol% Pb loaded ZnO enhances the transfer of photo excited carriers, decreases the recombination rate and shows more photo catalytic activities when compared to ZnO (Pavoski et al., 2019; Shanmugam & Jeyaperumal, 2018). Whereas the removal efficiency of about 97% was achieved by CuO-ZnO (20-80) after 85 min of irradiation, whereas CuO-ZnO (20-80) nanocomposite achieved a maximum removal efficiency of 65% after 130 min (Fouda et al., 2020).

The absorption band is red shifted and reduced band gap is observed in cerium (Ce) doped ZnO NCs (Saad et al., 2020). Effect of Chitosan (CH) the natural polymer in the Ce-Zno composite is also reported by Zhong et al. (2020) in Methyl orange and Saad et al. (2020) in Malachite green. For the degradation of M.G by CH-ZnO, the complete removal of 5 mg/L M.G was achieved after 270 min of visible light illumination. Doping of ZnO with cerium in CH-Ce-ZnO reflected considerable enhancing the photocatalytic activity as the complete removal of the M.G dye (5 mg/L) was achieved after 90 min only. Additionally, the maximum degradation results which were obtained for 10 mg/L and 15 mg/L of M.G dye using CH-ZnO are 83% and 76%, respectively while the application of CH-Ce-ZnO resulted in complete degradation of MG for the same concentrations after 180 min and 210 min in the same order. Saad et al. (2020) showed that nZnO-CS possesses an UV-shielding rate of more than 70% at 220-380 nm and MO decolorization rate of more than 80%.

Asperagus racemosus root assisted ZnO nanoparticles (Ar-ZnO) were prepared by green synthesis and investigated for the degradation of malachite green (MG) dye solution under visible light at wavelength 617 nm. The removal efficiency of Ar-ZnO for MG dye was 93.20% on treatment for 3 hours was discussed by Pallela et al. (2020). Similarly, ZnO nanoflowers synthesised via green route using Panos extract shows good photocatalytic degradation against MB, EY and MG using UV as light source. ZnO-QNF exhibited >99% of degradation efficiency at 15 and 5 mg/L of all the three dyes, such as, MB dye degraded within 80 and 30 min, EY dye degraded within 90 and 35 min, and MG dye degraded within 110 and 40 min of contact time, respectively (Kaliraj et al., 2019).

During photo-degradation of water pollutant, ZnO nanoparticles tend to agglomerate and may also float on water surface resulting decrease in the photocatalytic efficiency (Kim et al., 2015). Presence of surface-active materials like zeolite, activated carbon, fly ash, biochar, natural clay etc with ZnO forms a composite. So, the photo generated electron-hole pairs of the nanocomposite photocatalyst remain separated, for photo-degradation of water pollutants (Thirumalai et al., 2016).

Sarkar et al. (2021) focused-on synthesis of ZnO nanoparticle and LD slag geopolymer composite photocatalyst by applying a facile hydrothermal method to increase the photo-degradation capacity under visible light irradiation by increasing its charge transfer capability. This ZnO-Geopolymer nanocomposite yields the maximum CR photo-degradation around 96.97% (initial CR concentration: 5 mg/L; pH: 6; photocatalyst dosage: 1 g/L) under visible light; and highest first order reaction rate constant (k) value as 0.0487 min^{-1} compared to other similar photocatalysts (Sarkar et al., 2021).

CONCLUSION

In this chapter we have discussed briefly about the various types of nanocomposites and its role in pollution prevention. As stated in the introduction, increasing population and rapid industrialization causes a serious threat to the environment by water pollution. The effluents from dye industries are a serious problem, which can be solved to some extent by science and technology developments. Various types of nanomaterials and nanocomposites play a very vital role in dye degradation. Nanocomposite preparation characterisation, degradation methodology and its efficiency towards degradation of various types of dyes are discussed in this chapter.

REFERENCES

Achouri, F., Corbel, S., Balan, L., Mozet, K., Girot, E., Medjahdi, G., Said, M. B., Ghrabi, A., & Schneider, R. (2016). Porous Mn-doped ZnO nanoparticles for enhanced solar and visible light photocatalysis. *Materials & Design*, *101*, 309–316. doi:10.1016/j.matdes.2016.04.015

Adeleye, A. S., Conway, J. R., Garner, K., Huang, Y., Su, Y., & Keller, A. A. (2016). Engineered nanomaterials for water treatment and remediation: Costs, benefits, and applicability. *Chemical Engineering Journal*, *286*, 640–662. doi:10.1016/j.cej.2015.10.105

Ahmad, M., Rehman, W., Khan, M. M., Qureshi, M. T., Gul, A., Haq, S., Ullah, R., Rab, A., & Menaa, F. (2021). Phytogenic fabrication of ZnO and gold decorated ZnO nanoparticles for photocatalytic degradation of Rhodamine B. *Journal of Environmental Chemical Engineering*, *9*(1), 104725. doi:10.1016/j.jece.2020.104725

Ahmed Quraishi, I., Pawar, R. A., Shinde, D. R., & Chaskar, M. G. (2020). Parametric Study on Photocatalytic Dye Degradation under Visible Light in Flat Slurry Reactor with nano-ZnO Photocatalyst. *Materials Today: Proceedings*, *23*, 410–422. doi:10.1016/j.matpr.2020.02.061

Alhebshi, N., Huang, H., Ghandour, R., Alghamdi, N. K., Alharbi, O., Aljurban, S., He, J.-H., & Al-Jawhari, H. (2020). Green synthesized CuxO@Cu nanocomposites on a Cu mesh with dual catalytic functions for dye degradation and hydrogen evaluation. *Journal of Alloys and Compounds*, *848*, 156284. doi:10.1016/j.jallcom.2020.156284

Aly, H. F., & Abd-Elhamid, A. I. (2018). Photocatalytic Degradation of Methylene Blue Dye Using Silica Oxide Nanoparticles as a Catalyst. *Water Environment Research*, *90*(9), 807–817. doi:10.2175/106143017X15131012187953 PMID:30208997

Aminuzzaman, M., Kei, L. M., & Liang, W. H. (2017). Green synthesis of copper oxide (CuO) nanoparticles using banana peel extract and their photocatalytic activities. *AIP Conference Proceedings*, *1828*(1), 020016. doi:10.1063/1.4979387

Andrade, G. R. S., Nascimento, C. C., Lima, Z. M., Teixeira-Neto, E., Costa, L. P., & Gimenez, I. F. (2017). Star-shaped ZnO/Ag hybrid nanostructures for enhanced photocatalysis and antibacterial activity. *Applied Surface Science*, *399*, 573–582. doi:10.1016/j.apsusc.2016.11.202

Atla, S. B., Lin, W.-R., Chien, T.-C., Tseng, M.-J., Shu, J.-C., Chen, C.-C., & Chen, C.-Y. (2018). Fabrication of Fe_3O_4/ZnO magnetite core shell and its application in photocatalysis using sunlight. *Materials Chemistry and Physics*, *216*, 380–386. doi:10.1016/j.matchemphys.2018.06.020

Atrak, K., Ramazani, A., & Taghavi Fardood, S. (2019). Eco-friendly synthesis of $Mg_{0.5}Ni_{0.5}Al_xFe_{2-x}O_4$ magnetic nanoparticles and study of their photocatalytic activity for degradation of direct blue 129 dye. *Journal of Photochemistry and Photobiology A Chemistry*, *382*, 111942. doi:10.1016/j.jphotochem.2019.111942

Aydoghmish, S. M., Hassanzadeh-Tabrizi, S. A., & Saffar-Teluri, A. (2019). Facile synthesis and investigation of NiO–ZnO–Ag nanocomposites as efficient photocatalysts for degradation of methylene blue dye. *Ceramics International*, *45*(12), 14934–14942. doi:10.1016/j.ceramint.2019.04.229

Bao, L., Zang, J., & Li, X. (2011). Flexible Zn_2SnO_4/MnO_2 Core/Shell Nanocable−Carbon Microfiber Hybrid Composites for High-Performance Supercapacitor Electrodes. *Nano Letters*, *11*(3), 1215–1220. doi:10.1021/nl104205s PMID:21306113

Barzegar, M. H., Ghaedi, M., Madadi Avargani, V., Sabzehmeidani, M. M., Sadeghfar, F., & Jannesar, R. (2019). Electrochemical synthesis and efficient photocatalytic degradation of azo dye alizarin yellow R by Cu/CuO nanorods under visible LED light irradiation using experimental design methodology. *Polyhedron*, *158*, 506–514. doi:10.1016/j.poly.2018.10.040

Basnet, P., Inakhunbi Chanu, T., Samanta, D., & Chatterjee, S. (2018). A review on bio-synthesized zinc oxide nanoparticles using plant extracts as reductants and stabilizing agents. *Journal of Photochemistry and Photobiology. B, Biology*, *183*, 201–221. doi:10.1016/j.jphotobiol.2018.04.036 PMID:29727834

Basnet, P., Samanta, D., Chanu, T. I., Mukherjee, J., & Chatterjee, S. (2019). Tea-phytochemicals functionalized Ag modified ZnO nanocomposites for visible light driven photocatalytic removal of organic water pollutants. *Materials Research Express*, *6*(8), 085095. doi:10.1088/2053-1591/ab234e

Basnet, P. C. S. (2019). Nanocomposites of ZnO for Water Remediation. In Composites for Environmental Engineering. doi:10.1002/9781119555346.ch7

Behjati, S., Sheibani, S., Herritsch, J., & Gottfried, J. M. (2020). Photodegradation of dyes in batch and continuous reactors by Cu_2O-CuO nano-photocatalyst on Cu foils prepared by chemical-thermal oxidation. *Materials Research Bulletin*, *130*, 110920. doi:10.1016/j.materresbull.2020.110920

Bhuyan, T., Mishra, K., Khanuja, M., Prasad, R., & Varma, A. (2015). Biosynthesis of zinc oxide nanoparticles from Azadirachta indica for antibacterial and photocatalytic applications. *Materials Science in Semiconductor Processing*, *32*, 55–61. doi:10.1016/j.mssp.2014.12.053

Borges, M. E., Sierra, M., Cuevas, E., García, R. D., & Esparza, P. (2016). Photocatalysis with solar energy: Sunlight-responsive photocatalyst based on TiO_2 loaded on a natural material for wastewater treatment. *Solar Energy*, *135*, 527–535. doi:10.1016/j.solener.2016.06.022

Bozetine, H., Wang, Q., Barras, A., Li, M., Hadjersi, T., Szunerits, S., & Boukherroub, R. (2016). Green chemistry approach for the synthesis of ZnO–carbon dots nanocomposites with good photocatalytic properties under visible light. *Journal of Colloid and Interface Science*, *465*, 286–294. doi:10.1016/j.jcis.2015.12.001 PMID:26674245

Chen, F., Tang, Y., Liu, C., Qian, J., Wu, Z., & Chen, Z. (2017). Synthesis of porous structured ZnO/Ag composite fibers with enhanced photocatalytic performance under visible irradiation. *Ceramics International*, *43*(16), 14525–14528. doi:10.1016/j.ceramint.2017.07.158

Choo, K.-H. (2018). Modeling Photocatalytic Membrane Reactors. In A. Basile, S. Mozia, & R. Molinari (Eds.), *Current Trends and Future Developments on (Bio-) Membranes* (pp. 297–316). Elsevier. doi:10.1016/B978-0-12-813549-5.00010-4

Crini, G., & Lichtfouse, E. (2019). Advantages and disadvantages of techniques used for wastewater treatment. *Environmental Chemistry Letters*, *17*(1), 145–155. doi:10.100710311-018-0785-9

Crini, G., Lichtfouse, E., Wilson, L. D., & Morin-Crini, N. (2019). Conventional and non-conventional adsorbents for wastewater treatment. *Environmental Chemistry Letters*, *17*(1), 195–213. doi:10.100710311-018-0786-8

Delnavaz, M., Farahbakhsh, J., & Mahdian, S. S. (2021). Photodegradation of reactive blue 19 dye using magnetic nanophotocatalyst α-Fe_2O_3/WO_3: A comparison study of α-Fe_2O_3/WO_3 and WO_3/NaOH. *Water Science and Engineering*, *14*(2), 119–128. doi:10.1016/j.wse.2021.06.007

Dhatshanamurthi, P., Shanthi, M., & Swaminathan, M. (2017). An efficient pilot scale solar treatment method for dye industry effluent using nano-ZnO. *Journal of Water Process Engineering*, *16*, 28–34. doi:10.1016/j.jwpe.2016.12.002

Dhivya, A., & Yadav, R. (2022). An Eco-approach synthesis of undoped and Mn doped ZnO nano-photocatalyst for prompt decoloration of methylene blue dye. *Materials Today: Proceedings*, *48*, 494–501. doi:10.1016/j.matpr.2021.02.751

Douafer, S., Lahmar, H., Benamira, M., Messaadia, L., Mazouzi, D., & Trari, M. (2019). Chromate reduction on the novel hetero-system $LiMn_2O_4$/SnO_2 catalyst under solar light irradiation. *Surfaces and Interfaces*, *17*, 100372. doi:10.1016/j.surfin.2019.100372

Ebrahimzadeh, M. A., Naghizadeh, A., Amiri, O., Shirzadi-Ahodashti, M., & Mortazavi-Derazkola, S. (2020). Green and facile synthesis of Ag nanoparticles using Crataegus pentagyna fruit extract (CP-AgNPs) for organic pollution dyes degradation and antibacterial application. *Bioorganic Chemistry*, *94*, 103425. doi:10.1016/j.bioorg.2019.103425 PMID:31740048

El-Bindary, A., Ismail, A., & Eladl, E. F. (2019). Photocatalytic degradation of reactive blue 21 using Ag doped ZnO nanoparticles. *Journal of Materials and Environmental Science*, *10*, 1258–1271.

Elfeky, A. S., Salem, S. S., Elzaref, A. S., Owda, M. E., Eladawy, H. A., Saeed, A. M., Awad, M. A., Abou-Zeid, R. E., & Fouda, A. (2020). Multifunctional cellulose nanocrystal /metal oxide hybrid, photo-degradation, antibacterial and larvicidal activities. *Carbohydrate Polymers*, *230*, 115711. doi:10.1016/j.carbpol.2019.115711 PMID:31887890

Fan, Y., Wu, D., Zhang, S., Zhang, L., Hu, W., Zhu, C., & Gong, X. (2021). *Effective photodegradation of 4-nitrophenol with CuO nano particles prepared by ionic liquids/water system.* Green Chemical Engineering. doi:10.1016/j.gce.2021.07.009

Fouda, A., Abdel-Maksoud, G., Abdel-Rahman, M. A., Eid, A. M., Barghoth, M. G., & El-Sadany, M. A.-H. (2019). Monitoring the effect of biosynthesized nanoparticles against biodeterioration of cellulose-based materials by Aspergillus niger. *Cellulose (London, England)*, *26*(11), 6583–6597. doi:10.100710570-019-02574-y

Fouda, A., Salem, S. S., Wassel, A. R., Hamza, M. F., & Shaheen, T. I. (2020). Optimization of green biosynthesized visible light active CuO/ZnO nano-photocatalysts for the degradation of organic methylene blue dye. *Heliyon, 6*(9), e04896. doi:10.1016/j.heliyon.2020.e04896 PMID:32995606

Gan, L., Xu, L., Shang, S., Zhou, X., & Meng, L. (2016). Visible light induced methylene blue dye degradation photo-catalyzed by WO_3/graphene nanocomposites and the mechanism. *Ceramics International, 42*(14), 15235–15241. doi:10.1016/j.ceramint.2016.06.160

Gao, M., Zhu, L., Ong, W. L., Wang, J., & Ho, G. (2015). Structural Design of TiO_2-based Photocatalyst for H2 Production and Degradation Applications. *Catalysis Science & Technology, 5*(10), 4703–4726. Advance online publication. doi:10.1039/C5CY00879D

George, A., Magimai Antoni Raj, D., Venci, X., Dhayal Raj, A., Albert Irudayaraj, A., Josephine, R. L., John Sundaram, S., Al-Mohaimeed, A. M., Al Farraj, D. A., Chen, T.-W., & Kaviyarasu, K. (2022). Photocatalytic effect of CuO nanoparticles flower-like 3D nanostructures under visible light irradiation with the degradation of methylene blue (MB) dye for environmental application. *Environmental Research, 203*, 111880. doi:10.1016/j.envres.2021.111880 PMID:34400161

Ghaffari, Y., Gupta, N. K., Bae, J., & Kim, K. S. (2020). One-step fabrication of Fe_2O_3/Mn_2O_3 nanocomposite for rapid photodegradation of organic dyes at neutral pH. *Journal of Molecular Liquids, 315*, 113691. doi:10.1016/j.molliq.2020.113691

Gu, C., Xiong, S., Zhong, Z., Wang, Y., & Xing, W. (2017). A promising carbon fiber-based photocatalyst with hierarchical structure for dye degradation. *RSC Advances, 7*(36), 22234-22242. doi:10.1039/C7RA02583A

Gudipati, T., Zaman, M. B., Singh, P., & Poolla, R. (2021). Enhanced photocatalytic activity of biogenically synthesized CuO nanostructures against xylenol orange and rhodamine B dyes. *Inorganic Chemistry Communications, 130*, 108677. doi:10.1016/j.inoche.2021.108677

Guo, N., Liang, Y., Lan, S., Liu, L., Zhang, J., Ji, G., & Gan, S. (2014). Microscale Hierarchical Three-Dimensional Flowerlike TiO_2/PANI Composite: Synthesis, Characterization, and Its Remarkable Photocatalytic Activity on Organic Dyes under UV-Light and Sunlight Irradiation. *The Journal of Physical Chemistry C, 118*(32), 18343–18355. doi:10.1021/jp5044927

Guo, N., Liu, H., Fu, Y., & Hu, J. (2020). Preparation of Fe_2O_3 nanoparticles doped with In2O3 and photocatalytic degradation property for rhodamine B. *Optik (Stuttgart), 201*, 163537. doi:10.1016/j.ijleo.2019.163537

Gupta, V. K., & Suhas. (2009). Application of low-cost adsorbents for dye removal—A review. *Journal of Environmental Management, 90*(8), 2313–2342. doi:10.1016/j.jenvman.2008.11.017 PMID:19264388

Hao, N., Nie, Y., Xu, Z., Jin, C., Fyda, T. J., & Zhang, J. X. J. (2020). Microfluidics-enabled acceleration of Fenton oxidation for degradation of organic dyes with rod-like zero-valent iron nanoassemblies. *Journal of Colloid and Interface Science, 559*, 254–262. doi:10.1016/j.jcis.2019.10.042 PMID:31634669

Hasan Khan Neon, M., & Islam, M. S. (2019). MoO_3 and Ag co-synthesized TiO_2 as a novel heterogeneous photocatalyst with enhanced visible-light-driven photocatalytic activity for methyl orange dye degradation. *Environmental Nanotechnology, Monitoring & Management, 12*, 100244. doi:10.1016/j.enmm.2019.100244

Hisatomi, T., Takanabe, K., & Domen, K. (2015). Photocatalytic Water-Splitting Reaction from Catalytic and Kinetic Perspectives. *Catalysis Letters, 145*(1), 95–108. doi:10.100710562-014-1397-z

Hu, W., Xie, L., & Zeng, H. (2020). Novel sodium alginate-assisted MXene nanosheets for ultrahigh rejection of multiple cations and dyes. *Journal of Colloid and Interface Science, 568*, 36–45. doi:10.1016/j.jcis.2020.02.028 PMID:32086010

Ihsanullah, D., Asmaly, H., Saleh, T., Laoui, T., Gupta, V., & Atieh, M. A. (2015). Enhanced adsorption of phenols from liquids by aluminum oxide/carbon nanotubes: Comprehensive study from synthesis to surface properties. *Journal of Molecular Liquids, 206*, 176–182, 182. doi:10.1016/j.molliq.2015.02.028

Jadhav, J., & Biswas, S. (2018). Hybrid ZnO:Ag core-shell nanoparticles for wastewater treatment: Growth mechanism and plasmonically enhanced photocatalytic activity. *Applied Surface Science, 456*, 49–58. doi:10.1016/j.apsusc.2018.06.028

Jayapriya, M., & Arulmozhi, M. (2021). Beta vulgaris peel extract mediated synthesis of Ag/TiO_2 nanocomposite: Characterization, evaluation of antibacterial and catalytic degradation of textile dyes-an electron relay effect. *Inorganic Chemistry Communications, 128*, 108529. doi:10.1016/j.inoche.2021.108529

Jung, S. M., Lee, J., Han, M., Choi, J., Kim, S., Seo, J., & Lim, H. (2010). Advanced photocatalytic activity using TiO_2/ceramic fiber-based honeycomb. *Studies in Surface Science and Catalysis, 175*, 441–444. doi:10.1016/S0167-2991(10)75080-1

Kahouli, M., Barhoumi, A., Bouzid, A., Al-Hajry, A., & Guermazi, S. (2015). Structural and optical properties of ZnO nanoparticles prepared by direct precipitation method. *Superlattices and Microstructures, 85*, 7–23. doi:10.1016/j.spmi.2015.05.007

Kajbafvala, A., Ghorbani, H., Paravar, A., Samberg, J., Kajbafvala, E., & Sadrnezhaad, S. K. (2012). Effects of morphology on photocatalytic performance of Zinc oxide nanostructures synthesized by rapid microwave irradiation methods. *Superlattices and Microstructures, 51*(4), 512–522. doi:10.1016/j.spmi.2012.01.015

Kaliraj, L., Ahn, J. C., Rupa, E. J., Abid, S., Lu, J., & Yang, D. C. (2019). Synthesis of panos extract mediated ZnO nano-flowers as photocatalyst for industrial dye degradation by UV illumination. *Journal of Photochemistry and Photobiology. B, Biology, 199*, 111588. doi:10.1016/j.jphotobiol.2019.111588 PMID:31450132

Kaushal, S., Kaur, N., Kaur, M., & Singh, P. P. (2020). Dual-Responsive Pectin/Graphene Oxide (Pc/GO) nano-composite as an efficient adsorbent for Cr (III) ions and photocatalyst for degradation of organic dyes in waste water. *Journal of Photochemistry and Photobiology A Chemistry, 403*, 112841. doi:10.1016/j.jphotochem.2020.112841

Keshavarz, H. R. P. M. H. (2011, November 2). Study of Congo red photodegradation kinetic catalyzed by $Zn_{1-X}Cu_XS$ and $Zn_{1-X}Ni_XS$ nanoparticles. *International Journal of Physical Sciences*, *6*(27), 6268–6279. doi:10.5897/IJPS09.251

Kim, H. J., Joshi, M. K., Pant, H. R., Kim, J. H., Lee, E., & Kim, C. S. (2015). One-pot hydrothermal synthesis of multifunctional Ag/ZnO/fly ash nanocomposite. *Colloids and Surfaces. A, Physicochemical and Engineering Aspects*, *469*, 256–262. doi:10.1016/j.colsurfa.2015.01.032

Kong, J.-Z., Li, A.-D., Zhai, H.-F., Gong, Y.-P., Li, H., & Wu, D. (2009). Preparation, characterization of the Ta-doped ZnO nanoparticles and their photocatalytic activity under visible-light illumination. *Journal of Solid State Chemistry*, *182*(8), 2061–2067. doi:10.1016/j.jssc.2009.03.022

Kubiak, A., Kubacka, M., Gabała, E., Dobrowolska, A., Synoradzki, K., Siwińska-Ciesielczyk, K., & Jesionowski, T. (2020). Hydrothermally Assisted Fabrication of TiO_2-Fe_3O_4 Composite Materials and Their Antibacterial Activity. *Materials (Basel)*, *13*(21), 4715. doi:10.3390/ma13214715 PMID:33105776

Kumar, S. S., Venkateswarlu, P., Rao, V. R., & Rao, G. N. (2013). Synthesis, characterization and optical properties of zinc oxide nanoparticles. *International Nano Letters*, *3*(1), 30. doi:10.1186/2228-5326-3-30

Kumaresan, S., Vallalperuman, K., Sathishkumar, S., Karthik, M., & SivaKarthik, P. (2017). Synthesis and systematic investigations of Al and Cu-doped ZnO nanoparticles and its structural, optical and photo-catalytic properties. *Journal of Materials Science Materials in Electronics*, *28*(13), 9199–9205. doi:10.100710854-017-6654-7

Li, D., & Haneda, H. (2003). Morphologies of zinc oxide particles and their effects on photocatalysis. *Chemosphere*, *51*(2), 129–137. doi:10.1016/S0045-6535(02)00787-7 PMID:12586145

Li, X., Mei, Q., Chen, L., Zhang, H., Dong, B., Dai, X., & Zhou, J. (2019). Enhancement in adsorption potential of microplastics in sewage sludge for metal pollutants after the wastewater treatment process. *Water Research*, *157*, 228–237. doi:10.1016/j.watres.2019.03.069 PMID:30954698

Li, Y., Zhang, B.-P., Zhao, J.-X., Ge, Z.-H., Zhao, X.-K., & Zou, L. (2013). ZnO/carbon quantum dots heterostructure with enhanced photocatalytic properties. *Applied Surface Science*, *279*, 367–373. doi:10.1016/j.apsusc.2013.04.114

Liu, C., Yu, Z., Neff, D., Zhamu, A., & Jang, B. Z. (2010). Graphene-based supercapacitor with an ultrahigh energy density. *Nano Letters*, *10*(12), 4863–4868. doi:10.1021/nl102661q PMID:21058713

Liu, H., Liu, H., Yang, J., Zhai, H., Liu, X., & Jia, H. (2019). Microwave-assisted one-pot synthesis of Ag decorated flower-like ZnO composites photocatalysts for dye degradation and NO removal. *Ceramics International*, *45*(16), 20133–20140. doi:10.1016/j.ceramint.2019.06.279

Liu, Y., Xu, C., Lu, J., Zhu, Z., Zhu, Q., Manohari, A. G., & Shi, Z. (2018). Template-free synthesis of porous ZnO/Ag microspheres as recyclable and ultra-sensitive SERS substrates. *Applied Surface Science*, *427*, 830–836. doi:10.1016/j.apsusc.2017.07.229

Manjari, G., Saran, S., Arun, T., Vijaya Bhaskara Rao, A., & Devipriya, S. P. (2017). Catalytic and recyclability properties of phytogenic copper oxide nanoparticles derived from Aglaia elaeagnoidea flower extract. *Journal of Saudi Chemical Society*, *21*(5), 610–618. doi:10.1016/j.jscs.2017.02.004

Mekasuwandumrong, O., Pawinrat, P., Praserthdam, P., & Panpranot, J. (2010). Effects of synthesis conditions and annealing post-treatment on the photocatalytic activities of ZnO nanoparticles in the degradation of methylene blue dye. *Chemical Engineering Journal, 164*(1), 77–84. doi:10.1016/j.cej.2010.08.027

Mendoza-Mendoza, E., Nuñez-Briones, A. G., García-Cerda, L. A., Peralta-Rodríguez, R. D., & Montes-Luna, A. J. (2018). One-step synthesis of ZnO and Ag/ZnO heterostructures and their photocatalytic activity. *Ceramics International, 44*(6), 6176–6180. doi:10.1016/j.ceramint.2018.01.001

Menon, S., Agarwal, H., & Kumar, V. (2021). Catalytical degradation of industrial dyes using biosynthesized selenium nanoparticles and evaluating its antimicrobial activities. *Sustainable Environment Research, 31*(1), 2. Advance online publication. doi:10.118642834-020-00072-6

Mishra, R. K., Choi, G. J., Sohn, Y., Lee, S. H., & Gwag, J. S. (2020). A novel RGO/N-RGO supercapacitor architecture for a wide voltage window, high energy density and long-life via voltage holding tests. *Chemical Communications, 56*(19), 2893-2896. doi:10.1039/D0CC00249F

Mostafa, A. M., Mwafy, E. A., & Toghan, A. (2021). ZnO nanoparticles decorated carbon nanotubes via pulsed laser ablation method for degradation of methylene blue dyes. *Colloids and Surfaces. A, Physicochemical and Engineering Aspects, 627*, 127204. doi:10.1016/j.colsurfa.2021.127204

Nabi, G., Majid, A., Riaz, A., Alharbi, T., Arshad Kamran, M., & Al-Habardi, M. (2021). Green synthesis of spherical TiO_2 nanoparticles using Citrus Limetta extract: Excellent photocatalytic water decontamination agent for RhB dye. *Inorganic Chemistry Communications, 129*, 108618. doi:10.1016/j.inoche.2021.108618

Neves, T., Frantz, T., Schenque, E., Gelesky, M., & Mortola, V. (2017). An investigation into an alternative photocatalyst based on CeO_2/Al_2O_3 in dye degradation. *Environmental Technology & Innovation, 8*, 349–359. Advance online publication. doi:10.1016/j.eti.2017.08.003

Nidheesh, P. V., Zhou, M., & Oturan, M. A. (2018). An overview on the removal of synthetic dyes from water by electrochemical advanced oxidation processes. *Chemosphere, 197*, 210–227. doi:10.1016/j.chemosphere.2017.12.195 PMID:29366952

Noreen, S., Mustafa, G., Ibrahim, S. M., Naz, S., Iqbal, M., Yaseen, M., & Nisar, J. (2020). Iron oxide (Fe_2O_3) prepared via green route and adsorption efficiency evaluation for an anionic dye: Kinetics, isotherms and thermodynamics studies. *Journal of Materials Research and Technology, 9*(3), 4206–4217. doi:10.1016/j.jmrt.2020.02.047

Norouzi, A., & Nezamzadeh-Ejhieh, A. (2020). α-Fe_2O_3/Cu_2O heterostructure: Brief characterization and kinetic aspect of degradation of methylene blue. *Physica B, Condensed Matter, 599*, 412422. doi:10.1016/j.physb.2020.412422

Nuengmatcha, P., Porrawatkul, P., Chanthai, S., Sricharoen, P., & Limchoowong, N. (2019). Enhanced photocatalytic degradation of methylene blue using Fe_2O_3/graphene/CuO nanocomposites under visible light. *Journal of Environmental Chemical Engineering, 7*(6), 103438. doi:10.1016/j.jece.2019.103438

Ong, C. B., Ng, L. Y., & Mohammad, A. W. (2018). A review of ZnO nanoparticles as solar photocatalysts: Synthesis, mechanisms and applications. *Renewable & Sustainable Energy Reviews, 81*, 536–551. doi:10.1016/j.rser.2017.08.020

Ong, W.-J., Yeong, J.-J., Tan, L.-L., Goh, B. T., Yong, S.-T., & Chai, S.-P. (2014). Synergistic effect of graphene as a co-catalyst for enhanced daylight-induced photocatalytic activity of $Zn_{0.5}Cd_{0.5}S$ synthesized via an improved one-pot co-precipitation-hydrothermal strategy. *RSC Advances, 4*(103), 59676-59685. doi:10.1039/C4RA10467F

Pallela, P. N. V. K., Ruddaraju, L. K., Veerla, S. C., Matangi, R., Kollu, P., Ummey, S., & Pammi, S. V. N. (2020). Synergetic antibacterial potential, dye degrading capability and biocompatibility of Asperagus racemosus root assisted ZnO nanoparticles. *Materials Today. Communications, 25*, 101574. doi:10.1016/j.mtcomm.2020.101574

Panchal, P., Paul, D. R., Sharma, A., Choudhary, P., Meena, P., & Nehra, S. P. (2020). Biogenic mediated Ag/ZnO nanocomposites for photocatalytic and antibacterial activities towards disinfection of water. *Journal of Colloid and Interface Science, 563*, 370–380. doi:10.1016/j.jcis.2019.12.079 PMID:31887701

Pavithra, M., & Jessie Raj, M. B. (2021). Influence of ultrasonication time on solar light irradiated photocatalytic dye degradability and antibacterial activity of Pb doped ZnO nanocomposites. *Ceramics International, 47*(22), 32324–32331. doi:10.1016/j.ceramint.2021.08.128

Pavoski, G., Baldisserotto, D. L. S., Maraschin, T., Brum, L. F. W., dos Santos, C., dos Santos, J. H. Z., Brandelli, A., & Galland, G. B. (2019). Silver nanoparticles encapsulated in silica: Synthesis, characterization and application as antibacterial fillers in the ethylene polymerization. *European Polymer Journal, 117*, 38–54. doi:10.1016/j.eurpolymj.2019.04.055

Pendolino, F., & Armata, N. (2017). *Graphene Oxide in Environmental Remediation Process*. doi:10.1007/978-3-319-60429-9

Phang, S. J., & Tan, L.-L. (2019). Recent advances in carbon quantum dot (CQD)-based two dimensional materials for photocatalytic applications. *Catalysis Science & Technology, 9*(21), 5882-5905. doi:10.1039/C9CY01452G

Prabhuraj, T., Prabhu, S., Dhandapani, E., Duraisamy, N., Ramesh, R., Kumar, K. A. R., & Maadeswaran, P. (2021). Bifunctional ZnO sphere/r-GO composites for supercapacitor and photocatalytic activity of organic dye degradation. *Diamond and Related Materials, 120*, 108592. doi:10.1016/j.diamond.2021.108592

Pragathiswaran, C., Smitha, C., Mahin Abbubakkar, B., Govindhan, P., & Anantha Krishnan, N. (2021). Synthesis and characterization of TiO_2/ZnO–Ag nanocomposite for photocatalytic degradation of dyes and anti-microbial activity. *Materials Today: Proceedings, 45*, 3357–3364. doi:10.1016/j.matpr.2020.12.664

Pugazhendhi, A., Prabhu, R., Muruganantham, K., Shanmuganathan, R., & Natarajan, S. (2019). Anticancer, antimicrobial and photocatalytic activities of green synthesized magnesium oxide nanoparticles (MgONPs) using aqueous extract of Sargassum wightii. *Journal of Photochemistry and Photobiology. B, Biology, 190*, 86–97. doi:10.1016/j.jphotobiol.2018.11.014 PMID:30504053

Rajendran, R., Vignesh, S., Sasireka, A., Priya, P., Suganthi, S., Raj, V., & AlFaify, S. (2021). Investigation on novel Cu_2O modified g-C_3N_4/ZnO heterostructures for efficient photocatalytic dye degradation performance under visible-light exposure. *Colloid and Interface Science Communications, 44*, 100480. doi:10.1016/j.colcom.2021.100480

Rashid Al-Mamun, M., Shofikul Islam, M., Rasel Hossain, M., Kader, S., Shahinoor Islam, M., & Zaved Hossain Khan, M. (2021). A novel and highly efficient Ag and GO co-synthesized ZnO nano photocatalyst for methylene blue dye degradation under UV irradiation. *Environmental Nanotechnology, Monitoring & Management*, *16*, 100495. doi:10.1016/j.enmm.2021.100495

Rodney, J. D., Deepapriya, S., Annie Vinosha, P., Krishnan, S., Janet Priscilla, S., Daniel, R., & Jerome Das, S. (2018). Photo-Fenton degradation of nano-structured La doped CuO nanoparticles synthesized by combustion technique. *Optik (Stuttgart)*, *161*, 204–216. doi:10.1016/j.ijleo.2018.01.125

Ruíz-Baltazar, Á. J. (2020). Green synthesis assisted by sonochemical activation of Fe_3O_4-Ag nano-alloys: Structural characterization and studies of sorption of cationic dyes. *Inorganic Chemistry Communications*, *120*, 108148. doi:10.1016/j.inoche.2020.108148

Saad, A. M., Abukhadra, M. R., Abdel-Kader Ahmed, S., Elzanaty, A. M., Mady, A. H., Betiha, M. A., Shim, J.-J., & Rabie, A. M. (2020). Photocatalytic degradation of malachite green dye using chitosan supported ZnO and Ce–ZnO nano-flowers under visible light. *Journal of Environmental Management*, *258*, 110043. doi:10.1016/j.jenvman.2019.110043 PMID:31929075

Saleh, R., & Djaja, N. F. (2014). Transition-metal-doped ZnO nanoparticles: Synthesis, characterization and photocatalytic activity under UV light. *Spectrochimica Acta. Part A: Molecular and Biomolecular Spectroscopy*, *130*, 581–590. doi:10.1016/j.saa.2014.03.089 PMID:24813289

Salem, S. S., & Fouda, A. (2021). Green Synthesis of Metallic Nanoparticles and Their Prospective Biotechnological Applications: An Overview. *Biological Trace Element Research*, *199*(1), 344–370. doi:10.100712011-020-02138-3 PMID:32377944

Sarkar, C., Basu, J. K., & Samanta, A. N. (2021). Synthesis of novel ZnO/Geopolymer nanocomposite photocatalyst for degradation of congo red dye under visible light. *Environmental Nanotechnology, Monitoring & Management*, *16*, 100521. doi:10.1016/j.enmm.2021.100521

Sayadi, M. H., Amadpour, N., & Homaeigohar, S. (2021). Photocatalytic and Antibacterial Properties of Ag-$CuFe_2O_4$@WO_3 Magnetic Nanocomposite. *Nanomaterials (Basel, Switzerland)*, *11*(2), 298. doi:10.3390/nano11020298 PMID:33498950

Shanmugam, V., & Jeyaperumal, K. S. (2018). Investigations of visible light driven Sn and Cu doped ZnO hybrid nanoparticles for photocatalytic performance and antibacterial activity. *Applied Surface Science*, *449*, 617–630. doi:10.1016/j.apsusc.2017.11.167

Sharma, M., Poddar, M., Gupta, Y., Nigam, S., Avasthi, D. K., Adelung, R., Abolhassani, R., Fiutowski, J., Joshi, M., & Mishra, Y. K. (2020). Solar light assisted degradation of dyes and adsorption of heavy metal ions from water by CuO–ZnO tetrapodal hybrid nanocomposite. *Materials Today. Chemistry*, *17*, 100336. doi:10.1016/j.mtchem.2020.100336

Sharma, S., Kumar, K., Thakur, N., Chauhan, S., & Chauhan, M. S. (2019). The effect of shape and size of ZnO nanoparticles on their antimicrobial and photocatalytic activities: A green approach. *Bulletin of Materials Science*, *43*(1), 20. doi:10.100712034-019-1986-y

Simamora, A. J., Hsiung, T. L., Chang, F. C., Yang, T. C., Liao, C. Y., & Wang, H. P. (2012). Photocatalytic splitting of seawater and degradation of methylene blue on CuO/nano TiO_2. *International Journal of Hydrogen Energy*, *37*(18), 13855–13858. doi:10.1016/j.ijhydene.2012.04.091

Singh, A., Ahmed, A., Sharma, A., Sharma, C., Paul, S., Khosla, A., & Arya, S. (2021). Promising photocatalytic degradation of methyl orange dye via sol-gel synthesized Ag–CdS@Pr-TiO_2 core/shell nanoparticles. *Physica B, Condensed Matter*, *616*, 413121. doi:10.1016/j.physb.2021.413121

Singh, J., Dutta, T., Kim, K.-H., Rawat, M., Samddar, P., & Kumar, P. (2018). 'Green' synthesis of metals and their oxide nanoparticles: Applications for environmental remediation. *Journal of Nanobiotechnology*, *16*(1), 84. doi:10.118612951-018-0408-4 PMID:30373622

Singhal, S., Dixit, S., & Shukla, A. K. (2018). Self-assembly of the Ag deposited ZnO/carbon nanospheres: A resourceful photocatalyst for efficient photocatalytic degradation of methylene blue dye in water. *Advanced Powder Technology*, *29*(12), 3483–3492. doi:10.1016/j.apt.2018.09.031

Srivastava, V., & Choubey, A. K. (2021). Investigation of adsorption of organic dyes present in wastewater using chitosan beads immobilized with biofabricated CuO nanoparticles. *Journal of Molecular Structure*, *1242*, 130749. doi:10.1016/j.molstruc.2021.130749

Stanley, R., Jebasingh, J. A., Stanley, P. K., Ponmani, P., Shekinah, M. E., & Vasanthi, J. (2021). Excellent Photocatalytic degradation of Methylene Blue, Rhodamine B and Methyl Orange dyes by Ag-ZnO nanocomposite under natural sunlight irradiation. *Optik (Stuttgart)*, *231*, 166518. doi:10.1016/j.ijleo.2021.166518

Szpyrkowicz, L., Juzzolino, C., & Kaul, S. N. (2001). A Comparative study on oxidation of disperse dyes by electrochemical process, ozone, hypochlorite and fenton reagent. *Water Research*, *35*(9), 2129–2136. doi:10.1016/S0043-1354(00)00487-5 PMID:11358291

Tantawy, H. R., Nada, A. A., Baraka, A., & Elsayed, M. A. (2021). Novel synthesis of bimetallic Ag-Cu nanocatalysts for rapid oxidative and reductive degradation of anionic and cationic dyes. *Applied Surface Science Advances*, *3*, 100056. doi:10.1016/j.apsadv.2021.100056

Thandapani, K., Kathiravan, M., Namasivayam, E., Padiksan, I. A., Natesan, G., Tiwari, M., Giovanni, B., & Perumal, V. (2018). Enhanced larvicidal, antibacterial, and photocatalytic efficacy of TiO2 nanohybrids green synthesized using the aqueous leaf extract of Parthenium hysterophorus. *Environmental Science and Pollution Research International*, *25*(11), 10328–10339. doi:10.100711356-017-9177-0 PMID:28537028

Thennarasu, G., & Sivasamy, A. (2016). Enhanced visible photocatalytic activity of cotton ball like nano structured Cu doped ZnO for the degradation of organic pollutant. *Ecotoxicology and Environmental Safety*, *134*, 412–420. doi:10.1016/j.ecoenv.2015.10.030 PMID:26560433

Thirumalai, K., Balachandran, S., & Swaminathan, M. (2016). Superior photocatalytic, electrocatalytic, and self-cleaning applications of Fly ash supported ZnO nanorods. *Materials Chemistry and Physics*, *183*, 191–200. doi:10.1016/j.matchemphys.2016.08.018

Udayabhanu, L. R., Lakshmana Reddy, N., Shankar, M. V., Sharma, S. C., & Nagaraju, G. (2020). One pot synthesis of Cu–TiO_2/CuO nanocomposite: Application to photocatalysis for enhanced H2 production, dye degradation & detoxification of Cr (VI). *International Journal of Hydrogen Energy, 45*(13), 7813–7828. doi:10.1016/j.ijhydene.2019.10.081

Ullah, R., & Dutta, J. (2008). Photocatalytic degradation of organic dyes with manganese-doped ZnO nanoparticles. *Journal of Hazardous Materials, 156*(1), 194–200. doi:10.1016/j.jhazmat.2007.12.033 PMID:18221834

Velumani, A., Sengodan, P., Arumugam, P., Rajendran, R., Santhanam, S., & Palanisamy, M. (2020). Carbon quantum dots supported ZnO sphere based photocatalyst for dye degradation application. *Current Applied Physics, 20*(10), 1176–1184. doi:10.1016/j.cap.2020.07.016

Vidya, C., Manjunatha, C., Chandraprabha, M. N., & Rajshekar, M., & Raj, M. (2017). Hazard free green synthesis of ZnO nano-photo-catalyst using Artocarpus Heterophyllus leaf extract for the degradation of Congo red dye in water treatment applications. *Journal of Environmental Chemical Engineering, 5*(4), 3172–3180. doi:10.1016/j.jece.2017.05.058

Vinay, S. P., & Chandrasekhar, N. (2019). Facile Green Chemistry Synthesis of Ag Nanoparticles Using Areca Catechu Extracts for the Antimicrobial Activity and Photocatalytic Degradation of Methylene Blue Dye. *Materials Today: Proceedings, 9*, 499–505. doi:10.1016/j.matpr.2018.10.368

Vu, A.-T., Xuan, T. N., & Lee, C.-H. (2019). Preparation of mesoporous $Fe_2O_3{\cdot}SiO_2$ composite from rice husk as an efficient heterogeneous Fenton-like catalyst for degradation of organic dyes. *Journal of Water Process Engineering, 28*, 169–180. doi:10.1016/j.jwpe.2019.01.019

Wan, H., Wang, F., Chen, Y., Zhao, Z., Zhang, G., Dou, M., & Xue, B. (2021). Enhanced Reactive Red 2 anaerobic degradation through improving electron transfer efficiency by nano-Fe_3O_4 modified granular activated carbon. *Renewable Energy, 179*, 696–704. doi:10.1016/j.renene.2021.07.046

Wan, X., Yang, J., Huang, X., Tie, S., & Lan, S. (2019). A high-performance room temperature thermo-catalyst Cu_2O/Ag_0@Ag-NPs for dye degradation under dark condition. *Journal of Alloys and Compounds, 785*, 398–409. doi:10.1016/j.jallcom.2019.01.215

Wang, K., Xu, X., Lu, L., Li, A., Han, X., Wu, Y., & Jiang, Y. (2019). Magnetically recoverable Ag/Bi2Fe4O9 nanoparticles as a visible-light-driven photocatalyst. *Chemical Physics Letters, 715*, 129–133. doi:10.1016/j.cplett.2018.11.021

Wang, P., Qi, C., Hao, L., Wen, P., & Xu, X. (2019). Sepiolite/Cu_2O/Cu photocatalyst: Preparation and high performance for degradation of organic dye. *Journal of Materials Science and Technology, 35*(3), 285–291. doi:10.1016/j.jmst.2018.03.023

Wang, Y., Liu, T., & Liu, J. (2020). Synergistically Boosted Degradation of Organic Dyes by CeO_2 Nanoparticles with Fluoride at Low pH. *ACS Applied Nano Materials, 3*(1), 842–849. doi:10.1021/acsanm.9b02356

Wongkaew, A., Kongsi, W., & Limsuwan, P. (2013). Physical Properties and Selective CO Oxidation of Coprecipitated CuO/CeO_2 Catalysts Depending on the CuO in the Samples. *Advances in Materials Science and Engineering, 2013*, 1–8. Advance online publication. doi:10.1155/2013/374080

Wu, L., Luo, Y., Zhou, S., Wu, Z., & Chen, X. (2021). Fabrication of Ag-TiO_2 functionalized activated carbon for dyes degradation based on tea residues. *Colloids and Surfaces. A, Physicochemical and Engineering Aspects*, *627*, 127130. doi:10.1016/j.colsurfa.2021.127130

Xu, X., Jia, K., Chen, S., Lang, D., Yang, C., Wang, L., & Wang, J. (2021). Ultra-fast degradation of phenolics and dyes by Cu_2O/Cu catalysts: Synthesis and degradation kinetics. *Journal of Environmental Chemical Engineering*, *9*(4), 105505. doi:10.1016/j.jece.2021.105505

Xu, Y., Wu, S., Li, X., Meng, H., Zhang, X., Wang, Z., & Han, Y. (2017). Ag nanoparticle-functionalized ZnO micro-flowers for enhanced photodegradation of herbicide derivatives. *Chemical Physics Letters*, *679*, 119–126. doi:10.1016/j.cplett.2017.04.091

Yang, M., Guo, Z., Qiu, K., Long, J., Yin, G., Guan, D., & Zhou, S. (2010). Synthesis and characterization of Mn-doped ZnO column arrays. *Applied Surface Science*, *256*(13), 4201–4205. doi:10.1016/j.apsusc.2010.01.125

Yang, Y., Sun, M., Zhou, J., Ma, J., & Komarneni, S. (2020). Degradation of orange II by Fe@Fe_2O_3 core shell nanomaterials assisted by $NaHSO_3$. *Chemosphere*, *244*, 125588. doi:10.1016/j.chemosphere.2019.125588 PMID:32050354

Zang, Z., & Tang, X. (2015). Enhanced fluorescence imaging performance of hydrophobic colloidal ZnO nanoparticles by a facile method. *Journal of Alloys and Compounds*, *619*, 98–101. doi:10.1016/j.jallcom.2014.09.072

Zhang, J., Li, L., Li, Y., & Yang, C. (2017). Microwave-assisted synthesis of hierarchical mesoporous nano-TiO_2/cellulose composites for rapid adsorption of Pb2+. *Chemical Engineering Journal*, *313*, 1132–1141. doi:10.1016/j.cej.2016.11.007

Zhang, X., Pan, J., Zhu, C., Sheng, Y., Yan, Z., Wang, Y., & Feng, B. (2015). The visible light catalytic properties of carbon quantum dots/ZnO nanoflowers composites. *Journal of Materials Science Materials in Electronics*, *26*(5), 2861–2866. doi:10.100710854-015-2769-x

Zhao, B., Li, H., Qin, X., Li, Z., Zhang, S., Wang, A., & Zhu, Z. (2021). Performance enhancement and catalytic mechanism identification of Cu-based composite for degradation of organic contaminants. *Powder Technology*, *389*, 11–20. doi:10.1016/j.powtec.2021.04.092

Zhong, R., Zhong, Q., Huo, M., Yang, B., & Li, H. (2020). Preparation of biocompatible nano-ZnO/chitosan microspheres with multi-functions of antibacterial, UV-shielding and dye photodegradation. *International Journal of Biological Macromolecules*, *146*, 939–945. doi:10.1016/j.ijbiomac.2019.09.217 PMID:31726126

Zyoud, A., Dwikat, M., Al-Shakhshir, S., Ateeq, S., Shteiwi, J., Zu'bi, A., & Hilal, H. S. (2016). Natural dye-sensitized ZnO nano-particles as photo-catalysts in complete degradation of E. coli bacteria and their organic content. *Journal of Photochemistry and Photobiology A Chemistry*, *328*, 207–216. doi:10.1016/j.jphotochem.2016.05.020

Chapter 14
Application of Carbon-Based Nanocomposite Materials for Wastewater Treatment

Arun Kant
Kirori Mal College, University of Delhi, India

Sudipta Ghosh
Kirori Mal College, University of Delhi, India

Gyanendra Kumar
Swami Shraddhanand College, India

M. Ramananda Singh
Kirori Mal College, University of Delhi, India

Mohd Ehtesham
Jamia Millia Islami University, India

Panmei Gaijon
Kirori Mal College, University of Delhi, India

ABSTRACT

Water is a vital component of life. It is naturally available as earth hydrosphere and plays an important role in the world economy, and it essential for balancing of the ecosystem. Numerous microbes and other toxins such as chemicals and heavy metals are integrated into rainwater and flowing water, resulting in water pollution. This chapter examines the numerous ways in which nanomaterials can be used to remove various kinds of contaminants from polluted water. In this chapter, carbon-based adsorbents material, that is, carbonaceous materials, has described. Carbonaceous materials such as stimulated carbon, carbon nanotubes, and graphene oxide have good performance and high adsorption value for medicinal active chemicals. In present-day investigations, researchers have found that carbon-based nanomaterials have been located progressively being applied in recycling of wastewater treatment research with overwhelmingly positive results.

INTRODUCTION

Water is an important natural resource of the earth on which all the living beings are depends. The water present on earth that cycling endlessly through the environment called hydrological cycle. Out of total water reserved of the earth, near about 97% present as salty and about 3% water present as fresh water.

DOI: 10.4018/978-1-6684-4553-2.ch014

About 9.86% of total freshwater resources of the earth are present as in the form of ground water. The water (Drinking or non-drinking) present in the nature, it has properties to dissolve many substances in it, therefore got polluted by different way, it is defined as alteration in chemical, physical and biological characteristics of water making it unsuitable for selected use in its normal state.

There are different sources that cause the water pollution, ground water polluted by textile industries, tanneries and chemical industries etc. In surface water pollution, the main role of marine water pollution, rivers bring pollutants from drainage basins, catchment area that is industries, hospitals waste, hotels; agricultural wastes directly come in this water. Oil drilling and shipment a big issue of marine water pollution because, petroleum, paint industries, ship-accidents add to marine water cause a pollution.

Due to because of the pollution, it necessary to create a mechanism to purify the water present on the earth, in which nano-composite and carbon-based material play a substantial role in wastewater treatment. The classification of wastewater treatment is well described by given pictorial charts shown in Figure 1.

The term "nano" is resultant from the Greek word nanus, which meaning "dwarf." Physics professor Richard Feynman initially introduced the concept of nanotechnology on December 29, (Feynman, 1960). The University of Tokyo was first devised the word "Nanotechnology" in 1974 year. Development in nanotechnology and nano-mediated domains has exploded in recent years, attracting academics, investors, governments, and the corporate sector. Quantum and nanodevices arose in the early 1980s as a result of significant advances in material modelling, as well as significant advances in characterisation, such as the scanning tunnelling microscope (STM) also atomic force microscope (AFM) (Capek, 2006). Quantum and functions are linked to the fabrication and characterisation of foundations based on size, shape, self-assembly, surface, and imperfection features using various nanotechnology-based physico-chemical characterization approaches. Each of these fields has its own unique applications based on the fundamental principles of nanotechnology. Their applications often include nanomedicine, nuclear physics, and nanocatalysis, among others (Guisbiers et al., 2012).

The ability to construct new nanostructures at the atomic scale too has resulted in novel materials and techniques with wide-ranging applications. In order to satisfy our ever-increasing energy needs, major advances must be made in the energy sector. Environment, food, health, water, and a range of other issues are all at risk in the modern world.

Wastewater obtained from different sources and its treatment could be divided into the following category such as chemical, physical, and biological treatments (El-Gendy et al., 2020; Russell, 2006). In case of Chemical method following steps involved like oxidation, ion exchange, photo-oxidation, coagulation and electrochemical processes, photo-oxidation further divided in two steps such as photolysis and photocatalysis; electrochemical processes are also further divided into electrocoagulation and electro degradation (Rajasulochana & Preethy, 2016). There are numerous applications for adsorption in wastewater treatment, as it is a physical/chemical remediation process that can handle practically any sort of contamination including organic chemicals, dyes, heavy metals, etc., (Das, & Poater, 2021). In this chapter, Authors have explained the variety of carbon-based adsorbent material used in remediation as well as purification of water. These types of materials are cost-effective and easily available.

Figure 1. Classification of wastewater treatment technologies

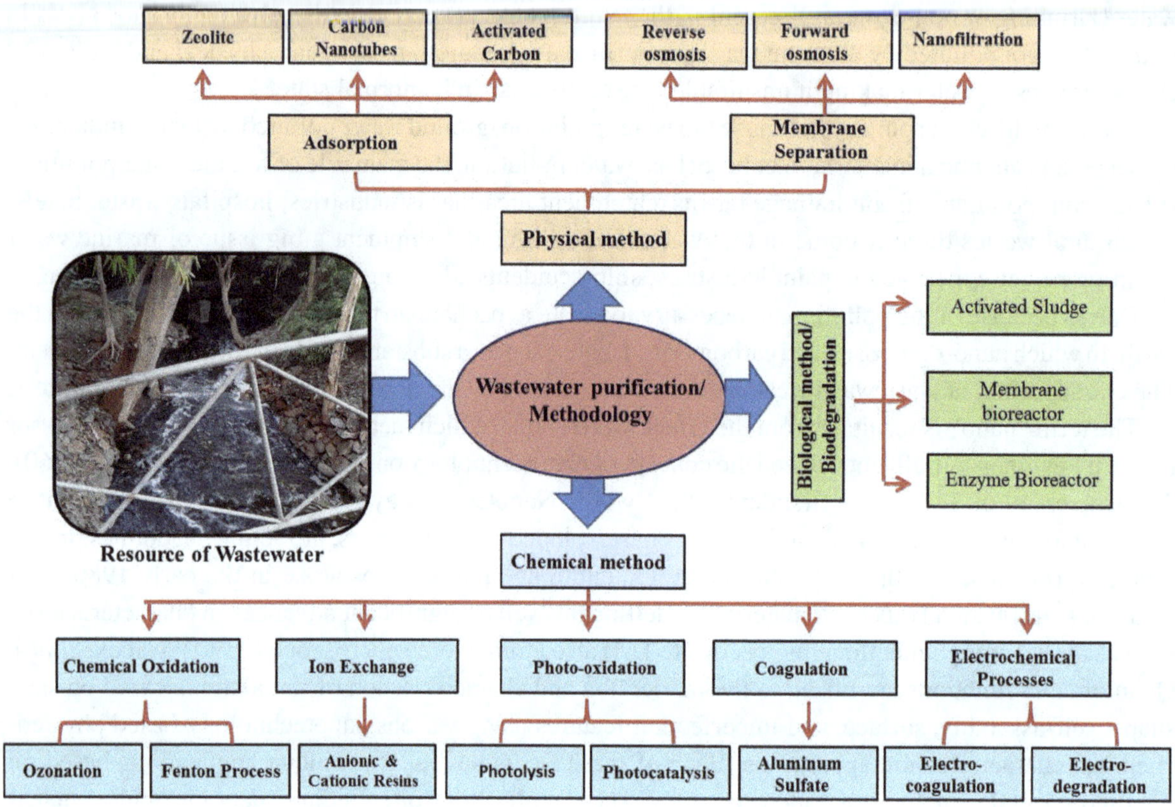

CARBON BASED MATERIALS

In the nature variety of carbon-based materials present that is naturally occurring as well as synthetic. Nano-composite and carbon-based material play the main role in purification of Wastewater. In detail we have discuss about Carbon based material as in following.

Activated Carbon

There are two most common methods such as filtration and purification, in which activated carbons (ACs) used as in adsorption processes. Moreover, stimulated carbon could be adsorbed unwanted metals and contaminants that have been suspended in the water. Because of its large superficial area, it is also used in commercial convenience and environmental applications. For example, ACs has been explored for the elimination of many pharmacological active substances. Such as tetracycline (an antibiotic medication) was removed from aqueous media using several forms of ACs. It was employed to remove three pharmaceutical active chemicals (tetracycline, penicillins, and quinolones) with a tetracycline adsorption. Among these it to remove tetracycline adsorption efficiency was found to be 455.33 mg /g (Wang et al., 2017b). In this adsorption processes following adsorption parameters including pH studies, contact time, as well as temperature were all examined.

Figure 2. Representation of the structure of carbon-based material like fullerene, Carbon nanotube and Graphite.

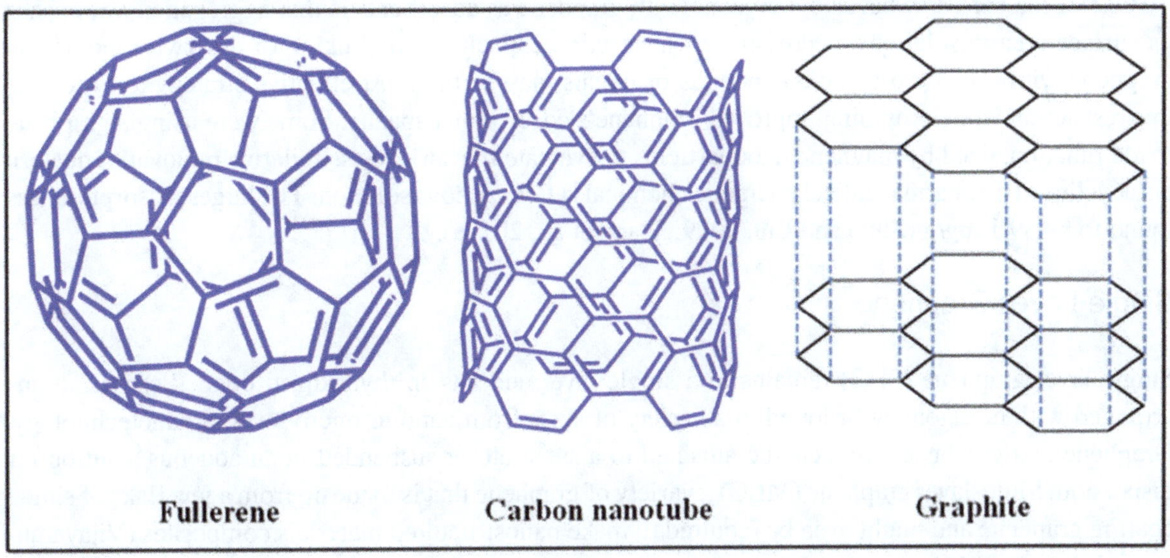

Carbon Nanotube

Carbon nanotube (CNTs) nanostructures have sp^2 hybridization with all carbon atoms. Further, CNTs are arranged in bundles to form a complex network by Van-der Waals interactions (Avcı et al., 2020). The electrical conductivity of hexagonal rings is regulated by their arrangement end-to-end on the tubular form of CNTs. Due to their special properties, which including high surface area, nanocavities, and electrical conductivity as a promising application that a diversity of nano technological significance of CNTs and its applications.

Graphene and Graphene Oxide

It is a carbonaceous single layer nanomaterial with a 2-dimensional structure made with sp^2 hybridized carbon atoms, it has high catalytic activity, and huge surface area is a few of the properties of graphene-based nanomaterial (Subodh et al., 2018; Baby et al., 2019). Various carbon-based material including fullerene, Carbon nanotube and Graphite is shown in Figure-2. These unexpected physical characteristics and they were recommended for a wide range of possible purposes as adsorptive material for remediation of toxic dye from wastewater.

Graphene/Graphene Oxide and Reduced Graphene Oxide

Exfoliations of graphite with chemical oxidation produce a single layer/multilayer graphene of highly oxygenated functional groups. RGO is identical to GO have functional groups are chemically, thermally, or physiologically reduced. Graphene nanoparticles and its functionalized forms have enormous surface area and excellent catalytic activity, could be used in a multitude of scenarios, including the eliminating of toxic dye (Subodh et al., 2018; Kumar & Masram, 2021). They demonstrated the adsorption of dye with the surface of graphene that cation interactions were responsible for the bulk of adsorption.

One of the core drawbacks of someone with graphene as adsorbents is that aggregation diminishes its surface area in solutions, decreasing its adsorption capacity. Crosslinking or otherwise modifying graphene with specific organic compounds or metals may be the most effective strategy to overcome this restriction while providing improved graphene's adsorption capacity. Some were using a graphene oxide functionalized by magnetic nanoparticles to evaluate the surface assimilative removal of quatern tetracycline (TC) pharmaceutical energetic chemicals after aqueous solution. The largest adsorption was found to be 39.1 mg/g (Zhao and Liu, 2009; Wang et al., 2017a).

Single Layer Graphene

Single layer graphene (SLG) remains in a single layer and has sp^2 hybridized bonded carbon atoms arranged with hexagonally. It found in a variety of sizes from nano to micro level in nanotechnology. Graphene and graphene oxide can be attached to a substrate or suspended in an aqueous solution for dispersion. Multi-layer graphene (MLG) a variety of graphene that is made up from a few flaks of single coating graphene and might even be required to make nanostructured materials composites (Vijayarangamuthu et al., 2016).

APPLICATIONS IN WASTEWATER TREATMENT

Including both organic and inorganic subtraction contaminants ion conversation, membrane purification, organic precipitation, chemical oxidation, adsorption, and photocatalysis have all been frequently employed in treatment of wastewater. Adsorption and photocatalysis have garnered significant consideration in the field of remediation of wastewater, among the approaches indicated above. As a result, the next section will focus on the usage of several categories of carbon-based nanomaterials in the adsorption-based experiment utilized it for remediation of several contaminants from wastewater.

Adsorption

Adsorption may be exact as the gathering of liquid solutes/gas upon the superficial of a solid adsorbent material. This methodology is recognised as a hygienic and multipurpose solution for the amputation of impurities from wastewater due to the very high skill; ease of process, and with low cost (Radaei et al., 2017). It frequently happens fashionable one of two ways reliant on how the interaction of adsorbent and adsorbates may be chemical adsorption or physisorption. Physical adsorption also known as physisorption is a reversible exothermic event that happens when the van der Waals forces of interaction are weak interactions between adsorbates and the adsorbent molecule. (Demirkıran et al., 2017). As opposed to electrostatic attraction, chemisorption sometimes referred to as chemical adsorption in which the formation of strong chemical bonds among adsorbate and adsorbent molecule, which results in an irreversible exothermic process. Some factor like Temperature, compression, type of adsorbate, and surface area of adsorbent can affect the chemisorption and physisorption (Access, 2018). The adsorption volume of the adsorbents utilised is right related to the efficiency at which pollutants are removed during the adsorption process (Thines et al., 2017). Surface properties like definite surface area, surface

Figure 3. Schematically representation of adsorption process.

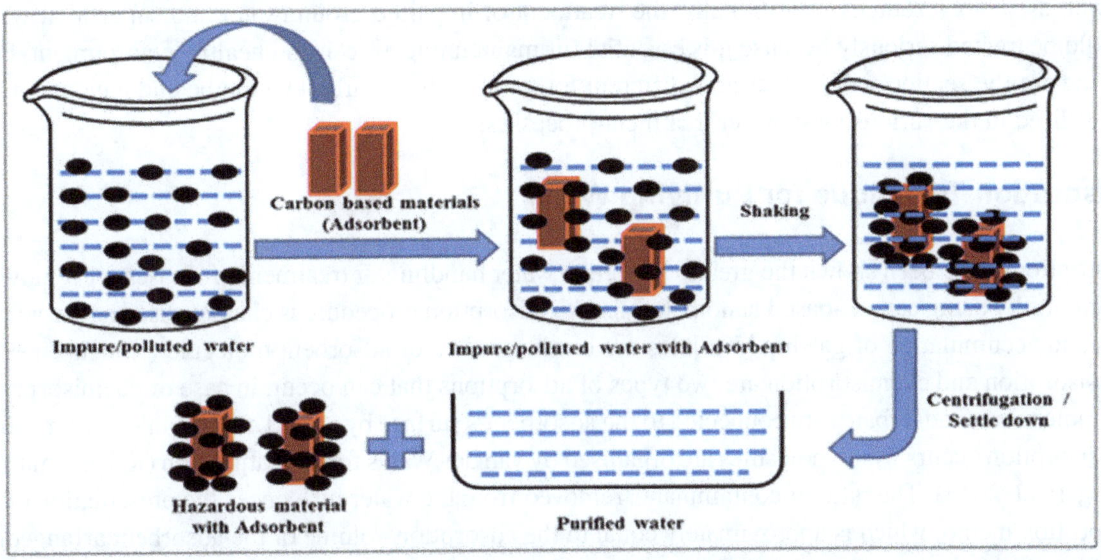

active sites, and affinity of adsorbent towards contaminants have a major impact on remediation of pollutant. Temperature, nature of adsorbates, adsorbents, pH of solution, and interaction time are tentative parameters in the adsorption process (Ali, 2010). Freundlich, Langmuir, Halsey, Henderson, Smith, and Elovich liquid flick dispersal are some of the isotherm models were commonly used to describe the adsorption mechanism (Ali, 2012). Furthermore, these models provide a good thoughtful of the communication between an adsorbent and the adsorbates on the surface. In the field of treatment of wastewater, Langmuir and Freundlich models were indeed frequently used to estimate the interaction of adsorbates such as chemicals, dye and other wastes material with adsorbents (Mubarak et al., 2017). In the case of monolayer coverage, the Langmuir adsorption isotherm models are generally used, but in the condition of heterogeneous surface coverage, the Freundlich isotherm method is frequently used (Anastopoulos et al., 2019). The kinetics of adsorption process was studied using the following models like pseudo-first-order, pseudo-second order, Elovich, and intra-atom diffusion models (Jun et al., 2018). A Schematically representation of adsorption process is show in Figure 3.

Application of Carbon Nanoparticles in the Remediation of Contaminants

Nanomaterials and polymeric nanoparticle often have size in range of 10-100 nm in diameter that assembled from a wide range of biodegradable such as chitosan polymeric nanoparticle. The behaviour and classification of nanomaterials have been influenced by the particle composition and surface possessions of the composite nanostructures. Furthermore, improving the chemical reactivity and functionalities of these carbon nanostructures by the improving their surface characteristics. It is possible to use these nanoparticles on their own or in combination with the additional of materials such as membranes or additional structural components.

Mining, battery manufacturing, galvanization and metal finishing are just some of the many industries that produce heavy metal ions containing wastewater and discharge it into nearby water sources (Mubarak et al., 2014). By the reprocessing the purified water after treatment, we can save both the environment as well as water resources. Additionally, the treatment of impaired groundwater and superficial water should be treated seriously because it is essential for maintaining a fresh and healthy environment. Following are the section will look at how different forms of carbon, carbon nanotube and nanomaterials are utilised in the various wastewater treatment processes.

Adsorption Technique for Purifying Water

Adsorptions have been shown the greatest effective water handling or treatment processes for removing pollutants by using carbon-based nanomaterials. An adsorption procedure is characterised as the attraction and accumulation of gas/liquid solutes are on the surface of adsorbent molecule (Kaneko, 1994). Physisorption and chemisorption are two types of adsorptions that can occur, in case of chemisorption, the molecules of adsorbates are connected to the adsorbent's surface by a solid organic interaction, while physisorption occurs when molecules are organised by van der Waals force of attraction (Krishnamurthy & Agarwal, 2013). The ratio of contaminants removed from the water influences the presentation of an adsorption metho, which is approximately equal to the adsorption volume of the adsorbent arranged. If Surface of adsorbent possess high definite surface area, active regions, and pollutant kinship are essential for an integral adsorbent's efficiency to adsorb contaminants. Carbon nanotubes (CNTs) impregnated on triggered carbon, and nanomaterials are efficient adsorbents because of their huge accurate superficial area and effective dynamic chemical properties (Ibrahim et al., 2016; Abdullah et al., 2014). The application of various forms of carbon nanomaterial as adsorbent will be examined in the future portion, which will investigate the adsorption of toxins such as contaminants present in wastewater.

With its high melting point and ability to form bonds with a wide variety of other elements, carbon is one of the periodic table's most multidimensional elements. CNTs are first revealed in 1991 (Goel et al., 2005), had been widely used in research to demonstrate their capacity to purify the water. CNTs are cylinder in shape macromolecules with radiuses as tiny as a rare nanometres and lengths as long as 20 cm. identified that the principal component of the tube walls is a hexagonal arrangement of carbon particles, which corresponds to the atomic planes of graphite. CNTs are two types such as single walled CNTs (SWCNTs) and many walled CNTs (MWCNTs). MWCNTs consisting of hundreds of concentric tubes and SWCNTs containing of a single layer of carbon particles (Hirsch and Vostrowsky, 2007). MWCNT's could have sizes of up to the range 100 nm, whereas SWCNTs have diameters ranging from 0.4 to 3 nm. SWCNT's have an advantage over MWCNTs in that they can be wrapped up into dissimilar types of graphene pieces to make a variety of CNTs. They have been regarded as one of the greatest promising nanostructured resources, and the prospect of mass production of CNTs has piqued the interest of all researchers (Luo et al., 2000). CNTs a novel member of the carbon family, have been shown to have exceptional properties, leading to a wide range of applications including hydrogen storing (Cheng et al., 2001)., organic sensors (Kong et al., 2000)., separation and sanitization (Mubarak et al., 2010, 2011), and compound material reinforcing (Dalton et al., 2003). On the other hand, CNTs have a hollow and covered structure as well as a huge exact surface zone (Li et al., 2006), which makes them ideal for eliminating heavy metals from wastewater using the adsorption approach (AlSaadi et al., 2016; Mubarak et al., 2016; AlOmar et al., 2016). CNTs can be functionalized in the influence of acid and alkali solution to improve their performance. Because of the impurities introduced during CNT formation, clean CNTs do not always deliver the best efficiency in the field of adsorption. The occurrence of these contaminants on the surface of CNTs would result in a change in the CNTs' vital properties. As a result, pollutants will be removed from CNTs using an acid or alkaline strategy to prevent the characteristics from being altered. Functionalization, in other words, is the process of adding a new functional group to CNTs' surface. By swapping weak interactions by considerable electrostatic forces, a negatively charged functional set introduced to the exposed end or sidewall of CNTs would improve their solubility in any solvent and improve their surface area (Georgakilas et al., 2002; Hu et al., 2009). Nitric acid (Yang et al., 2009), sulphuric acid and potassium permanganate are commonly acidic and alkaline solutions used to functionalize CNTs (Lu and Chiu, 2008). Further, this functionalization approach incorporates functional groups (–COOH, -OH, and –C=O) at the opening end of CNTs (Krishnamurthy and Agarwal, 2013; Mubarak et al., 2015).

One of the wide-ranging interferences for the groundwater environmental problem is the deposition of heavy metal ions, obviously or covertly, to any local water properties from firms including mining, batteries manufacture, and metal finishing. Along with its relatively cheap price, productivity, and flexibility, adsorption has been highlighted as a multifunction technique (Li et al., 2005). Mubarak et al. (2012) just published an experiment comparing non-functionalized and derivatized CNTs in the aqueous environment for Cu^{2+} adsorption. As a consequence, CNTs were chosen as an adsorbent in order to get a more attractive and effective alternatives it may be required to extract metal ions also including Cd^{2+}, Cu^{2+}, Ni^{2+}, and Pb^{2+} from aqueous solution, however according (Rao et al., 2007) discovered that metal ion adsorption to dissimilar kinds of CNTs followed basically the direction $Pb^{2+} > Ni^{2+} > Zn^{2+} > Cu^{2+} > Cd^{2+}$, with surface oxidised CNTs through HNO_3, NaOCl, as well as $KMnO_4$ having a greater adsorption volume than raw CNTs. Both the raw and chemically modified CNTs are becoming more prominent in the removal of impurities containing divalent metal ions. In addition to heavy metals, toxic amounts of expected various organic matter in drinking water, such as trihalomethanes (THMs), have been shown to stimulate bacterial growth and biofilm development in the water distribution system when Organic matter levels exceed the permissible limit (Lu and Chu, 2005). And a need for organic matter removal from the environment has increased as a result of these chemicals' harmful effects on consumer health and aesthetics. Researchers Su and Lu (2007) showed that after being introduced to 10 cycles of water treatment and reactivation, CNTs demonstrated better adsorption capacity and lost less weight. While CNTs are more expensive in terms of its synthesis, they have the potential to be an even better Organic matter adsorbent, preventing microorganisms in drinking water from deteriorating, culminating in a higher-quality also healthier water supply. Dyes have been recognized as one of the hazardous chemical compounds found in manufacturing waste, and their contamination of water has a detrimental impact since it reduces light dispersion, prohibiting aquatic plants from photosynthesis. Colors in drinking water induce allergic responses, skin rashes, and cancer (Alves de Lima et al., 2007; Carneiro et al., 2010) and inversion in people, according to the statement. MWCNTs destroyed at least 98.7% of the dye in a solution including high salt attentions (Machado et al., 2011). Methanol and NaOH (4 mol/L) could this be used to recover the MWCNTs from oversensitive Red M-2BE dye-loaded MWCNTs. As according (Wu & co-worker's, 2007), CNTs were particularly successful in decreasing Procion Bloodshot MX-5B from aqueous solution. The excellent adsorption capacity was reached by applying 0.25 g/L of CNTs, which reduced the dye's adsorption at high pH and temperature. Excess adsorption volume was introduced into the Langmuir isotherm model in order to fit the adsorption data at 321 K. On the same note, the pseudo second order model accurately matched the adsorption kinetic data. According to (Kuo et al., 2008), investigated and described that CNTs were used to successfully remove direct dyes from aqueous solutions.

Fullerenes being discovered and synthesised before CNTs were discovered and synthesised. In fact, the identification of fullerenes is often considered as a critical juncture in the creation of a wide variety of carbon-based nanomaterials. Fullerenes, often notorious as bucky balls, are the third kind of carbon allotrope between diamond and graphite. The primary distinction among CNTs and fullerenes is the figure of the carbons, with CNTs consisting of nanotubes and fullerenes consisting of a cage-like structure with 5-member rings and indeterminate number of 6-member rings. Fullerenes are hexagonal rings in which the carbon atoms are arranged. Weak sp^3 bonding, more strain, and more subtle carbon sites are all well-known benefits of structures with fewer hexagons. Moreover, isomers featuring adjoining pentagons generally have lesser stability and virtual abundance than isomers having isolated pentagons (Kroto, 1987; Campbell et al., 1996), delocalize aromatic bonds over the fullerene structures. C-60 is a fullerene that has been icosahedral symmetric (Johnson et al., 1990) with the use of resonance structures that support a carbon atom with something like a power conditioning state and bonding geometry (Johnson et al., 1992). Because of its comparable stability, C-60 has become a stand-alone starting material for chemical processes. Transformations that have been covalent, supramolecular and endohedral also allow for manipulation of substances and the production of polymeric materials for specialised eco-friendly applications. The features of fullerene have been successfully utilised in order to expand its use in the environmental domain. A number of research have been conducted out using fullerene to absorb heavy metal ions from wastewater as an adsorbent. Yang et al. (2006) asserted that all of the adsorption isotherms are nonlinear and well represented by the Polanyi–Manes model. CNTs adsorb second to four orders of magnitude stronger phenanthrene than fullerene, with SWCNTs greater then MWCNTs greater than fullerene has always been the direction. SWCNT showed highest adsorption performance relatively large, larger surface and reduced particle size which strengthened their applicability in environmental risk supervision and even had a considerable influence on the ultimate fate of PAHs once unrestricted into the environment. CNTs adsorb second to four orders of magnitude greater phenanthrene than fullerene, SWCNTs > MWCNTs > fullerene seems to be the ranking. SWCNT showed highest adsorption capability due to the greater porous structure as well as reduced particle size.

Activated Nanoporous Carbon in the Purification of Water

The carbon-based porous materials should be synthesised and used as adsorbent that have application in wastewater treatment. Carbon-based Nanoporous materials are used as adsorbents such as dust (Ansari and Pornahad, 2010), gripe shells (Pradhan et al., 2005), petroleum by-products (Rao et al., 2002), walnut stone waste (Fiol et al., 2006), and adsorbents fabric (Babić et al., 2002). Even though, all of these carbon-based adsorbents demonstrated as high catalytic activity for removing pollutants from wastewater. As a by-product, the adsorption volume of this activated carbon may be employed to improve the higher surface area. In other words, this stimulated carbon could well be turned into nanoporous stimulated charcoal and surface functionalization, which can substantially adsorb the pollutants. According to Xiao and Thomas (2005), synthesised stimulated carbon was created by varying the nitrogen concentration throughout invention and were used to assess the adsorption of metals. The surfaces of these carbons are wrapped in oxygen functional groups. The fragmentation of surface oxygen functional groups causes a reduction in oxygen content and an improvement in the carbon series. Moreover, the quantity of separate transition metals adsorbed on the adsorbent declines during competitive adsorptions between these ions as compared to a single ion system adsorption (Xiao & Thomas, 2004). When it came to ion adsorption, the order was $Hg^{2+} > Pb^{2+} > Cd^{2+} > Ca^{2+}$ for Hg^{2+}, Pb^{2+}. Electrostatic influence, numerous ion species in solution play a role in metal ion adsorption onto the adsorbents. Mangun et al. (201) utilized manufactured nanoporous stimulated carbon fibres (SCFs) with a normal pore size and surface areas ranging from 171 to 483 m^2/g to investigate the adsorption capacity of toluene, ethylbenzene and pxylene. The adsorption information obtained perfectly linked the Freundlich adsorption isotherm, and SCFs have been shown to have good adsorption than granular stimulated carbon. Takeuchi et al. (1999) evaluated and demonstrated the use an effective adsorbent for PAH components such as fluorine, anthracene, naphthalene, and also pyrene in aqueous solution. Pyrene, acenaphthene, anthracene, fluorine, fluoranthene, and naphthalene all grew in hydrophobicity in a logical sequence, with pyrrole at the top. Few studies have employed activated charcoal granules or carbon fibres to establish the adsorption capability for chlorinated compounds from water, and the results have indeed been outstanding (Sakoda et al., 1987). Takeuchi and companions decided to carry out a successful research project on the production of stimulated carbon with a high specific surface area of 3000 m^2/g in granular maintained on honeycomb board, and this stimulated carbon demonstrated great adsorption volume of pentachlorophenol vapours including such as trichloroethylene (TCE), methylchloroform (MC) and also tetrachloroethylene (TCE) (O'Mahony et al., 2002). The adsorbate and adsorbent interact more effectively with a high-surface-area material, resulting in increased adsorption capacity. If dye exceeds the allowable limits in any water resource, as environmental disclosure in this text, it is considered harmful. Azo dyes are some kinds of dye that has at least having one azo bond (-N=N-) with aromatic rings making them a significant component in the international dyestuffs market (Yang and Al-Duri, 2001). One form of azo dye is reactive dyes, which are employed for their vivid colours and high colourfastness (O'Mahony et al., 2002). Some azo dyes are proven to be harmful to aquatic species and may be used to kill them directly. These azo dyes, which make up the majority of oversensitive dyes, are typically azo compounds joined together by an azo bridge (Cohen, 1947). On the other hand, standard physicochemical and biological treatment procedures are unable to remove these reactive colours from wastewater. Adsorption of these colours from industrial wastewater has been demonstrated to be highly successful by a few researchers who used carbon sources in the adsorption approach (Ahmad and Hameed, 2010). Adsorption of basic dyes from aqueous solution onto granular stimulated carbon was also reported by Meshko et al. (2001). Özacar and Şengil (2002) established that dye from manufacturing effluent may very well be obtained using activated charcoal as the adsorbent.

Graphene in the Water Treatment Industry

Due to its distinctive 2-dimensional structure which is made up of a rare atomic layered that have special mechanical, thermal, and electrical abilities have been used to research as well in as in the field of adsorption for wastewater treatment (Su et al., 2009). Graphene oxide nanosheets produced from graphite powder with the improved properties were used as the adsorbent for the exclusion of Cd^{2+} and Co^{2+} from huge capacities of aqueous solutions (Zhao et al., 2011). The effects of pH, surface charge, and carbonaceous acid on Cd^{2+} also Co^{2+} demonstrate that pH has a major impact on metal ion absorption on graphene oxide nanosheets, although ionic strength has a slight impact. Cd^{2+} and Co^{2+} adsorption on graphene oxide nanosheets at pH 8 was hindered by humic acid. For Cd^{2+} and Co^{2+} at pH 6.0, the greatest sorption strengths were found to be 106.3 and 68.2 mg/g for graphene oxide nanosheets at 303 K. Furthermore, graphene oxide nanosheets would have been the ideal material for purifying water if mass-produced at a low-priced rate. According to Deng et al. (2010) used potassium treatment and prevention solution as the electrolyte in a modest and low-cost electrolysis to produce synthesised graphene. The adsorption isotherm data of both the isotherm such as Langmuir and Freundlich isotherm models is good agreement. Pb^{2+} was completely removed from an aqueous solution using graphene nanosheets (GNSs) at room temperature (Huang et al., 2011). The concomitant adsorption of Pb^{2+} ions also protons on GNSs strengthens overall Lewis's basicity, which is further explained by the high-vacuum thermal treatment. Permanent magnet monolayer hybrids been employed to create a chemical reaction with 10 nm-sized magnetite (Chandra et al., 2010) and by using magnetic graphene hybrids, the elimination ratio of arsenic compound within 1 ppb reached 99.9%. Along with the adsorption of heavy metals, a number of studies have shown that graphene may be used for the adsorption of textile colours, resulting in a cleaner water supply. After oxidising graphite with both the Hummers–Offeman reaction, Bradder et al. (2011) successfully implemented it for dye adsorption from aqueous solution. The presence of negatively charged with functional groups on the adsorbent's surface, adsorption has also been characterized as electrostatic attraction. Researchers have reported hydroxy graphite has also been used to eliminate lubricating oil and colours from wastewater. Exfoliated graphite with maximum adsorption volume for methyl orange has been generated from residual sulphuric acid compounds in natural graphite (Tryba et al., 2003). From the (Zhao and Liu, 2009) discovered that modulated expansion graphite residue might be used as a permeable adsorbent to adsorb cationic dyes also including methylene blue from acidic suspension.

Disinfection

In addition to adsorption have introduced by disinfection and was discovered to be an amazing technology for removing toxins from several wastewater. Water treatment agencies are under more pressure to eradicate pathogens in raw waters as a result of the improved value of drinking water that is required (Kfir et al., 1995). An issue through pathogen removal is the inconstant concentration and variety of pathogens present fashionable influent water quality. According to (LeChevallier et al., 2004). investigation, the clearance effectiveness of microbiological particles which have varied from 0.04 to 5.5 logs. According to research conducted by Upadhyayula et al. (2009), when water is treated with microbes such as cyanobacteria it releases poisonous microcystins. According to Krasner et al. (2006), the synthesis of disinfection by products DBPs is facilitated by the presence of some chemical oxidants which are found in water or in the water treatment. Some diseases such as Cryptosporidium and Giardia are very resistant to disinfectants, which necessary to remove an increased dosage of disinfectants in order to control this pathogen. The existence of microorganisms that are resistant to disinfection has been classified as an important problem in water treatment. As a result, nanotechnology was adopted as the solution worker for this problem in order to gap the production of disinfection by products DBPs and advance the value of treated wastewater (Li et al., 2008).

Recently, a number of synthetic nanomaterials included silver nanoparticles (n-Ag) (Morones et al., 2005; Badireddy et al., 2007), n-C60 fullerene nanoparticles (Lyon et al., 2006), and CNTs (Kang et al., 2007), have been shown to be effective disinfectants. These synthetic nanoparticles are highly effective against microorganisms. Due to their safety and lack of production of hazardous disinfection by products DBPs, these nanomaterials have been proven to be ideal disinfectants in the wastewater treatment.

Carbon Nanotubes used in Disinfection

CNTs are composed of carbon sheets and folded upon the layer of tubes, resulting it have two form such as SWCNTs and MWCNTs. There have been few studies on the influence of carbon nanotubes on the elimination of bacteria and virus activity from water. Apart from heavy metals emitted by industry, one of the biggest contaminants in the water is the result of bacteria and virus activity. According to Wick et al. (2007), with toxicity levels of water reduces by treating with SWCNT, MWCNT, and C60 nanomaterials (Jia et al., 2001). Microbial adhesion and befouling development on the surface of CNTs were then inhibited by CNTs. To maximise the CNTs performance as a disinfectant, the matching stabilising affected by organic matter (Hyung et al., 2007) degree of accumulation (Kang et al., 2007) and bioavailability of the CNTs (Brunet et al., 2008) have to be active in another investigation. According to (Lyon et al., 2006) have as long as the first direct proof that extremely pure SWCNTs have a substantial antibacterial action and show that direct contact with SWCNTs causes severe cell membrane damage. However, CNTs have been discovered to have trouble dispersing in water, resulting in deficient contact between CNT and microorganisms for decontamination. Supplementary processes such as functionalization and covering can improve the disinfectant efficacy of CNTs by immobilising pristine SWCNTs on a membrane filter surface. From the (Brady-Estévez et al., 2008) covered a thin layer of SWCNTs with microporous membrane and achieved effective E. coli inactivation and bacteriophage eradication. CNTs can also be integrated into hollow fibres to effectively inactivate E. coli and poliovirus, according to (Srivastava, 2004). According to (Kfir et al., 1995) investigated the sporicidal effects of SWCNTs and SWCNTs in combination with oxidising antimicrobial compounds like H_2O_2 and NaOCl on Bacillus antharacis (B. antharacis) spores. Antimicrobial compounds were also utilised, which rendered the spores more permeable to the chemicals. Arias and Yang (2009) investigated the antibacterial properties of both SWCNTs and MWCNs linked to bacterial pathogens such as Gram-negative versus Gram-positive types. When compared to SWCNTs with functional groups of $-NH_2$, SWCNTs with surface functional groups of –OH and –COOH had higher antibacterial action against both Gram-positive and Gram-negative bacterial cells. In the case of SWCNTs-NH_2, the presence of functional groups in a lengthy chain will impact the connections between the SWCNTs-NH_2 and bacterial cells. Since the cylinder structure of CNTs may not be in direct hint with the bacteria cell walls, resulting in a reduction in SWCNTs-NH_2 antimicrobial movement. As a result of the decreased toxicity level of MWCNTs compared to SWCNTs, they were found to have reduced antibacterial activity.

Fullerenes as a Disinfectant

The use of fullerenes as an antibacterial agent is still relatively new, and many academics are just beginning to explore the possibility of using fullerene as a disinfectant. Despite their well-known antibacterial properties, many fullerenes did not produce extremely effective antibacterial movement. When fullerene was examined as antiseptic for mammalian cells, Lyon et al. (2008) found that it has neither high toxicity nor not show antibacterial action. C_{60} molecules can be produced to collective into steady fullerene water suspension (FWS) with properties different from bulk solids of C_{60} by four unlike techniques, (Lyon et al., 2005). The antibacterial activities of these created FWS, which were made without the use of any intermediary solvents, are being investigated further. In accumulation to surface area, the morphologies of these interruptions were discovered to have an impact on antibacterial activity. Both crystalline and amorphous collections make up the separated C_{60} suspensions. These suspensions were centrifuged, resulting in a slighter size fraction with bigger amorphous aggregates than the greater size fraction. Because the aggregates with a lower proportion lack the recurring structure that determines crystallinity, they seem more amorphous.

Nanoporous Activated Carbon as Antiseptic

Activated carbon is familiar for its incredible as adsorbent and it is also operating as disinfect into research reveal antimicrobial effects by minimizing its surface area. Using Granular activated carbon (GAC) as an antiseptic to examine effluent from a predicted water purification pilot plant. Stewart & Wolfe (1990) found that GAC had a positive influence on the water's microbial quality. Samples including both disinfected and non-disinfected industrial effluent were examined for microorganisms after being visible to chloramines and chlorine for 1 hour. Bacteria such as Flavobacterium spp. were bred using chloramines and chlorine, respectively. Stimulated carbon fibres (ACF) were used as electrodes to disinfect the microorganisms such as Escherichia coli, E. coli in consumption water reported by Stewart & Wolfe (1994). Rather than carbon fabric or rough stimulated carbon, E. coli was successfully absorbed onto ACF. Furthermore, E. coli cells that had been adsorbed on the ACF were electrochemically done when a voltage of 0.8 V was given to a saturated calomel electrode (SCE). This study established that the drinking water is become disinfected without chlorination.

Membrane's Process

Membrane's process has been used in the elimination of pollutants from water. It is classified as a physical procedure that works by causing atoms to move depending on a varying concentration equal or a variance in atom size between the two sides of the membrane. An automated and adaptable design is provided by a membrane, which uses relatively little land and chemicals (Qu et al., 2013). A membrane's total performance is influenced by its two critical factor such selectivity and permeability. When it comes to defining permeability and selectivity, permeability and selectivity are exact as the invention of flow and membrane width alienated by the pressure differential diagonally the membrane (Freeman, 1999). A simplified definition of permeability is the capacity of water-borne contaminants to pass through the membrane and there are two types of membrane (Hsu, 2004). As a rule, a membrane with high permeability and exercising judgment is the best choice in most cases, when the permeability of the membrane is improved; less membrane area is needed to treat the water, lowering its initial investment cost. The composition of a membrane's substance determines its overall performance. One of the most effective materials to be put into the membrane is one that can provide high permeability, high selectivity, and low energy consumption. On that topic, effective research has shown that incorporating carbon nanotubes into membranes improves permeability, stability, and fouling resistance by leaps and bounds.

Membranes with Carbon Nanotubes and Companies

CNTs will often be mixed with other biodegradable polymers as well as nanomaterials to generate nanocomposite membranes. Shawky et al. (2011) has proposed the polymer grafting procedure to combine MWNCTs with aromatic polyamide to create nanocomposite membranes. MWCNTs were originated to be well diffused in the polyamide matrix, and an enhancement in MWCNT content increased the membrane's roughness, and structural rigidity. The solid interaction becomes stronger as the MWCNT level rises, resulting in greater permeability and salt rejection (Lee et al., 2007). MWCNTs can be used in conjunction with polyethersulfone (C/P) to balance sheaths, (Celik et al., 2011). The C/P mix membranes were originated to be extra hydrophilic and to have larger clean water fluidity than the polyethersulfone (PES) membranes. The volume of MWCNTs used in the blend membranes was shown to have a significant impact on the shape and permeability properties of the membranes. However, due to the hydrophobic landscape of CNT pore walls, aligned CNTs combined into mix membranes will cause an in-height fouling (Kim and Van der Bruggen, 2010). The obtained results reveal that the flow of the bare PES membrane dropped by up to 43 percent as compared to the operation time (Fan & Harris, 2001). The SWCNTs were effectively covalently bound to the surface of thin film membranes (TFMs) (Tiraferri et al., 2011).

Membranes Made of Fullerenes

Fullerenes were similarly used to integrate into membranes, and the results of the study reveal that they are effective at removing toxins from water. Jin et al. (2007) testified the development of a novel polymer membrane founded on the aquaphobic polymer poly (2, 6-dimethyl-1, 4-phenylene oxide) (PPO) improved by C_{60}, which was used to investigate the adsorption behaviour of estrogenic chemicals found in superficial water and filtered wastewater. Membranes were made by combining sulfonated polystyrene and also with fullerene reported by Saga et al. (2008). The addition of fullerene in the polyelectrolyte membrane increased the membrane's oxidation resistance and reduced the methanol crossing. Another study found that adding fullerene to the membrane did not improve its mechanical strength. The membrane's Young's modulus, rupture pressure, and extreme stress all dropped as the fullerene content in the membranes enlarged. The MeOH permeability of the fullerene composite membranes was 50% lower than that of the PS membrane, with the MeOH penetrability of the 1.4 percent influenza-phosphalidylserine Flu-PS and 2.8 percent Flu-PS membranes being 50% lesser than that of the PS membrane. This study shows the Fullerenes incorporated membranes, has been shown more effective for removing the toxins material form wastewater.

CONCLUSION

There are different sources of polluted water such as textile industries, tanneries and chemical industries, Industrial effluents, Sewage water and agrochemicals. Marine water got polluted from; wastewater of rivers brings pollutants from drainage, Catchment area that is industries, agricultural wastes. Oil drilling and shipment a big issue of marine water pollution because washing of ship and ship-accidents add to marine water cause pollution. It necessary to create a mechanism to purify the water present on the earth, in which carbon-based material play an important role in Wastewater treatment. There are numerous types of nanomaterials that have tremendous potential for effectively treating polluted water (including metal toxins and various organic and inorganic contaminants) due to their unique qualities such as increased surface area, high porosity, and ability to work at low concentrations. Although technologies utilizing nanostructured catalytic membranes, nanosorbents, and nanocomposites to remove contaminants from wastewater are environmentally friendly and effective to purify the wastewater.

REFERENCES

Abdullah, A. M., Mohammed, A. A.-S., Suleyman, A. A. M., & Zahangir, Md. (2014). A, iis S. *Kinetics of Cadmium Adsorption by CNTS Grown on PACs*, *7*(7), 25–30.

Access, O. (2018). We are IntechOpen, the world's leading publisher of Open Access books Built by scientists, for scientists TOP 1%. *Long-Haul Travel Motiv by Int Tour to Penang*, *i*, 13.

Ahmad, A. A., & Hameed, B. H. (2010). Fixed-bed adsorption of reactive azo dye onto granular activated carbon prepared from waste. *Journal of Hazardous Materials*, *175*(1–3), 298–303. https://doi.org/10.1016/j.jhazmat.2009.10.003

Ahmed, Y. M., Al-Mamun, A., Al Khatib, M. F. R., Jameel, A. T., & AlSaadi, M. A. H. A. R. (2015). Efficient lead sorption from wastewater by carbon nanofibers. *Environmental Chemistry Letters*, *13*(3), 341–346. https://doi.org/10.1007/s10311-015-0509-3

Ali, I. (2010). The quest for active carbon adsorbent substitutes: Inexpensive adsorbents for toxic metal ions removal from wastewater. *Separation and Purification Reviews*, *39*(3–4), 95–171. https://doi.org/10.1080/15422119.2010.527802

Ali, I. (2012). New generation adsorbents for water treatment. *Chemical Reviews*, *112*(10), 5073–5091. https://doi.org/10.1021/cr300133d

AlOmar, M. K., Alsaadi, M. A., Hayyan, M., Akib, S., Ibrahim, R. K., & Hashim, M. A. (2016). Lead removal from water by choline chloride based deep eutectic solvents functionalized carbon nanotubes. *Journal of Molecular Liquids*, *222*, 883–894. https://doi.org/10.1016/j.molliq.2016.07.074

AlSaadi, M. A., Al Mamun, A., Alam, M. Z., Amosa, M. K., & Atieh, M. A. (2016). Removal of cadmium from water by CNT–PAC composite: Effect of functionalization. *Nano*, *11*(1), 1650011. https://doi.org/10.1142/S1793292016500119

Alves de Lima, R. O., Bazo, A. P., Salvadori, D. M. F., Rech, C. M., de Palma Oliveira, D., & de Aragão Umbuzeiro, G. (2007). Mutagenic and carcinogenic potential of a textile azo dye processing plant effluent that impacts a drinking water source. *Mutation Research*, *626*(1–2), 53–60. https://doi.org/10.1016/j.mrgentox.2006.08.002

Anastopoulos, I., Robalds, A., Tran, H. N., Mitrogiannis, D., Giannakoudakis, D. A., Hosseini-Bandegharaei, A., & Dotto, G. L. (2019). Removal of heavy metals by leaves-derived biosorbents. *Environmental Chemistry Letters*, *17*(2), 755–766. https://doi.org/10.1007/s10311-018-00829-x

Ansari, R., & Pornahad, A. (2010). Removal of cerium (IV) *ion from aqueous solutions using sawdust as a very low cost bioadsorbent. Journal of Applied Sciences in Environmental Sanitation*, *5*(3).

Arias, L. R., & Yang, L. (2009). Inactivation of bacterial pathogens by carbon nanotubes in suspensions. *Langmuir*, *25*(5), 3003–3012. https://doi.org/10.1021/la802769m

Avcı, A., İnci, İ., & Baylan, N. (2020). Adsorption of ciprofloxacin hydrochloride on multiwall carbon nanotube. *Journal of Molecular Structure*, *1206*, 127711.

Babić, B. M., Milonjić, S. K., Polovina, M. J., Čupić, S., & Kaludjerović, B. V. (2002). Cupi ˇ c¢ S, Kaludjerovic´ BV. *Carbon*, *40*(7), 1109–1115. https://doi.org/10.1016/S0008-6223(01)00256-1

Baby, R., Saifullah, B., & Hussein, M. Z. (2019). Carbon nanomaterials for the treatment of heavy metal-contaminated water and environmental remediation. *Nanoscale Research Letters*, *14*(1), 341. https://doi.org/10.1186/s11671-019-3167-8

Badireddy, A. R., Hotze, E. M., Chellam, S., Alvarez, P., & Wiesner, M. R. (2007). Inactivation of bacteriophages via photosensitization of fullerol nanoparticles. *Environmental Science & Technology*, *41*(18), 6627–6632. https://doi.org/10.1021/es0708215

Bradder, P., Ling, S. K., Wang, S., & Liu, S. (2011). Dye adsorption on layered graphite oxide. *Journal of Chemical & Engineering Data*, *56*(1), 138–141. https://doi.org/10.1021/je101049g

Brady-Estévez, A. S., Kang, S., & Elimelech, M. (2008). A single-walled-carbon-nanotube filter for removal of viral and bacterial pathogens. *Small*, *4*(4), 481–484. https://doi.org/10.1002/smll.200700863

Brunet, L., Lyon, D. Y., Zodrow, K., Rouch, J.-C., Caussat, B., Serp, P., Remigy, J.-C., Wiesner, M. R., & Alvarez, P. J. J. (2008). Properties of membranes containing semi-dispersed carbon nanotubes. *Environmental Engineering Science*, *25*(4), 565–576. https://doi.org/10.1089/ees.2007.0076

Campbell, E. E. B., Fowler, P. W., Mitchell, D., & Zerbetto, F. (1996). Increasing cost of pentagon adjacency for larger fullerenes. *Chemical Physics Letters*, *250*(5–6), 544–548. https://doi.org/10.1016/0009-2614(96)00055-3

Capek, I. (2006). Chapter 1. In Nanotechnology and nanomaterials (pp. 1–69). Elsevier.

Carneiro, P. A., Umbuzeiro, G. A., Oliveira, D. P., & Zanoni, M. V. B. (2010). Assessment of water contamination caused by a mutagenic textile effluent/dyehouse effluent bearing disperse dyes. *Journal of Hazardous Materials*, *174*(1–3), 694–699. https://doi.org/10.1016/j.jhazmat.2009.09.106

Celik, E., Park, H., Choi, H., & Choi, H. (2011). Carbon nanotube blended polyethersulfone membranes for fouling control in water treatment. *Water Research*, *45*(1), 274–282. https://doi.org/10.1016/j.watres.2010.07.060

Chandra, V., Park, J., Chun, Y., Lee, J. W., Hwang, I. C., & Kim, K. S. (2010). Water-dispersible magnetite-reduced graphene oxide composites for arsenic removal. *ACS Nano*, *4*(7), 3979–3986. https://doi.org/10.1021/nn1008897

Cheng, H.-M., Yang, Q.-H., & Liu, C. (2001). Hydrogen storage in carbon nanotubes. *Carbon*, *39*(10), 1447–1454. https://doi.org/10.1016/S0008-6223(00)00306-7

Cohen, S. G. (1947). Encyclopedia of chemical technology, I. Journal of Polymer Science, 3(3), 463.

Dalton, A. B., Collins, S., Muñoz, E., Razal, J. M., Ebron, V. H., Ferraris, J. P., Coleman, J. N., Kim, B. G., & Baughman, R. H. (2003). Super- tough carbon-nanotube fibres. *Nature*, *423*(6941), 703. https://doi.org/10.1038/423703a

Das, T. K., & Poater, A. (2021). Review on the use of heavy metal deposits from water treatment waste towards catalytic chemical syntheses. *International Journal of Molecular Sciences*, *22*(24), 13383. https://doi.org/10.3390/ijms222413383

Demirkıran, N., Turhan Özdemir, G. D. T., Saraç, M., & Dardağan, M. (2017). Adsorption of methylene blue from aqueous solutions by pyrolusite ore. *Mongolian Journal of Chemistry*, *18*(44), 5–11. https://doi.org/10.5564/mjc.v18i44.880

Deng, X., Lü, L., Li, H., & Luo, F. (2010). The adsorption properties of Pb(II) and Cd(II) on functionalized graphene prepared by electrolysis method. *Journal of Hazardous Materials*, *183*(1–3), 923–930. https://doi.org/10.1016/j.jhazmat.2010.07.117

El-Gendy, A., El-Shafie, A. S., Issa, A., Al-Meer, S., Al-Saad, K., & El-Azazy, M. (2020). Carbon-based materials (CBMS) for determination and remediation of antimicrobials in different substrates: Wastewater and infant foods as examples. In Carbon-Based Material for Environmental Protection and Remediation. IntechOpen.

Fan, L., Harris, J. L., Roddick, F. A., & Booker, N. A. (2001). Influence of the characteristics of natural organic matter on the fouling of microfiltration membranes. *Water Research*, *35*(18), 4455–4463. https://doi.org/10.1016/s0043-1354(01)00183-x

Feynman, R. P. (1960). There's plenty of room at the bottom. California Institute of Technology Caltech Engineering and Science, 23(5), 22–36.

Fiol, N., Villaescusa, I., Martínez, M., Miralles, N., Poch, J., & Serarols, J. (2006). Sorption of Pb(II), Ni(II), Cu(II) and Cd(II) from aqueous solution by olive stone waste. *Separation and Purification Technology*, *50*(1), 132–140. https://doi.org/10.1016/j.seppur.2005.11.016

Freeman, B. D. (1999). Basis of permeability/selectivity tradeoff relations in polymeric gas separation membranes. *Macromolecules*, *32*(2), 375–380. https://doi.org/10.1021/ma9814548

Georgakilas, V., Kordatos, K., Prato, M., Guldi, D. M., Holzinger, M., & Hirsch, A. (2002). Organic functionalization of carbon nanotubes. *Journal of the American Chemical Society*, *124*(5), 760–761. https://doi.org/10.1021/ja016954m

Goel, J., Kadirvelu, K., Rajagopal, C., & Kumar Garg, V. (2005). Removal of lead(II) by adsorption using treated granular activated carbon: Batch and column studies. *Journal of Hazardous Materials*, *125*(1–3), 211–220. https://doi.org/10.1016/j.jhazmat.2005.05.032

Guisbiers, G., Mejía-Rosales, S., & Deepak, F. L. (2012). Nanomaterial properties: Size and shape dependencies. *Journal of Nanomaterials*, *2012*, 2.

Hirsch, A., & Vostrowsky, O. (2007). Functionalization of carbon nanotubes. In Functional organic materials (pp. 1–57). Wiley-VCH Verlag GmbH & Co. KGaA.

Hsu, V. R. (2004). *Ion transport through biological cell Membranes: From electro-diffusion to Hodgkin-Huxley via a quasi steady-state approach*. University of Washington.

Hu, C. Y., Xu, Y. J., Duo, S. W., Zhang, R. F., & Li, M. S. (2009). Non-covalent functionalization of carbon nanotubes with surfactants and polymers. *Journal of the Chinese Chemical Society (Taipei)*, *56*(2), 234–239. https://doi.org/10.1002/jccs.200900033

Hu, X., Zhao, Y., Wang, H., Tan, X., Yang, Y., & Liu, Y. (2017). Efficient removal of tetracycline from aqueous media with a Fe3O4 Nanoparticles@graphene oxide nanosheets assembly. *International Journal of Environmental Research and Public Health*, *14*(12), 1495. https://doi.org/10.3390/ijerph14121495

Huang, Z. H., Zheng, X., Lv, W., Wang, M., Yang, Q. H., & Kang, F. (2011). Adsorption of lead (II). *Langmuir*, *27*(12), 7558–7562. https://doi.org/10.1021/la200606r

Hyung, H., Fortner, J. D., Hughes, J. B., & Kim, J. H. (2007). Natural organic matter stabilizes carbon nanotubes in the aqueous phase. *Environmental Science & Technology*, *41*(1), 179–184. https://doi.org/10.1021/es061817g

Ibrahim, R. K., Hayyan, M., AlSaadi, M. A., Hayyan, A., & Ibrahim, S. (2016). Environmental application of nanotechnology: Air, soil, and water. *Environmental Science and Pollution Research International*, *23*(14), 13754–13788. https://doi.org/10.1007/s11356-016-6457-z

Jia, Z., Shen, D., & Xu, W. (2001). Synthesis and antibacterial activities of quaternary ammonium salt of chitosan. *Carbohydrate Research*, *333*(1), 1–6. https://doi.org/10.1016/s0008-6215(01)00112-4

Jin, X., Hu, J. Y., Tint, M. L., Ong, S. L., Biryulin, Y., & Polotskaya, G. (2007). Estrogenic compounds removal by fullerene-containing membranes. *Desalination*, *214*(1–3), 83–90. https://doi.org/10.1016/j.desal.2006.10.019

Johnson, R. D., Bethune, D. S., & Yannoni, C. S. (1992). Fullerene structure and dynamics: A magnetic resonance potpourri. *Accounts of Chemical Research*, *25*(3), 169–175. https://doi.org/10.1021/ar00015a011

Johnson, R. D., Meijer, G., & Bethune, D. S. (1990). C60 has icosahedral symmetry. *Journal of the American Chemical Society*, *112*(24), 8983–8984. https://doi.org/10.1021/ja00180a055

Jun, L. Y., Mubarak, N. M., Yee, M. J., Yon, L. S., Bing, C. H., Khalid, M., & Abdullah, E. C. (2018). An overview of functionalised carbon nanomaterial for organic pollutant removal. *Journal of Industrial and Engineering Chemistry*, *67*, 175–186. https://doi.org/10.1016/j.jiec.2018.06.028

Kaneko, K. (1994). Determination of pore size and pore size distribution: 1. Adsorbents and catalysts. *Journal of Membrane Science*, *96*(1–2), 59–89.

Kang, S., Pinault, M., Pfefferle, L. D., & Elimelech, M. (2007). Single-walled carbon nanotubes exhibit strong antimicrobial activity. *Langmuir*, *23*(17), 8670–8673. https://doi.org/10.1021/la701067r

Kfir, R., Hilner, C., du Preez, M., & Bateman, B. (1995). Studies evaluating the applicability of utilising the same concentration techniques for the detection of protozoan parasites and viruses in water. *Water Science and Technology*, *31*(5–6), 417–423. https://doi.org/10.2166/wst.1995.0651

Kim, J., & Van der Bruggen, B. (2010). The use of nanoparticles in polymeric and ceramic membrane structures: Review of manufacturing procedures and performance improvement for water treatment. *Environmental Pollution*, *158*(7), 2335–2349. https://doi.org/10.1016/j.envpol.2010.03.024

Kong, J., Franklin, N. R., Zhou, C., Chapline, M. G., Peng, S., Cho, K., & Dai, H. (2000). Nanotube molecular wires as chemical sensors. *Science*, *287*(5453), 622–625. https://doi.org/10.1126/science.287.5453.622

Krasner, S. W., Weinberg, H. S., Richardson, S. D., Pastor, S. J., Chinn, R., Sclimenti, M. J., Onstad, G. D., & Thruston, A. D. (2006). Occurrence of a new generation of disinfection byproducts. *Environmental Science & Technology*, *40*(23), 7175–7185. https://doi.org/10.1021/es060353j

Krishnamurthy, G., & Agarwal, S. (2013). Optimization of reaction conditions for high yield synthesis of carbon nanotube bundles by low-temperature solvothermal process and study of their H_2Storage capacity. *Bulletin of the Korean Chemical Society*, *34*(10), 3046–3054. https://doi.org/10.5012/bkcs.2013.34.10.3046

Kroto, H. W. (1987). The stability of the fullerenes Cn, with n=24, 28, 32, 36, 50. *Nature*, *329*(6139), 529–531. doi:10.1038/329529a0

Kumar, G., & Masram, D. T. (2021). Sustainable synthesis of MOF-5@GO nanocomposites for efficient removal of rhodamine b from water. *ACS Omega*, *6*(14), 9587–9599. https://doi.org/10.1021/acsomega.1c00143

Kumar, G., & Masram, D. T. (2021). Sustainable synthesis of MOF-5@GO nanocomposites for efficient removal of rhodamine b from water. *ACS Omega*, *6*(14), 9587–9599. https://doi.org/10.1021/acsomega.1c00143

Kumar, G., Mogha, N. K., & Kumar, M. (2020). NiO nanocomposites/rGO as a heterogeneous catalyst for imidazole scaffolds with applications in inhibiting the DNA binding activity. *Dalton Transactions*, *49*(6), 1963–1974. doi:10.1039/c9dt04416g

Kuo, C. Y., Wu, C. H., & Wu, J. Y. (2008). Adsorption of direct dyes from aqueous solutions by carbon nanotubes: Determination of equilibrium, kinetics and thermodynamics parameters. *Journal of Colloid and Interface Science*, *327*(2), 308–315. https://doi.org/10.1016/j.jcis.2008.08.038

LeChevallier, M. W. KwokKeung, A., & Au, K.-K. (2004). Water treatment and pathogen control: Process efficiency in achieving safe drinking-water. IWA Publishing.

Lee, S. Y., Kim, H. J., Patel, R., Im, S. J., Kim, J. H., & Min, B. R. (2007). Silver nanoparticles immobilized on thin film composite polyamide membrane: Characterization, nanofiltration, antifouling properties. *Polymers for Advanced Technologies*, *18*(7), 562–568. https://doi.org/10.1002/pat.918

Li, Q., Mahendra, S., Lyon, D. Y., Brunet, L., Liga, M. V., Li, D., & Alvarez, P. J. (2008). Antimicrobial nanomaterials for water disinfection and microbial control: Potential applications and implications. *Water Research*, *42*(18), 4591–4602. https://doi.org/10.1016/j.watres.2008.08.015

Li, Y. H., Di, Z., Ding, J., Wu, D., Luan, Z., & Zhu, Y. (2005). Adsorption thermodynamic, kinetic and desorption studies of Pb2+ on carbon nanotubes. *Water Research*, *39*(4), 605–609. https://doi.org/10.1016/j.watres.2004.11.004

Li, Y.-H., Zhu, Y., Zhao, Y., Wu, D., & Luan, Z. (2006). Different morphologies of carbon nanotubes effect on the lead removal from aqueous solution. *Diamond and Related Materials*, *15*(1), 90–94. https://doi.org/10.1016/j.diamond.2005.07.004

Lu, C., & Chiu, H. (2008). Chemical modification of multiwalled carbon nanotubes for sorption of Zn2+ from aqueous solution. *Chemical Engineering Journal*, *139*(3), 462–468. https://doi.org/10.1016/j.cej.2007.08.013

Lu, C., & Chu, C. (2005). Effects of acetic acid on the regrowth of heterotrophic bacteria in the drinking water distribution system. *World Journal of Microbiology & Biotechnology*, *21*(6–7), 989–998. https://doi.org/10.1007/s11274-004-7554-6

Luo, J. Z., Gao, L. Z., Leung, Y. L., & Au, C. T. (2000). The decomposition of NO on CNTs and 1 wt% Rh/CNTs. *Catalysis Letters*, *66*(1–2), 91–97.

Lyon, D. Y., Adams, L. K., Falkner, J. C., & Alvarezt, P. J. J. (2006). Antibacterial activity of fullerene water suspensions: Effects of preparation method and particle size. *Environmental Science & Technology, 40*(14), 4360–4366. https://doi.org/10.1021/es0603655

Lyon, D. Y., Brown, D. A., & Alvarez, P. J. J. (2008). Implications and potential applications of bactericidal fullerene water suspensions: Effect of nc60 concentration, exposure conditions and shelf life. *Water Science and Technology, 57*(10), 1533–1538. https://doi.org/10.2166/wst.2008.282

Lyon, D. Y., Fortner, J. D., Sayes, C. M., Colvin, V. L., & Hughe, J. B. (2005). Bacterial cell association and antimicrobial activity of a C60 water suspension. *Environmental Toxicology and Chemistry, 24*(11), 2757–2762. https://doi.org/10.1897/04-649r.1

Machado, F. M., Bergmann, C. P., Fernandes, T. H. M., Lima, E. C., Royer, B., Calvete, T., & Fagan, S. B. (2011). Adsorption of Reactive Red M-2BE dye from water solutions by multi-walled carbon nanotubes and activated carbon. *Journal of Hazardous Materials, 192*(3), 1122–1131. https://doi.org/10.1016/j.jhazmat.2011.06.020

Mangun, C. L., Yue, Z., Economy, J., Maloney, S., Kemme, P., & Cropek, D. (2001). Adsorption of organic contaminants from water using tailored ACFs. *Chemistry of Materials, 13*(7), 2356–2360. https://doi.org/10.1021/cm000880g

Matsunaga, T., Nakasono, S., Kitajima, Y., & Horiguchi, K. (1994). Electrochemical disinfection of bacteria in drinking water using activated carbon fibers. *Biotechnology and Bioengineering, 43*(5), 429–433. https://doi.org/10.1002/bit.260430511

Meshko, V., Markovska, L., Mincheva, M., & Rodrigues, A. E. (2001). Adsorption of basic dyes on granular activated carbon and natural zeolite. *Water Research, 35*(14), 3357–3366. https://doi.org/10.1016/s0043-1354(01)00056-2

Morones, J. R., Elechiguerra, J. L., Camacho, A., Holt, K., Kouri, J. B., Ramírez, J. T., & Yacaman, M. J. (2005). The bactericidal effect of silver nanoparticles. *Nanotechnology, 16*(10), 2346–2353. https://doi.org/10.1088/0957-4484/16/10/059

Mubarak, N., Daniel, S., Khalid, M., & Tan, J. (2012). Comparative study of functionalize and non-functionalized carbon nanotube for Removal of Copper from polluted water. *Int. J. Chem. Environ. Eng., 3*(5).

Mubarak, N. M., Sahu, J. N., Abdullah, E. C., & Jayakumar, N. S. (2014). Removal of heavy metals from wastewater using carbon nanotubes. *Separation and Purification Reviews, 43*(4), 311–338. https://doi.org/10.1080/15422119.2013.821996

Mubarak, N. M., Sahu, J. N., Abdullah, E. C., Jayakumar, N. S., & Ganesan, P. (2016). Microwave-assisted synthesis of multi-walled carbon nanotubes for enhanced removal of Zn(II) from wastewater. *Research on Chemical Intermediates, 42*(4), 3257–3281. https://doi.org/10.1007/s11164-015-2209-9

Mubarak, N. M., Sahu, J. N., Wong, J. R., Jayakumar, N. S., Ganesan, P., & Abdullah, E. C. (2015). *Overview on the functionalization of carbon nanotubes. Chemical functionalization of carbon nanomaterials*. CRC Press.

Mubarak, N. M., Sazila, N., & Nizamuddin, S. (2017). Adsorptive removal of phenol from aqueous solution by using carbon nanotubes and magnetic biochar. *NanoWorld J*. doi:10.17756/nwj.2017-043

Mubarak, N. M., Yusof, F., & Alkhatib, M. F. (2011). The production of carbon nanotubes using two-stage chemical vapor deposition and their potential use in protein purification. *Chemical Engineering Journal, 168*(1), 461–469. https://doi.org/10.1016/j.cej.2011.01.045

Mubarak, N. M., Yusof, F., Alkhatib, M. F., Ameen, E., Khalid, M., Mohammed, A. S., Muataz, A., Qudsieh, I. Y., & Rashmi, W. (2010). Optimization of CNTs production using full factorial design and its advanced application in protein purification. *International Journal of Nanoscience, 09*(3), 181–192. https://doi.org/10.1142/S0219581X10006648

O'Mahony, T., Guibal, E., & Tobin, J. M. (2002). Reactive dye biosorption by Rhizopus arrhizus biomass. *Enzyme and Microbial Technology, 31*(4), 456–463. https://doi.org/10.1016/S0141-0229(02)00110-2

Özacar, M., & Şengil, İ. A. (2002). engil ˙ IA. *Adsorption, 8*(4), 301–308. https://doi.org/10.1023/A:1021585413857

Pradhan, S., Shukla, S. S., & Dorris, K. L. (2005). Removal of nickel from aqueous solutions using crab shells. *Journal of Hazardous Materials, 125*(1–3), 201–204. https://doi.org/10.1016/j.jhazmat.2005.05.029

Qu, X., Brame, J., Li, Q., & Alvarez, P. J. (2013). Nanotechnology for a safe and sustainable water supply: Enabling integrated water treatment and reuse. *Accounts of Chemical Research, 46*(3), 834–843. https://doi.org/10.1021/ar300029v

Radaei, E., Moghaddam, M. R. A., & Arami, M. (2017). Application of artifcial neural network on modeling of reactive blue 19 removal by modified pomegranate residual. *Environmental Engineering and Management Journal, 16*(9), 2113–2122. https://doi.org/10.30638/eemj.2017.218

Rajasulochana, P., & Preethy, V. (2016). Comparison on efficiency of various techniques in treatment of waste and sewage water – A comprehensive review. *Resource-Efficient Technologies, 2*(4), 175–184. doi:10.1016/j.reffit.2016.09.004

Rao, G. P., Lu, C., & Su, F. (2007). Sorption of divalent metal ions from aqueous solution by carbon nanotubes: A review. *Separation and Purification Technology, 58*(1), 224–231. https://doi.org/10.1016/j.seppur.2006.12.006

Rao, M., Parwate, A. V., & Bhole, A. G. (2002). Removal of Cr6+ and Ni2+ from aqueous solution using bagasse and fly ash. *Waste Management, 22*(7), 821–830. doi:10.1016/s0956-053x(02)00011-9

Russell, D. L. (2006). *Practical wastewater treatment*. John Wiley & Sons, Inc.

Saga, S., Matsumoto, H., Saito, K., Minagawa, M., & Tanioka, A. (2008). Polyelectrolyte membranes based on hydrocarbon polymer containing fullerene. *Journal of Power Sources, 176*(1), 16–22. https://doi.org/10.1016/j.jpowsour.2007.10.017

Sakoda, A., Kawazoe, K., & Suzuki, M. (1987). Adsorption of tri- and tetra-chloroethylene from aqueous solutions on activated carbon fibers. *Water Research, 21*(6), 717–722. https://doi.org/10.1016/0043-1354(87)90084-4

Shawky, H. A., Chae, S.-R., Lin, S., & Wiesner, M. R. (2011). Synthesis and characterization of a carbon nanotube/polymer nanocomposite membrane for water treatment. *Desalination*, *272*(1–3), 46–50. https://doi.org/10.1016/j.desal.2010.12.051

Srivastava, A., Srivastava, O. N., Talapatra, S., Vajtai, R., & Ajayan, P. M. (2004). Carbon nanotube filters. *Nature Materials*, *3*(9), 610–614. https://doi.org/10.1038/nmat1192

Stewart, M. H., Wolfe, R. L., & Means, E. G. (1990). Assessment of the bacteriological activity associated with granular activated carbon treatment of drinking water. *Applied and Environmental Microbiology*, *56*(12), 3822–3829. https://doi.org/10.1128/aem.56.12.3822-3829.1990

Su, F., & Lu, C. (2007). Adsorption kinetics, thermodynamics and desorption of natural dissolved organic matter by multiwalled carbon nanotubes. *Journal of Environmental Science and Health. Part A, Toxic/Hazardous Substances & Environmental Engineering*, *42*(11), 1543–1552. https://doi.org/10.1080/10934520701513381

Su, Q., Pang, S., Alijani, V., Li, C., Feng, X., & Müllen, K. (2009). Composites of graphene with large aromatic molecules. *Advanced Materials*, *21*(31), 3191–3195. https://doi.org/10.1002/adma.200803808

Subodh, N. K., Mogha, N. K., Chaudhary, K., Kumar, G., & Masram, D. T. (2018). Fur-Imine-Functionalized Graphene Oxide-Immobilized Copper Oxide Nanoparticle Catalyst for the Synthesis of Xanthene Derivatives. *ACS Omega, 3*(11), 16377–16385. doi:10.1021/acsomega.8b01781

Takeuchi, Y., Hino, M., Yoshimura, Y., Otowa, T., Izuhara, H., & Nojima, T. (1999). Removal of single component chlorinated hydrocarbon vapor by activated carbon of very high surface area. *Separation and Purification Technology*, *15*(1), 79–90. https://doi.org/10.1016/S1383-5866(98)00082-3

Thines, R. K., Mubarak, N. M., Nizamuddin, S., Sahu, J. N., Abdullah, E. C., & Ganesan, P. (2017). Application potential of carbon nanomaterials in water and wastewater treatment: A review. *Journal of the Taiwan Institute of Chemical Engineers*, *72*, 116–133. https://doi.org/10.1016/j.jtice.2017.01.018

Tiraferri, A., Vecitis, C. D., & Elimelech, M. (2011). Covalent binding of single-walled carbon nanotubes to polyamide membranes for antimicrobial surface properties. *ACS Applied Materials & Interfaces*, *3*(8), 2869–2877. https://doi.org/10.1021/am200536p

Tryba, B., & Morawski, A. W., & Kalenczuk. (2003). Exfoliated graphite as a new sorbent for removal of engine oils from wastewater. *Spill Science & Technology Bulletin*, *8*(5–6), 569–571.

Upadhyayula, V. K. K., Deng, S., Mitchell, M. C., & Smith, G. B. (2009). Application of carbon nanotube technology for removal of contaminants in drinking water: A review. *The Science of the Total Environment*, *408*(1), 1–13. https://doi.org/10.1016/j.scitotenv.2009.09.027

Vijayarangamuthu, K., Ahn, S., Seo, H., Yoon, S.-H., Park, C.M., & Jeon, K. J. (2016). Temporospatial control of graphene wettability. *Advanced Materials, 28*(4), 661–667. doi:10.1002/adma.201503444

Wang, B., Jiang, Y. S., Li, F. Y., & Yang, D. Y. (2017b). Preparation of biochar by simultaneous carbonization, magnetization and activation for norfloxacin removal in water. *Bioresource Technology*, *233*, 159–165. https://doi.org/10.1016/j.biortech.2017.02.103

Wang, T., Pan, X., Ben, W., Wang, J., Hou, P., & Qiang, Z. (2017a). Adsorptive removal of antibiotics from water using magnetic ion exchange resin. *Journal of Environmental Sciences (China)*, *52*, 111–117. https://doi.org/10.1016/j.jes.2016.03.017

Wick, P., Manser, P., Limbach, L. K., Dettlaff-Weglikowska, U., Krumeich, F., Roth, S., Stark, W. J., & Bruinink, A. (2007). The degree and kind of agglomeration affect carbon nanotube cytotoxicity. *Toxicology Letters*, *168*(2), 121–131. https://doi.org/10.1016/j.toxlet.2006.08.019

Wu, C. H. (2007). Adsorption of reactive dye onto carbon nanotubes: Equilibrium, kinetics and thermodynamics. *Journal of Hazardous Materials*, *144*(1–2), 93–100. https://doi.org/10.1016/j.jhazmat.2006.09.083

Xiao, B., & Thomas, K. M. (2004). Competitive adsorption of aqueous metal ions on an oxidized nanoporous activated carbon. *Langmuir*, *20*(11), 4566–4578. https://doi.org/10.1021/la049712j

Xiao, B., & Thomas, K. M. (2005). Adsorption of aqueous metal ions on oxygen and nitrogen functionalized nanoporous activated carbons. *Langmuir*, *21*(9), 3892–3902. https://doi.org/10.1021/la047135t

Yang, K., Zhu, L., & Xing, B. (2006). Adsorption of polycyclic aromatic hydrocarbons by carbon nanomaterials. *Environmental Science & Technology*, *40*(6), 1855–1861. https://doi.org/10.1021/es052208w

Yang, S., Li, J., Shao, D., Hu, J., & Wang, X. (2009). Adsorption of Ni(II) on oxidized multi- -walled carbon nanotubes: Effect of contact time, pH, foreign ions and PAA. *Journal of Hazardous Materials*, *166*(1), 109–116. https://doi.org/10.1016/j.jhazmat.2008.11.003

Yang, X.-Y., & Al-Duri, B. (2001). Application of branched pore diffusion model in the adsorption of reactive dyes on activated carbon. *Chemical Engineering Journal*, *83*(1), 15–23. https://doi.org/10.1016/S1385-8947(00)00233-3

Zhao, G., Li, J., Ren, X., Chen, C., & Wang, X. (2011). Few-layered graphene oxide nanosheets as superior sorbents for heavy metal ion pollution management. *Environmental Science & Technology*, *45*(24), 10454–10462. https://doi.org/10.1021/es203439v

Zhao, M., & Liu, P. (2009). Adsorption of methylene blue from aqueous solutions by modified expanded graphite powder. *Desalination*, *249*(1), 331–336. https://doi.org/10.1016/j.desal.2009.01.037

Chapter 15
Synthetic Methods of Nanomaterials

Deepak Vasant Nagarale
VVM's S.G. Patil Arts, Science, and Commerce College, India

ABSTRACT

The word "nano" is from the Greek word "nanos" meaning "dwarf." It is a prefix used to describe "one billionth" of something. A nanometer (nm) is a billionth of a meter or a millionth of a millimeter. This chapter started with an introduction to nanoscience followed by what nanostructure is and its applications of nanotechnology (basic idea) and various size-dependent properties of nanomaterials. In this chapter, some unique properties like 1) semiconducting nanoparticles and 2) metallic nanoparticles are explained with examples. Synthesis aspects of nanomaterials also need to be understood using bottom-up and top-down approaches including mechanical alloying and shape and size control of nanomaterials. In the current scenario, the research and development of nanotechnology is very active globally, and nanotechnologies are already used in many products. Further, nanotechnologies are also being developed for use in environmental applications (e.g., clean-up of environmental pollutants).

INTRODUCTION TO NANO SCIENCE

Nanoscience is a combine word consisting with two parts, nano and science. Although nano is a refer as in nanometer, nanoampere, nanosecond, etc. in nanoscience, it refers specifically to nanometer. Therefore, nanoscience is the science of objects in the size regime of nanometers. The 'nano' word basically is a Greek prefix meaning dwarf or something very small and depicts one billionth (10^{-9}) of a unit Nanomaterials, therefore, refer to the class of materials with at least one of the dimensions in the nanometric range. For an immediate comparison, a nanometer represents a dimension about a few tens of thousand times thinner than human hair. Table 1 gives an idea of the scale of different objects, from macroscale to nanoscale. In the case of polycrystalline materials, the grain size is typically of the order of 1 – 100 microns (1 micron = 10^{-6} m). Nanocrystalline materials have a grain size of the order of 1 – 100 nm and are therefore 100 – 1000 times smaller than conventional grain dimensions. However,

DOI: 10.4018/978-1-6684-4553-2.ch015

Table 1. The list of small dimensions

Number	Name	Symbol
0.1	deci	d
0.01	centi	c
0.001	milli	m
0.000 001	micro	μ
0.000 000 001	nano	n
0.000 000 000 001	pico	p
0.000 000 000 000 001	femto	F
0.000 000 000 000 000 001	atto	a
0.000 000 000 000 000 000 001	zepto	z
0.000 000 000 000 000 000 000 001	yocto	y

compared to the size of an atom (0.2 – 0.4 nm in diameter), nanocrystalline grains are still significantly large. (Mansoori and Soelaiman, 2005).

Applications of Nanoscience

Nanoscience is the study of structures and materials on an ultra-small scale, and the unique and interesting properties these materials demonstrate. Nanoscience is cross-disciplinary, meaning scientists from a range of fields including chemistry, physics, biology, medicine, computing, materials science and engineering are studying it and using it for following stream areas of society as shown in Figure 1.

1. Computers
2. Clothing
3. Furniture
4. Adhesives
5. Coatings for car paintwork
6. Sports
7. Cosmetics
8. Pharmaceutical etc.

Nanostructure

Hundreds of years ago nanomaterials have been produced and used by humans on daily basis. on the other hand, the understanding of certain materials as nanostructured materials is relatively current, made possible by the advent of advanced tools that are capable of resolving information at nanoscale. Nanostructured materials have generated extensive interest among chemists in recent years because of their physicochemical and plasmonic properties and potential applications. The extraordinary physicochemical and biological properties of materials at the nanoscale enable new applications ranging from

Figure 1. Nanoscience and Society

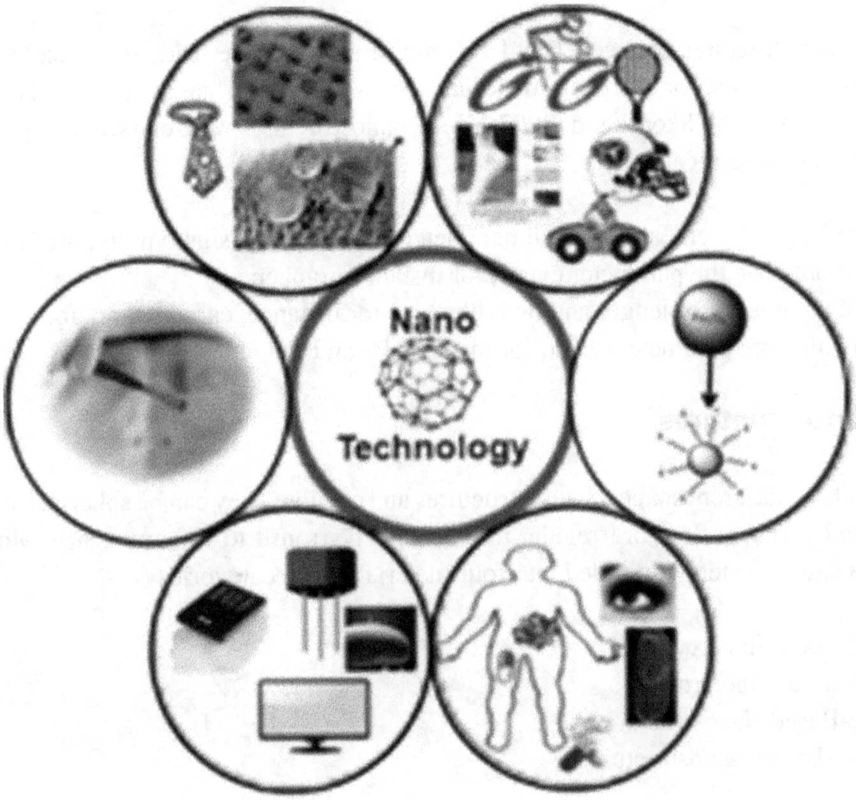

energy conservation and structural strength enhancement to antimicrobial characteristics and self-cleaning surfaces. The discovery of new materials and structures at the nanoscale, and the development of new theoretical and experimental methods for research, can provide novel opportunities for the development and improvement of hitherto unknown nanomaterials.

Nanostructures have obtained prominence in technological developments owing to their tunable physicochemical properties including their melting points, electrical and thermal conductivities, light absorption and scattering properties, optical sensitivity, (photo) catalytic activity, and wettability resulting in their significantly enhanced performance over their bulk counterparts.

The International Organization for Standardization (ISO)

This defines a nanostructured material as a "material with any external nanoscale dimension or having the internal nanoscale surface structure". Similarly, the European Commission describes nanostructures as a "manufactured or natural material that possesses unbounded aggregated or agglomerated particles with external dimensions between 1-100 nm size range". (Atkins and Overton, 2010)

Classification of Nanostructures Based on Dimension

Nanostructures are structures between 1 and 100 nm in size that are made up of carbon, composite, metal, metal oxide, organic, or inorganic material. A new scheme of nanostructure classification was provided by Pokropivny and Skorokhod (2007). Here, nanostructures were classified depending on their dimensions into one of four categories:

1. 0D, where length, height, and breadth parameters are fixed at a single point, for instance at a dot.
2. 1D, where only one the parameter exists, for instance graphene,
3. 2D, where parameters of length and breadth exist, for instance, carbon nanotubes.
4. 3D, where all three parameters exist, for instance Pd and ZnO NPs.

Types of Nanostructures

Nanostructures have different shapes, sizes, structures and origins. They can be spherical, conical, spiral, cylindrical, tubular, flat, hollow, or irregular in shape and be from 1 to 100 nm in size. Most nanostructured materials can be generally divided into four material-based categories

1. Carbon-Based Nanostructures
2. Organic-Based Nanostructures
3. Inorganic-Based Nanostructures
4. Composite-Based Nanostructures

Carbon-Based Nanostructures

Nanostructures made of carbon are known as carbon-based nanostructures. They can have different morphologies, such as ellipsoid, hollow tube, or sphere.

- Generally, these nanostructures can be classified into diamonds, fullerenes (C_{60}, C_{80}, and C_{240}), carbon nanotubes (CNTs), graphene, and carbon nanofibers.
- Eg. Fullerene (C_{60}) is a carbon-based molecule that is spherical in morphology and made up of carbon atoms held together via sp^2 hybridization.
- Generally, the other fullerenes (0D), such as C_{76}, C_{80}, C_{240}, etc, are synthesized from larger numbers of carbon atoms.
- Fullerenes are comprised of between 28 and 1500 carbon atoms that form spherical structures.
- Single-layer fullerenes have diameters up to 8.2 nm while multilayer fullerenes have diameters of between 4 and 36 nm.
- Over the past few years, solar cells have attracted much attention due to their important role in the production of energy.

Organic-Based Nanostructures

Dendrimers, liposomes, micelles, polymer NPs, etc., are usually known as organic nanostructures or polymers.

- These include nanostructures made mostly from organic material, excluding carbon-based or inorganic-based nanostructures.
- These nanostructures are nontoxic, biodegradable, and some of their structures,
- e.g., liposomes and micelles, have hollow cores (also known as nanocapsules).

Inorganic-Based Nanostructures

Inorganic nanostructures and nanoparticles are structures and particles that are not made from carbon-based or organic-based NPs.

- These nanostructures include metal and metal oxide NPs.
- Metal-based and metal oxide-based NPs are commonly categorized as inorganic nanostructures.
- These nanostructures can be synthesized into metal NPs, such as Pd or Au, metal oxide NPs like TiO_2, and also semiconductors, such as ceramics and silicon.

Composite-Based Nanostructures

Nanocomposites can be described as multiphase nanostructures with one phase being at the nanoscale dimension.

- They can either combine nanostructures with other NPs or NPs with bulk-type or larger materials (e.g., hybrid nonporous materials) or more complicated structures.
- Nanocomposites can be any combination of metal-based, carbon-based, or organic-based nanostructures with any form of ceramic, metal, or polymer bulk materials. (Murty et al., 2013)

NANOTECHNOLOGY

The coming decades could well be dominated by nanotechnology ('Nano Age'), a deviation from the practice of identification of an era based on materials to one based on technology. The term 'nanotechnology' was first coined by Norio Taniguchi in 1974 to describe semiconductor processes such as thin film deposition and ion beam milling, where the features can be controlled at the nanometric level.

The creation of functional materials, devices and systems through control of matter on the nanometer length scale (1–100 nm), and exploitation of novel phenomena and properties (physical, chemical, biological) at that length scale (Dino, 2008).

Uses

1. Nanotechnology is being used in developing countries to help treat disease and prevent health issues.
2. The umbrella term for this kind of nanotechnology is Nano-medicine.
3. Nanotechnology is also being applied to or developed for application to a variety of industrial and purification processes

4. Food and agricultural systems, such as food security.
5. Disease treatment delivery methods,
6. Molecular and cellular biology,
7. Materials for pathogen detection
8. Protection of the environment

SIZE-DEPENDENT PROPERTIES OF NANOMATERIALS

For a science that is all about size, one of the most interesting aspects of nanoscience is that properties of nanoparticles change with size.

It has been shown that many fundamental properties are size dependent on the nanoscale.

For example, the most stable crystalline phase of a material is size dependent. From thermodynamic considerations, the total free energy is a sum of the free energy of the bulk and the surface of the nanoparticle.

$$G_{nanoparticle} = G_{surface} + G_{bulk}$$

For nanoparticles, $G_{surface}$ is no longer a minor component but in fact becomes a large component of the total free energy. Surface free energies and surface stress are important components to the overall phase stability of nanoparticles. Titanium dioxide is interesting as anatase becomes more stable than rutile (Anatase and rutile two forms of Titanium dioxide TiO_2) for a particle size below 14 nm. (Weller et al., 2014)

PHYSICAL PROPERTIES OF NANOPARTICLES

Physical properties of nanoparticles are dependent on:

- Size
- Shape (spheres, rods, platelets, etc.)
- Composition
- Crystal Structure (FCC, BCC, etc.)
- Surface ligands or capping agents
- The medium in which they are dispersed
- As the percentage of atoms at the surface increases, the mechanical, optical, electrical, chemical, and magnetic properties change.
- For example, optical properties (color) of gold and silver change, when the spatial dimensions are reduced and the concentration is changed.

Semiconducting Nanoparticles

Quantum dots (QDs) are a particular kind of semiconducting crystalline nanostructure containing a metallic semiconductor core that is most commonly coated in a shell to improve its optical behavior.

- A quantum dot has a discrete quantized energy spectrum.
- Typical quantum dot sizes range between 2 and 20 nm. The color they glow depends on the size of the nanoparticle.
- The capability to tune the size of Quantum dots is advantageous for many applications.
- Due to their unique properties, such as bright fluorescence, broad UV excitation, narrow emission, and high photostability.
- Quantum dots have an enormous impact on pharmaceutical research and drug development.
- Physics can be used to explain the basis of Quantum dots performance. When an atom is exposed to energy it absorbs some of it, causing an electron to move to a higher energy level. When the electron falls back to its lower level a photon is emitted with the same energy as that absorbed by the original atom.
- However, different atoms emit different colors. This is possible because the energy levels in atoms have set values; on the other hand, they are quantized.
- Two important effects occur in semiconductors when electrons are confined to tiny regions. First, the HOMO-LUMO energy gap increases from the value observed in bulk crystals. Second, the energy levels of electrons in the LUMOs (and holes-the absence of electrons in the HOMOs) are quantized, like those of a particle in a box. Both effects play an important role in determining the optical properties of Quantum dots. (Sengupta and Sarkar, 2015)

Metallic Nanoparticles

The optical properties of metallic nanoparticles arise from a complex electrodynamic effect that is strongly influenced by the surrounding dielectric medium. Light impinging on metallic particles causes optical excitations of their electrons.

- The principal type of optical excitation that occurs is the collective oscillation of electrons in the valence band of the metal. Such coherent oscillations occur at the interface of a metal with a dielectric medium and are called surface plasmons.
- In bulk particles, the surface plasmons are travelling waves and are characterized by a linear momentum.
- To excite plasmons using photons in bulk metals, the momenta of the plasmon and the photon must match.
- This matching is possible only for very specific geometries of the interaction between light and matter and is a weak contributor to the optical properties of the metal.
- In nanoparticles, however, the surface plasmons are localized and have no characteristic momentum.
- As a result, the momenta of the plasmon and the photon do not need to match and plasmon excitation occurs with a greater intensity.
- The peak intensity of the surface plasmon absorption for gold and silver occurs in the optical region of the spectrum, and so these metallic nanoparticles are useful as pigments.
- The characteristics of plasmon absorption depend strongly on the metal and the dielectric surroundings as well as on the size and shape of the nanoparticle.
- To control the dielectric surroundings, so-called core–shell composite nanoparticles have been designed in which metallic shells of nanometer-scale thickness encapsulate a dielectric nanoparticle.

- Metallic nanoparticles and metallic nano-shells are used as dielectric sensors because their optical properties change when they come into contact with different dielectric materials.
- In particular, biological sensing is of interest because biological analytes can bind to the surface of the nanoparticle, causing a detectable shift in the plasmon absorption band.
- Gold nanoparticles are common examples of metallic nanoparticles and have found practical applications as biological and chemical sensors, 'smart bombs' for cancer therapy, and optical switching and fluorescent display materials. (Kuno, 2005)

SYNTHESIS ROUTE OF NANOMATERIALS

Nanoparticle synthesis refers to methods for creating nanoparticles. Nanoparticles can be derived from larger molecules or synthesized by 'bottom-up' methods or Top-down approaches. For example, nucleate and grow particles from fine molecular distributions in liquid or vapour phase. (Klabunde and Sergeev, 2013.)

Synthesis can also include functionalization by conjugation to bioactive molecules.

Bottom-Up Approaches

The bottom-up approach consists of the building up of nanomaterials from smaller building blocks. This approach has given rise to nano-chemistry, typically with the use of self-assembly methods for the formation of highly ordered two- and three-dimensional nanoscale structures. The idea of building nanomaterials atom-by-atom was first popularized by Drexler in the mid-1980s, who envisioned the construction of 'nanorobots'

Chemical Vapor Deposition (CVD)

Chemical vapor deposition (CVD) is a famous thin film deposition technique in which vapors condense into solid phase onto the surface of a substrate as a result of chemical reaction. The conventional CVD is also known as thermal CVD in which the deposition is carried out in a furnace usually at high temperature to vaporize the precursor material and promote the chemical reaction. A typical CVD process involves several steps. The precursors placed in the reaction chamber/furnace are evaporated and the reactant vapors are transported towards the substrate with the gas flow. The reactants are adsorbed on the surface of the substrate followed by diffusion, nucleation, and chemical reaction which leads to formation of deposition of material of interest in the form of thin film. In parallel to this, The precursor vapors can also go through gas phase reaction to produce products and by-products in gaseous phase. These products can adsorb on the substrate surface or ejected out of the furnace with gas flow. A conventional CVD reactor consists of several components which basically include:

1. Reactor chamber or furnace
2. Substrate holder
3. Carrier gas as H, N, Ar, or mixture of these gases
4. Heating source
5. Boats to load precursors in the chamber

Figure 2. Chemical vapor deposition (CVD)

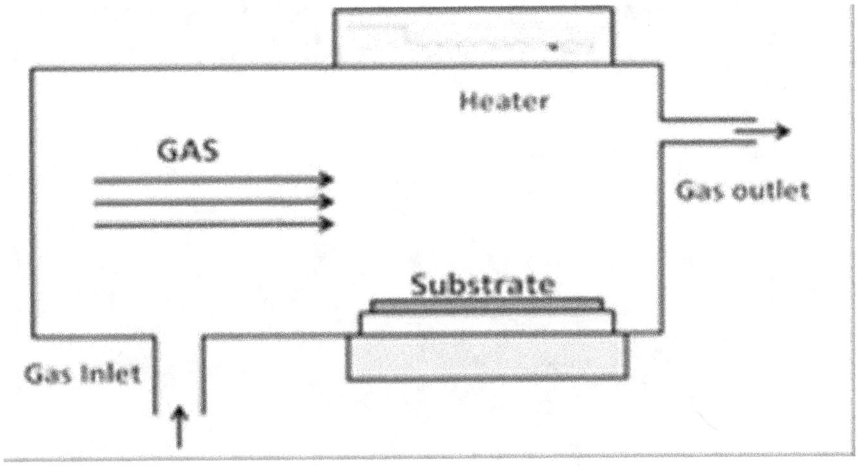

6. Exhaust system
7. Vacuum system.

The schematic diagram of a conventional CVD reactor is given in **Figure 2** The conventional CVD process involving vapor–solid mechanism is not efficient for deposition of nanomaterials (e.g., NWs) because direct adsorption of vapors on solid is a slow process. The synthesis stages include:

1. Loading of the precursor reactants and inert gases into the chamber,
2. Shifting of vapor species to substrate.

Spray Pyrolysis

The spray pyrolysis deposition (SPD) is a simple and cost-effective deposition technique based on chemical synthesis method in which layer of thin film is deposited on the surface of substrate by spraying the solution. This technique is used for deposition of mono or multilayer thin films having high density, controlled thickness, and usually porous structure. A speciality of this technique is the preparation of the product material in the form of porous or dense powder comprising of ultrafine grains. This capability of spray pyrolysis makes it a suitable technique for synthesis of nanomaterials.

The processing via this technique basically consists of spraying the precursor solution onto the surface of heated substrate. For controlled deposition of the desired products, the growth process is carried out in thermally insulated chamber equipped with necessary components. In this technique, the initial solution is prepared by melting the metal salt into a solvent. The small droplets of solution are then atomized into the furnace where the velocity and concentration of these droplets establish the morphology and size of the grown NPs. In this method, the characteristic solvent evaporation time and characteristic solute diffusion time are important and define the deposition rate and quality of grown layers. The product composition can be controlled by composition of reactants in precursor solution, growth parameters, and dynamics of the chemical reaction. The product particle will be a porous material or a hollow particle, the morphology of the particles is decided by the characteristic times. Temperature plays an important

Figure 3. Schematic Diagram of Spray Pyrolysis Process

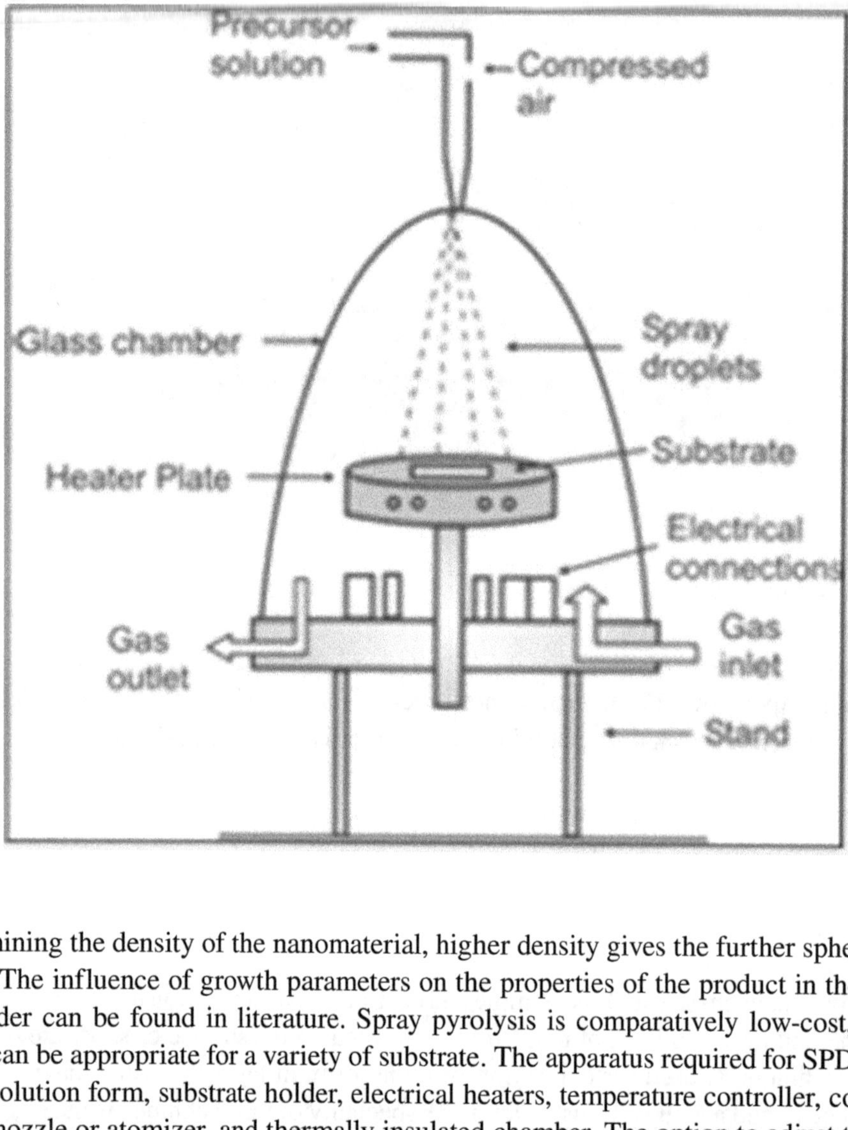

role in determining the density of the nanomaterial, higher density gives the further spherical shape of the particles. The influence of growth parameters on the properties of the product in the form of thin films or powder can be found in literature. Spray pyrolysis is comparatively low-cost, eco-friendly, scalable and can be appropriate for a variety of substrate. The apparatus required for SPD comprises of precursor in solution form, substrate holder, electrical heaters, temperature controller, cooling system, filters, spray nozzle or atomizer, and thermally insulated chamber. The option to adjust the opening of nozzle tip offers the control on particle size of the product which points to suitability of this technique for growth of NPs. The deposition rate and quality of the product strongly depend upon the processes, equipment, and growth parameters. Though, several atomizers have been tested but air blast, ultrasonic, and electrostatic type atomizers have been commonly utilized. The schematic diagram showing spray pyrolysis process is sketched in **Figure 3.** (Brechignac and Houdy, 2007)

Sol-Gel Process

A description of the sol-gel process can be formation of an oxide network through polycondensation reactions of a molecular precursor in a liquid. In general, in this process, several stages are identified, starting with a silicate solution and then forming a sol, which will then be transformed into a gel, and finally, a dry gel is obtained which is generally formed by a three-dimensional network of silica, with

Figure 4. Schematic of Sol-gel Process

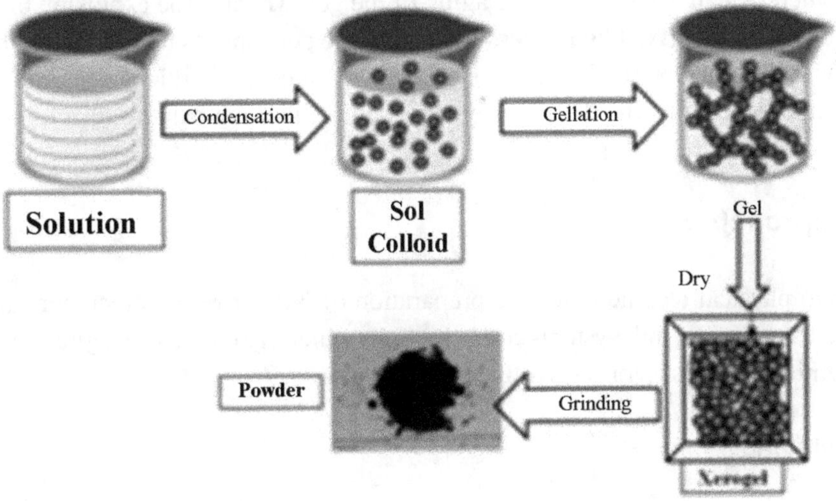

numerous pores of various sizes interconnected. **Figure 4** presents an outline of the routes of this mechanism Among the advantages of using the sol-gel process in the synthesis is because it can be carried out at room temperature, it allows us to produce a wide range of novel and functional materials, with potential applications in different areas; and finally, it is really attractive compared to other methods, due to its low production costs.

Sol-gel samples can be designed with a wide variety of morphologies, such as monoliths, films, fibers, and powders. In particular, films are the most important from the technological point of view. The process begins with the formation of a "sol," which is a stable dispersion of colloidal particles (amorphous or crystalline) or polymers in a solvent. A "gel" is formed by a three-dimensional continuous network, which contains a liquid phase, or by the joining of polymer chains. In a colloidal gel, the network is built from agglomerates of colloidal particles. Generally, van der Waals forces or hydrogen bonds dominate the interactions between the sol's particles. During synthesis, in most gel systems, covalent-type interactions dominate, and the gel process is irreversible. The gelation process may be reversible if there are other interactions involved. The purpose behind the sol-gel synthesis is to dissolve a compound in a liquid to obtain a solid controlling the factors of given synthesis. Using a controlled stoichiometry, sols of different reagents can be mixed to prepare multicomponent compounds. The sol-gel method prevents the problems with coprecipitation, which may be inhomogeneous, as it is a gelation reaction. It allows mixing at an atomic level to form small particles, which are easily sinter able. Typically, in the sol-gel chemistry, there is a reaction of an organometallic compound, which is generally an alkoxide, nitrate, or chloride under aqueous conditions to form a solid product. This product can be a dense glass monolith, a high surface area molecular filter, an aerogel to a metal oxide, a nitride coating, or nanoparticle. The process begins with reactions of hydrolysis and condensation of a precursor to form a gel followed by aging, solvent extraction, and finally drying. These reactions may be catalyzed by the addition of an acid or a base, which will produce dense or diffuse networks, respectively, by altering the hydrolysis kinetics. The selection of the precursor and catalyst depends ultimately on what you would like to make. **In the gelation step**, condensations are produced from the gel precursors in aqueous solution which are hydrolyzed and polymerized through alcohol or water. When starting the gelation, when the average size of the conglomerate is very small, they are best modeled with an approximation at the atomic level. In the next stage, syneresis can occur during the aging of the gel, which is the expulsion of solvent due to the contraction of the gel matrix. **The process of drying the gel** consists in eliminating the water from the gel system, with simultaneous collapse of the gel structure, under conditions of constant temperature, pressure, and humidity Usually, the dry gel is given a calcination treatment to turn it into a crystalline material.

Top-Down Approaches

There is top-down physical technique for the preparation of NPs in which precursors are grinded via milling systems. The grinding mill systems comprise balls which are used as the agitation media to carry out mechanochemical processes for the synthesis of NPs

Mechanical Alloying

The bulk material is placed inside the milling jar and grinding balls are used in synthesis process of nanomaterials. During this high-energy ball milling process (**Figure 5**), there is a choice to add a surfactant. Without a surfactant, the aggregation process takes place because of high surface energy of the particles, which results in the formation of larger NPs during mechanical grinding. If the surfactant is added, then the surfactant molecules form an organic layer on the surface of particles. The adsorption of these molecules lowers the surface energy of particles, and consequently no agglomeration takes place and NPs are produced with smaller size range and desired surface properties. The rotation speed of the ball's collision determines the efficiency of the grinding mechanism. The grinding capability also depends upon the number of balls and ball diameter, as the ball diameter is inversely proportional to the frequency of the ball's collisions. The structure, shape, and morphology of nanomaterials strongly depend upon parameters of bill milling strategy. The variety of milling parameters, growth conditions, and growth velocity directions produce nanomaterials in the form of diverse shapes like particles, rods, cubes, fibers, etc. In this method, it is very easy to introduce the impurities with less effort, high uniformity, and high yield. The top-down techniques are inexpensive, well-established, and traditional methods for the preparation of large-scale nanomaterials. However, the synthesized nanomaterials are often irregular in shapes and may have defects.

Mechanical Alloying (MA) is described as a high energy milling process in which powder particles are subjected to repeated cold welding, fracturing, and rewelding. The transfer of mechanical energy to the powder particles results in introduction of strain into the powder through generation of dislocations and other defects which act as fast diffusion paths. Additionally, refinement of particle and grain sizes occurs, and consequently the diffusion distances are reduced. Further, a slight rise in powder temperature occurs during milling. All these effects lead to alloying of the blended elemental powders during the milling process. The result could be constitutional changes formation of solid solutions (both equilibrium and supersaturated), intermetallic phases (equilibrium, metastable and quasi-crystalline), and amorphous phases or microstructural changes leading to development of ultrafine-grained and nanostructured phases

Figure 5. Ball Mailing

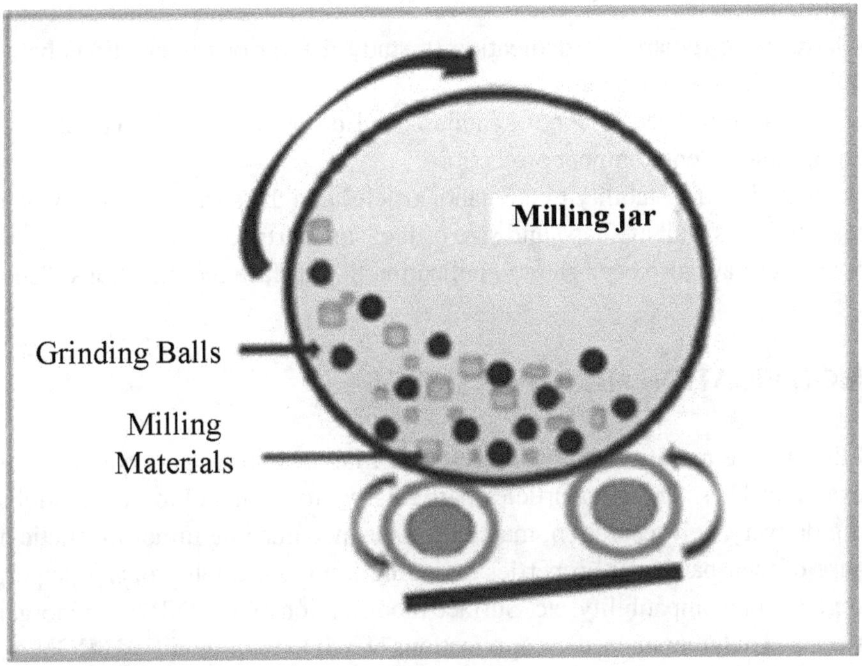

ROLE OF SURFACTANT IN SHAPE AND SIZE CONTROL OF NANOMATERIALS

Surfactants can be categorized according to the charge present in the hydrophilic portion of the molecule (after dissociation in aqueous solution):

1. Anionic surfactants
2. Nonionic surfactants
3. Cationic surfactants
4. Amphoteric surfactants

SURFACTANTS PLAY MAJOR ROLES IN THE FORMATION OF NANOEMULSIONS

By lowering the interfacial tension, Laplace pressure P (the difference in pressure between inside and outside the droplet) is reduced and hence the stress needed to break up a drop is reduced. (Grassian, 2008)

- Surfactants prevent coalescence of newly formed drops
- Surfactants play an important role in microemulsion reactions by lowering the tension among microemulsion and excess phases and to prepare the smaller NPs with narrow size distribution.
- They facilitate the synthesis of NPs in limited space by acting as a reagent to make the system stable and it can also merge with a co-surfactant.

- The shapes and sizes are very diverse and which consist of spheres, cubes, nanorods, nanowires, nanopyramids, etc
- Chemists have made remarkable dedication to study the various interactions between nanoparticles and the surface- active agents.
- These findings were utilized to control fundamental properties of nanoparticles. like shape and size, and morphology, environment protection.
- Surfactants provide extra stability to the nanoparticles and the surface forces, which play a major role in the stability as well as shape and size of the Nanoparticles.
- The nanoparticles have also been shows application like surface modifications. (Ling et al., 2019)

SURFACE MODIFICATION NANOPARTICLES

To understand the surface modification, we have taken Magnetic iron oxide nanoparticles (MIONPs) as a standard example: These are nanoparticles particularly attractive in biosensor, antibacterial activity, targeted drug delivery, cell separation, magnetic resonance imaging tumor magnetic hyperthermia, and so on because of their particular properties including Immobilisation property, superparamagnetic behavior, low toxicity, biocompatibility, etc. Surface modification of MIONPs with inorganic materials, organic molecules, and polymer molecules; applications of MIONPs or modified MIONPs; the technical challenges of synthesizing MIONPs; and their limitations in biomedical applications were described in this review to provide the theoretical and technological guidance for their future applications. (Schmidt and Brown, 2017).

Preparation of Nanoparticles Using Surfactants

The synthesis procedure for the preparation of metal nanoparticles from water-in oil microemulsions is simple. Two microemulsions are formulated, one with a metal salt or a metal complex dissolved in the water pools and one with a reducing agent, such as sodium borohydride or hydrazine. The reaction leading to the solid materials, such as a reduction of a metal salt to the metal. The overall kinetics of the system are usually such that the particles cease to grow when they have reached a size comparable with the size of starting microemulsion. Thus, for systems involving fast reaction steps within the droplets, the overall reaction kinetics are mainly governed by transport of species through the hydrocarbon domain and by droplet fusion, which governs the size and the size distribution of the nanoparticles. The water droplets of the starting water in-oil microemulsion should not be seen as a mold that is being filled with product during the course of the reaction.

CONCLUSION

In this chapter, We have gave synthestic aspects of nanomaterials and need to understand methods like Bottom-up and Top-down approaches include mechanical alloying, shape and size control of nanomaterials. By understanding the current scenario, the research and development of nanotechnology is very active globally, and nanotechnologies are already used in many products, In this chapter importance of nanotechnologies ho threy being developed for use in environmental applications were discuss.

Figure 6. Synthesis of surface modification of MIONPs

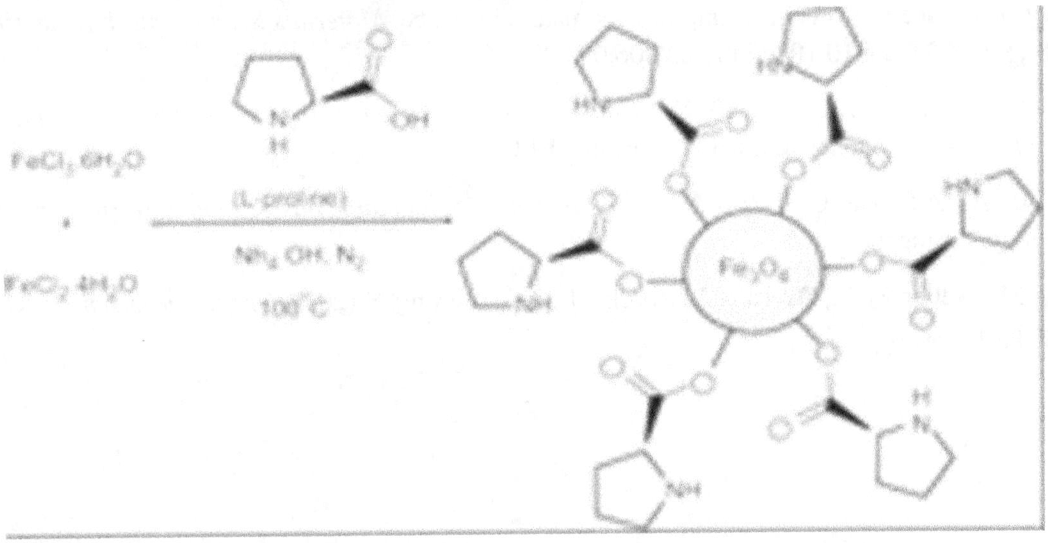

REFERENCES

Atkins, P., & Overton, T. (2010). *Shriver and Atkins' inorganic chemistry*. Oxford University Press.

Brechignac, C. P., & Houdy, M. L. (2007). *Nanomaterials and Nanochemistry*. Springer. doi:10.1007/978-3-662-47314-6

Dino, J. (2008, March 29). *NASA Ames Research Center Public Affairs Office*. NASA. Retrieved from https://www.nasa.gov/centers/ames/news/releases/2002/02images/nanogear/nanogears.html#:~:text=Nanotechnology%20is%20the%20creation%20of,biological)%20at%20that%20length%20scale

Grassian, V. H. (2008). When size really matters: Size-dependent properties and surface chemistry of metal and metal oxide nanoparticles in gas and liquid phase environments. *The Journal of Physical Chemistry C, 112*(47), 18303–18313. doi:10.1021/jp806073t

Klabunde, K. J., & Sergeev, G. B. (2013). *Nanochemistry*. Newnes.

Kuno, M. (2005). *Introduction to nanoscience and nanotechnology: A workbook*. University of Notre Dame.

Ling, W., Wang, M., Xiong, C., Xie, D., Chen, Q., Chu, X., Qiu, X., Li, Y., & Xiao, X. (2019). Synthesis, surface modification, and applications of magnetic iron oxide nanoparticles. *Journal of Materials Research, 34*(11), 1828–1844. doi:10.1557/jmr.2019.129

Mansoori, G. A., & Soelaiman, T. F. (2005). *Nanotechnology—An introduction for the standards community*. ASTM International.

Murty, B. S., Shankar, P., & Baldev Raj, R. B. (2013). *James Murdas. Text book of nano science and nano technology*. Academic Press.

Pokropivny, V. V., & Skorokhod, V. V. (2007). Classification of nanostructures by dimensionality and concept of surface forms engineering in nanomaterial science. *Materials Science and Engineering C*, *27*(5-8), 990–993. doi:10.1016/j.msec.2006.09.023

Schmidt, N. A., & Brown, J. M. (2017). *Evidence-based practice for nurses: Appraisal and application of research* (4th ed.). Jones & Bartlett Learning, LLC.

Sengupta, A., & Sarkar, C. K. (Eds.). (2015). *Introduction to nano: Basics to nanoscience and nano-technology*. Springer.

Weller, M., Weller, M. T., Overton, T., Rourke, J., & Armstrong, F. (2014). *Inorganic chemistry*. Oxford University Press.

Compilation of References

Abdel Aal, A., Mahmoud, S. A., & Aboul-Gheit, A. K. (2009). Sol–Gel and Thermally Evaporated Nanostructured Thin ZnO Films for Photocatalytic Degradation of Trichlorophenol. *Nanoscale Research Letters*, *4*, 627.

Abdel-Ghani, N. T., Hefny, M., & El-Chaghaby, G. A. F. (2007). Removal of lead from aqueous solution using low cost abundantly available adsorbents. *International Journal of Environmental Science and Technology*, *4*(1), 67–73. doi:10.1007/BF03325963

Abdullah, A. M., Mohammed, A. A.-S., Suleyman, A. A. M., & Zahangir, Md. (2014). A, iis S. *Kinetics of Cadmium Adsorption by CNTS Grown on PACs*, *7*(7), 25–30.

Abdulrazzak, F. (2018). Preparation and Characterization of Silver Oxide Nanoparticles AgNPs and Evaluation the Ratios of Oxides. *Journal of Engineering and Applied Sciences (Asian Research Publishing Network)*, *14*. Advance online publication. doi:10.36478/jeasci.2019.9177.9184

Abou El-Nour, K. M. M., Eftaiha, A., Al-Warthan, A., & Ammar, R. A. A. (2010). Synthesis and applications of silver nanoparticles. *Arab. J. Chem.*, *3*(3), 135–140. https://www.sciencedirect.com/science/article/pii/S1878535210000377

Abraham, J. K., Philip, B., Witchurch, A., Varadan, V. K., & Reddy, C. C. (2004). A compact wireless gas sensor using a carbon nanotube/PMMA thin film chemiresistor. *Smart Materials and Structures*, *13*(5), 1045–1049. https://doi.org/10.1088/0964-1726/13/5/010

Abrahamson, J., Wiles, P. G., & Rhoades, B. L. (1999). Structure of carbon fibers found on carbon arc anodes. *Carbon*, *37*(11), 1873–1874. doi:10.1016/S0008-6223(99)00199-2

Access, O. (2018). We are IntechOpen, the world's leading publisher of Open Access books Built by scientists, for scientists TOP 1%. *Long-Haul Travel Motiv by Int Tour to Penang*, *i*, 13.

Achouri, F., Corbel, S., Balan, L., Mozet, K., Girot, E., Medjahdi, G., Said, M. B., Ghrabi, A., & Schneider, R. (2016). Porous Mn-doped ZnO nanoparticles for enhanced solar and visible light photocatalysis. *Materials & Design*, *101*, 309–316. doi:10.1016/j.matdes.2016.04.015

Adams, L. K., Lyon, D. Y., & Alvarez, P. J. J. (2006). Comparative eco-toxicity of nanoscale TiO2, SiO2, and ZnO water suspensions. *Water Research*, *40*(19), 3527–3532. doi:10.1016/j.watres.2006.08.004 PMID:17011015

Adeleye, A. S., Conway, J. R., Garner, K., Huang, Y., Su, Y., & Keller, A. A. (2016). Engineered nanomaterials for water treatment and remediation: Costs, benefits, and applicability. *Chemical Engineering Journal*, *286*, 640–662. doi:10.1016/j.cej.2015.10.105

Aderhold, D., Williams, C. J., & Edyvean, R. G. J. (1996). The removal of heavy-metal ions by seaweeds and their derivatives. *Bioresource Technology*, *58*(1), 1–6. doi:10.1016/S0960-8524(96)00072-7

Adesola Babarinde N.A., Oyebamiji Babalola, J., & Adebowale Sanni, R. (2006). Biosorption of lead ions from aqueous solution by maize leaf. *Int. J. Phy. Sci.*, *1*(1), 23–26.

Adnan, M. A. M., Phoon, B. L., & Julkapli, N. M. (2020). Mitigation of pollutants by chitosan/metallic oxide photocatalyst: A review. *Journal of Cleaner Production*, *261*, 121190. Advance online publication. doi:10.1016/j.jclepro.2020.121190

Agarwal, M., & Singh, K. (2017). Heavy metal removal from wastewater using various adsorbents: A review. *Journal of Water Reuse and Desalination*, *7*(4), 387–419.

Ahalya, N., Ramachandra, T. V., & Kanamandi, R. D. (2003). Biosorption of heavy metals. *Research Journal of Chemistry and Environment*, *7*, 71–78.

Ahmad, A. A., & Hameed, B. H. (2010). Fixed-bed adsorption of reactive azo dye onto granular activated carbon prepared from waste. *Journal of Hazardous Materials*, *175*(1–3), 298–303. https://doi.org/10.1016/j.jhazmat.2009.10.003

Ahmad, A., Mohd-Setapar, S. H., Chuong, C. S., Khatoon, A., Wani, W. A., Kumar, R., & Rafatullah, M. (2015). Recent advances in new generation dye removal technologies: Novel search for approaches to reprocess wastewater. *RSC Advances*, *5*(39), 30801–30818. doi:10.1039/C4RA16959J

Ahmad, M., Ahmed, S., Swami, B. L., & Ikram, S. (2015). Adsorption of heavy metal ions: Role of chitosan and cellulose for water treatment. *Langmuir*, *79*, 109–155.

Ahmad, M., Rehman, W., Khan, M. M., Qureshi, M. T., Gul, A., Haq, S., Ullah, R., Rab, A., & Menaa, F. (2021). Phytogenic fabrication of ZnO and gold decorated ZnO nanoparticles for photocatalytic degradation of Rhodamine B. *Journal of Environmental Chemical Engineering*, *9*(1), 104725. doi:10.1016/j.jece.2020.104725

Ahmed Quraishi, I., Pawar, R. A., Shinde, D. R., & Chaskar, M. G. (2020). Parametric Study on Photocatalytic Dye Degradation under Visible Light in Flat Slurry Reactor with nano-ZnO Photocatalyst. *Materials Today: Proceedings*, *23*, 410–422. doi:10.1016/j.matpr.2020.02.061

Ahmed, B., Ojha, A. K., Singh, A., Hirsch, F., Fischer, I., Patrice, D., & Materny, A. (2018). Well-controlled in-situ growth of 2D WO3 rectangular sheets on reduced graphene oxide with strong photocatalytic and antibacterial properties. *Journal of Hazardous Materials*, *347*, 266–278. doi:10.1016/j.jhazmat.2017.12.069 PMID:29329009

Ahmed, S., Ahmad, M., Swami, B. L., & Ikram, S. (2016). A review on plants extract mediated synthesis of silver nanoparticles for antimicrobial applications: A green expertise. *Journal of Advanced Research*, *7*(1), 17–28. doi:10.1016/j.jare.2015.02.007 PMID:26843966

Ahmed, Y. M., Al-Mamun, A., Al Khatib, M. F. R., Jameel, A. T., & AlSaadi, M. A. H. A. R. (2015). Efficient lead sorption from wastewater by carbon nanofibers. *Environmental Chemistry Letters*, *13*(3), 341–346. https://doi.org/10.1007/s10311-015-0509-3

Ahuja, S. (2015). Overview of Global Water Challenges and Solutions. *ACS Symposium Series*, *1206*, 1–25. doi:10.1021/bk-2015-1206.ch001

Ajayan, P. M. (1999). Nanotubes from carbon. *Chemical Reviews*, *99*(7), 1787–1800. https://doi.org/10.1021/cr970102g

Ajmal, M., Rao, R. A. K., Ahmad, R., & Ahmad, J. (2000). Adsorption studies on Citrus reticulata (fruit peel of orange): Removal and recovery of Ni (II) from electroplating wastewater. *Journal of Hazardous Materials*, *79*(1-2), 117–131. doi:10.1016/S0304-3894(00)00234-X PMID:11040390

Ajmal, M., Rao, R. A. K., Anwar, S., Ahmad, J., & Ahmad, R. (2003). Adsorption studies on rice husk: Removal and recovery of Cd (II) from wastewater. *Bioresource Technology*, *86*(2), 147–149. doi:10.1016/S0960-8524(02)00159-1 PMID:12653279

Akarslan, F., & Demiralay, H. (2018). Effective of textile materials harmful to human health. *Acta Physica Polonica*, *128*, B-407–B-408.

Akinpelu, A. A., Ali, M. E., Johan, M. R., Saidu, R., Qurban, M. A., & Saleh, T. A. (2019). Polycyclic aromatic hydrocarbons extraction and removal from wastewater by carbon nanotubes: A review of the current technologies, challenges and prospects. *Process Safety and Environmental Protection*, *122*, 68–82. doi:10.1016/j.psep.2018.11.006

Akola, J., Walter, M., Whetten, R. L., Hakkinen, H., & Gronbeck, H. (2008). On the structure of thiolate-protected Au25. *Journal of the American Chemical Society*, *130*(12), 3756–3757. doi:10.1021/ja800594p PMID:18321117

Al-Asheh, S., & Duvnjak, Z. (1998). Binary metal sorption by pine bark: Study of equilibria and mechanisms. *Separation Science and Technology*, *33*(9), 1303–1329. doi:10.1080/01496399808544985

Albelda, J. A. V., Uzunoglu, A., Santos, G. N. C., & Stanciu, L. A. (2017). Graphene-titanium dioxide nanocomposite based hypoxanthine sensor for assessment of meat freshness. *Biosensors & Bioelectronics*, *89*(1), 518–524. https://doi.org/10.1016/j.bios.2016.03.041

Al-Bsoul, A., Al-Shannag, M., Tawalbeh, M., Al-Taani, A. A., Lafi, W. K., Al-Othman, A., & Alsheyab, M. (2020). Optimal conditions for olive mill wastewater treatment using ultrasound and advanced oxidation processes. *The Science of the Total Environment*, *700*, 134576.

Alhebshi, N., Huang, H., Ghandour, R., Alghamdi, N. K., Alharbi, O., Aljurban, S., He, J.-H., & Al-Jawhari, H. (2020). Green synthesized CuxO@Cu nanocomposites on a Cu mesh with dual catalytic functions for dye degradation and hydrogen evaluation. *Journal of Alloys and Compounds*, *848*, 156284. doi:10.1016/j.jallcom.2020.156284

Al-Hussaini, A. S., Eltabie, K. R., & Rashad, M. E. E. (2016). One-pot modern fabrication and characterization of TiO2@ terpoly(aniline, anthranilic acid and o-phenylenediamine) core-shell nanocomposites via polycondensation. *Polymer*, *101*, 328–337. https://doi.org/10.1016/j.polymer.2016.08.104

Ali, A., Ambreen, S., Maqbool, Q., Naz, S., Shams, M. F., Ahmad, M., Phull, A. R., & Zia, M. (2016). Zinc impregnated cellulose nanocomposites: Synthesis, characterization and applications. *Journal of Physics and Chemistry of Solids*, *98*, 174–182.

Ali, H., Guler, A. C., Masar, M., Urbanek, P., Urbanek, M., Skoda, D., Suly, P., Machovsky, M., Galusek, D., & Kuritka, I. (2021). Solid-State Synthesis of Direct Z-Scheme Cu2O/WO3 Nanocomposites with Enhanced Visible-Light Photocatalytic Performance. *Catalysts*, *11*(2), 293. doi:10.3390/catal11020293

Ali, I. (2010). The quest for active carbon adsorbent substitutes: Inexpensive adsorbents for toxic metal ions removal from wastewater. *Separation and Purification Reviews*, *39*(3–4), 95–171. https://doi.org/10.1080/15422119.2010.527802

Ali, I. (2012). New generation adsorbents for water treatment. *Chemical Reviews*, *112*(10), 5073–5091. https://doi.org/10.1021/cr300133d

Ali, I., & Gupta, V. K. (2007). Advances in water treatment by adsorption technology. *Nature Protocols*, *1*(6), 2661–2667. doi:10.1038/nprot.2006.370 PMID:17406522

Alkaim, A. F., Sadik, Z., Mahdi, D. K., Alshrefi, S. M., Al-Sammarraie, A. M., Alamgir, F. M., Singh, P. M., & Aljeboree, A. M. (2015). Preparation, structure and adsorption properties of synthesized multiwall carbon nanotubes for highly effective removal of maxilon blue dye. *Korean Journal of Chemical Engineering*, *32*(12), 2456–2462. doi:10.100711814-015-0078-y

Allhoff, F. (2007). On the autonomy and justification of nanoethics. *NanoEthics*, *1*(3), 185–210. doi:10.100711569-007-0018-3

Alluri, H. K., Ronda, S. R., Settalluri, V. S., Bondili, J. S., Suryanarayana, V., & Venkateshwar, P. (2007). Biosorption: An eco-friendly alternative for heavy metal removal. *African Journal of Biotechnology*, *6*(25).

Almomani, F., Al Ketife, A., Judd, S., Shurair, M., Bhosale, R. R., Znad, H., & Tawalbeh, M. (2019). Impact of CO2 concentration and ambient conditions on microalgal growth and nutrient removal from wastewater by a photobioreactor. *The Science of the Total Environment*, *662*, 662–671.

AlOmar, M. K., Alsaadi, M. A., Hayyan, M., Akib, S., Ibrahim, R. K., & Hashim, M. A. (2016). Lead removal from water by choline chloride based deep eutectic solvents functionalized carbon nanotubes. *Journal of Molecular Liquids*, *222*, 883–894. https://doi.org/10.1016/j.molliq.2016.07.074

Al-Qodah, Z., Tawalbeh, M., Al-Shannag, M., Al-Anber, Z., & Bani-Melhem, K. (2020). Combined electrocoagulation processes as a novel approach for enhanced pollutants removal: A state-of-the-art review. *The Science of the Total Environment*, *744*, 140806.

AlSaadi, M. A., Al Mamun, A., Alam, M. Z., Amosa, M. K., & Atieh, M. A. (2016). Removal of cadmium from water by CNT–PAC composite: Effect of functionalization. *Nano*, *11*(1), 1650011. https://doi.org/10.1142/S1793292016500119

Alves de Lima, R. O., Bazo, A. P., Salvadori, D. M. F., Rech, C. M., de Palma Oliveira, D., & de Aragão Umbuzeiro, G. (2007). Mutagenic and carcinogenic potential of a textile azo dye processing plant effluent that impacts a drinking water source. *Mutation Research*, *626*(1–2), 53–60. https://doi.org/10.1016/j.mrgentox.2006.08.002

Aly, H. F., & Abd-Elhamid, A. I. (2018). Photocatalytic Degradation of Methylene Blue Dye Using Silica Oxide Nanoparticles as a Catalyst. *Water Environment Research*, *90*(9), 807–817. doi:10.2175/106143017X15131012187953 PMID:30208997

Amarnath, K., Kumar, J., Reddy, T., Mahesh, V., Ayyappan, S. R., & Nellore, J. (2012). Synthesis and characterization of chitosan and grape polyphenols stabilized palladium nanoparticles and their antibacterial activity. *Colloid Surf. B*, *92*, 254–261. doi:10.1016/j.colsurfb.2011.11.049 PMID:22225943

Ameen, S., Akhtar, M. S., Kim, Y. S., & Shin, H. S. (2011). Nanocomposites of poly(1-naphthylamine)/SiO2 and poly(1-naphthylamine)/TiO2: Comparative photocatalytic activity evaluation towards methylene blue dye. *Applied Catalysis B: Environmental*, *103*(1–2), 136–142. https://doi.org/10.1016/j.apcatb.2011.01.019

Aminuzzaman, M., Kei, L. M., & Liang, W. H. (2017). Green synthesis of copper oxide (CuO) nanoparticles using banana peel extract and their photocatalytic activities. *AIP Conference Proceedings*, *1828*(1), 020016. doi:10.1063/1.4979387

Anasori, B., Lukatskaya, M. R., & Gogotsi, Y. (2017). 2D metal carbides and nitrides (MXenes) for energy storage. *Nature Reviews Materials*, *2*(2), 16098. Advance online publication. doi:10.1038/natrevmats.2016.98

Anastas, P. T., & Warner, J. C. (1998). *12 principles of green chemistry. Green chemistry: theory and practice*. Oxford University Press.

Anastopoulos, I., Robalds, A., Tran, H. N., Mitrogiannis, D., Giannakoudakis, D. A., Hosseini-Bandegharaei, A., & Dotto, G. L. (2019). Removal of heavy metals by leaves-derived biosorbents. *Environmental Chemistry Letters*, *17*(2), 755–766. https://doi.org/10.1007/s10311-018-00829-x

Anaya-Esparza, L. M., Ruvalcaba-Gómez, J. M., Maytorena-Verdugo, C. I., González-Silva, N., Romero-Toledo, R., Aguilera-Aguirre, S., ... Montalvo-González, E. (2020). Chitosan-TiO2: A versatile hybrid composite. *Materials (Basel)*, *13*(4), 811.

Andrade, G. R. S., Nascimento, C. C., Lima, Z. M., Teixeira-Neto, E., Costa, L. P., & Gimenez, I. F. (2017). Star-shaped ZnO/Ag hybrid nanostructures for enhanced photocatalysis and antibacterial activity. *Applied Surface Science, 399*, 573–582. doi:10.1016/j.apsusc.2016.11.202

Andrade, N. F., Martinez, D. S. T., Paula, A. J., Silveira, J. V., Alves, O. L., & Fiho, A. G. S. (2013). Temperature effects on the nitric acid oxidation of industrial grade multi-walled carbon nanotubes. *Journal of Nanoparticle Research, 15*(7), 1767. doi:10.100711051-013-1761-8

Andrei, V., Hoye, R. L. Z., Quesada, M. C., Bajada, M., Ahmad, S., Volder, M. D., Friend, R., & Reisner, E. (2018). Scalable triple cation mixed halide perovskite-$BiVO_4$ tandems for bias-free water splitting. *Advanced Energy Materials, 8*(25), 1801403. doi:10.1002/aenm.201801403

An, H. K., Park, B. Y., & Kim, D. S. (2001). Crab shell for the removal of heavy metals from aqueous solution. *Water Research, 35*(15), 3551–3556. doi:10.1016/S0043-1354(01)00099-9 PMID:11561614

Ansari, R., & Pornahad, A. (2010). Removal of cerium (IV) *ion from aqueous solutions using sawdust as a very low cost bioadsorbent. Journal of Applied Sciences in Environmental Sanitation, 5*(3).

Antony, R. P., Bassi, P. S., Abdi, F. F., Chiam, S. Y., Ren, Y., Barber, J., Loo, J. S. C., & Wong, L. H. (2016). Electrospun Mo-$BiVO_4$ for efficient photoelectrochemical water oxidation: Direct evidence of improved hole diffusion length and charge separation. *Electrochimica Acta, 211*, 173–182. doi:10.1016/j.electacta.2016.06.008

Anzar, N., Hasan, R., Tyagi, M., Yadav, N., & Narang, J. (2020). Carbon nanotube - A review on synthesis, properties and plethora of applications in the biomedical science. *Sensors International, 1*, 100003. doi:10.1016/j.sintl.2020.100003

Arenas, M. C., Rodríguez-Núnez, L. F., Rangel, D., Martínez-Álvarez, O., Martínez-Alonso, C., & Castaño, V. M. (2013). Simple one-step ultrasonic synthesis of anatase Titania/polypyrrole nanocomposites. *Ultrasonics Sonochemistry, 20*(2), 777–784. doi:10.1016/j.ultsonch.2012.09.009 PMID:23099056

Arepalli, S. (2004). Laser ablation process for single-walled carbon nanotube production. *Journal of Nanoscience and Nanotechnology, 4*(4), 317–325. doi:10.1166/jnn.2004.072 PMID:15296222

Arias, L. R., & Yang, L. (2009). Inactivation of bacterial pathogens by carbon nanotubes in suspensions. *Langmuir, 25*(5), 3003–3012. https://doi.org/10.1021/la802769m

Aromal, S. A., & Philip, D. (2012). Green synthesis of gold nanoparticles using Trigonella foenum-graecum and its size dependent catalytic activity. *Spectrochimica Acta. Part A: Molecular Spectroscopy, 97*, 1–5. doi:10.1016/j.saa.2012.05.083 PMID:22743607

Arora, R., Mandal, U. K., Sharma, P., & Srivastav, A. (2014). Effect of fabrication technique on microstructure and electrical conductivity of polyaniline-TiO2 -PVA composite material. *Procedia Materials Science, 6*, 238–243. https://doi.org/10.1016/j.mspro.2014.07.029

Arora, N., & Sharma, N. (2014). Arc discharge synthesis of carbon nanotubes. A comprehensive review. *Diamond and Related Materials, 50*, 135–150. doi:10.1016/j.diamond.2014.10.001

Arora, R., Mandal, U. K., Sharma, P., & Srivastav, A. (2015). Article. *Materials Today: Proceedings, 2*, 2215–2225.

Arvand, M., Latify, L., Tajmehri, H., Yagubov, A. I., Nuriyev, A. N., Pourhabib, A., ... Abolhassani, M. R. (2009). Comparative study for the removal of oxadiazon from aqueous solutions by adsorption on chitosan and activated carbon. *Analytical Letters, 42*(6), 856–869.

Ashkanani, A., Almomani, F., Khraisheh, M., Bhosale, R., Tawalbeh, M., & AlJaml, K. (2019). Bio-carrier and operating temperature effect on ammonia removal from secondary wastewater effluents using moving bed biofilm reactor (MBBR). *The Science of the Total Environment*, *693*, 133425.

Ashkenazy, R., Gottlieb, L., & Yannai, S. (1997). Characterization of acetone-washed yeast biomass functional groups involved in lead biosorption. *Biotechnology and Bioengineering*, *55*(1), 1–10. doi:10.1002/(SICI)1097-0290(19970705)55:1<1::AID-BIT1>3.0.CO;2-H PMID:18636438

Asuri, P., Bale, S. S., Pangule, R. C., Shah, D. A., Kane, R. S., & Dordick, J. S. (2007). Structure, function, and stability of enzymes covalently attached to single-walled carbon nanotubes. *Langmuir*, *23*(24), 12318–12321. https://doi.org/10.1021/la702091c

Asuri, P., Karajanagi, S. S., Yang, H., Yim, T. J., Kane, R. S., & Dordick, J. S. (2006). Increasing protein stability through control of the nanoscale environment. *Langmuir*, *22*(13), 5833–5836. https://doi.org/10.1021/la0528450

Atkins, P., & Overton, T. (2010). *Shriver and Atkins' inorganic chemistry*. Oxford University Press.

Atla, S. B., Lin, W.-R., Chien, T.-C., Tseng, M.-J., Shu, J.-C., Chen, C.-C., & Chen, C.-Y. (2018). Fabrication of Fe_3O_4/ZnO magnetite core shell and its application in photocatalysis using sunlight. *Materials Chemistry and Physics*, *216*, 380–386. doi:10.1016/j.matchemphys.2018.06.020

Atrak, K., Ramazani, A., & Taghavi Fardood, S. (2019). Eco-friendly synthesis of $Mg_{0.5}Ni_{0.5}Al_xFe_{2-x}O_4$ magnetic nanoparticles and study of their photocatalytic activity for degradation of direct blue 129 dye. *Journal of Photochemistry and Photobiology A Chemistry*, *382*, 111942. doi:10.1016/j.jphotochem.2019.111942

Avcı, A., İnci, İ., & Baylan, N. (2020). Adsorption of ciprofloxacin hydrochloride on multiwall carbon nanotube. *Journal of Molecular Structure*, *1206*, 127711.

Axtell, N. R., Sternberg, S. P., & Claussen, K. (2003). Lead and nickel removal using Microspora and Lemna minor. *Bioresource Technology*, *89*(1), 41–48. doi:10.1016/S0960-8524(03)00034-8 PMID:12676499

Aydoghmish, S. M., Hassanzadeh-Tabrizi, S. A., & Saffar-Teluri, A. (2019). Facile synthesis and investigation of NiO–ZnO–Ag nanocomposites as efficient photocatalysts for degradation of methylene blue dye. *Ceramics International*, *45*(12), 14934–14942. doi:10.1016/j.ceramint.2019.04.229

Azizi, S., Alloin, F., & Dufresne, A. (2005). Review of recent research into cellulosic whiskers, their properties and their application in nanocomposite field. *Biomacromolecules*, *6*, 612–626.

Azmi, W., Sani, R. K., & Banerjee, U. C. (1998). Biodegradation of triphenylmethane dyes. *Enzyme and Microbial Technology*, *22*(3), 185–191. doi:10.1016/S0141-0229(97)00159-2 PMID:9463944

Ba-Abbad, M. M., Kadhum, A. A. H., Mohamad, A. B., Takriff, M. S., & Sopian, K. (2012). Synthesis and catalytic activity of TiO_2 nanoparticles for photochemical oxidation of concentrated chlorophenols under direct solar radiation. *International Journal of Electrochemical Science*, *7*, 4871–4888.

Babić, B. M., Milonjić, S. K., Polovina, M. J., Čupić, S., & Kaludjerović, B. V. (2002). Cupi ˇ c¢ S, Kaludjerovic´ BV. *Carbon*, *40*(7), 1109–1115. https://doi.org/10.1016/S0008-6223(01)00256-1

Baby, R., Saifullah, B., & Hussein, M. Z. (2019). Carbon nanomaterials for the treatment of heavy metal-contaminated water and environmental remediation. *Nanoscale Research Letters*, *14*(1), 341. https://doi.org/10.1186/s11671-019-3167-8

Badireddy, A. R., Hotze, E. M., Chellam, S., Alvarez, P., & Wiesner, M. R. (2007). Inactivation of bacteriophages via photosensitization of fullerol nanoparticles. *Environmental Science & Technology*, *41*(18), 6627–6632. https://doi.org/10.1021/es0708215

Baek, Y. W., & An, Y. J. (2011). Microbial toxicity of metal oxide nanoparticles (CuO, NiO, ZnO, and Sb2O3) to Escherichia coli, Bacillus subtilis, and Streptococcus aureus. *The Science of the Total Environment*, *409*(8), 1603–1608. doi:10.1016/j.scitotenv.2011.01.014 PMID:21310463

Bagheri, M., & Mirbagheri, S. A. (2018). Critical review of fouling mitigation strategies in membrane bioreactors treating water and wastewater. *Bioresource Technology*, *258*, 318–334. doi:10.1016/j.biortech.2018.03.026 PMID:29548641

Baig, T. H., Garcia, A. E., Tiemann, K. J., & Gardea-Torresdey, J. L. (1999). Adsorption of heavy metal ions by the biomass of Solanum elaeagnifolium (Silverleaf night-shade). In *Proceedings of the 1999 conference on Hazardous Waste Research, (131)*. Academic Press.

Baig, N., Kammakakam, I., & Falath, W. (2021). Nanomaterials: A review of synthesis methods, properties, recent progress, and challenges. *Mater. Adv.*, *2*(6), 1821–1871. doi:10.1039/D0MA00807A

Baillie, C. (2004). Green composites. In Polymer Composites and the Environment. CRC Press.

Bai, S., Wang, L., Chen, X., Du, J., & Xiong, Y. (2015). Chemically exfoliated metallic MoS2 nanosheets: A promising supporting co-catalyst for enhancing the photocatalytic performance of TiO2 nanocrystals. *Nano Research*, *8*(1), 175–183. doi:10.100712274-014-0606-9

Bakhsh, E. M., Khan, S. A., Marwani, H. M., Danish, E. Y., Asiri, A. M., & Khan, S. B. (2018). Performance of cellulose acetate-ferric oxide nanocomposite supported metal catalysts toward the reduction of environmental pollutants. *International Journal of Biological Macromolecules*, *107*, 668–677.

Balasubramanian, K., & Burghard, M. (2005). Chemically functionalized carbon nanotubes. *Small*, *1*(2), 180–192. doi:10.1002mll.200400118 PMID:17193428

Balasubramanian, K., & Burghard, M. (2008). Electrochemically functionalized carbon nanotubes for device applications. *Journal of Materials Chemistry*, *18*(26), 3071–3083. doi:10.1039/b718262g

Banat, F., Al-Asheh, S., & Al-Makhadmeh, L. (2003). Kinetics and equilibrium study of cadmium ion sorption onto date pits—An agricultural waste. *Adsorption Science and Technology*, *21*(3), 245–260. doi:10.1260/026361703322404395

Banks, C. E., & Compton, R. G. (2006). New electrodes for old: From carbon nanotubes to edge plane pyrolytic graphite. *Analyst (London)*, *131*(1), 15–21. doi:10.1039/B512688F PMID:16425467

Bansal, R. C., & Goyal, M. (2005). *Activated carbon adsorption*. Taylor & Francis Group.

Bansode, S. S., Banarjee, S. K., Gaikwad, D. D., Jadhav, S. L., & Thorat, R. M. (2010). Microencapsulation: A Review. *International Journal of Pharmaceutical Sciences Review and Research*, *1*(2).

Bao, L., Zang, J., & Li, X. (2011). Flexible Zn_2SnO_4/MnO_2 Core/Shell Nanocable–Carbon Microfiber Hybrid Composites for High-Performance Supercapacitor Electrodes. *Nano Letters*, *11*(3), 1215–1220. doi:10.1021/nl104205s PMID:21306113

Barakat, M. A. (2011). New trends in removing heavy metals from industrial wastewater. *Arabian Journal of Chemistry*, *4*(4), 361–377.

Baran, W., Makowski, A., & Wardas, W. (2003). The influence of $FeCl_3$ on the photocatalytic degradation of dissolved azo dyes in aqueous TiO_2 suspensions. *Chemosphere*, *53*(1), 87–95. doi:10.1016/S0045-6535(03)00435-1 PMID:12892670

Barisci, J. N., Wallace, G. G., & Baughman, R. H. (2000). Electrochemical studies of single-wall carbon nanotubes in aqueous solutions. *Journal of Electroanalytical Chemistry (Lausanne, Switzerland)*, *488*(2), 92–98. doi:10.1016/S0022-0728(00)00179-0

Barsoum, M. W., & Radovic, M. (2011). Elastic and Mechanical Properties of the MAX Phases. *Annual Review of Materials Research*, *41*(1), 195–227. doi:10.1146/annurev-matsci-062910-100448

Barzegar, M. H., Ghaedi, M., Madadi Avargani, V., Sabzehmeidani, M. M., Sadeghfar, F., & Jannesar, R. (2019). Electrochemical synthesis and efficient photocatalytic degradation of azo dye alizarin yellow R by Cu/CuO nanorods under visible LED light irradiation using experimental design methodology. *Polyhedron*, *158*, 506–514. doi:10.1016/j.poly.2018.10.040

Bashir, S., Liu, J., Zhang, H., Sun, X., & Guo, J. (2013). Band gap evaluations of metal-inserted titania Nanomaterials. *Journal of Nanoparticle Research*, *15*(4), 1572. doi:10.100711051-013-1572-y

Basnet, P. C. S. (2019). Nanocomposites of ZnO for Water Remediation. In Composites for Environmental Engineering. doi:10.1002/9781119555346.ch7

Basnet, P., Inakhunbi Chanu, T., Samanta, D., & Chatterjee, S. (2018). A review on bio-synthesized zinc oxide nanoparticles using plant extracts as reductants and stabilizing agents. *Journal of Photochemistry and Photobiology. B, Biology*, *183*, 201–221. doi:10.1016/j.jphotobiol.2018.04.036 PMID:29727834

Basnet, P., Samanta, D., Chanu, T. I., Mukherjee, J., & Chatterjee, S. (2019). Tea-phytochemicals functionalized Ag modified ZnO nanocomposites for visible light driven photocatalytic removal of organic water pollutants. *Materials Research Express*, *6*(8), 085095. doi:10.1088/2053-1591/ab234e

Basso, M. C., Cerrella, E. G., & Cukierman, A. L. (2002). Activated carbons developed from a rapidly renewable biosource for removal of cadmium (II) and nickel (II) ions from dilute aqueous solutions. *Industrial & Engineering Chemistry Research*, *41*(2), 180–189. doi:10.1021/ie010664x

Basu, A., Kumar, S., & Mukherjee, S. (2003). Arsenic reduction from aqueous environment by water lettuce (Pistia stratiotes L.). *Indian Journal of Environmental Health*, *45*(2), 143–150. PMID:15270347

Bauschlicher, C. W., & Ricca, A. (2004). Binding of NH3 to graphite and to a (9,0) carbon nanotube. *Physical Review. Part B*, *70*(11), 115409–115413. doi:10.1103/PhysRevB.70.115409

Bayramoglu, G., Kunduzcu, G., & Arica, M. Y. (2020). Preparation and characterization of strong cation exchange terpolymer resin as effective adsorbent for removal of disperse dyes. *Polymer Engineering and Science*, *60*(1), 192–201. doi:10.1002/pen.25272

Behjati, S., Sheibani, S., Herritsch, J., & Gottfried, J. M. (2020). Photodegradation of dyes in batch and continuous reactors by Cu_2O-CuO nano-photocatalyst on Cu foils prepared by chemical-thermal oxidation. *Materials Research Bulletin*, *130*, 110920. doi:10.1016/j.materresbull.2020.110920

Bekele, E. T., Gonfa, B. A., Zelekew, O. A., Belay, H. H., & Sabir, F. K. (2020). Synthesis of Titanium Oxide Nanoparticles Using Root Extract of Kniphofia foliosa as a Template, Characterization, and Its Application on Drug Resistance Bacteria. *Journal of Nanomaterials*. . doi:10.1155/2020/2817037

Bell, J., Plumb, J. J., Buckley, C. A., & Stuckey, D. C. (2000). Treatment and decolorization of dyes in an anaerobic baffled reactor. *Journal of the Environmental Engineering Division*, *126*, 1026–1032.

Benkhaya, S., M'rabet, S., & El Harfi, A. (2020). A review on classifications, recent synthesis and applications of textile dyes. *Inorganic Chemistry Communications*, *115*, 107891.

Bera, R., Kundu, S., & Patra, A. (2015). 2D Hybrid Nanostructure of Reduced Graphene Oxide–CdS Nanosheet for Enhanced Photocatalysis. *ACS Applied Materials & Interfaces*, *7*(24), 13251–13259. doi:10.1021/acsami.5b03800 PMID:26029992

Berber, M. R. (2020). Current advances of polymer composites for water treatment and desalination. *Journal of Chemistry*.

Berglund, S. P., Rettie, A. J. E., Hoang, S., & Mullins, C. B. (2012). Incorporation of Mo and W into nanostructured $BiVO_4$ films for efficient photoelectrochemical water oxidation. *Physical Chemistry Chemical Physics*, *14*(19), 7065–7075. doi:10.1039/c2cp40807d PMID:22466715

Bharagava, R. N., & Chandra, R. (2010). Biodegradation of the major color containing compounds in distillery wastewater by an aerobic bacterial culture and characterization of their metabolites. *Biodegradation*, *21*(5), 703–711. doi:10.100710532-010-9336-1 PMID:20146090

Bharagava, R. N., Saxena, G., Mulla, S. I., & Patel, D. K. (2018). Characterization and identification of recalcitrant organic pollutants (ROPs) in tannery wastewater and its phytotoxicity evaluation for environmental safety. *Archives of Environmental Contamination and Toxicology*, *75*(2), 259–272. doi:10.100700244-017-0490-x PMID:29243159

Bhatia, D., Datta, D., Joshi, A., Gupta, S., & Gote, Y. (2019). Adsorption of isonicotinic acid from aqueous solution using multi-walled carbon nanotubes/Fe_3O_4. *Journal of Molecular Liquids*, *276*, 163–169. doi:10.1016/j.molliq.2018.11.127

Bhatnagar, A., & Sillanpää, M. (2009). Applications of chitin-and chitosan-derivatives for the detoxification of water and wastewater—A short review. *Advances in Colloid and Interface Science*, *152*(1-2), 26–38.

Bhat, S. S. M., Lee, S. A., Suh, J. M., Hong, S. P., & Jang, H. W. (2018). Triple planar hetero-junction of SnO_2/WO_3/$BiVO_4$ with enhanced photoelectrochemical performance under front illumination. *Applied Sciences (Basel, Switzerland)*, *8*(10), 1765. doi:10.3390/app8101765

Bhuiyan, M., Miah, M. Y., Paul, S. C., Aka, T. D., Saha, O., Rahaman, M. M., Sharif, M., Habiba, O., & Ashaduzzaman, M. (2020). Green synthesis of iron oxide nanoparticle using Carica papaya leaf extract: application for photocatalytic degradation of remazol yellow RR dye and antibacterial activity. *Heliyon*, *6*(8). doi:10.1016/j.heliyon.2020.e04603

Bhuyan, T., Mishra, K., Khanuja, M., Prasad, R., & Varma, A. (2015). Biosynthesis of zinc oxide nanoparticles from Azadirachta indica for antibacterial and photocatalytic applications. *Materials Science in Semiconductor Processing*, *32*, 55–61. doi:10.1016/j.mssp.2014.12.053

Bian, Z. Y., Zhu, Y. Q., Zhang, J. X., Ding, A. Z., & Wang, H. (2014). Visible-light driven degradation of ibuprofen using abundant metal-loaded $BiVO_4$ photocatalysts. *Chemosphere*, *117*, 527–531. doi:10.1016/j.chemosphere.2014.09.017 PMID:25268078

Bibi, I., Nazar, N., Ata, S., Sultan, M., Ali, A., Abbas, A., Jilani, K., Kamal, S., Sarim, F. M., Khan, M. I., Jalal, F., & Iqbal, M. (2019). Green synthesis of iron oxide nanoparticles using pomegranate seeds extract and photocatalytic activity evaluation for the degradation of textile dye. *Journal of Materials Research and Technology*, *8*(6), 6115–6124. doi:10.1016/j.jmrt.2019.10.006

Bienfait, M., Zeppenfeld, P., Dupont-Pavlovsky, N. D., Muris, M., Johnson, M. R., Wilson, T., DePies, M., & Vilches, O. E. (2004). Thermodynamics and structure of hydrogen, methane, argon, oxygen, and carbon dioxide adsorbed on single-wall carbon nanotube bundles. *Physical Review. Part B*, *70*(3), 035410–035419. doi:10.1103/PhysRevB.70.035410

Biliuta, G., & Coseri, S. (2019). Cellulose: A ubiquitous platform for ecofriendly metal nanoparticles preparation. *Coordination Chemistry Reviews*, *383*, 155–173.

Bitton, G. (2005). *Wastewater Microbiology*. Wiley-Liss. doi:10.1002/0471717967

Boddu, V. M., Abburi, K., Talbott, J. L., & Smith, E. D. (2003). Removal of hexavalent chromium from wastewater using a new composite chitosan biosorbent. *Environmental Science & Technology*, *37*(19), 4449–4456. doi:10.1021/es021013a PMID:14572099

Bogue, R. W. (2004). Nanotechnology: What are the prospects for sensors? *Sensor Review*, *24*(3), 253–260. https://doi.org/10.1108/02602280410545362

Bohonak, D. M., & Zydney, A. L. (2005). Compaction and permeability effects with virus filtration membranes. *Journal of Membrane Science*, *254*(1–2), 71–79. https://doi.org/10.1016/j.memsci.2004.12.035

Bolan, N., Kunhikrishnan, A., Thangarajan, R., Kumpiene, J., Park, J., Makino, T., ... Scheckel, K. (2014). Remediation of heavy metal (loid) s contaminated soils–to mobilize or to immobilize? *Journal of Hazardous Materials*, *266*, 141–166.

Bondavalli, P., Legagneux, P., & Pribat, D. (2009). Carbon nanotubes based transistors as gas sensors: State of the art and critical review. *Sensors and Actuators. Part B*, *140*(1), 304–318. doi:10.1016/j.snb.2009.04.025

Booshehri, A. Y., Goh, S. C. K., Hong, J., Jiang, R., & Xu, R. (2014). Effect of depositing silver nanoparticles on $BiVO_4$ in enhancing visible light photocatalytic inactivation of bacteria in water. *Journal of Materials Chemistry. A, Materials for Energy and Sustainability*, *2*(17), 6209–6217. doi:10.1039/C3TA15392D

Borges, M. E., Sierra, M., Cuevas, E., García, R. D., & Esparza, P. (2016). Photocatalysis with solar energy: Sunlight-responsive photocatalyst based on TiO_2 loaded on a natural material for wastewater treatment. *Solar Energy*, *135*, 527–535. doi:10.1016/j.solener.2016.06.022

Borgohain, K., & Mahamuni, S. (2002). Formation of single-phase CuO quantum particles. *Journal of Materials Research*, *17*(5), 1220–1223. doi:10.1557/JMR.2002.0180

Borsagli, F. G. M., Mansur, A. A., Chagas, P., Oliveira, L. C., & Mansur, H. S. (2015). O-carboxymethyl functionalization of chitosan: Complexation and adsorption of Cd (II) and Cr (VI) as heavy metal pollutant ions. *Reactive & Functional Polymers*, *97*, 37–47.

Borstlap, A. C., & Schuurmans, J. (1988). Kinetics of L-valine uptake in tobacco leaf discs. Comparison of wild-type, the digenic mutant Valr-2, and its monogenic derivatives. *Planta*, *176*(1), 42–50. doi:10.1007/BF00392478 PMID:24220733

Bozetine, H., Wang, Q., Barras, A., Li, M., Hadjersi, T., Szunerits, S., & Boukherroub, R. (2016). Green chemistry approach for the synthesis of ZnO–carbon dots nanocomposites with good photocatalytic properties under visible light. *Journal of Colloid and Interface Science*, *465*, 286–294. doi:10.1016/j.jcis.2015.12.001 PMID:26674245

Bradder, P., Ling, S. K., Wang, S., & Liu, S. (2011). Dye adsorption on layered graphite oxide. *Journal of Chemical & Engineering Data*, *56*(1), 138–141. https://doi.org/10.1021/je101049g

Brady-Estévez, A. S., Kang, S., & Elimelech, M. (2008). A single-walled-carbon-nanotube filter for removal of viral and bacterial pathogens. *Small*, *4*(4), 481–484. https://doi.org/10.1002/smll.200700863

Brechignac, C. P., & Houdy, M. L. (2007). *Nanomaterials and Nanochemistry*. Springer. doi:10.1007/978-3-662-47314-6

Briggs, A. M., Cross, M. J., Hoy, D. G., Blyth, F. H., Woolf, A. D., & March, L. (2016). Musculoskeletal Health Conditions Represent a Global Threat to Healthy Aging: A Report for the 2015World Health Organization World Report on Ageing and Health. *The Gerontologist*, *56*(Suppl 2), 243–255. doi:10.1093/geront/gnw002 PMID:26994264

Brunet, L., Lyon, D. Y., Zodrow, K., Rouch, J.-C., Caussat, B., Serp, P., Remigy, J.-C., Wiesner, M. R., & Alvarez, P. J. J. (2008). Properties of membranes containing semi-dispersed carbon nanotubes. *Environmental Engineering Science*, *25*(4), 565–576. https://doi.org/10.1089/ees.2007.0076

Cai, T., Wang, L., Liu, Y., Zhang, S., Dong, W., Chen, H., & Luo, S. (2018). Ag3PO4/Ti3C2 MXene interface materials as a Schottky catalyst with enhanced photocatalytic activities and anti-photocorrosion performance. *Applied Catalysis B: Environmental*, *239*, 545–554. doi:10.1016/j.apcatb.2018.08.053

Calderón, J., Navarro, M. E., Jimenez-Capdeville, M. E., Santos-Diaz, M. A., Golden, A., Rodriguez-Leyva, I., Borja-Aburto, V., & Díaz-Barriga, F. (2001). Exposure to arsenic and lead and neuropsychological development in Mexican children. *Environmental Research*, *85*(2), 69–76. https://doi.org/10.1006/enrs.2000.4106

Campbell, E. E. B., Fowler, P. W., Mitchell, D., & Zerbetto, F. (1996). Increasing cost of pentagon adjacency for larger fullerenes. *Chemical Physics Letters*, *250*(5–6), 544–548. https://doi.org/10.1016/0009-2614(96)00055-3

Cantalini, C., Valentini, L., Lozzi, L., Armentano, I., Kenny, J. M., & Santucci, S. (2003). NO2 gas sensitivity of carbon nanotubes obtained by plasma enhanced chemical vapor deposition. *Sensors and Actuators. Part B*, *93*(1–3), 333–337. doi:10.1016/S0925-4005(03)00224-7

Cao, S., Shen, B., Tong, T., Fu, J., & Yu, J. (2018). 2D/2D Heterojunction of Ultrathin MXene/Bi2WO6 Nanosheets for Improved Photocatalytic CO2 Reduction. *Advanced Functional Materials*, *28*(21), 1800136. doi:10.1002/adfm.201800136

Cao, W.-T., Chen, F.-F., Zhu, Y.-J., Zhang, Y.-G., Jiang, Y.-Y., Ma, M.-G., & Chen, F. (2018). Binary Strengthening and Toughening of MXene/Cellulose Nanofiber Composite Paper with Nacre-Inspired Structure and Superior Electromagnetic Interference Shielding Properties. *ACS Nano*, *12*(5), 4583–4593. doi:10.1021/acsnano.8b00997 PMID:29709183

Capek, I. (2006). Chapter 1. In Nanotechnology and nanomaterials (pp. 1–69). Elsevier.

Carabineiro, S. A. C., Thavorn-Amornsri, T., Pereira, M. F. R., & Figueiredo, J. L. (2011). Adsorption of ciprofloxacin on surface-modified carbon materials. *Water Research*, *45*(15), 4583–4591. doi:10.1016/j.watres.2011.06.008 PMID:21733541

Carneiro, P. A., Umbuzeiro, G. A., Oliveira, D. P., & Zanoni, M. V. B. (2010). Assessment of water contamination caused by a mutagenic textile effluent/dyehouse effluent bearing disperse dyes. *Journal of Hazardous Materials*, *174*(1–3), 694–699. https://doi.org/10.1016/j.jhazmat.2009.09.106

Carolin, C. F., Kumar, P. S., Saravanan, A., Joshiba, G. J., & Naushad, M. (2017). Efficient techniques for the removal of toxic heavy metals from aquatic environment: A review. *Journal of Environmental Chemical Engineering*, *5*(3), 2782–2799. doi:10.1016/j.jece.2017.05.029

Carpenter, A. W., de Lannoy, C. F., & Wiesner, M. R. (2015). Cellulose nanomaterials in water treatment technologies. *Environmental Science & Technology*, *49*(9), 5277–5528.

Celik, E., Park, H., Choi, H., & Choi, H. (2011). Carbon nanotube blended polyethersulfone membranes for fouling control in water treatment. *Water Research*, *45*(1), 274–282. https://doi.org/10.1016/j.watres.2010.07.060

Chai, B., Liu, C., Yan, J., Ren, Z., & Wang, Z. (2018). In-situ synthesis of WO3 nanoplates anchored on g-C3N4 Z-scheme photocatalysts for significantly enhanced photocatalytic activity. *Applied Surface Science*, *448*, 1–8. doi:10.1016/j.apsusc.2018.04.116

Chala, S., Wetchakun, K., Phanichphant, S., Inceesungvorn, B., & Wetchakun, N. (2014). Enhanced visible-light-response photocatalytic degradation of methylene blue on Fe-loaded $BiVO_4$ photocatalyst. *Journal of Alloys and Compounds*, *597*, 129–135. doi:10.1016/j.jallcom.2014.01.130

Chandran, S. P., Chaudhary, M., Pasricha, R., Ahmad, A., & Sastry, M. (2006). Synthesis of gold nanotriangles and silver nanoparticles using Aloe vera plant extract. *Biotechnology Progress*, *22*(2), 577–583. doi:10.1021/bp0501423 PMID:16599579

Chandra, V., Park, J., Chun, Y., Lee, J. W., Hwang, I. C., & Kim, K. S. (2010). Water-dispersible magnetite-reduced graphene oxide composites for arsenic removal. *ACS Nano*, *4*(7), 3979–3986. https://doi.org/10.1021/nn1008897

Chang, Y., & Zeng, H. C. (2004). Controlled synthesis and self-assembly of single-crystalline CuO Nanorods and Nanoribbons. *Crystal Growth & Design*, *4*(2), 397–402. doi:10.1021/cg034127m

Chaudhary, L. B., Rana, T. S., & Anand, K. K. (2008). Current status of the systematics of Astragalus L. (Fabaceae) with special reference to the Himalayan species in India. *Taiwania, 53*, 338–355.

Che, G., Lakshmi, B. B., Fisher, E. R., & Martin, C. R. (1998). Carbon nanotubule membranes for electrochemical energy storage and production. *Nature, 393*(6683), 346–349. https://doi.org/10.1038/30694

Chen, C. C., & Lu, C. S. (2007). Photocatalytic degradation of Basic violet 4: Degradation efficiency, product distribution, and mechanisms. *The Journal of Physical Chemistry C, 111*(37), 13922–13932. doi:10.1021/jp0738964

Chen, C., Li, X., Zhao, D., Tan, X., & Wang, X. (2007). Adsorption kinetic, thermodynamic and desorption studies of Th(IV) on oxidized multi-wall carbon nanotubes. *Colloids and Surfaces. A, Physicochemical and Engineering Aspects, 302*(1–3), 449–454. https://doi.org/10.1016/j.colsurfa.2007.03.007

Chen, C., & Wang, X. (2006). Adsorption of Ni(II) from aqueous solution using oxidized multiwall carbon nanotubes. *Industrial & Engineering Chemistry Research, 45*(26), 9144–9149. https://doi.org/10.1021/ie060791z

Chen, D., & Ray, A. K. (1999). Photocatalytic kinetics of phenol and its derivatives over UV irradiated TiO_2. *Applied Catalysis B: Environmental, 23*(2–3), 143–157. doi:10.1016/S0926-3373(99)00068-5

Chen, F., Tang, Y., Liu, C., Qian, J., Wu, Z., & Chen, Z. (2017). Synthesis of porous structured ZnO/Ag composite fibers with enhanced photocatalytic performance under visible irradiation. *Ceramics International, 43*(16), 14525–14528. doi:10.1016/j.ceramint.2017.07.158

Cheng, H.-M., Yang, Q.-H., & Liu, C. (2001). Hydrogen storage in carbon nanotubes. *Carbon, 39*(10), 1447–1454. https://doi.org/10.1016/S0008-6223(00)00306-7

Cheng, J., Feng, J., & Pan, W. (2015). Enhanced Photocatalytic Activity in Electrospun Bismuth Vanadate Nano-fibers with Phase Junction. *ACS Applied Materials & Interfaces, 7*(18), 9638–9644. doi:10.1021/acsami.5b01305 PMID:25856118

Cheng, X., Zu, L., Jiang, Y., Shi, D., Cai, X., Ni, Y., & Qin, Y. (2018). A titanium-based photo-Fenton bifunctional catalyst of mp-MXene/TiO2−x nanodots for dramatic enhancement of catalytic efficiency in advanced oxidation processes. *Chemical Communications, 54*(82), 11622–11625. doi:10.1039/C8CC05866K PMID:30264081

Chen, L., He, F., Zhao, N., & Guo, R. (2017). Fabrication of 3D quasi-hierarchical Z-scheme RGO-Fe2O3-MoS2 nanoheterostructures for highly enhanced visible-light-driven photocatalytic degradation. *Applied Surface Science, 420*, 669–680. doi:10.1016/j.apsusc.2017.05.099

Chen, L., Huang, R., Ma, Y. J., Luo, S. L., Au, C. T., & Yin, S. F. (2013). Controllable synthesis of hollow and porous Ag/$BiVO_4$ composites with enhanced visible-light photocatalytic performance. *RSC Advances, 3*(46), 24354–24361. doi:10.1039/c3ra43691h

Chen, P., Wu, X., Lin, J., & Tan, K. L. (1999). High H2 uptake by alkali-doped carbon nanotubes under ambient pressure and moderate temperatures. *Science, 285*(5424), 91–93. https://doi.org/10.1126/science.285.5424.91

Chen, X., Li, L., Yi, T., Zhang, W. Z., Zhang, X., & Wang, L. (2015). Microwave assisted synthesis of sheet-like Cu/$BiVO_4$ and its activities of various photocatalytic conditions. *Journal of Solid State Chemistry, 229*, 141–149. doi:10.1016/j.jssc.2015.05.026

Chen, X., & Mao, S. S. (2007). Titanium dioxide nanomaterials: Synthesis, properties, modifications, and applications. *Chemical Reviews, 107*(7), 2891–2959. doi:10.1021/cr0500535 PMID:17590053

Chen, X., Xia, X., Wang, X., Qiao, J., & Chen, H. (2011). A comparative study on sorption of perfluorooctane sulfonate (PFOS) by chars, ash and carbon nanotubes. *Chemosphere, 83*(10), 1313–1319. doi:10.1016/j.chemosphere.2011.04.018 PMID:21531440

Chen, Z., Pronkin, S., Fellinger, T.-P., Kailasam, K., Vilé, G., Albani, D., Krumeich, F., Leary, R., Barnard, J., Thomas, J. M., Pérez-Ramírez, J., Antonietti, M., & Dontsova, D. (2016). Merging Single-Atom-Dispersed Silver and Carbon Nitride to a Joint Electronic System via Copolymerization with Silver Tricyanomethanide. *ACS Nano*, *10*(3), 3166–3175. doi:10.1021/acsnano.5b04210 PMID:26863408

Cheremisinoff, N. P. (2002). *Handbook of Water and Wastewater Treatment Technologies*. Butterworth-Heinemann.

Chhabra, T., Kumar, A., Bahuguna, A., & Krishnan, V. (2019). Reduced graphene oxide supported MnO2 nanorods as recyclable and efficient adsorptive photocatalysts for pollutants removal. *Vacuum*, *160*, 333–346. doi:10.1016/j.vacuum.2018.11.053

Cho, W. S., Moon, S. I., Paek, K. K., Lee, Y. H., Park, J. H., & Ju, B. K. (2006). Patterned multiwall carbon nanotube films as materials of NO2 gas sensors. *Sensors and Actuators. Part B*, *119*(1), 180–185. doi:10.1016/j.snb.2005.12.004

Cho, H.-H., Huang, H., & Schwab, K. (2011). Effects of solution chemistry on the adsorption of ibuprofen and triclosan onto carbon nanotubes. *Langmuir*, *27*(21), 12960–12967. doi:10.1021/la202459g PMID:21913654

Choo, K.-H. (2018). Modeling Photocatalytic Membrane Reactors. In A. Basile, S. Mozia, & R. Molinari (Eds.), *Current Trends and Future Developments on (Bio-) Membranes* (pp. 297–316). Elsevier. doi:10.1016/B978-0-12-813549-5.00010-4

Cho, S. K., Park, H. S., Lee, H. C., Nam, K. M., & Bard, A. J. (2013). Metal doping of $BiVO_4$ by composite electrodeposition with improved photoelectrochemical water oxidation. *The Journal of Physical Chemistry C*, *117*(44), 23048–23056. doi:10.1021/jp408619u

Chowdhury, S., & Balasubramanian, R. (2014). Article. *Applied Catalysis B: Environmental*, *160–161*, 307–324.

Clarke, E., & Anliker, R. (1980). Organic Dyes and Pigments. The Handbook of Environmental Chemistry, 3, 181-215.

Clausen, C. A. (2000). Isolating metal-tolerant bacteria capable of removing copper, chromium, and arsenic from treated wood. *Waste Management & Research*, *18*(3), 264–268. doi:10.1177/0734242X0001800308

Clifford, D. M., Castano, C. E., & Rojas, J. V. (2017). Supported transition metal nanomaterials: Nanocomposites synthesized by ionizing radiation. *Radiation Physics and Chemistry*, *132*, 52–64. https://doi.org/10.1016/j.radphyschem.2016.12.001

Cohen, S. G. (1947). Encyclopedia of chemical technology, I. Journal of Polymer Science, 3(3), 463.

Collins, P. G., Bradley, K., Ishigami, M., & Zettl, A. (2000). Extreme oxygen sensitivity of electronic properties of carbon nanotubes. *Science*, *287*(5459), 1801–1804. https://doi.org/10.1126/science.287.5459.1801

Collins, P. G., Zettl, A., Bando, H., Thess, A., & Smalley, R. E. (1997). Nanotube nanodevice. *Science*, *278*(5335), 100–102. https://doi.org/10.1126/science.278.5335.100

Córdoba, R., Ibarra, A., Mailly, D., & De Teresa, J. M. (2018). Vertical Growth of Superconducting Crystalline Hollow Nanowires by He+ Focused Ion Beam Induced Deposition. *Nano Letters*, *18*(2), 1379–1386. doi:10.1021/acs.nanolett.7b05103 PMID:29357248

Cortes, P., Deng, S., & Smith, G. B. (2009). The adsorption properties of bacillus atrophaeus spores on singlewall carbon nanotubes. *Journal of Sensors*, *2009*, 1–6. https://doi.org/10.1155/2009/131628

Coupal, B., & Lalancette, J. M. (1976). The treatment of waste waters with peat moss. *Water Research*, *10*(12), 1071–1076. doi:10.1016/0043-1354(76)90038-5

Crabtree, G. W., Dresselhaus, M. S., & Buchanan, M. V. (2004). The hydrogen economy. *Physics Today*, *57*(12), 39–45. doi:10.1063/1.1878333

Cravanzola, S., Cesano, F., Magnacca, G., Zecchina, A., & Scarano, D. (2016). Designing rGO/MoS2 hybrid nanostructures for photocatalytic applications. *RSC Advances*, *6*(64), 59001–59008. doi:10.1039/C6RA08633K

Crini, G. (2005). Recent developments in polysaccharide-based materials used as adsorbents in wastewater treatment. *Progress in Polymer Science*, *30*(1), 38–70. doi:10.1016/j.progpolymsci.2004.11.002

Crini, G., & Lichtfouse, E. (2019). Advantages and disadvantages of techniques used for wastewater treatment. *Environmental Chemistry Letters*, *17*(1), 145–155. doi:10.100710311-018-0785-9

Crini, G., & Lichtfouse, E. (2019). Advantages and disadvantages of techniques used in wastewater treatment. *Environmental Chemistry Letters*, *17*, 145–155.

Crini, G., Lichtfouse, E., Wilson, L. D., & Morin-Crini, N. (2019). Conventional and non-conventional adsorbents for wastewater treatment. *Environmental Chemistry Letters*, *17*(1), 195–213. doi:10.100710311-018-0786-8

Cruz, A. M., & Pérez, U. M. G. (2010). Photocatalytic properties of $BiVO_4$ prepared by the co-precipitation method: Degradation of rhodamine B and possible reaction mechanisms under visible irradiation. *Materials Research Bulletin*, *45*(2), 135–141. doi:10.1016/j.materresbull.2009.09.029

Cui, C., Guo, R., Xiao, H., Ren, E., Song, Q., Xiang, C., & Jiang, S. (2020). Bi2WO6/Nb2CTx MXene hybrid nanosheets with enhanced visible-light-driven photocatalytic activity for organic pollutants degradation. *Applied Surface Science*, *505*, 144595. doi:10.1016/j.apsusc.2019.144595

Cullity, B. D., & Stock, S. R. (2001). Elements of X-ray diffraction (3rd ed.). Prentice-Hall, Inc.

Dadigala, R., Bandi, R., Gangapuram, B. R., Dasari, A., Belay, H. H., & Guttena, V. (2019). Fabrication of novel 1D/2D V2O5/g-C3N4 composites as Z-scheme photocatalysts for CR degradation and Cr (VI) reduction under sunlight irradiation. *Journal of Environmental Chemical Engineering*, *7*(1), 102822. doi:10.1016/j.jece.2018.102822

Dagher, S., Haik, Y., Ayesh, A.I., & Tit, N. (2014). Synthesis and optical properties of colloidal CuO nanoparticles. *J. Lumin.*, *151*, 149–154. .02.015 doi:10.1016/j.jlumin.2014

Dai, L., & Mau, A. W. H. (2001). Controlled synthesis and modification of carbon nanotubes and C60: Carbon nanostructures for advanced polymeric composite materials. *Advanced Materials*, *13*(12–13), 899–913. https://doi.org/10.1002/1521-4095(200107)13:12/13<899::AID-ADMA899>3.0.CO;2-G

Dalton, A. B., Collins, S., Muñoz, E., Razal, J. M., Ebron, V. H., Ferraris, J. P., Coleman, J. N., Kim, B. G., & Baughman, R. H. (2003). Super– tough carbon-nanotube fibres. *Nature*, *423*(6941), 703. https://doi.org/10.1038/423703a

Dalton, A. B., Collins, S., Muñoz, E., Razal, J. M., Ebron, V. H., Ferraris, J. P., Coleman, J. N., Kim, B. G., & Baughman, R. H. (2003). Super-tough carbonnanotube fibers. *Nature*, *423*(6941), 703. https://doi.org/10.1038/423703a

Darwent, J. R., & Mills, A. (1982). Photo-oxidation of water sensitized by WO_3powder. *Journal of the Chemical Society. Faraday Transactions II*, *78*(2), 359–367. doi:10.1039/f29827800359

Das, R. K., Gogoi, N., & Bora, U. (2011). Green synthesis of gold nanoparticles using Nyctanthes arbortristis flower extract. Bioprocess. *Biosystems Engineering*, *34*, 615–619. PMID:21229266

Das, R., Leo, B. F., & Murphy, F. (2018). The Toxic Truth About Carbon Nanotubes in Water Purification: A Perspective View. *Nanoscale Research Letters*, *13*(1), 183. doi:10.118611671-018-2589-z PMID:29915874

Das, R., Vecitis, C. D., Schulze, A., Cao, B., Ismail, A. F., Lu, X., & Ramakrishna, S. (2017). Recent advances in nanomaterials for water protection and monitoring. *Chemical Society Reviews*, *46*(22), 6946–7020. doi:10.1039/C6CS00921B PMID:28959815

Das, T. K., & Poater, A. (2021). Review on the use of heavy metal deposits from water treatment waste towards catalytic chemical syntheses. *International Journal of Molecular Sciences*, *22*(24), 13383. https://doi.org/10.3390/ijms222413383

de Heer, W. A. D., Châtelain, A., & Ugarte, D. (1995). A carbon nanotube fieldemission electron source. *Science*, *270*(5239), 1179–1180. https://doi.org/10.1126/science.270.5239.1179

De la Rosa, G., Gardea-Torresdey, J. L., Peralta-Videa, J. R., Herrera, I., & Contreras, C. (2003). Use of silica-immobilized humin for heavy metal removal from aqueous solution under flow conditions. *Bioresource Technology*, *90*(1), 11–17. doi:10.1016/S0960-8524(03)00099-3 PMID:12835051

Delnavaz, M., Farahbakhsh, J., & Mahdian, S. S. (2021). Photodegradation of reactive blue 19 dye using magnetic nanophotocatalyst α-Fe_2O_3/WO_3: A comparison study of α-Fe_2O_3/WO_3 and WO_3/NaOH. *Water Science and Engineering*, *14*(2), 119–128. doi:10.1016/j.wse.2021.06.007

Demirkıran, N., Turhan Özdemir, G. D. T., Saraç, M., & Dardağan, M. (2017). Adsorption of methylene blue from aqueous solutions by pyrolusite ore. *Mongolian Journal of Chemistry*, *18*(44), 5–11. https://doi.org/10.5564/mjc.v18i44.880

Deng, D., Novoselov, K. S., Fu, Q., Zheng, N., Tian, Z., & Bao, X. (2016). Catalysis with two-dimensional materials and their heterostructures. *Nature Nanotechnology*, *11*(3), 218–230. doi:10.1038/nnano.2015.340 PMID:26936816

Deng, F., Li, Y., Luo, X., Yang, L., & Tu, X. (2012). Preparation of conductive polypyrrole/TiO2 nanocomposite via surface molecular imprinting technique and its photocatalytic activity under simulated solar light irradiation. *Colloids and Surfaces. A, Physicochemical and Engineering Aspects*, *395*, 183–189. doi:10.1016/j.colsurfa.2011.12.029

Deng, F., Min, L., Luo, X., Wu, S., & Luo, S. (2013). Visible-light photocatalytic degradation performances and thermal stability due to the synergetic effect of TiO2 with conductive copolymers of polyaniline and polypyrrole. *Nanoscale*, *5*(18), 8703–8710. doi:10.1039/c3nr02502k PMID:23900296

Deng, J., Shao, Y., Gao, N., Deng, Y., Tan, C., Zhou, S., & Hu, X. (2012). Multiwalled carbon nanotubes as adsorbents for removal of herbicide diuron from aqueous solution. *Chemical Engineering Journal*, *193–194*, 339–347. doi:10.1016/j.cej.2012.04.051

Deng, S., Yang, Z., Lv, G., Zhu, Y., Li, H., Wang, F., & Zhang, X. (2019). WO3 nanosheets/g-C3N4 nanosheets' nanocomposite as an effective photocatalyst for degradation of rhodamine B. *Applied Physics. A, Materials Science & Processing*, *125*(1), 44. doi:10.100700339-018-2331-9

Deng, X., Lü, L., Li, H., & Luo, F. (2010). The adsorption properties of Pb(II) and Cd(II) on functionalized graphene prepared by electrolysis method. *Journal of Hazardous Materials*, *183*(1–3), 923–930. https://doi.org/10.1016/j.jhazmat.2010.07.117

Devanesan, S., AlSalh, M. S., Balaji, R. V., Ranjitsingh, A. J. A., Ahamed, A., Alfuraydi, A. A., AlQahtani, F. Y., Aleanizy, F. S., & Othman, A. H. (2018). Antimicrobial and cytotoxicity e_ects of synthesized silver nanoparticles from Punica granatum peel extract. *Nanoscale Research Letters*, *13*(1), 315. doi:10.118611671-018-2731-y PMID:30288618

Devanesan, S., Jayamala, M., AlSalhi, M. S., Umamaheshwari, S., & Ranjitsingh, A. J. A. (2021, May). Antimicrobial and anticancer properties of Carica papaya leaves derived di-methyl flubendazole mediated silver nanoparticles. *Journal of Infection and Public Health*, *14*(5), 577–587. doi:10.1016/j.jiph.2021.02.004 PMID:33848887

Devatha, C. P., Thalla, A., & Katte, S. (2016). Green synthesis of iron nanoparticles using different leaf extracts for treatment of domestic wastewater. *Journal of Cleaner Production*, *139*, 1425–1435. Advance online publication. doi:10.1016/j.jclepro.2016.09.019

Devi, H. S., & Singh, T. D. (2014). Synthesis of copper oxide nanoparticles by a novel method and its application in the degradation of methyl orange. *Adv Electron Electr Eng.*, *4*, 83–88.

Dhatshanamurthi, P., Shanthi, M., & Swaminathan, M. (2017). An efficient pilot scale solar treatment method for dye industry effluent using nano-ZnO. *Journal of Water Process Engineering*, *16*, 28–34. doi:10.1016/j.jwpe.2016.12.002

Dhivya, A., & Yadav, R. (2022). An Eco-approach synthesis of undoped and Mn doped ZnO nano-photocatalyst for prompt decoloration of methylene blue dye. *Materials Today: Proceedings*, *48*, 494–501. doi:10.1016/j.matpr.2021.02.751

Di Francia, G., Alfano, B., & La Ferrara, V. (2009). Conductometric gas nanosensors. *Journal of Sensors*, *2009*, 1–18. https://doi.org/10.1155/2009/659275

Di, J., Xiong, J., Li, H., & Liu, Z. (2018). Ultrathin 2D Photocatalysts: Electronic-Structure Tailoring, Hybridization, and Applications. *Advanced Materials*, *30*(1), 1704548. doi:10.1002/adma.201704548 PMID:29178550

Dillon, A. C., Jones, K. M., Bekkedahl, T. A., Kiang, C. H., Bethune, D. S., & Heben, M. J. (1997). storage of hydrogen in single walled carbon nanotubes. *Nature*, *386*(6623), 377–379. https://doi.org/10.1038/386377a0

Ding, X., Li, Y., Li, C., Wang, W., Wang, L., Feng, L., & Han, D. (2019). 2D visible-light-driven TiO2@Ti3C2/g-C3N4 ternary heterostructure for high photocatalytic activity. *Journal of Materials Science*, *54*(13), 9385–9396. doi:10.100710853-018-03289-4

Ding, Y., Zhou, Y., Nie, W., & Chen, P. (2015). MoS2–GO nanocomposites synthesized via a hydrothermal hydrogel method for solar light photocatalytic degradation of methylene blue. *Applied Surface Science*, *357*, 1606–1612. doi:10.1016/j.apsusc.2015.10.030

Dino, J. (2008, March 29). *NASA Ames Research Center Public Affairs Office*. NASA. Retrieved from https://www.nasa.gov/centers/ames/news/releases/2002/02images/nanogear/nanogears.html#:~:text=Nanotechnology%20is%20the%20creation%20of,biological)%20at%20that%20length%20scale

Dixit, R., Malaviya, D., Pandiyan, K., Singh, U. B., Sahu, A., Shukla, R., ... Paul, D. (2015). Bioremediation of heavy metals from soil and aquatic environment: An overview of principles and criteria of fundamental processes. *Sustainability*, *7*(2), 2189–2212.

Doan, V. D., Luc, V. S., Nguyen, T. L. H., Nguyen, T. D., & Nguyen, T. D. (2020). Utilizing waste corn-cob in biosynthesis of noble metallic nanoparticles for antibacterial effect and catalytic degradation of contaminants. *Environmental Science and Pollution Research International*, *27*(6), 6148–6162. doi:10.100711356-019-07320-2 PMID:31863387

Dong, Q., Wang, J., Duan, X., Tan, X., Liu, S., & Wang, S. (2019). Self-assembly of 3D MnO2/N-doped graphene hybrid aerogel for catalytic degradation of water pollutants: Structure-dependent activity. *Chemical Engineering Journal*, *369*, 1049–1058. doi:10.1016/j.cej.2019.03.139

Douafer, S., Lahmar, H., Benamira, M., Messaadia, L., Mazouzi, D., & Trari, M. (2019). Chromate reduction on the novel hetero-system $LiMn_2O_4/SnO_2$ catalyst under solar light irradiation. *Surfaces and Interfaces*, *17*, 100372. doi:10.1016/j.surfin.2019.100372

Dreaden, E. C., Alkilany, A. M., Huang, X., Murphy, C. J., & El-Sayed, M. A. (2012). The golden age: Gold nanoparticles for biomedicine. *Chemical Society Reviews*, *41*(7), 2740–2779. doi:10.1039/C1CS15237H PMID:22109657

Dreyer, G., & Tillmanns, E. (1981). NeuesJahrb. Mineral. *Monatshe,* 151-154.

Drisya, K. T., López, M. S., Ramírez, J. J. R., Álvarez, J. C. D., Rousseau, A., Velumani, S., Asomoza, R., Kassiba, A., Jantrania, A., & Castaneda, H. (2020). Electronic and optical competence of $TiO_2/BiVO_4$ nanocomposites in the photocatalytic processes. *Scientific Reports*, *10*(1), 13507. doi:10.103841598-020-69032-9 PMID:32782289

Dubey, R. (2009). Microencapsulation technology and applications. *Defence Science Journal*, *59*(1), 82.

Dubey, S. P., Lahtinen, M., Särkkä, H., & Sillanpää, M. (2010). Bioprospecting of Sorbus aucuparia leaf extract in development of silver and gold nanocolloids. *Colloid Surf. B*, *80*(1), 26–33. doi:10.1016/j.colsurfb.2010.05.024 PMID:20620889

Duc Hoa, N., Van Quy, N., Suk Cho, Y., & Kim, D. (2007). Nanocomposite of SWNTs and SnO2 fabricated by soldering process for ammonia gas sensor application. *Physica Status Solidi*, *204*(6), 1820–1824. https://doi.org/10.1002/pssa.200675318

Du, C., Song, Y., Shi, S., Jiang, B., Yang, J., & Xiao, S. (2020). Preparation and characterization of a novel Fe3O4-graphene-biochar composite for crystal violet adsorption. *The Science of the Total Environment*, *711*, 134662. doi:10.1016/j.scitotenv.2019.134662 PMID:31831251

Dutta, S., Gupta, B., Srivastava, S. K., & Gupta, A. K. (2021). Recent advances on the removal of dyes from wastewater using various adsorbents: A critical review. *Materials Advances*, *2*, 4497–4531.

Ebrahimzadeh, M. A., Naghizadeh, A., Amiri, O., Shirzadi-Ahodashti, M., & Mortazavi-Derazkola, S. (2020). Green and facile synthesis of Ag nanoparticles using Crataegus pentagyna fruit extract (CP-AgNPs) for organic pollution dyes degradation and antibacterial application. *Bioorganic Chemistry*, *94*, 103425. doi:10.1016/j.bioorg.2019.103425 PMID:31740048

Eichhorn, S. J., Dufresne, A., & Aranguren, M. (2010). Review: Current international research into cellulose nanofibres and nanocomposites. *Journal of Materials Science*, *45*, 1–33.

Eisenberg, D., Ahn, H. S., & Bard, A. J. (2014). Enhancedphotoelectrochemical water oxidation on bismuth vanadate by electrodeposition of amorphous titanium dioxide. *Journal of the American Chemical Society*, *136*(40), 14011–14014. doi:10.1021/ja5082475 PMID:25243345

El Saeed, A. M., El-Fattah, M. A., & Azzam, A. M. (2015). Synthesis of ZnO nanoparticles and studying its influence on the antimicrobial, anticorrosion and mechanical behavior of polyurethane composite for surface coating. *Dyes and Pigments*, *121*, 282–289. doi:10.1016/j.dyepig.2015.05.037

El-Bindary, A., Ismail, A., & Eladl, E. F. (2019). Photocatalytic degradation of reactive blue 21 using Ag doped ZnO nanoparticles. *Journal of Materials and Environmental Science*, *10*, 1258–1271.

Elfeky, A. S., Salem, S. S., Elzaref, A. S., Owda, M. E., Eladawy, H. A., Saeed, A. M., Awad, M. A., Abou-Zeid, R. E., & Fouda, A. (2020). Multifunctional cellulose nanocrystal /metal oxide hybrid, photo-degradation, antibacterial and larvicidal activities. *Carbohydrate Polymers*, *230*, 115711. doi:10.1016/j.carbpol.2019.115711 PMID:31887890

El-Gendy, A., El-Shafie, A. S., Issa, A., Al-Meer, S., Al-Saad, K., & El-Azazy, M. (2020). Carbon-based materials (CBMS) for determination and remediation of antimicrobials in different substrates: Wastewater and infant foods as examples. In Carbon-Based Material for Environmental Protection and Remediation. IntechOpen.

Endo, M., Strano, M. S., & Ajayan, P. M. (2007). Potential applications of carbon nanotubes. *Topics in Applied Physics*, 13–62. doi:10.1007/978-3-540-72865-8_2

Endo, M., Takeuchi, K., Igarashi, S., Kobori, K., Shiraishi, M., & Kroto, H. W. (1993). The production and structure of pyrolytic carbon nanotubes (PCMTs). *Journal of Physics and Chemistry of Solids*, *54*(12), 1841–1848. doi:10.1016/0022-3697(93)90297-5

Engates, K., & Shipley, H. (2011). Adsorption of Pb, cd, cu, Zn, and Ni to titanium dioxide nanoparticles: Effect of particle size, solid concentration, and exhaustion. *Environmental Science and Pollution Research International*, *18*(3), 386–395. doi:10.100711356-010-0382-3 PMID:20694836

Epling, G. A., & Lin, C. (2002). Photoassisted bleaching of dyes utilizing TiO2 and visible light. *Chemosphere*, *46*(4), 561–570. doi:10.1016/S0045-6535(01)00173-4 PMID:11838435

Eriksson, P. (1988). Nanofiltration extends the range of membrane ultrafiltration. *Environment and Progress*, *7*, 58–62.

Ersan, G., Apul, O. G., Perreault, F., & Karanfil, T. (2017). Adsorption of organic contaminants by graphene nanosheets: A review. *Water Research*, *126*, 385–398. doi:10.1016/j.watres.2017.08.010 PMID:28987890

Esmailzadeh, H., Sangpour, P., Shahraz, F., Hejazi, J., & Khaksar, R. (2016). Effect of nanocomposite packaging containing ZnO on growth of Bacillus subtilis and Enterobacter aerogenes. *Materials Science and Engineering C*, *58*, 1058–1063. doi:10.1016/j.msec.2015.09.078 PMID:26478403

Espinosa, E. H., Ionescu, R., Chambon, B., Bedis, G., Sotter, E., Bittencourt, C., Felten, A., Pireaux, J. J., Correig, X., & Llobet, E. (2007). Hybrid metal oxide and multiwall carbon nanotube films for low temperature gas sensing. *Sensors and Actuators. Part B*, *127*(1), 137–142. doi:10.1016/j.snb.2007.07.108

Essam, J. (2012 April 12). *Microencapsulation methods* [PowerPoint slides]. https://www.slideshare.net/JehanEssam/microencapsulation-methods-12259395

Ezugbe, E. O., & Rathilal, S. (2020). Membrane technologies in wastewater treatment: A review. *Membranes*, •••, 10.

Fagan, S. B., Santos, E. J. G., Souza Filho, A. G., Mendes Filho, J., & Fazzio, A. (2007). Ab initio study of 2,3,7,8-tetrachlorinated dibenzo-p-dioxin adsorption on single wall carbon nanotubes. *Chemical Physics Letters*, *437*(1–3), 79–82. https://doi.org/10.1016/j.cplett.2007.01.071

Fang, H., Pan, Y., Yin, M., & Pan, C. (2020). Enhanced photocatalytic activity and mechanism of Ti3C2–OH/Bi2WO6:Yb3+, Tm3+ towards degradation of RhB under visible and near infrared light irradiation. *Materials Research Bulletin*, *121*, 110618. doi:10.1016/j.materresbull.2019.110618

Fang, H., Pan, Y., Yin, M., Xu, L., Zhu, Y., & Pan, C. (2019). Facile synthesis of ternary Ti3C2–OH/ln2S3/CdS composite with efficient adsorption and photocatalytic performance towards organic dyes. *Journal of Solid State Chemistry*, *280*, 120981. doi:10.1016/j.jssc.2019.120981

Fang, Y., Cao, Y., & Chen, Q. (2019). Synthesis of an Ag2WO4/Ti3C2 Schottky composite by electrostatic traction and its photocatalytic activity. *Ceramics International*, *45*(17, 17, Part A), 22298–22307. doi:10.1016/j.ceramint.2019.07.256

Fan, L., Harris, J. L., Roddick, F. A., & Booker, N. A. (2001). Influence of the characteristics of natural organic matter on the fouling of microfiltration membranes. *Water Research*, *35*(18), 4455–4463. https://doi.org/10.1016/s0043-1354(01)00183-x

Fan, Y., Wu, D., Zhang, S., Zhang, L., Hu, W., Zhu, C., & Gong, X. (2021). *Effective photodegradation of 4-nitrophenol with CuO nano particles prepared by ionic liquids/water system*. Green Chemical Engineering. doi:10.1016/j.gce.2021.07.009

Farhadian, M., Sangpour, P., & Hosseinzadeh, G. (2015). Morphology dependent photocatalytic activity of WO3 nanostructures. *Journal of Energy Chemistry*, *24*(2), 171–177. doi:10.1016/S2095-4956(15)60297-2

Fayazi, M. (2021). Preparation and characterization of carbon nanotubes/pyrite nanocomposite for degradation of methylene blue by heterogeneous Fenton reaction. *Journal of the Taiwan Institute of Chemical Engineers*, *120*, 229–235. doi:10.1016/j.jtice.2021.03.033

Feng, R., Lei, W., Sui, X., Liu, X., Qi, X., Tang, K., Liu, G., & Liu, M. (2018). Anchoring black phosphorus quantum dots on molybdenum disulfide nanosheets: A 0D/2D nanohybrid with enhanced visible–and NIR –light photoactivity. *Applied Catalysis B: Environmental*, *238*, 444–453. doi:10.1016/j.apcatb.2018.07.052

Feng, X., Sheng, N., Liu, Y., Chen, X., Chen, D., Yang, C., & Zhou, X. (2017). Simultaneously Enhanced Stability and Selectivity for Propene Epoxidation with H2 and O2 on Au Catalysts Supported on Nano-Crystalline Mesoporous TS-1. *ACS Catalysis*, *7*(4), 2668–2675. doi:10.1021/acscatal.6b03498

Ferreira, C. A., Domenech, S. C., & Lacaze, P. C. (2001). Synthesis and characterization of polypyrrole/TiO2 composites on mild steel. *Journal of Applied Electrochemistry*, *31*(1), 49–56. doi:10.1023/A:1004149421649

Feynman, R. P. (1960). There's plenty of room at the bottom. California Institute of Technology Caltech Engineering and Science, 23(5), 22–36.

Fibre2Fashion. (n.d.). *Microencapsulation of fragrances in textiles*. Fibre2Fashion. Retrieved from https://www.fibre2fashion.com/industry-article/6962/microencapsulation-of-fragrances-in-textiles

Fiol, N., Villaescusa, I., Martínez, M., Miralles, N., Poch, J., & Serarols, J. (2006). Sorption of Pb(II), Ni(II), Cu(II) and Cd(II) from aqueous solution by olive stone waste. *Separation and Purification Technology*, *50*(1), 132–140. https://doi.org/10.1016/j.seppur.2005.11.016

Fischback, M., Youn, J., Zhao, X., Wang, P., Park, H., Chang, H., Kim, J., & Ha, S. (2006). Miniature biofuel cells with improved stability under continuous operation. *Electroanalysis*, *18*(19–20), 2016–2022. https://doi.org/10.1002/elan.200603626

Foo, K. Y., & Hameed, B. H. (2010). An overview of dye removal via activated carbon adsorption process. *Desalination and Water Treatment*, *19*, 255–274.

Fornasiero, F., Park, H. G., Holt, J. K., Stadermann, M., Grigoropoulos, C. P., Noy, A., & Bakajin, O. (2008). Ion exclusion by sub-2-nm carbon nanotube pores. *Proceedings of the National Academy of Sciences of the United States of America*, *105*(45), 17250–17255. https://doi.org/10.1073/pnas.0710437105

Foster, H. A., Ditta, I. B., Varghese, S., & Steele, A. (2011). Photocatalytic disinfection using titanium dioxide: Spectrum and mechanism of antimicrobial activity. *Applied Microbiology and Biotechnology*, *90*(6), 1847–1868. doi:10.100700253-011-3213-7 PMID:21523480

Fouda, A., Abdel-Maksoud, G., Abdel-Rahman, M. A., Eid, A. M., Barghoth, M. G., & El-Sadany, M. A.-H. (2019). Monitoring the effect of biosynthesized nanoparticles against biodeterioration of cellulose-based materials by Aspergillus niger. *Cellulose (London, England)*, *26*(11), 6583–6597. doi:10.100710570-019-02574-y

Fouda, A., Salem, S. S., Wassel, A. R., Hamza, M. F., & Shaheen, T. I. (2020). Optimization of green biosynthesized visible light active CuO/ZnO nano-photocatalysts for the degradation of organic methylene blue dye. *Heliyon*, *6*(9), e04896. doi:10.1016/j.heliyon.2020.e04896 PMID:32995606

Fox, D. F., & Duxbury, D. F. (1993). The photochemistry and photophysics of triphenylmethane dyes in solid and liquid media. *Chemical Reviews*, *93*(1), 381–433. doi:10.1021/cr00017a018

Franco, A., Celma de Oliveira Lima, E., Novak, M. A., & Wells, P. R. Jr. (2007). Synthesis of nanoparticles of $Co_xFe_{(3-x)}O_4$ by combustion reaction method. *Journal of Magnetism and Magnetic Materials*, *308*(2), 198–202. doi:10.1016/j.jmmm.2006.05.022

Freeman, B. D. (1999). Basis of permeability/selectivity tradeoff relations in polymeric gas separation membranes. *Macromolecules*, *32*(2), 375–380. https://doi.org/10.1021/ma9814548

Friberg, L., Nordberg, G. F., & Vouk, V. B. (1979). *Handbook on the toxicology of metals*. Elsevier/North-Holland.

Fu, F., & Wang, Q. (2011). Removal of heavy metal ions from wastewaters: A review. *Journal of Environmental Management*, *92*(3), 407–418.

Fujishima, A., & Honda, K. (1972). Electrochemical photolysis of water at a semiconductor electrode. *Nature*, *238*(5358), 37–38. doi:10.1038/238037a0 PMID:12635268

Gandhi, M. R., Kousalya, G., & Meenakshi, S. (2011). Removal of copper (II) using chitin/chitosan nano-hydroxyapatite composite. *International Journal of Biological Macromolecules*, *48*(1), 119–124.

Ganeshbabu, M., Kannan, N., Venkatesh, P. S., Paulraj, G., Jeganathan, K., & Ali, D. M. (2020). Synthesis and characterization of $BiVO_4$ nanoparticles for environmental applications. *RSC Advances*, *10*(31), 18315–18322. doi:10.1039/D0RA01065K PMID:35517221

Gan, L., Xu, L., Shang, S., Zhou, X., & Meng, L. (2016). Visible light induced methylene blue dye degradation photo-catalyzed by WO_3/graphene nanocomposites and the mechanism. *Ceramics International*, *42*(14), 15235–15241. doi:10.1016/j.ceramint.2016.06.160

Gan, P. P., Ng, S. H., Huang, Y., & Li, S. F. Y. (2012). Green synthesis of gold nanoparticles using palm oil mill effluent (POME): A low-cost and eco-friendly viable approach. *Bioresource Technology*, *113*, 132–135. doi:10.1016/j.biortech.2012.01.015 PMID:22297042

Gao, H., Yang, H., Xu, J., Zhang, S., & Li, J. (2018). Strongly Coupled g-C3N4 Nanosheets-Co3O4 Quantum Dots as 2D/0D Heterostructure Composite for Peroxymonosulfate Activation. *Small, 14*(31), 1801353.

Gao, B., Iftekhar, S., Srivastava, V., Doshi, B., & Sillanpää, M. (2018). Insights into the generation of reactive oxygen species (ROS) over polythiophene/ZnIn 2 S 4 based on different modification processing. *Catalysis Science & Technology*, *8*(8), 2186–2194. doi:10.1039/C8CY00303C

Gao, L., Liu, G., Zamyadi, A., Wang, Q., & Li, M. (2021). Life-cycle cost analysis of a hybrid algae-based biological desalination–low pressure reverse osmosis system. *Water Research*, *195*, 116957.

Gao, M., Zhu, L., Ong, W. L., Wang, J., & Ho, G. (2015). Structural Design of TiO_2-based Photocatalyst for H2 Production and Degradation Applications. *Catalysis Science & Technology*, *5*(10), 4703–4726. Advance online publication. doi:10.1039/C5CY00879D

Gao, X., Wang, Z., Zhai, X., Fu, F., & Li, W. (2015). Thesynthesize of lanthanide doped $BiVO_4$ and its enhanced photocatalytic activity. *Journal of Molecular Liquids*, *211*, 25–30. doi:10.1016/j.molliq.2015.06.058

Gao, Z., Bandosz, T. J., Zhao, Z., Han, M., Liang, C., & Qiu, J. (2008). Investigation of the role of surface chemistry and accessibility of cadmium adsorption sites on open-surface carbonaceous materials. *Langmuir*, *24*(20), 11701–11710. https://doi.org/10.1021/la703638h

Gardea-Torresdey, J. L., Becker-Hapak, M. K., Hosea, J. M., & Darnall, D. W. (1990). Effect of chemical modification of algal carboxyl groups on metal ion binding. *Environmental Science & Technology*, *24*(9), 1372–1378. doi:10.1021/es00079a011

Gardea-Torresdey, J. L., De La Rosa, G., & Peralta-Videa, J. R. (2004). Use of phytofiltration technologies in the removal of heavy metals: A review. *Pure and Applied Chemistry*, *76*(4), 801–813. doi:10.1351/pac200476040801

Gardea-Torresdey, J. L., Tang, L., & Salvador, J. M. (1996a). Copper adsorption by esterified and unesterified fractions of Sphagnum peat moss and its different humic substances. *Journal of Hazardous Materials*, *48*(1-3), 191–206. doi:10.1016/0304-3894(95)00156-5

Gardea-Torresdey, J. L., Tiemann, K. J., Dokken, K., & Gamez, G. (1998). Investigation of metal binding in alfalfa biomass through chemical modification of amino and sulfhydryl ligands. In *Proceedings of the 1998 Conference on Hazardous Waste Research* (pp. 111-121). Snowbird.

Gardea-Torresdey, J. L., Tiemann, K. J., Gonzalez, J. H., Henning, J. A., & Townsend, M. S. (1996b). Ability of silica-immobilized Medicago sativa (alfalfa) to remove copper ions from solution. *Journal of Hazardous Materials*, *48*(1-3), 181–190. doi:10.1016/0304-3894(95)00155-7

Garg, A., Chhipa, K., & Kumar, L. (2018). Microencapsulation techniques in pharmaceutical formulation. *European Journal Pharmaceutical and Medical Research*, *5*(3), 199–206.

Garibo, D., Borbón-Nuñez, H. A., de León, J. N. D., García Mendoza, E., Estrada, I., Toledano-Magaña, Y., Tiznado, H., Ovalle-Marroquin, M., Soto-Ramos, A. G., Blanco, A., Rodríguez, J. A., Romo, O. A., Chávez-Almazán, L. A., & Susarrey-Arce, A. (2020). Green synthesis of silver nanoparticles using Lysiloma acapulcensis exhibit high-antimicrobial activity. *Scientific Reports*, *10*(1), 12805. doi:10.103841598-020-69606-7 PMID:32732959

Garusinghe, U. M., Raghuwanshi, V. S., Batchelor, W., & Garnier, G. (2018). Water Resistant Cellulose – Titanium Dioxide Composites for Photocatalysis. *Scientific Reports*, *8*, 2306.

Gavrilescu, M., Demnerová, K., Aamand, J., Agathos, S., & Fava, F. (2015). Emerging pollutants in the environment: Present and future challenges in biomonitoring, ecological risks and bioremediation. *New Biotechnology*, *32*(1), 147–156. doi:10.1016/j.nbt.2014.01.001 PMID:24462777

Ge, L. (2008). Novel Pd/$BiVO_4$ composite photocatalysts for efficient degradation of methyl orange under visible light irradiation. *Materials Chemistry and Physics*, *107*(2-3), 465–470. doi:10.1016/j.matchemphys.2007.08.016

Ge, L., Peng, Z., Wang, W., Tan, F., Wang, X., Su, B., Qiao, X., & Wong, P. K. (2018). g-C3N4/MgO nanosheets: Light-independent, metal-poisoning-free catalysts for the activation of hydrogen peroxide to degrade organics. *Journal of Materials Chemistry. A, Materials for Energy and Sustainability*, *6*(34), 16421–16429. doi:10.1039/C8TA05488F

Geng, Y., Zhang, P., Li, N., & Sun, Z. (2015). Synthesis of Co doped $BiVO_4$ with enhanced visible-light photocatalytic activities. *Journal of Alloys and Compounds*, *651*, 744–748. doi:10.1016/j.jallcom.2015.08.123

Georgakilas, V., Kordatos, K., Prato, M., Guldi, D. M., Holzinger, M., & Hirsch, A. (2002). Organic functionalization of carbon nanotubes. *Journal of the American Chemical Society*, *124*(5), 760–761. https://doi.org/10.1021/ja016954m

George, A., Magimai Antoni Raj, D., Venci, X., Dhayal Raj, A., Albert Irudayaraj, A., Josephine, R. L., John Sundaram, S., Al-Mohaimeed, A. M., Al Farraj, D. A., Chen, T.-W., & Kaviyarasu, K. (2022). Photocatalytic effect of CuO nanoparticles flower-like 3D nanostructures under visible light irradiation with the degradation of methylene blue (MB) dye for environmental application. *Environmental Research*, *203*, 111880. doi:10.1016/j.envres.2021.111880 PMID:34400161

Ghaedi, M., & Mosallanejad, N. (2013). Removal of heavy metal ions from polluted waters by using of low cost adsorbents. *Journal of Chemical Health Risks*, *3*, 7–21.

Ghaffari, Y., Gupta, N. K., Bae, J., & Kim, K. S. (2020). One-step fabrication of Fe_2O_3/Mn_2O_3 nanocomposite for rapid photodegradation of organic dyes at neutral pH. *Journal of Molecular Liquids*, *315*, 113691. doi:10.1016/j.molliq.2020.113691

Ghidiu, M., Naguib, M., Shi, C., Mashtalir, O., Pan, L. M., Zhang, B., & Barsoum, M. W. (2014). Synthesis and characterization of two-dimensional Nb4C3 (MXene). *Chemical Communications*, *50*(67), 9517–9520. doi:10.1039/C4CC03366C PMID:25010704

Ghimire, K. N., Inoue, K., Makino, K., & Miyajima, T. (2002). Adsorptive removal of arsenic using orange juice residue. *Separation Science and Technology*, *37*(12), 2785–2799. doi:10.1081/SS-120005466

Ghobadifard, M., & Mohebbi, S. (2018). Novel nanomagnetic Ag/β-Ag2WO4/CoFe2O4 as a highly efficient photocatalyst under visible light irradiation. *New Journal of Chemistry*, *42*(12), 9530–9542. doi:10.1039/C8NJ00834E

Ghosh, S. K. (2006). Functional coatings and microencapsulation: a general perspective. *Functional Coatings: By Polymer Microencapsulation*, 1-28.

Ginley, D., Green, M. A., & Collins, R. (2008). Solar energy conversion toward 1 terawatt. *MRS Bulletin*, *33*(4), 355–364. doi:10.1557/mrs2008.71

Gnanasangeetha, D., & Saralathambavani, D. (2014). Biogenic production of zinc oxide nanoparticles using Acalypha indica. *Journal of Chemical, Biological and Physical Sciences*, *4*, 238–246.

Goel, J., Kadirvelu, K., Rajagopal, C., & Kumar Garg, V. (2005). Removal of lead(II) by adsorption using treated granular activated carbon: Batch and column studies. *Journal of Hazardous Materials*, *125*(1–3), 211–220. https://doi.org/10.1016/j.jhazmat.2005.05.032

Gole, J. L., Stout, J. D., Burda, C., Lou, Y., & Chen, X. (2004). Highly efficient formation of visible light tunable TiO2-XnX photocatalysts and their transformation at the nanoscale. *The Journal of Physical Chemistry B*, *108*(4), 1230–1240. doi:10.1021/jp030843n

Goncalves, A. G., Orfao, J. J. M., & Pereira, M. F. R. (2012). Catalytic ozonation of sulphamethoxazole in the presence of carbon materials: Catalytic performance and reaction pathways. *Journal of Hazardous Materials*, *239-240*, 167–174. doi:10.1016/j.jhazmat.2012.08.057 PMID:23009796

Gondal, M. A., Dastageer, M. A., Khalil, A., Hayat, K., & Yamani, Z. H. (2011). Nanostructured ZnO synthesis and its application for effective disinfection of Escherichia coli microorganism in water. *Journal of Nanoparticle Research*, *13*(8), 3423–3430. doi:10.100711051-011-0264-8

Gong, J., Sun, J., & Chen, Q. (2008). Micromachined solgel carbon nanotube/SnO2 nanocomposite hydrogen sensor. *Sensors and Actuators. Part B*, *130*(2), 829–835. doi:10.1016/j.snb.2007.10.051

Gong, K., Yan, Y., Zhang, M., Su, L., Xiong, S., & Mao, L. (2005). Electrochemistry and electroanalytical applications of carbon nanotubes: A review. *Analytical Sciences*, *21*(12), 1383–1393. doi:10.2116/analsci.21.1383 PMID:16379375

Goodwin, T. W., & Mercer, E. I. (1972). *Introduction to plant biochemistry*. Academic Press.

Gopinath, V., MubarakAli, D., Priyadarshini, S., Priyadharsshini, N. M., Thajuddin, N., & Velusamy, P. (2012). Biosynthesis of silver nanoparticles from Tribulus terrestris and its antimicrobial activity: A novel biological approach. *Colloid Surf. B*, *96*, 69–74. doi:10.1016/j.colsurfb.2012.03.023 PMID:22521683

Gorityala, B., Ma, J., Wang, X., Chen, P., & Liu, X. (2010). Carbohydrate Functionalized Carbon Nanotubes and Their Applications. *Chemical Society Reviews*, *39*(8), 2925–2934. doi:10.1039/b919525b PMID:20585681

Govardhan, C. P. (1999). Crosslinking of enzymes for improved stability and performance. *Current Opinion in Biotechnology*, *10*(4), 331–335. https://doi.org/10.1016/S0958-1669(99)80060-3

Goyal, P., Sharma, P., Srivastava, S., & Srivastava, M. M. (2008). Saraca indica leaf powder for decontamination of Pb: Removal, recovery, adsorbent characterization and equilibrium modeling. *International Journal of Environmental Science and Technology*, *5*(1), 27–34. doi:10.1007/BF03325994

Grassian, V. H. (2008). When size really matters: Size-dependent properties and surface chemistry of metal and metal oxide nanoparticles in gas and liquid phase environments. *The Journal of Physical Chemistry C*, *112*(47), 18303–18313. doi:10.1021/jp806073t

Grossiord, N., Loos, J., & Koning, C. E. (2005). Strategies for dispersing carbon nanotubes in highly viscous polymers. *Journal of Materials Chemistry*, *15*(24), 2349. https://doi.org/10.1039/b501805f

Gu, C., Xiong, S., Zhong, Z., Wang, Y., & Xing, W. (2017). A promising carbon fiber-based photocatalyst with hierarchical structure for dye degradation. *RSC Advances, 7*(36), 22234-22242. doi:10.1039/C7RA02583A

Guan, G., & Han, M.-Y. (2019). Functionalized Hybridization of 2D Nanomaterials. *Advancement of Science*, *6*(23), 1901837. PMID:31832321

Guan, G., Liu, S., Cheng, Y., Zhang, Y.-W., & Han, M.-Y. (2018). BSA-caged metal clusters to exfoliate MoS2 nanosheets towards their hybridized functionalization. *Nanoscale*, *10*(23), 10911–10917. doi:10.1039/C8NR02121J PMID:29850713

Guan, G., Xia, J., Liu, S., Cheng, Y., Bai, S., Tee, S. Y., Zhang, Y.-W., & Han, M.-Y. (2017). Electrostatic-Driven Exfoliation and Hybridization of 2D Nanomaterials. *Advanced Materials*, *29*(32), 1700326. doi:10.1002/adma.201700326 PMID:28640388

Gudipati, T., Zaman, M. B., Singh, P., & Poolla, R. (2021). Enhanced photocatalytic activity of biogenically synthesized CuO nanostructures against xylenol orange and rhodamine B dyes. *Inorganic Chemistry Communications*, *130*, 108677. doi:10.1016/j.inoche.2021.108677

Guesh, K., Mayoral, Á., Márquez-Álvarez, C., Chebude, Y., & Díaz, I. (2016). Enhanced photocatalytic activity of TiO2 supported on zeolites tested in real wastewaters from the textile industry of Ethiopia. *Microporous and Mesoporous Materials*, *225*, 88–97. doi:10.1016/j.micromeso.2015.12.001

Guettaı, N., & Ait Amar, H. A. (2005). Photocatalytic oxidation of methyl orange in presence of titanium dioxide in aqueous suspension. Part II: Kinetics study. *Desalination*, *185*(1–3), 439–448. doi:10.1016/j.desal.2005.04.049

Guisbiers, G., Mejía-Rosales, S., & Deepak, F. L. (2012). Nanomaterial properties: Size and shape dependencies. *Journal of Nanomaterials*, *2012*, 2.

Gunalan, S., Sivaraj, R., & Rajendran, V. (2012). Green synthesized ZnO nanoparticles against bacterial and fungal pathogens. *Progress in Natural Science*, *22*(6), 693–700. doi:10.1016/j.pnsc.2012.11.015

Gunay, M. (Ed.). (2013). *Eco-friendly textile dyeing and finishing*. BoD–Books on Demand. doi:10.5772/3436

Guo, J. (2009). Interface science in nanoparticles: An electronic structure view of photon-in/photon-out soft-X-ray spectroscopy. *International Journal of Quantum Chemistry*, *109*(12), 2714–2721. doi:10.1002/qua.22174

Guo, M., Wang, J., Strong, P. J., Jiang, P., Ok, Y. S., & Wang, H. (2019). Carbon nanotube grafted chitosan and its adsorption capacity for phenol in aqueous solution. *The Science of the Total Environment*, *682*, 340–347. doi:10.1016/j.scitotenv.2019.05.148 PMID:31125747

Guo, N., Liang, Y., Lan, S., Liu, L., Zhang, J., Ji, G., & Gan, S. (2014). Microscale Hierarchical Three-Dimensional Flowerlike TiO_2/PANI Composite: Synthesis, Characterization, and Its Remarkable Photocatalytic Activity on Organic Dyes under UV-Light and Sunlight Irradiation. *The Journal of Physical Chemistry C*, *118*(32), 18343–18355. doi:10.1021/jp5044927

Guo, N., Liu, H., Fu, Y., & Hu, J. (2020). Preparation of Fe_2O_3 nanoparticles doped with In2O3 and photocatalytic degradation property for rhodamine B. *Optik (Stuttgart)*, *201*, 163537. doi:10.1016/j.ijleo.2019.163537

Guo, X., Du, Y., Chen, F., Park, H. S., & Xie, Y. (2007). Mechanism of removal of arsenic by bead cellulose loaded with iron oxyhydroxide (β-FeOOH): EXAFS study. *Journal of Colloid and Interface Science*, *314*, 427–433.

Guo, Y., Zhou, J., Song, Y., & Zhang, L. (n.d.). An efficient and environmentally friendly method for the synthesis of cellulose carbamate by microwave heating. *Macromolecular Rapid Communications*, *30*, 1504–1508.

Guo, Z., Shin, K., Karki, A. B., Young, D. P., Kaner, R. B., & Hahn, H. T. (2009). Fabrication and characterization of iron oxide nanoparticles filled polypyrrole, nanocomposites. *Journal of Nanoparticle Research*, *11*(6), 1441–1452. doi:10.100711051-008-9531-8

Gupta, V. (2009). Application of Low-Cost Adsorbents for Dye Removal: A Review. *Journal of* Environmental Management, 90, 2313–2342.

Gupta, V. K., & Suhas. (2009). Application of low-cost adsorbents for dye removal—A review. *Journal of Environmental Management*, *90*(8), 2313–2342. doi:10.1016/j.jenvman.2008.11.017 PMID:19264388

Gusain, R., Kumar, N., Fosso-Kankeu, E., & Ray, S. S. (2019). Efficient removal of Pb(II) and Cd(II) for industrial mine water by a hierarchical MoS_2/SH-MWCNT nanocomposite. *ACS Omega*, *4*(9), 13922–13935. doi:10.1021/acsomega.9b01603 PMID:31497710

Gusain, R., Kumar, N., & Ray, S. S. (2020). Recent advances in carbon nanomaterials-based adsorbents for water purification. *Coordination Chemistry Reviews*, *405*, 213111. doi:10.1016/j.ccr.2019.213111

Habibi, Y. (2014). Key advances in the chemical modification of nanocelluloses. *Macromolecules*, *45*, 9220.

Habibi, Y., Lucia, L. A., & Rojas, O. J. (2010). Cellulose nanocrystals: Chemistry, self-assembly, and applications. *Chemical Reviews*, 3479–3500.

Han, Y., Yang, Y., Zhao, J., Yin, X., & Que, W. (2018). Facile One-Pot Synthesis of Ternary Copper-Tin-Chalcogenide Quantum Dots on Reduced Graphene Oxide for Enhanced Photocatalytic Activity. *Catalysis Letters*, *148*(10), 3112–3118. doi:10.100710562-018-2525-y

Hao, N., Nie, Y., Xu, Z., Jin, C., Fyda, T. J., & Zhang, J. X. J. (2020). Microfluidics-enabled acceleration of Fenton oxidation for degradation of organic dyes with rod-like zero-valent iron nanoassemblies. *Journal of Colloid and Interface Science*, *559*, 254–262. doi:10.1016/j.jcis.2019.10.042 PMID:31634669

Haque, F., Daeneke, T., Kalantar-zadeh, K., & Ou, J. Z. (2017). Two-Dimensional Transition Metal Oxide and Chalcogenide-Based Photocatalysts. *Nano-Micro Letters*, *10*(2), 23. doi:10.100740820-017-0176-y PMID:30393672

Hasan Khan Neon, M., & Islam, M. S. (2019). MoO_3 and Ag co-synthesized TiO_2 as a novel heterogeneous photocatalyst with enhanced visible-light-driven photocatalytic activity for methyl orange dye degradation. *Environmental Nanotechnology, Monitoring & Management*, *12*, 100244. doi:10.1016/j.enmm.2019.100244

Hasija, V., Raizada, P., Sudhaik, A., Sharma, K., Kumar, A., Singh, P., Jonnalagadda, S. B., & Thakur, V. K. (2019). Recent advances in noble metal free doped graphitic carbon nitride based nanohybrids for photocatalysis of organic contaminants in water: A review. *Applied Materials Today*, *15*, 494–524. doi:10.1016/j.apmt.2019.04.003

Heinlaan, M., Ivask, A., Blinova, I., Dubourguier, H.-C., & Kahru, A. (2008). Toxicity of nanosized and bulk ZnO, CuO and TiO2 to bacteria Vibrio fischeri and crustaceans Daphnia magna and Thamnocephalus platyurus. *Chemosphere*, *71*(7), 1308–1316. doi:10.1016/j.chemosphere.2007.11.047 PMID:18194809

Hema, M., Arasi, A. Y., Tamilselvi, P., & Anbarasan, R. (2013). Titania nanoparticles synthesized by sol–gel technique. *Chemical Science Transactions*, *2*(1), 239–245. doi:10.7598/cst2013.344

Hernándeza, S., Gerardia, G., Bejtkab, K., Finaa, A., & Russo, N. (2016). Evaluation of the charge transfer kinetics of spin-coated $BiVO_4$ thin films for sun-driven water photoelectrolysis. *Applied Catalysis B: Environmental*, *190*, 66–74. doi:10.1016/j.apcatb.2016.02.059

Herschy, R. W. (2012). Earth Sci. Ser.: Vol. 876–883. *Water quality for drinking: WHO guidelines, Encycl.* doi:10.1007/978-1-4020-4410-6_184

Hirakawa, T., & Nosaka, Y. (2002). Properties of O2•- and OH• Formed in TiO2 Aqueous Suspensions by Photocatalytic Reaction and the Influence of H2O2 and Some Ions. *Langmuir, 18*(8), 3247–3254. doi:10.1021/la015685a

Hirsch, A., & Vostrowsky, O. (2007). Functionalization of carbon nanotubes. In Functional organic materials (pp. 1–57). Wiley-VCH Verlag GmbH & Co. KGaA.

Hisatomi, T., Takanabe, K., & Domen, K. (2015). Photocatalytic Water-Splitting Reaction from Catalytic and Kinetic Perspectives. *Catalysis Letters, 145*(1), 95–108. doi:10.100710562-014-1397-z

Hong, W., Wyatt, B. C., Nemani, S. K., & Anasori, B. (2020). Double transition-metal MXenes: Atomistic design of two-dimensional carbides and nitrides. *MRS Bulletin, 45*(10), 850–861. doi:10.1557/mrs.2020.251

Houa, Y., Yan, S., Huang, G., Yang, Q., Huang, S., & Cai, J. (2020). Fabrication of N-doped carbons from waste bamboo shoot shell with high removal efficiency of organic dyes from water. *Bioresource Technology, 303*, 122939. doi:10.1016/j.biortech.2020.122939 PMID:32045864

Hou, J., Cao, S., Wu, Y., Liang, F., Sun, Y., Lin, Z., & Sun, L. (2017). Simultaneously efficient light absorption and charge transport of phosphate and oxygen-vacancy confined in bismuth tungstate atomic layers triggering robust solar CO2 reduction. *Nano Energy, 32*, 359–366. doi:10.1016/j.nanoen.2016.12.054

Hsu, V. R. (2004). *Ion transport through biological cell Membranes: From electro-diffusion to Hodgkin- Huxley via a quasi steady-state approach*. University of Washington.

Huang, C., Li, C., & Shi, G. (2012). Graphene based catalysts. *Energy & Environmental Science, 5*(10), 8848–8868. doi:10.1039/c2ee22238h

Huang, H., Gan, M., Ma, L., Yu, L., Hu, H., Yang, F., Li, Y., & Ge, C. (2015). Fabrication of Polyaniline/Graphen oxide/titania nanotube arrays nanocomposites and their application in supercapacitors. *Journal of Alloys and Compounds, 630*, 214–221. doi:10.1016/j.jallcom.2015.01.059

Huang, H., Song, Y., Li, N., Chen, D., Xu, Q., Li, H., & Lu, J. (2019). One-step in-situ preparation of N-doped TiO2@C derived from Ti3C2 MXene for enhanced visible-light driven photodegradation. *Applied Catalysis B: Environmental, 251*, 154–161. doi:10.1016/j.apcatb.2019.03.066

Huang, K., Li, Z., Lin, J., Han, G., & Huang, P. (2018). Two-dimensional transition metal carbides and nitrides (MXenes) for biomedical applications. *Chemical Society Reviews, 47*(14), 5109–5124. doi:10.1039/C7CS00838D PMID:29667670

Huang, Q., Liu, Y., Cai, T., & Xia, X. (2019). Simultaneous removal of heavy metal ions and organic pollutant by BiOBr/Ti3C2 nanocomposite. *Journal of Photochemistry and Photobiology A Chemistry, 375*, 201–208. doi:10.1016/j.jphotochem.2019.02.026

Huang, Z. H., Zheng, X., Lv, W., Wang, M., Yang, Q. H., & Kang, F. (2011). Adsorption of lead (II). *Langmuir, 27*(12), 7558–7562. https://doi.org/10.1021/la200606r

Hu, C. Y., Xu, Y. J., Duo, S. W., Zhang, R. F., & Li, M. S. (2009). Non-covalent functionalization of carbon nanotubes with surfactants and polymers. *Journal of the Chinese Chemical Society (Taipei), 56*(2), 234–239. https://doi.org/10.1002/jccs.200900033

Hu, J., Tong, Z., Hu, Z., Chen, G., & Chen, T. (2012). Adsorption of roxarsone from aqueous solution by multi-walled carbon nanotubes. *Journal of Colloid and Interface Science, 377*(1), 355–361. doi:10.1016/j.jcis.2012.03.064 PMID:22513167

Hummer, G., Rasaiah, J. C., & Noworyta, J. P. (2001). Water conduction through the hydrophobic channel of a carbon nanotube. *Nature, 414*(6860), 188–190. https://doi.org/10.1038/35102535

Hunger, K. (2003). *Industrial Dyes-Chemistry, Properties, Application*. Wiley VCH.

Hunge, Y. M., Mohlte, V. S., Kumbhar, S. S., Rajpure, K. V., Moholkar, A. V., & Bhosale, C. H. (2015). Photoelctrocatalytic degradation of methyl red using sprayed WO_3 thin films under visible light irradiation. *Journal of Materials Science, 26*, 8404–8412.

Huo, J., Yuan, C., & Wang, Y. (2019). Nanocomposites of Three-Dimensionally Ordered Porous TiO2 Decorated with Pt and Reduced Graphene Oxide for the Visible-Light Photocatalytic Degradation of Waterborne Pollutants. *ACS Applied Nano Materials, 2*(5), 2713–2724. doi:10.1021/acsanm.9b00215

Hu, T., Wang, J., Zhang, H., Li, Z., Hu, M., & Wang, X. (2015). Vibrational properties of Ti3C2 and Ti3C2T2 (T = O, F, OH) monosheets by first-principles calculations: A comparative study. *Physical Chemistry Chemical Physics, 17*(15), 9997–10003. doi:10.1039/C4CP05666C PMID:25785395

Hu, W., Xie, L., & Zeng, H. (2020). Novel sodium alginate-assisted MXene nanosheets for ultrahigh rejection of multiple cations and dyes. *Journal of Colloid and Interface Science, 568*, 36–45. doi:10.1016/j.jcis.2020.02.028 PMID:32086010

Hu, X., Deng, F., Huang, W., Zeng, G., Luo, X., & Dionysiou, D. D. (2018). The band structure control of visible-light-driven rGO/ZnS-MoS2 for excellent photocatalytic degradation performance and long-term stability. *Chemical Engineering Journal, 350*, 248–256. doi:10.1016/j.cej.2018.05.182

Hu, X., Zhao, Y., Wang, H., Tan, X., Yang, Y., & Liu, Y. (2017). Efficient removal of tetracycline from aqueous media with a Fe3O4 Nanoparticles@graphene oxide nanosheets assembly. *International Journal of Environmental Research and Public Health, 14*(12), 1495. https://doi.org/10.3390/ijerph14121495

Hu, Z., Cai, X., Wang, Z., Li, S., Wang, Z., & Xie, X. (2019). Construction of carbon-doped supramolecule-based g-C3N4/TiO2 composites for removal of diclofenac and carbamazepine: A comparative study of operating parameters, mechanisms, degradation pathways. *Journal of Hazardous Materials, 380*, 120812. doi:10.1016/j.jhazmat.2019.120812 PMID:31326838

Hyung, H., Fortner, J. D., Hughes, J. B., & Kim, J. H. (2007). Natural organic matter stabilizes carbon nanotubes in the aqueous phase. *Environmental Science & Technology, 41*(1), 179–184. https://doi.org/10.1021/es061817g

Ibrahim, R. K., Hayyan, M., AlSaadi, M. A., Hayyan, A., & Ibrahim, S. (2016). Environmental application of nanotechnology: Air, soil, and water. *Environmental Science and Pollution Research International, 23*(14), 13754–13788. https://doi.org/10.1007/s11356-016-6457-z

Ida, S., Kim, N., Ertekin, E., Takenaka, S., & Ishihara, T. (2015). Photocatalytic Reaction Centers in Two-Dimensional Titanium Oxide Crystals. *Journal of the American Chemical Society, 137*(1), 239–244. doi:10.1021/ja509970z PMID:25479408

Ihsanullah, A. A., Al-Amer, A. M., Laoui, T., Al-Marri, M. J., Nasser, M. S., Khraisheh, M., & Atieh, M. A. (2016). Heavy metal removal from aqueous solution by advanced carbon nanotubes: A critical review of adsorption applications. *Separation and Purification Technology, 157*, 141–161. doi:10.1016/j.seppur.2015.11.039

Ihsanullah, D., Asmaly, H., Saleh, T., Laoui, T., Gupta, V., & Atieh, M. A. (2015). Enhanced adsorption of phenols from liquids by aluminum oxide/carbon nanotubes: Comprehensive study from synthesis to surface properties. *Journal of Molecular Liquids, 206*, 176–182, 182. doi:10.1016/j.molliq.2015.02.028

Iijama, S. (1991). Helical microtubules of graphitic carbon. *Nature, 354*(6348), 56–58. doi:10.1038/354056a0

Ikeda, S., Tanaka, A., Shinohara, K., Hara, M., Kondo, J. N., Maruya, K., & Domen, K. (1997). Heterogeneous photocatalyst materials for water splitting. *Microporous and Mesoporous Materials, 9*, 253–258. doi:10.1016/S0927-6513(96)00112-5

Ikram, M., Hassan, J., Imran, M., Haider, J., Ul-Hamid, A., Shahzadi, I., Ikram, M., Raza, A., Qumar, U., & Ali, S. (2020). 2D chemically exfoliated hexagonal boron nitride (hBN) nanosheets doped with Ni: Synthesis, properties and catalytic application for the treatment of industrial wastewater. *Applied Nanoscience*, *10*(9), 3525–3528. doi:10.100713204-020-01439-2

Ikram, M., Hassan, J., Raza, A., Haider, A., Naz, S., Ul-Hamid, A., Haider, J., Shahzadi, I., Qamar, U., & Ali, S. (2020). Photocatalytic and bactericidal properties and molecular docking analysis of TiO2 nanoparticles conjugated with Zr for environmental remediation. *RSC Advances*, *10*(50), 30007–30024. doi:10.1039/D0RA05862A PMID:35518250

Ikram, M., Khan, M. I., Raza, A., Imran, M., Ul-Hamid, A., & Ali, S. (2020). Outstanding performance of silver-decorated MoS2 nanopetals used as nanocatalyst for synthetic dye degradation. *Physica E, Low-Dimensional Systems and Nanostructures*, *124*, 114246. doi:10.1016/j.physe.2020.114246

Ikram, M., Umar, E., Raza, A., Haider, A., Naz, S., Ul-Hamid, A., Haider, J., Shahzadi, I., Hassan, J., & Ali, S. (2020). Dye degradation performance, bactericidal behavior and molecular docking analysis of Cu-doped TiO2 nanoparticles. *RSC Advances*, *10*(41), 24215–24233. doi:10.1039/D0RA04851H PMID:35516171

Inoue, Y., Kohno, M., Kaneko, T., Ogura, S., & Sato, K. (1998). Dispersion of ruthenium oxide on barium titanates ($Ba_6Ti_{17}O_{40}$, $Ba_4Ti_{13}O_{30}$, $BaTi_4O_9$ and $Ba_2Ti_9O_{20}$) and photocatalytic activity for water decomposition. *Journal of the Chemical Society, Faraday Transactions*, *94*(1), 89–94. doi:10.1039/a704947a

Intaphong, P., Phuruangrat, A., & Pookmanee, P. (2016). Synthesis and characterization of $BiVO_4$ photocatalyst by microwave method. *Integrated Ferroelectrics*, *175*(1), 51–58. doi:10.1080/10584587.2016.1200910

International Agency of Research on Cancer. (1976). *Cadmium and Cadmium Compounds. Monographs on the Evaluation of Carcinogenic Risk of Chemicals to Man* (Vol. 11). International Agency for Research on Cancer.

Iqbal, M., Saeed, A., & Akhtar, N. (2002). Petiolar felt-sheath of palm: A new biosorbent for the removal of heavy metals from contaminated water. *Bioresource Technology*, *81*(2), 151–153. doi:10.1016/S0960-8524(01)00126-2 PMID:11762907

Islam, M. A., Church, J., Han, C., Chung, H.-S., Ji, E., Kim, J. H., Choudhary, N., Lee, G.-H., Lee, W. H., & Jung, Y. (2017). Noble metal-coated MoS2 nanofilms with vertically-aligned 2D layers for visible light-driven photocatalytic degradation of emerging water contaminants. *Scientific Reports*, *7*(1), 14944. doi:10.103841598-017-14816-9 PMID:29097721

Izadi, A., Mohebbi, A., Amiri, M., & Izadi, N. (2017). Removal of iron ions from industrial copper raffinate and electrowinning electrolyte solutions by chemical precipitation and ion exchange. *Minerals Engineering*, *113*, 23–35.

Izvestiya Akademii Nauk, S. S. S. R. (1982). Article. *Metals*, *3*, 12–17.

Jadhav, J., & Biswas, S. (2018). Hybrid ZnO:Ag core-shell nanoparticles for wastewater treatment: Growth mechanism and plasmonically enhanced photocatalytic activity. *Applied Surface Science*, *456*, 49–58. doi:10.1016/j.apsusc.2018.06.028

Jadupati, M., Tanmay, D., & Souvik, G. (2012). Microencapsulation: An indispensable technology for drug delivery system. *Int. Res. J. Pharm*, *3*(4), 8–13.

Jahurul Islam, M., Amaranatha Reddy, D., Han, N. S., Choi, J., Song, J. K., & Kim, T. K. (2016). An oxygen-vacancy rich 3D novel hierarchical MoS2/BiOI/AgI ternary nanocomposite: Enhanced photocatalytic activity through photogenerated electron shuttling in a Z-scheme manner. *Physical Chemistry Chemical Physics*, *18*(36), 24984–24993. doi:10.1039/C6CP02246D PMID:27722571

Jamshaid, A., Hamid, A., Muhammad, N., Naseer, A., Ghauri, M., Iqbal, J., ... Shah, N. S. (2017). Cellulose-based Materials for the Removal of Heavy Metals from Wastewater–An Overview. *ChemBioEng Reviews*, *4*(4), 240–256.

Jasper, J. T., Yang, Y., & Hoffmann, M. R. (2017). Toxic Byproduct Formation during Electrochemical Treatment of Latrine Wastewater. *Environmental Science & Technology*, *51*(12), 7111–7119. doi:10.1021/acs.est.7b01002 PMID:28538093

Jassby, D., Cath, T. Y., & Buisson, H. (2018). The role of nanotechnology in industrial water treatment. *Nature Nanotechnology*, *13*(8), 670–672. doi:10.103841565-018-0234-8 PMID:30082807

Javey, A., Guo, J., Wang, Q., Lundstrom, M., & Dai, H. J. (2003). Ballistic carbon nanotube field-effect transistors. *Nature*, *424*(6949), 654–657. https://doi.org/10.1038/nature01797

Jayapriya, M., & Arulmozhi, M. (2021). Beta vulgaris peel extract mediated synthesis of Ag/TiO_2 nanocomposite: Characterization, evaluation of antibacterial and catalytic degradation of textile dyes-an electron relay effect. *Inorganic Chemistry Communications*, *128*, 108529. doi:10.1016/j.inoche.2021.108529

Jiang, H., Dai, H., Meng, X., Zhang, L., Deng, J., Liu, Y., & Au, C. T. (2012). Hydrothermal fabrication and visible-light-driven photocatalytic properties of bismuth vanadate with multiple morphologies and/or porous structures for Methyl Orange degradation. *Journal of Environmental Sciences (China)*, *24*(3), 449–457. doi:10.1016/S1001-0742(11)60793-6 PMID:22655358

Jiang, H., Zang, C., Zhang, Y., Wang, W., Yang, C., Sun, B., Shen, Y., & Bian, F. (2020). 2D MXene-derived Nb2O5/C/Nb2C/g-C3N4 heterojunctions for efficient nitrogen photofixation. *Catalysis Science & Technology*, *10*(17), 5964–5972. doi:10.1039/D0CY00656D

Jiang, L., Yuan, X., Pan, Y., Liang, J., Zeng, G., Wu, Z., & Wang, H. (2017). Doping of graphitic carbon nitride for photocatalysis: A reveiw. *Applied Catalysis B: Environmental*, *217*, 388–406. doi:10.1016/j.apcatb.2017.06.003

Jiang, X., Kuklin, A. V., Baev, A., Ge, Y., Ågren, H., Zhang, H., & Prasad, P. N. (2020). Two-dimensional MXenes: From morphological to optical, electric, and magnetic properties and applications. *Physics Reports*, *848*, 1–58. doi:10.1016/j.physrep.2019.12.006

Jia, Z., Shen, D., & Xu, W. (2001). Synthesis and antibacterial activities of quaternary ammonium salt of chitosan. *Carbohydrate Research*, *333*(1), 1–6. https://doi.org/10.1016/s0008-6215(01)00112-4

Ji, H., Zeng, W., & Li, Y. (2019). Gas sensing mechanisms of metal oxide semiconductors: A focus review. *Nanoscale*, *11*(47), 22664–22684. doi:10.1039/C9NR07699A PMID:31755888

Ji, L., Zhou, L., Bai, X., Shao, Y., Zhao, G., Qu, Y., Wang, C., & Li, Y. (2012). Facile synthesis of multiwall carbon nanotubes/iron oxides for removal of tetrabromobisphenol A and Pb(II). *Journal of Materials Chemistry*, *22*(31), 15853–15862. doi:10.1039/c2jm32896h

Jing, S., Cao, X., Zhong, L., Peng, X., Sun, R., & Liu, J. (2018). Effectively enhancing conversion of cellulose to HMF by combining in-situ carbonic acid from CO2 and metal oxides. *Industrial Crops and Products*, *126*, 151–157.

Jin, S., Fallgren, P. H., Morris, J. M., & Chen, Q. (2007). Removal of bacteria and viruses from waters using layered double hydroxide nanocomposites. *Science and Technology of Advanced Materials*, *8*(1–2), 67–70. https://doi.org/10.1016/j.stam.2006.09.003

Jin, X., Hu, J. Y., Tint, M. L., Ong, S. L., Biryulin, Y., & Polotskaya, G. (2007). Estrogenic compounds removal by fullerene-containing membranes. *Desalination*, *214*(1–3), 83–90. https://doi.org/10.1016/j.desal.2006.10.019

Joerger, R., Klaus, T., & Granqvist, C. G. (2000). Biologically produced silver–carbon composite materials for optically functional thin-film coatings. *Advanced Materials*, *12*(6), 407–409. doi:10.1002/(SICI)1521-4095(200003)12:6<407::AID-ADMA407>3.0.CO;2-O

Johari, K., Saman, N., Song, S. T., Chin, C. S., Kong, H., & Mat, H. (2016). Adsorption enhancement of elemental mercury by various surface modified coconut husk as eco-friendly low-cost adsorbents. *International Biodeterioration & Biodegradation, 109*, 45–52.

Johnson, R. D., Bethune, D. S., & Yannoni, C. S. (1992). Fullerene structure and dynamics: A magnetic resonance potpourri. *Accounts of Chemical Research, 25*(3), 169–175. https://doi.org/10.1021/ar00015a011

Johnson, R. D., Meijer, G., & Bethune, D. S. (1990). C60 has icosahedral symmetry. *Journal of the American Chemical Society, 112*(24), 8983–8984. https://doi.org/10.1021/ja00180a055

Johnston, H. J., Hutchison, G. R., Christensen, F. M., Peters, S., Hankin, S., Aschberger, K., & Stone, V. (2010). A critical review of the biological mechanisms underlying the in vivo and in vitro toxicity of carbon nanotubes: The contribution of physico-chemical characteristics. *Nanotoxicology, 4*(2), 207–246. doi:10.3109/17435390903569639 PMID:20795897

Joo, E. J., Park, G., Gwak, J. S., Seo, J. H., Jang, K. Y., Oh, K. H., & Nam, K. M. (2016). Efficient photoelectrochemical water oxidation by metal-doped bismuth vanadate photoanode with Iron oxyhydroxideelectrocatalyst. *Journal of Nanomaterials*, 1827151.

Jordao, C. P., Pereira, M. D. G., Einloft, R., Santana, M. B., Bellato, C. R., & Vargas de Mello, J. W. (2002). Removal of Cu, Cr, Ni, Zn, and Cd from electroplating wastes and synthetic solutions by vermicompost of cattle manure. *Journal of Environmental Science and Health. Part A, Toxic/Hazardous Substances & Environmental Engineering, 37*(5), 875–892. doi:10.1081/ESE-120003594 PMID:12049122

Jo, W. J., Jang, J. W., Kong, K. J., Kang, H. J., Kim, J. Y., Jun, H., Parmar, K. P. S., & Lee, J. S. (2012). Phosphatedoping into monoclinic $BiVO_4$ for enhanced photoelectrochemical water oxidation activity. *Angewandte Chemie International Edition, 51*(13), 3147–3151. doi:10.1002/anie.201108276 PMID:22344930

Jo, W.-K., & Tonda, S. (2019). Novel CoAl-LDH/g-C3N4/RGO ternary heterojunction with notable 2D/2D/2D configuration for highly efficient visible-light-induced photocatalytic elimination of dye and antibiotic pollutants. *Journal of Hazardous Materials, 368*, 778–787. doi:10.1016/j.jhazmat.2019.01.114 PMID:30739031

Jung, S. M., Lee, J., Han, M., Choi, J., Kim, S., Seo, J., & Lim, H. (2010). Advanced photocatalytic activity using TiO_2/ceramic fiber-based honeycomb. *Studies in Surface Science and Catalysis, 175*, 441–444. doi:10.1016/S0167-2991(10)75080-1

Jun, L. Y., Mubarak, N. M., Yee, M. J., Yon, L. S., Bing, C. H., Khalid, M., & Abdullah, E. C. (2018). An overview of functionalised carbon nanomaterial for organic pollutant removal. *Journal of Industrial and Engineering Chemistry, 67*, 175–186. https://doi.org/10.1016/j.jiec.2018.06.028

Jüttner, K., Galla, U., & Schmieder, H. (2000). Electrochemical approaches to environmental problems in the process industry. *Electrochimica Acta, 45*, 2575–2594.

Kahouli, M., Barhoumi, A., Bouzid, A., Al-Hajry, A., & Guermazi, S. (2015). Structural and optical properties of ZnO nanoparticles prepared by direct precipitation method. *Superlattices and Microstructures, 85*, 7–23. doi:10.1016/j.spmi.2015.05.007

Kajbafvala, A., Ghorbani, H., Paravar, A., Samberg, J., Kajbafvala, E., & Sadrnezhaad, S. K. (2012). Effects of morphology on photocatalytic performance of Zinc oxide nanostructures synthesized by rapid microwave irradiation methods. *Superlattices and Microstructures, 51*(4), 512–522. doi:10.1016/j.spmi.2012.01.015

Kalanura, S. S., Yoo, I. L., Park, H. J., & Seo, H. (2017). Insights into the electronic bands ofWO_3/$BiVO_4$/TiO_2, revealing high solar water splitting efficiency. *Journal of Materials Chemistry. A, Materials for Energy and Sustainability, 5*(4), 1455–1461. doi:10.1039/C6TA07592D

Kaliraj, L., Ahn, J. C., Rupa, E. J., Abid, S., Lu, J., & Yang, D. C. (2019). Synthesis of panos extract mediated ZnO nano-flowers as photocatalyst for industrial dye degradation by UV illumination. *Journal of Photochemistry and Photobiology. B, Biology*, *199*, 111588. doi:10.1016/j.jphotobiol.2019.111588 PMID:31450132

Kamali, M., Persson, K. M., Costa, M. E., & Capela, I. (2019). Sustainability criteria for assessing nanotechnology applicability in industrial wastewater treatment: Current status and future outlook. *Environment International*, *125*, 261–276. doi:10.1016/j.envint.2019.01.055 PMID:30731376

Kandah, M. I., & Meunier, J. L. (2007). Removal of nickel ions from water by multi-walled carbon nanotubes. *Journal of Hazardous Materials*, *146*(1–2), 283–288. https://doi.org/10.1016/j.jhazmat.2006.12.019

Kaneko, K. (1994). Determination of pore size and pore size distribution: 1. Adsorbents and catalysts. *Journal of Membrane Science*, *96*(1–2), 59–89.

Kang, S., Pinault, M., Pfefferle, L. D., & Elimelech, M. (2007). Single-walled carbon nanotubes exhibit strong antimicrobial activity. *Langmuir*, *23*(17), 8670–8673. https://doi.org/10.1021/la701067r

Kansal, S. K., Singh, M., & Sud, D. (2007). Studies on photodegradation of two commercial dyes in aqueous phase using different photocatalysts. *Journal of Hazardous Materials*, *141*(3), 581–590. doi:10.1016/j.jhazmat.2006.07.035

Kansal, S. K., Singh, M., & Sud, D. (2007). Studies on photo degradation of two commercial dyes in aqueous phase using different photocatalyst. *Journal of Hazardous Materials*, *141*, 581–590.

Kar, A., Smith, Y. R., & Subramanian, V. R. (2009). Improved photocatalytic degradation of, textile dye using titanium dioxide nanotubes formed over titanium wires. *Environmental Science & Technology*, *43*(9), 3260–3265. doi:10.1021/es8031049 PMID:19534144

Kardam, A., Raj, K. R., Arora, J. K., & Srivastava, S. (2012). Artificial neural network modeling for biosorption of Pb (II) ions on nanocellulose fibers. *BioNanoScience*, *2*(3), 153–160.

Kar, R. N., Sahoo, B. N., & Sukla, C. B. (1992). Removal of heavy metals from pure water using sulphate-reducing bacteria (SRB). *Pollution Research*, *11*, 1–13.

Kartal, S. N., & Imamura, Y. (2005). Removal of copper, chromium, and arsenic from CCA-treated wood onto chitin and chitosan. *Bioresource Technology*, *96*(3), 389–392. doi:10.1016/j.biortech.2004.03.004 PMID:15474943

Karthikeyan, P., Banu, H., & Meenakshi, S. (2019). Synthesis and characterization of metal-loaded chitosan-alginate biopolymeric hybrid beads for the efficient removal of phosphate and nitrate ions from aqueous solution. *International Journal of Biological Macromolecules*, *130*, 407–418. doi:10.1016/j.ijbiomac.2019.02.059 PMID:30802518

Karthikeyan, P., Elanchezhiyan, S. S. D., Preethi, J., Meenakshi, S., & Park, C. M. (2020). Mechanistic performance of polyaniline-substituted hexagonal boron nitride composite as a highly efficient adsorbent for the removal of phosphate, nitrate, and hexavalent chromium ions from an aqueous environment. *Applied Surface Science*, *511*, 145543. doi:10.1016/j.apsusc.2020.145543

Karthikeyan, P., & Meenakshi, S. (2019). In-situ fabrication of cerium incorporated chitosan-β- cyclodextrin microspheres as an effective adsorbent for toxic anions removal. *Environmental Nanotechnology, Monitoring & Management*, *12*, 100272. doi:10.1016/j.enmm.2019.100272

Karunakaran, C., & Kalaivani, S. (2014). Enhanced visible light-photocatalysisby hydrothermally synthesized thallium-doped bismuth vanadate nanoparticles. *Materials Science in Semiconductor Processing*, *27*, 352–361. doi:10.1016/j.mssp.2014.07.004

Kasiri, M. B. (2018) Application of chitosan derivatives as promising adsorbents for treatment of textile wastewater. *The Impact and Prospects of Green Chemistry for Textile Technology*, 417-469.

Katal, R., Masudy-Panah, S., Tanhaei, M., Farahani, M. H. D. A., & Jiangyong, H. (2020). A review on the synthesis of the various types of anatase TiO2 facets and their applications for photocatalysis. *Chemical Engineering Journal, 384*, 123384. doi:10.1016/j.cej.2019.123384

Kato, H., & Kudo, A. (2003). Photocatalytic water splitting into H_2 and O_2 over various tantalite photocatalysts. *Catalysis Today, 78*(1-4), 561–569. doi:10.1016/S0920-5861(02)00355-3

Kaur, J., & Singhal, S. (2014). Heterogeneous photocatalytic degradation of rose bengal: Effect of operational parameters. *Physica B, Condensed Matter, 450*, 49–53. doi:10.1016/j.physb.2014.05.069

Kaushal, S., Kaur, N., Kaur, M., & Singh, P. P. (2020). Dual-Responsive Pectin/Graphene Oxide (Pc/GO) nano-composite as an efficient adsorbent for Cr (III) ions and photocatalyst for degradation of organic dyes in waste water. *Journal of Photochemistry and Photobiology A Chemistry, 403*, 112841. doi:10.1016/j.jphotochem.2020.112841

Keshavarz, H. R. P. M. H. (2011, November 2). Study of Congo red photodegradation kinetic catalyzed by $Zn_{1-X}Cu_XS$ and $Zn_{1-X}Ni_XS$ nanoparticles. *International Journal of Physical Sciences, 6*(27), 6268–6279. doi:10.5897/IJPS09.251

Kfir, R., Hilner, C., du Preez, M., & Bateman, B. (1995). Studies evaluating the applicability of utilising the same concentration techniques for the detection of protozoan parasites and viruses in water. *Water Science and Technology, 31*(5–6), 417–423. https://doi.org/10.2166/wst.1995.0651

Khalid, N., Rahman, A., Ahmad, S., Kiani, S. N., & Ahmed, J. (1998). Adsorption of cadmium from aqueous solutions on rice husk. *Radiochimica Acta, 83*(3), 157–162. doi:10.1524/ract.1998.83.3.157

Khan, I., Saeed, K., & Khan, I. (2019). Nanoparticles: Properties, applications and toxicities. *Arabian Journal of Chemistry, 12*(7), 908–931. doi:10.1016/j.arabjc.2017.05.011

Kim, H. C., & Yu, M. J. (2005). Characterization of natural organic matter in conventional water treatment processes for selection of treatment processes focused on DBPs control. *Water Research, 39*, 4779–4789.

Kim, H. J., Joshi, M. K., Pant, H. R., Kim, J. H., Lee, E., & Kim, C. S. (2015). One-pot hydrothermal synthesis of multifunctional Ag/ZnO/fly ash nanocomposite. *Colloids and Surfaces. A, Physicochemical and Engineering Aspects, 469*, 256–262. doi:10.1016/j.colsurfa.2015.01.032

Kim, J. S., Kuk, E., Yu, K. N., Jong-Ho, K., Park, S. J., Lee, H. J., & Kim, S. H. (2007). Antimicrobial effects of silver nanoparticles. *Nanomedicine (London), 3*(1), 95–101. doi:10.1016/j.nano.2006.12.001 PMID:17379174

Kim, J., Jia, H., & Wang, P. (2006). Challenges in biocatalysis for enzyme-based biofuel cells. *Biotechnology Advances, 24*(3), 296–308. https://doi.org/10.1016/j.biotechadv.2005.11.006

Kim, J., & Van der Bruggen, B. (2010). The use of nanoparticles in polymeric and ceramic membrane structures: Review of manufacturing procedures and performance improvement for water treatment. *Environmental Pollution, 158*(7), 2335–2349. https://doi.org/10.1016/j.envpol.2010.03.024

Kishor, R., Bharagava, R. N., & Saxena, G. (2018). Industrial wastewaters: the major sources of dye contamination in the environment, ecotoxicological effects, and bioremediation approaches. In *Recent advances in environmental management* (pp. 1–25). CRC Press. doi:10.1201/9781351011259-1

Klabunde, K. J., & Sergeev, G. B. (2013). *Nanochemistry*. Newnes.

Kobya, M. (2004). Removal of Cr (VI) from aqueous solutions by adsorption onto hazelnut shell activated carbon: Kinetic and equilibrium studies. *Bioresource Technology*, *91*(3), 317–321. doi:10.1016/j.biortech.2003.07.001 PMID:14607493

Kobya, M., Demirbas, E., Senturk, E., & Ince, M. (2005). Adsorption of Heavy Metal Ions from Aqueous Solutions by Activated Carbon Prepared from Apricot Stone. *Bioresource Technology*, *96*, 1518–1521.

Kong, D., Qi, J., Liu, D., Zhang, X., Pan, L., & Zou, J. (2019). Ni-doped $BiVO_4$ with V^{4+} species and oxygen vacancies for efficient photoelectrochemical water splitting. *Transactions of Tianjin University*, *25*(4), 340–347. doi:10.100712209-019-00202-1

Kong, J., Franklin, N. R., Zhou, C., Chapline, M. G., Peng, S., Cho, K., & Dai, H. (2000). Nanotube molecular wires as chemical sensors. *Science*, *287*(5453), 622–625. https://doi.org/10.1126/science.287.5453.622

Kong, J.-Z., Li, A.-D., Zhai, H.-F., Gong, Y.-P., Li, H., & Wu, D. (2009). Preparation, characterization of the Ta-doped ZnO nanoparticles and their photocatalytic activity under visible-light illumination. *Journal of Solid State Chemistry*, *182*(8), 2061–2067. doi:10.1016/j.jssc.2009.03.022

Kong, X., Gao, P., Jiang, R., Feng, J., Yang, P., Gai, S., Chen, Y., Chi, Q., Xu, F., & Ye, W. (2020). Orderly layer-by-layered TiO2/carbon superstructures based on MXene's defect engineeringfor efficient hydrogen evolution. *Applied Catalysis A, General*, *590*, 117341. doi:10.1016/j.apcata.2019.117341

Kong, X., Xu, Y., Cui, Z., Li, Z., Liang, Y., Gao, Z., Zhu, S., & Yang, X. (2018). Defect enhances photocatalytic activity of ultrathin TiO2 (B) nanosheets for hydrogen production by plasma engraving method. *Applied Catalysis B: Environmental*, *230*, 11–17. doi:10.1016/j.apcatb.2018.02.019

Konstantinou, I. K., & Albanis, T. A. (2004). TiO_2-assisted photocatalytic degradation of azo dyes in aqueous solution: Kinetic and mechanistic investigations. *Applied Catalysis B: Environmental*, *49*(1), 1–14. doi:10.1016/j.apcatb.2003.11.010

Kordouli, E., Bourikas, K., Lycourghiotis, A., & Kordulis, C. (2015). The mechanism of azo-dyes adsorption on the titanium dioxide surface and their photocatalytic degradation over samples with various anatase/rutile ratios. *Catalysis Today*, *252*, 128–135. doi:10.1016/j.cattod.2014.09.010

Krasner, S. W., Weinberg, H. S., Richardson, S. D., Pastor, S. J., Chinn, R., Sclimenti, M. J., Onstad, G. D., & Thruston, A. D. (2006). Occurrence of a new generation of disinfection byproducts. *Environmental Science & Technology*, *40*(23), 7175–7185. https://doi.org/10.1021/es060353j

Kratochvil, D., Pimentel, P., & Volesky, B. (1998). Removal of trivalent and hexavalent chromium by seaweed biosorbent. *Environmental Science & Technology*, *32*(18), 2693–2698. doi:10.1021/es971073u

Krishnamurthy, G., & Agarwal, S. (2013). Optimization of reaction conditions for high yield synthesis of carbon nanotube bundles by low-temperature solvothermal process and study of their H_2Storage capacity. *Bulletin of the Korean Chemical Society*, *34*(10), 3046–3054. https://doi.org/10.5012/bkcs.2013.34.10.3046

Krishnan, K. A., & Anirudhan, T. S. (2003). Removal of cadmium (II) from aqueous solutions by steam-activated sulphurised carbon prepared from sugar-cane bagasse pith: Kinetics and equilibrium studies. *Water S.A.*, *29*(2), 147–156. doi:10.4314/wsa.v29i2.4849

Krivoshapkin, P. V., Ivanets, A. I., Torlopov, M. A., Mikhaylov, V. I., Srivastava, V., Sillanpää, M., Prozorovich, V. G., Kouznetsova, T. F., Koshevaya, E. D., & Krivoshapkina, E. F. (2019). Nanochitin/manganese oxide-biodegradable hybrid sorbent for heavy metal ions. *Carbohydrate Polymers*, *210*, 135–143.

Kroto, H. W. (1987). The stability of the fullerenes Cn, with n=24, 28, 32, 36, 50. *Nature*, *329*(6139), 529–531. doi:10.1038/329529a0

Kubiak, A., Kubacka, M., Gabała, E., Dobrowolska, A., Synoradzki, K., Siwińska-Ciesielczyk, K., & Jesionowski, T. (2020). Hydrothermally Assisted Fabrication of TiO_2-Fe_3O_4 Composite Materials and Their Antibacterial Activity. *Materials (Basel)*, *13*(21), 4715. doi:10.3390/ma13214715 PMID:33105776

Kudo, A., Omori, K., & Kato, H. (1999). A novel aqueous process for preparation of crystal form-controlled and highly crystalline $BiVO_4$ powder from layered vanadates at room temperature and its photocatalytic and photo physical properties. *Journal of the American Chemical Society*, *121*(49), 11459–11467. doi:10.1021/ja992541y

Kudo, A., Sayama, K., Tanaka, A., Asakura, K., Domen, K., Maruya, K., & Onishi, T. (1989). Nickel-loaded $K_4Nb_6O_{17}$photocatalyst in the decomposition of H_2O into H_2 and O_2: Structure and reaction mechanism. *Journal of Catalysis*, *120*(2), 337–352. doi:10.1016/0021-9517(89)90274-1

Kudo, A., Ueda, K., Kato, H., & Mikami, I. (1998). Photocatalytic O_2 evolution under visible light irradiation on $BiVO_4$ in aqueous $AgNO_3$ solution. *Catalysis Letters*, *53*(3/4), 229–230. doi:10.1023/A:1019034728816

Kumar, G., Mogha, N. K., & Kumar, M. (2020). NiO nanocomposites/rGO as a heterogeneous catalyst for imidazole scaffolds with applications in inhibiting the DNA binding activity. *Dalton Transactions*, *49*(6), 1963–1974. doi:10.1039/c9dt04416g

Kumaresan, S., Vallalperuman, K., Sathishkumar, S., Karthik, M., & SivaKarthik, P. (2017). Synthesis and systematic investigations of Al and Cu-doped ZnO nanoparticles and its structural, optical and photo-catalytic properties. *Journal of Materials Science Materials in Electronics*, *28*(13), 9199–9205. doi:10.100710854-017-6654-7

Kumar, G., & Masram, D. T. (2021). Sustainable synthesis of MOF-5@GO nanocomposites for efficient removal of rhodamine b from water. *ACS Omega*, *6*(14), 9587–9599. https://doi.org/10.1021/acsomega.1c00143

Kumar, S. S., Venkateswarlu, P., Rao, V. R., & Rao, G. N. (2013). Synthesis, characterization and optical properties of zinc oxide nanoparticles. *International Nano Letters*, *3*(1), 30. doi:10.1186/2228-5326-3-30

Kumar, S., Ahlawat, W., Bhanjana, G., Heydarifard, S., Nazhad, M. M., & Dilbaghi, N. (2014). Nanotechnology-based water treatment strategies. *Journal of Nanoscience and Nanotechnology*, *14*(2), 1838–1858. doi:10.1166/jnn.2014.9050 PMID:24749460

Kuno, M. (2005). *Introduction to nanoscience and nanotechnology: A workbook.* University of Notre Dame.

Kuo, C. Y., Wu, C. H., & Wu, J. Y. (2008). Adsorption of direct dyes from aqueous solutions by carbon nanotubes: Determination of equilibrium, kinetics and thermodynamics parameters. *Journal of Colloid and Interface Science*, *327*(2), 308–315. https://doi.org/10.1016/j.jcis.2008.08.038

Kuppusamy, P., Yusoff, M. M., Maniam, G. P., & Govindan, N. (2016). Biosynthesis of metallic nanoparticles using plant derivatives and their new avenues in pharmacological applications – An updated report. *Saudi Pharmaceutical Journal*, *24*(4), 473–484. doi:10.1016/j.jsps.2014.11.013 PMID:27330378

Kurniawan, T. A., Sillanpää, M. E., & Sillanpää, M. (2012). Nanoadsorbents for Remediation of Aquatic Environment: Local and Practical Solutions for Global Water Pollution Problems. *Critical Reviews in Environmental Science and Technology*, *42*(12), 1233–1295. doi:10.1080/10643389.2011.556553

Kurtoglu, M., Naguib, M., Gogotsi, Y., & Barsoum, M. W. (2012). First principles study of two-dimensional early transition metal carbides. *MRS Communications*, *2*(4), 133–137. doi:10.1557/mrc.2012.25

Kwon, J. D., Kim, P. H., Keum, J. H., & Kim, J. S. (2004). Polypyrrole/titania hybrids: Synthetic variation and test for the photovoltaic materials. *Solar Energy Materials and Solar Cells*, *83*(2–3), 311–321. doi:10.1016/j.solmat.2004.02.033

Lachheb, H., Puzenat, E., Houas, A., Ksibi, M., Elaloui, E., Guillard, C., & Herrmann, J. M. (2002). Photocatalytic degradation of various types of dyes (alizarin S, Crocein orange G, methyl Red, Congo red, methylene blue) in water by UV-irradiated titania. *Applied Catalysis B: Environmental*, *39*(1), 75–90. doi:10.1016/S0926-3373(02)00078-4

Lade, H., Kadam, A., Paul, D., & Govindwar, S. (2015). Biodegradation and detoxification of textile azo dyes by bacterial consortium under sequential microaerophilic/aerobic processes. *EXCLI Journal*, *14*, 158.

Lai, C., Li, G. R., Dou, Y. Y., & Gao, X. P. (2010). Mesoporous polyaniline or polypyrrole/anatase TiO2 nanocomposite as anode materials for lithium-ion batteries. *Electrochimica Acta*, *55*(15), 4567–4572. https://doi.org/10.1016/j.electacta.2010.03.010

Lam, C. W., James, J. T., McCluskey, R., Arepalli, S., & Hunter, R. L. (2006). A review of carbon nanotube toxicity and assessment of potential occupational and environmental health risks. *Critical Reviews in Toxicology*, *36*(3), 189–217. doi:10.1080/10408440600570233 PMID:16686422

Lam, E., Male, K. B., Chong, J. H., Leung, A. C., & Luong, J. H. (2012). Applications of functionalized and nanoparticle-modified nanocrystalline cellulose. *Trends in Biotechnology*, *30*(5), 283–290.

LeChevallier, M. W. KwokKeung, A., & Au, K.-K. (2004). Water treatment and pathogen control: Process efficiency in achieving safe drinking-water. IWA Publishing.

Lee, J.-S., Shevchenko, E. V., & Talapin, D. V. (2008). Au−PbS Core−Shell Nanocrystals: Plasmonic Absorption Enhancement and Electrical Doping via Intra-particle Charge Transfer. *Journal of the American Chemical Society*, *130*(30), 9673–9675. doi:10.1021/ja802890f PMID:18597463

Lee, S. M., & Lee, Y. H. (2000). Hydrogen storage in single-walled carbon nanotubes. *Applied Physics Letters*, *76*(20), 2877–2879. https://doi.org/10.1063/1.126503

Lee, S. Y., Kim, H. J., Patel, R., Im, S. J., Kim, J. H., & Min, B. R. (2007). Silver nanoparticles immobilized on thin film composite polyamide membrane: Characterization, nanofiltration, antifouling properties. *Polymers for Advanced Technologies*, *18*(7), 562–568. https://doi.org/10.1002/pat.918

Lefatshe, K., Muiva, C. M., & Kebaabetswe, L. P. (2017). Extraction of nanocellulose and in-situ casting of ZnO/cellulose nanocomposite with enhanced photocatalytic and antibacterial activity. *Polymers*, *164*, 301–308.

Lei, B. X., Zeng, L. L., Zhang, P., Sun, Z. F., Sun, W., & Zhang, X. X. (2014). Hydrothermal synthesis and photocatalytic properties of visible-light induced $BiVO_4$ with different morphologies. *Advanced Powder Technology*, *25*(3), 946–951. doi:10.1016/j.apt.2014.01.014

Lei, F., Sun, Y., Liu, K., Gao, S., Liang, L., Pan, B., & Xie, Y. (2014). Oxygen Vacancies Confined in Ultrathin Indium Oxide Porous Sheets for Promoted Visible-Light Water Splitting. *Journal of the American Chemical Society*, *136*(19), 6826–6829. doi:10.1021/ja501866r PMID:24773473

Lei, J.-C., Zhang, X., & Zhou, Z. (2015). Recent advances in MXene: Preparation, properties, and applications. *Frontiers in Physics*, *10*(3), 276–286. doi:10.100711467-015-0493-x

Lei, Y., Chen, F., Luo, Y., & Zhang, L. (2014). Three-dimensional magnetic graphene oxide foam/Fe3O4 nanocomposite as an efficient absorbent for Cr (VI) removal. *Journal of Materials Science*, *49*(12), 4236–4245. doi:10.100710853-014-8118-2

Li, X., Yu, J., Wageh, S., Al-Ghamdi, A. A., & Xie, J. (2016). Graphene in Photocatalysis: A Review. *Small, 12*(48), 6640-6696.

Liang, C., Niu, C.-G., Zhang, L., Wen, X.-J., Yang, S.-F., Guo, H., & Zeng, G.-M. (2019). Construction of 2D heterojunction system with enhanced photocatalytic performance: Plasmonic Bi and reduced graphene oxide co-modified Bi5O7I with high-speed charge transfer channels. *Journal of Hazardous Materials*, *361*, 245–258. doi:10.1016/j.jhazmat.2018.08.099 PMID:30199824

Liang, C., Zhang, L., Guo, H., Niu, C.-G., Wen, X.-J., Tang, N., Liu, H.-Y., Yang, Y.-Y., Shao, B.-B., & Zeng, G.-M. (2019). Photo-removal of 2,2¢4,4¢-tetrabromodiphenyl ether in liquid medium by reduced graphene oxide bridged artificial Z-scheme system of Ag@Ag3PO4/g-C3N4. *Chemical Engineering Journal*, *361*, 373–386. doi:10.1016/j.cej.2018.12.092

Liang, H. C., & Li, X. Z. (2009). Visible-induced photocatalytic reactivity of polymer–sensitized titania nanotube films. *Applied Catalysis B: Environmental*, *86*(1–2), 8–17. doi:10.1016/j.apcatb.2008.07.015

Liang, L., Li, X., Zhang, J., Ling, P., Sun, Y., Wang, C., Zhang, Q., Pan, Y., Xu, Q., Zhu, J., Luo, Y., & Xie, Y. (2020). Efficient infrared light induced CO2 reduction with nearly 100% CO selectivity enabled by metallic CoN porous atomic layers. *Nano Energy*, *69*, 104421. doi:10.1016/j.nanoen.2019.104421

Lian, P., Dong, Y., Wu, Z.-S., Zheng, S., Wang, X., Sen, W., ... Bao, X. (2017). Alkalized Ti3C2 MXene nanoribbons with expanded interlayer spacing for high-capacity sodium and potassium ion batteries. *Nano Energy*, *40*, 1–8. doi:10.1016/j.nanoen.2017.08.002

Li, D., & Haneda, H. (2003). Morphologies of zinc oxide particles and their effects on photocatalysis. *Chemosphere*, *51*(2), 129–137. doi:10.1016/S0045-6535(02)00787-7 PMID:12586145

Li, H., Shang, J., Ai, Z., & Zhang, L. (2015). Efficient Visible Light Nitrogen Fixation with BiOBr Nanosheets of Oxygen Vacancies on the Exposed {001} Facets. *Journal of the American Chemical Society*, *137*(19), 6393–6399. doi:10.1021/jacs.5b03105 PMID:25874655

Li, J., Chen, Y., Chen, C., & Wang, S. (2019). Solid-phase synthesis of visible-light-driven $BiVO_4$ photocatalyst and photocatalytic reduction of aqueous Cr (VI). *Bulletin of Chemical Reaction Engineering & Catalysis*, *14*(2), 336–344. doi:10.9767/bcrec.14.2.3182.336-344

Li, J.-Y., Li, Y.-H., Zhang, F., Tang, Z.-R., & Xu, Y.-J. (2020). Visible-light-driven integrated organic synthesis and hydrogen evolution over 1D/2D CdS-Ti3C2Tx MXene composites. *Applied Catalysis B: Environmental*, *269*, 118783. doi:10.1016/j.apcatb.2020.118783

Li, N., Liu, Z., Liu, M., Xue, C., Chang, Q., Wang, H., Li, Y., Song, Z., & Hu, S. (2019). Facile Synthesis of Carbon Dots@2D MoS2 Heterostructure with Enhanced Photocatalytic Properties. *Inorganic Chemistry*, *58*(9), 5746–5752. doi:10.1021/acs.inorgchem.9b00111 PMID:30950600

Li, N., Tian, Y., Zhao, J., Zhang, J., Zuo, W., Kong, L., & Cui, H. (2018). Z-scheme 2D/3D g-C3N4@ZnO with enhanced photocatalytic activity for cephalexin oxidation under solar light. *Chemical Engineering Journal*, *352*, 412–422. doi:10.1016/j.cej.2018.07.038

Ling, W., Wang, M., Xiong, C., Xie, D., Chen, Q., Chu, X., Qiu, X., Li, Y., & Xiao, X. (2019). Synthesis, surface modification, and applications of magnetic iron oxide nanoparticles. *Journal of Materials Research*, *34*(11), 1828–1844. doi:10.1557/jmr.2019.129

Lin, H. F., Ravikrishna, R., & Valsaraj, K. T. (2002). Reusable adsorbents for dilute solution separation. 6. Batch and continuous reactors for the adsorption and degradation of 1,2- dichlorobenzene from dilute wastewater streams using titania as a photocatalyst. *Separation and Purification Technology*, *28*(2), 87–102. https://doi.org/10.1016/S1383-5866(02)00017-5

Lin, K., Xu, Y., He, G., & Wang, X. (2006). The kinetic and thermodynamic analysis of Li ion in multi-walled carbon nanotubes. *Materials Chemistry and Physics*, *99*(2–3), 190–196. https://doi.org/10.1016/j.matchemphys.2005.09.035

Linsebigler, A. L., Lu, G., & Yates, J. T. Jr. (1995). Photocatalysis on TiO_2 surfaces: Principles, mechanisms, and selected results. *Chemical Reviews*, *95*(3), 735–758. doi:10.1021/cr00035a013

Lin, Y., Tian, Z., Zhang, L., Ma, J., Jiang, Z., Deibert, B. J., Ge, R., & Chen, L. (2019). Chromium-ruthenium oxide solid solution electrocatalyst for highly efficient oxygen evolution reaction in acidic media. *Nature Communications*, *10*(1), 162. doi:10.103841467-018-08144-3 PMID:30635581

Lin, Z., Shao, H., Xu, K., Taberna, P.-L., & Simon, P. (2020). MXenes as High-Rate Electrodes for Energy Storage. *Trends in Chemistry*, *2*(7), 654–664. doi:10.1016/j.trechm.2020.04.010

Li, Q., Mahendra, S., Lyon, D. Y., Brunet, L., Liga, M. V., Li, D., & Alvarez, P. J. J. (2008). Antimicrobial nanomaterials for water disinfection and microbial control: Potential applications and implications. *Water Research*, *42*(18), 4591–4602. https://doi.org/10.1016/j.watres.2008.08.015

Li, Q., Yu, J., Zhou, F., & Jiang, X. (2015). Synthesis and characterization of dithiocarbamate carbon nanotubes for removal of heavy metal ions from aqueous solutions. *Colloids and Surfaces. A, Physicochemical and Engineering Aspects*, *482*, 306–314. doi:10.1016/j.colsurfa.2015.06.034

Li, R., Chen, Z., Ren, N., Wang, Y., Wang, Y., & Yu, F. (2019). Biosynthesis of silver oxide nanoparticles and their photocatalytic and antimicrobial activity evaluation for wound healing applications in nursing care. J. Photochem. Photobiol. *Boletín Biológico*, *199*, 111593. doi:10.1016/j.jphotobiol.2019.111593 PMID:31505420

Li, S. M., Dong, Y. Y., Ma, M. G., Fu, L. H., Sun, R. C., & Xu, F. (2013). Hydrothermal synthesis, characterization, and bactericidal activities of hybrid from cellulose and TiO_2. *Carbohydrate Polymers*, *96*, 15–20.

Liu, C.H., Li, J.J., Zhang, H.L., Li, B. R., & Guo, Y. (2008). Structure dependent interaction between organic dyes and carbon nanotubes. *Colloid Surface, A, 313*, 9-12.

Liu, X., Zhang, S., & Pan, B. (2012). Potential of Carbon Nanotubes in Water Treatment. INTECH. doi:10.5772/51332

Liu, C., Dong, S., & Chen, Y. (2019). Enhancement of visible-light-driven photocatalytic activity of carbon plane/g-C3N4/TiO2 nanocomposite by improving heterojunction contact. *Chemical Engineering Journal*, *371*, 706–718. doi:10.1016/j.cej.2019.04.089

Liu, C., Fan, Y. Y., Liu, M., Cong, H. T., Cheng, H. M., & Dresselhaus, M. S. (1999). Hydrogen storage in single-walled carbon nanotube at room temperature. *Science*, *286*(5442), 1127–1129. https://doi.org/10.1126/science.286.5442.1127

Liu, C., Yu, Z., Neff, D., Zhamu, A., & Jang, B. Z. (2010). Graphene-based supercapacitor with an ultrahigh energy density. *Nano Letters*, *10*(12), 4863–4868. doi:10.1021/nl102661q PMID:21058713

Liu, C., Zhu, H., Zhu, Y., Dong, P., Hou, H., Xu, Q., Chen, X., Xi, X., & Hou, W. (2018). Ordered layered N-doped KTiNbO5/g-C3N4 heterojunction with enhanced visible light photocatalytic activity. *Applied Catalysis B: Environmental*, *228*, 54–63. doi:10.1016/j.apcatb.2018.01.074

Liu, H., Liu, H., Yang, J., Zhai, H., Liu, X., & Jia, H. (2019). Microwave-assisted one-pot synthesis of Ag decorated flower-like ZnO composites photocatalysts for dye degradation and NO removal. *Ceramics International*, *45*(16), 20133–20140. doi:10.1016/j.ceramint.2019.06.279

Liu, N., Huang, W., Tang, M., Yin, C., Gao, B., Li, Z., Tang, L., Lei, J., Cui, L., & Zhang, X. (2019). In-situ fabrication of needle-shaped MIL-53(Fe) with 1T-MoS2 and study on its enhanced photocatalytic mechanism of ibuprofen. *Chemical Engineering Journal*, *359*, 254–264. doi:10.1016/j.cej.2018.11.143

Liu, N., Lu, N., Yu, H., Chen, S., & Quan, X. (2020). Efficient day-night photocatalysis performance of 2D/2D Ti3C2/Porous g-C3N4 nanolayers composite and its application in the degradation of organic pollutants. *Chemosphere*, *246*, 125760. doi:10.1016/j.chemosphere.2019.125760 PMID:31901663

Liu, X., & Chen, C. (2020). Mxene enhanced the photocatalytic activity of ZnO nanorods under visible light. *Materials Letters*, *261*, 127127. doi:10.1016/j.matlet.2019.127127

Liu, X., Hu, Q., Fang, Z., Zhang, X., & Zhang, B. (2008). Magnetic chitosan nanocomposites: A useful recyclable tool for heavy metal ion removal. *Langmuir*, *25*(1), 3–8.

Liu, X., Shao, X., Li, F., & Zhao, M. (2018). Anchoring effects of S-terminated Ti2C MXene for lithium-sulfur batteries: A first-principles study. *Applied Surface Science*, *455*, 522–526. doi:10.1016/j.apsusc.2018.05.200

Liu, X., Xing, Z., Zhang, H., Wang, W., Zhang, Y., Li, Z., Wu, X., Yu, X., & Zhou, W. (2016). Fabrication of 3 D Mesoporous Black TiO2/MoS2/TiO2 Nanosheets for Visible-Light-Driven Photocatalysis. *ChemSusChem*, *9*(10), 1118–1124. doi:10.1002/cssc.201600170 PMID:27111114

Liu, X., Yan, Y., Da, Z., Shi, W., Ma, C., Lv, P., Tang, Y., Yao, G., Wu, Y., Huo, P., & Yan, Y. (2014). Significantly enhanced photocatalytic performance of CdS coupled WO3 nanosheets and the mechanism study. *Chemical Engineering Journal*, *241*, 243–250. doi:10.1016/j.cej.2013.12.058

Liu, Y., Xu, C., Lu, J., Zhu, Z., Zhu, Q., Manohari, A. G., & Shi, Z. (2018). Template-free synthesis of porous ZnO/Ag microspheres as recyclable and ultra-sensitive SERS substrates. *Applied Surface Science*, *427*, 830–836. doi:10.1016/j.apsusc.2017.07.229

Liu, Y., Zeng, X., Hu, X., Hu, J., & Zhang, X. (2019). Two-dimensional nanomaterials for photocatalytic water disinfection: Recent progress and future challenges. *Journal of Chemical Technology and Biotechnology (Oxford, Oxfordshire)*, *94*(1), 22–37. doi:10.1002/jctb.5779

Liu, Z. Q., Ma, J., Cui, Y. H., Zhao, L., & Zhang, B. P. (2011). Factors affecting the catalytic activity of multiwalled carbon nanotubes for ozonation of oxalic acid. *Separation and Purification Technology*, *78*(2), 147–153. doi:10.1016/j.seppur.2011.01.034

Li, W., Xiao, L., & Qin, C. (2010). The characterization and thermal investigation of chitosan-Fe3O4 nanoparticles synthesized via a novel one-step modifying process. *Journal of Macromolecular Science. Part A*, *48*(1), 57–64.

Li, X., Liu, G., & Zhao, J. (1999). Two competitive primary processes in the photodegradation of cationic triaryl methane dyes under visible irradiation in TiO_2 dispersions. *New Journal of Chemistry*, *23*(12), 1193–1196. doi:10.1039/a906765e

Li, X., Mei, Q., Chen, L., Zhang, H., Dong, B., Dai, X., & Zhou, J. (2019). Enhancement in adsorption potential of microplastics in sewage sludge for metal pollutants after the wastewater treatment process. *Water Research*, *157*, 228–237. doi:10.1016/j.watres.2019.03.069 PMID:30954698

Li, X., Qu, L., Dong, Y., Han, L., Liu, E., Fang, S., Zhang, Y., & Wang, T. (2014). A review of recent research progress on the Astragalus genus. *Molecules (Basel, Switzerland)*, *19*(11), 18850–18880. doi:10.3390/molecules191118850 PMID:25407722

Li, X., Zhou, H., Yu, P., Su, L., Ohsaka, T., & Mao, L. (2008b). A Miniature glucose/O2 biofuel cell with single-walled carbon nanotubes-modified carbon fiber microelectrodes as the substrate. *Electrochemistry Communications*, *10*(6), 851–854. https://doi.org/10.1016/j.elecom.2008.03.019

Li, Y. H., Ding, J., Luan, Z., Di, Z., Zhu, Y., Xu, C., Wu, D., & Wei, B. (2003c). Competitive adsorption of Pb2+, Cu2+ and CD2+ ions from aqueous solutions by multiwalled carbon nanotubes. *Carbon, 41*(14), 2787–2792. https://doi.org/10.1016/S0008-6223(03)00392-0

Li, Y. H., Di, Z., Ding, J., Wu, D., Luan, Z., & Zhu, Y. (2005). Adsorption thermodynamic, kinetic and desorption studies of Pb2+ on carbon nanotubes. *Water Research, 39*(4), 605–609. https://doi.org/10.1016/j.watres.2004.11.004

Li, Y. H., Luan, Z., Xiao, X., Zhou, X., Xu, C., Wu, D., & Wei, B. (2003b). Removal Cu^{2+} ions from aqueous solutions by carbon nanotubes. *Adsorption Science and Technology, 21*, 475–485.

Li, Y. H., Wang, S. G., Cao, A. Y., Zhao, D., Zhang, X. F., Xu, C. L., Luan, Z. K., Ruan, D. B., Liang, J., Wu, D. H., & Wei, B. Q. (2001). Adsorption of fluoride from water by amorphous alumina supported on carbon nanotubes. *Chemical Physics Letters, 350*(5-6), 412–416. doi:10.1016/S0009-2614(01)01351-3

Li, Y. H., Wang, S. G., Wei, J. Q., Zhang, X. F., Xu, C. L., Luan, Z. K., Wu, D. H., & Wei, B. Q. (2002). Lead adsorption on carbon nanotubes. *Chemical Physics Letters, 357*(3-4), 263–266. doi:10.1016/S0009-2614(02)00502-X

Li, Y. H., Wang, S. W., Luan, Z. L., Ding, J. D., Xu, C. X., & Wu, D. (2003a). Adsorption of cadmium (II). *Carbon, 41*(5), 1057–1062. https://doi.org/10.1016/S0008-6223(02)00440-2

Li, Y. H., Wang, S., Cao, A., Zhao, D., Zhang, X., Xu, C., Luan, Z., Ruan, D., Liang, J., Wu, D., & Wei, B. (2001). Adsorption of fluoride fromwater by amorphous alumina supported on carbon nanotubes. *Chemical Physics Letters, 350*(5–6), 412–416. https://doi.org/10.1016/S0009-2614(01)01351-3

Li, Y. H., Wang, S., Zhang, X., Wei, J., Xu, C., Luan, Z., & Wu, D. (2003d). Adsorption of fluoride from water by aligned carbon nanotubes. *Materials Research Bulletin, 38*(3), 469–476. https://doi.org/10.1016/S0025-5408(02)01063-2

Li, Y. H., Zhu, Y., Zhao, Y., Wu, D., & Luan, Z. (2006). Different morphologies of carbon nanotubes effect on the lead removal from aqueous solution. *Diamond and Related Materials, 15*(1), 90–94. https://doi.org/10.1016/j.diamond.2005.07.004

Li, Y., Chen, X., Sun, Y., Meng, X., Dall'Agnese, Y., Chen, G., Dall'Agnese, C., Ren, H., Sasaki, S., Tamiaki, H., & Wang, X.-F. (2020). Chlorosome-Like Molecular Aggregation of Chlorophyll Derivative on Ti3C2Tx MXene Nanosheets for Efficient Noble Metal-Free Photocatalytic Hydrogen Evolution. *Advanced Materials Interfaces, 7*(8), 1902080. doi:10.1002/admi.201902080

Li, Y., Deng, X., Tian, J., Liang, Z., & Cui, H. (2018). Ti3C2 MXene-derived Ti3C2/TiO2 nanoflowers for noble-metal-free photocatalytic overall water splitting. *Applied Materials Today, 13*, 217–227. doi:10.1016/j.apmt.2018.09.004

Li, Y., Ding, L., Yin, S., Liang, Z., Xue, Y., Wang, X., Cui, H., & Tian, J. (2019). Photocatalytic H2 Evolution on TiO2 Assembled with Ti3C2 MXene and Metallic 1T-WS2 as Co-catalysts. *Nano-Micro Letters, 12*(1), 6. doi:10.100740820-019-0339-0 PMID:34138075

Li, Y., Guo, S., Zhu, Y., Yan, H., Qian, D. W., Wang, H. Q., Yu, J. Q., & Duan, J. A. (2019). Comparative analysis of twenty-five compounds in different parts of Astragalus membranaceus var. mongholicus and Astragalus membranaceus by UPLC-MS/MS. *Journal of Pharmaceutical Analysis, 9*(6), 392–399. doi:10.1016/j.jpha.2019.06.002 PMID:31890338

Li, Y.-H., Ding, J., Luan, Z., Di, Z., Zhu, Y., Xu, C., Wu, D., & Wei, B. (2003). Competitive adsorption of Pb(II)· Cu(II), Cd^{2+} ions from aqueous solutions by multiwalled carbon nanotubes. *Carbon, 41*(14), 2787–2792. doi:10.1016/S0008-6223(03)00392-0

Li, Y., Yin, Z., Ji, G., Liang, Z., Xue, Y., Guo, Y., Tian, J., Wang, X., & Cui, H. (2019). 2D/2D/2D heterojunction of Ti3C2 MXene/MoS2 nanosheets/TiO2 nanosheets with exposed (001) facets toward enhanced photocatalytic hydrogen production activity. *Applied Catalysis B: Environmental, 246*, 12–20. doi:10.1016/j.apcatb.2019.01.051

Li, Y., Yu, Y., Wu, L., & Zhi, J. (2013). Processable Polyaniline/Titania nanocomposites with good photocatalytic and conductivity properties prepared via peroxo-titanium complex catalyzed emulsion polymerization approach. *Applied Surface Science*, *273*, 135–143. doi:10.1016/j.apsusc.2013.01.213

Li, Y., Yu, Y., Wu, L., & Zhi, J. (2013). Processable polyaniline/titania nanocomposites with good photocatalytic and conductivity properties prepared via peroxo-titanium complex catalyzed emulsion polymerization approach. *Applied Surface Science*, *273*, 135–143. https://doi.org/10.1016/j.apsusc.2013.01.213

Li, Y., Zhang, B.-P., Zhao, J.-X., Ge, Z.-H., Zhao, X.-K., & Zou, L. (2013). ZnO/carbon quantum dots heterostructure with enhanced photocatalytic properties. *Applied Surface Science*, *279*, 367–373. doi:10.1016/j.apsusc.2013.04.114

Li, Z., Sellaoui, L., Franco, D., Netto, M. S., Georgin, J., Dotto, G. L., & Li, Q. (2020). Adsorption of hazardous dyes in functionalized multiwalled carbon nanotubes in single and binary systems. Experimental study and physicochemical interpretation of adsorption mechanism. *Chemical Engineering Journal*, *389*, 124467. doi:10.1016/j.cej.2020.124467

Logan, B. E., Hamelers, B., Rozendal, R., Schröder, U., Keller, J., Freguia, S., Aelterman, P., Verstraete, W., & Rabaey, K. (2006). Microbial fuel cells: Methodology and technology. *Environmental Science & Technology*, *40*(17), 5181–5192. https://doi.org/10.1021/es0605016

Long, M., Brame, J., Qin, F., Bao, J., Li, Q., & Alvarez, P. J. J. (2017). Phosphate Changes Effect of Humic Acids on TiO2 Photocatalysis: From Inhibition to Mitigation of Electron–Hole Recombination. *Environmental Science & Technology*, *51*(1), 514–521. doi:10.1021/acs.est.6b04845 PMID:27982576

Long, R. Q., & Yang, R. T. (2001). Carbon nanotubes as superior sorbent for dioxin removal. *Journal of the American Chemical Society*, *123*(9), 2058–2059. https://doi.org/10.1021/ja003830l

Low, J., Cheng, B., & Yu, J. (2017). Surface modification and enhanced photocatalytic CO2 reduction performance of TiO2: A review. *Applied Surface Science*, *392*, 658–686. doi:10.1016/j.apsusc.2016.09.093

Low, K. S., & Lee, C. K. (1991). Cadmium uptake by the moss, Calymperes delessertii, Besch. *Bioresource Technology*, *38*(1), 1–6. doi:10.1016/0960-8524(91)90214-5

Lu, J., Kumar, B., Castro, M., & Feller, J. F. (2009). Vapour sensing with conductive polymer nanocomposites (CPC): Polycarbonate-carbon nanotubes transducers with hierarchical structure processed by spray layer by layer. *Sensors and Actuators. Part B*, *140*(2), 451–460. doi:10.1016/j.snb.2009.05.006

Lu, C., & Chiu, H. (2006). Adsorption of zinc (II). *Chemical Engineering Science*, *61*(4), 1138–1145. https://doi.org/10.1016/j.ces.2005.08.007

Lu, C., & Chiu, H. (2008). Chemical modification of multiwalled carbon nanotubes for sorption of Zn2+ from aqueous solution. *Chemical Engineering Journal*, *139*(3), 462–468. https://doi.org/10.1016/j.cej.2007.08.013

Lu, C., Chiu, H., & Liu, C. (2006). Removal of zinc (II). *Industrial & Engineering Chemistry Research*, *45*(8), 2850–2855. https://doi.org/10.1021/ie051206h

Lu, C., & Chu, C. (2005). Effects of acetic acid on the regrowth of heterotrophic bacteria in the drinking water distribution system. *World Journal of Microbiology & Biotechnology*, *21*(6–7), 989–998. https://doi.org/10.1007/s11274-004-7554-6

Lu, C., & Liu, C. (2006). Removal of nickel (II). *Journal of Chemical Technology and Biotechnology (Oxford, Oxfordshire)*, *81*(12), 1932–1940. https://doi.org/10.1002/jctb.1626

Lu, C., Liu, C., & Rao, G. P. (2008). Comparisons of sorbent cost for the removal of Ni^{2+} from aqueous solution by carbon nanotubes and granular activated carbon. *Journal of Hazardous Materials*, *151*(1), 239–246. https://doi.org/10.1016/j.jhazmat.2007.05.078

Lu, C., Tranca, D., Zhang, J., Rodríguez Hernández, F., Su, Y., Zhuang, X., Zhang, F., Seifert, G., & Feng, X. (2017). Molybdenum Carbide-Embedded Nitrogen-Doped Porous Carbon Nanosheets as Electrocatalysts for Water Splitting in Alkaline Media. *ACS Nano*, *11*(4), 3933–3942. doi:10.1021/acsnano.7b00365 PMID:28291319

Lukatskaya, M. R., Mashtalir, O., & Chang, R. E., Dall'Agnese, Y., Rozier, P., Taberna Pierre, L., . . . Gogotsi, Y. (2013). Cation Intercalation and High Volumetric Capacitance of Two-Dimensional Titanium Carbide. *Science*, *341*(6153), 1502–1505. Advance online publication. doi:10.1126cience.1241488 PMID:24072919

Luo, H., Shi, Z., Li, N., Gu, Z., & Zhuang, Q. (2001). Investigation of the electrochemical and electro catalytic behaviour of single-wall carbon nanotube film on a glassy carbon electrode. *Analytical Chemistry*, *73*(5), 915–920. doi:10.1021/ac000967l PMID:11289436

Luo, J. Z., Gao, L. Z., Leung, Y. L., & Au, C. T. (2000). The decomposition of NO on CNTs and 1 wt% Rh/CNTs. *Catalysis Letters*, *66*(1–2), 91–97.

Luo, W., Yang, Z., Li, Z., Zhang, J., Liu, J., Zhao, Z., Wang, Z., Yan, S., Yu, T., & Zou, Z. (2011). Solar hydrogen generation from seawater with a modified $BiVO_4$ photoanode. *Energy & Environmental Science*, *4*(10), 4046–4051. doi:10.1039/c1ee01812d

Luo, Y., Tan, G., Dong, G., Ren, H., & Xi, A. (2016). A comprehensive investigation of tetragonal Gd-doped $BiVO_4$ with enhanced photocatalytic performance under sun-light. *Applied Surface Science*, *364*, 156–165. doi:10.1016/j.apsusc.2015.12.100

Lu, W., Li, J., Sheng, Y., Zhang, X., You, J., & Chen, L. (2017). One-pot synthesis of magnetic iron oxide nanoparticle-multiwalled carbon nanotube composites for enhanced removal of Cr(VI) from aqueous solution. *Journal of Colloid and Interface Science*, *505*, 1134–1146. doi:10.1016/j.jcis.2017.07.013 PMID:28709373

Lv, X., Zhang, Y., Wang, Y., Zhang, G., & Liu, J. (2021). Formation of carbon nanofibres/nanotubes by chemical vapor deposition using Al_2O_3/KOH. *Diamond and Related Materials*, *113*, 108265. doi:10.1016/j.diamond.2021.108265

Lyon, D. Y., Adams, L. K., Falkner, J. C., & Alvarezt, P. J. J. (2006). Antibacterial activity of fullerene water suspensions: Effects of preparation method and particle size. *Environmental Science & Technology*, *40*(14), 4360–4366. https://doi.org/10.1021/es0603655

Lyon, D. Y., Brown, D. A., & Alvarez, P. J. J. (2008). Implications and potential applications of bactericidal fullerene water suspensions: Effect of nc60 concentration, exposure conditions and shelf life. *Water Science and Technology*, *57*(10), 1533–1538. https://doi.org/10.2166/wst.2008.282

Lyon, D. Y., Fortner, J. D., Sayes, C. M., Colvin, V. L., & Hughe, J. B. (2005). Bacterial cell association and antimicrobial activity of a C60 water suspension. *Environmental Toxicology and Chemistry*, *24*(11), 2757–2762. https://doi.org/10.1897/04-649r.1

Lytle, D. A., Sorg, T., Wang, L., & Chen, A. (2014). The accumulation of radioactive contaminants in drinking water distribution systems. *Water Research*, *50*, 396–407. doi:10.1016/j.watres.2013.10.050 PMID:24275108

Machado, F. M., Bergmann, C. P., Fernandes, T. H. M., Lima, E. C., Royer, B., Calvete, T., & Fagan, S. B. (2011). Adsorption of Reactive Red M-2BE dye from water solutions by multi-walled carbon nanotubes and activated carbon. *Journal of Hazardous Materials*, *192*(3), 1122–1131. https://doi.org/10.1016/j.jhazmat.2011.06.020

Maensiri, S., Laokul, P., & Klinkaewnarong, J. (2008). Indium oxide (in 2O3) nanoparticles using aloe vera plant extract: Synthesis and optical properties. *Journal of Optoelectronics and Advanced Materials*, *10*, 161–165.

Mahadeva, S. K., & Kim, J. (2009). Hybrid nanocomposite based on cellulose and tin oxide: Growth, structure, tensile and electrical characteristics. *Science and Technology of Advanced Materials*, *12*, 055006.

Mahmoud, A. M., Ibrahim, F. A., Shaban, S. A., & Youssef, N. A. (2015). Adsorption of heavy metal ion from aqueous solution by nickel oxide nano catalyst prepared by different methods. *Egyptian Journal of Petroleum*, *24*, 27–35.

Mai, F. D., Lu, C. S., Wu, C. W., Huang, C. H., Chen, J. Y., & Chen, C. C. (2008). Mechanisms of photocatalytic degradation of Victoria Blue R using Nano-TiO_2. *Separation and Purification Technology*, *62*(2), 423–436. doi:10.1016/j.seppur.2008.02.006

Ma, J., Sun, Z., Wang, Z., & Zhou, X. (2016). Preparation of ZnO–cellulose nanocomposites by different cellulose solution systems with a colloid mill. *Cellulose (London, England)*, *23*, 3703–3715.

Makkar, H. P. S., & Becker, K. (1997). Nutrients and antiquality factors in different morphological parts of the Moringa oleifera tree. *The Journal of Agricultural Science*, *128*(3), 311–322. doi:10.1017/S0021859697004292

Maldonado, M. I., Saggioro, E., Peral, J., Rodríguez-Castellón, E., Jiménez-Jiménez, J., & Malato, S. (2019). Hydrogen generation by irradiation of commercial CuO + TiO2 mixtures at solar pilot plant scale and in presence of organic electron donors. *Applied Catalysis B: Environmental*, *257*, 117890. doi:10.1016/j.apcatb.2019.117890

Mallakpour, S., & Khadem, E. (2016). Carbon nanotube-metal oxide nanocomposites: Fabrication, properties and applications. *Chemical Engineering Journal*, *302*, 344–367. doi:10.1016/j.cej.2016.05.038

Mangun, C. L., Yue, Z., Economy, J., Maloney, S., Kemme, P., & Cropek, D. (2001). Adsorption of organic contaminants from water using tailored ACFs. *Chemistry of Materials*, *13*(7), 2356–2360. https://doi.org/10.1021/cm000880g

Manikandan, V. (2017). Green synthesis of silver oxide nanoparticles and its antibacterial activity against dental pathogens. *Biotech.*, *3*(7). . doi:10.1007/s13205-017-0670-4

Manjari, G., Saran, S., Arun, T., Vijaya Bhaskara Rao, A., & Devipriya, S. P. (2017). Catalytic and recyclability properties of phytogenic copper oxide nanoparticles derived from Aglaia elaeagnoidea flower extract. *Journal of Saudi Chemical Society*, *21*(5), 610–618. doi:10.1016/j.jscs.2017.02.004

Manju, G. N., & Anirudhan, T. S. (1997). Use of coconut fibre pith-based pseudo-activated carbon for chromium (VI) removal. *Indian Journal of Environmental Health*, *39*, 289–298.

Mansoori, G. A., & Soelaiman, T. F. (2005). *Nanotechnology—An introduction for the standards community*. ASTM International.

Marchiol, L. (2012). Synthesis of metal nanoparticles in living plants. *Italian Journal of Agronomy*, *7*(3), 274–282. doi:10.4081/ija.2012.e37

Martinez, M., & Silley, P. (2010). *Antimicrobial Drug Resistance*. In F. Cunningham, J. Elliott, & P. Lees (Eds.), *Comparative and Veterinary Pharmacology* (pp. 227–264). Springer Berlin Heidelberg.

Martins, N. C. T., Freire, C. S. R., Neto, C. P., Silvestre, A. J. D., Causio, J., Baldi, G., Sadocco, P., & Trindade, T. (2013). Antibacterial paper based on composite coatings of nanofibrillated cellulose and ZnO. *Colloids and Surfaces. A, Physicochemical and Engineering Aspects*, *417*, 111–119.

Martins, R. J., Pardo, R., & Boaventura, R. A. (2004). Cadmium (II) and zinc (II) adsorption by the aquatic moss Fontinalis antipyretica: Effect of temperature, pH and water hardness. *Water Research*, *38*(3), 693–699. doi:10.1016/j.watres.2003.10.013 PMID:14723939

Maryam, A. T., & Torai, M. (2011). Permanent hard water softening using carbon nanotube sheets. *Desalination, 268*(1-3), 208–213. doi:10.1016/j.desal.2010.10.028

Masciangioli, T., & Zhang, W. X. (2003). Environmental technologies at the nanoscale. *Environmental Science & Technology, 37*(5), 102A–108A. doi:10.1021/es0323998 PMID:12666906

Masenelli-Varlot, K., McRae, E., & Dupont-Pavlovsky, N. (2002). Comparative adsorption of simple molecules on carbon nanotubes. *Applied Surface Science, 196*(1–4), 209–215. doi:10.1016/S0169-4332(02)00059-4

Matlock, M. M., Howerton, B. S., & Atwood, D. A. (2002). Chemical precipitation of heavy metals from acid mine drainage. *Water Research, 36*, 4757–4764.

Matsunaga, T., Nakasono, S., Kitajima, Y., & Horiguchi, K. (1994). Electrochemical disinfection of bacteria in drinking water using activated carbon fibers. *Biotechnology and Bioengineering, 43*(5), 429–433. https://doi.org/10.1002/bit.260430511

Matthews, R. W. (1988). Kinetics of photocatalytic oxidation of organic solutes over titanium dioxide. *Journal of Catalysis, 111*(2), 264–272. doi:10.1016/0021-9517(88)90085-1

Ma, W., Wang, N., Du, Y., Tong, T., Zhang, L., Andrew Lin, K.-Y., & Han, X. (2019). One-step synthesis of novel Fe3C@nitrogen-doped carbon nanotubes/graphene nanosheets for catalytic degradation of Bisphenol A in the presence of peroxymonosulfate. *Chemical Engineering Journal, 356*, 1022–1031. doi:10.1016/j.cej.2018.09.093

Mekasuwandumrong, O., Pawinrat, P., Praserthdam, P., & Panpranot, J. (2010). Effects of synthesis conditions and annealing post-treatment on the photocatalytic activities of ZnO nanoparticles in the degradation of methylene blue dye. *Chemical Engineering Journal, 164*(1), 77–84. doi:10.1016/j.cej.2010.08.027

Mendoza-Mendoza, E., Nuñez-Briones, A. G., García-Cerda, L. A., Peralta-Rodríguez, R. D., & Montes-Luna, A. J. (2018). One-step synthesis of ZnO and Ag/ZnO heterostructures and their photocatalytic activity. *Ceramics International, 44*(6), 6176–6180. doi:10.1016/j.ceramint.2018.01.001

Meng, N., Zhou, Y., Nie, W., & Chen, P. (2016). Synthesis of CdS-decorated RGO nanocomposites by reflux condensation method and its improved photocatalytic activity. *Journal of Nanoparticle Research, 18*(8), 241. doi:10.100711051-016-3522-y

Menon, S., Agarwal, H., & Kumar, V. (2021). Catalytical degradation of industrial dyes using biosynthesized selenium nanoparticles and evaluating its antimicrobial activities. *Sustainable Environment Research, 31*(1), 2. Advance online publication. doi:10.118642834-020-00072-6

Merkoçi, A. (2006). Carbon nanotubes in analytical sciences. *Mikrochimica Acta, 152*(3–4), 157–174. doi:10.100700604-005-0439-z

Meshko, V., Markovska, L., Mincheva, M., & Rodrigues, A. E. (2001). Adsorption of basic dyes on granular activated carbon and natural zeolite. *Water Research, 35*(14), 3357–3366. https://doi.org/10.1016/s0043-1354(01)00056-2

Meunier, N., Laroulandie, J., Blais, J. F., & Tyagi, R. D. (2003). Cocoa shells for heavy metal removal from acidic solutions. *Bioresource Technology, 90*(3), 255–263. doi:10.1016/S0960-8524(03)00129-9 PMID:14575948

Micro-encapsulation. (2022, April 28). In *Wikipedia.* https://en.wikipedia.org/wiki/Micro-encapsulation

Minteer, S. D., Liaw, B. Y., & Cooney, M. J. (2007). Enzyme-based biofuel cells. *Current Opinion in Biotechnology, 18*(3), 228–234. https://doi.org/10.1016/j.copbio.2007.03.007

Min, W., Yinsheng, C., Chao, N., Mingyan, D., & Duo, D. (2013). Lanthanum and boron co-doped $BiVO_4$ with enhanced visible light photocatalytic activity for degradation of methyl orange. *Journal of Rare Earths*, *31*(9), 878–884. doi:10.1016/S1002-0721(12)60373-1

Mishra, R. K., Choi, G. J., Sohn, Y., Lee, S. H., & Gwag, J. S. (2020). A novel RGO/N-RGO supercapacitor architecture for a wide voltage window, high energy density and long-life via voltage holding tests. *Chemical Communications*, *56*(19), 2893-2896. doi:10.1039/D0CC00249F

Mishra, G., & Tripathy, M. (1993). A critical review of the treatment. for decolorization of textile effluent. *Colourage*, *40*, 35–38.

Mittal, A. K., Chisti, Y., & Banerjee, U. C. (2013). Synthesis of metallic nanoparticles using plant extracts. *Biotechnology Advances*, *31*(2), 346–356. doi:10.1016/j.biotechadv.2013.01.003 PMID:23318667

Moghaddam, H. M., & Nasirian, S. (2014). Article. *Applied Surface Science*, *317*, 117–124.

Mohamed, R. R., Abu Elella, M. H., Sabaa, M. W., & Saad, G. R. (2018). Synthesis of an efficient adsorbent hydrogel based on biodegradable polymers for removing crystal violet dye from aqueous solution. *Cellulose (London, England)*, *25*(11), 6513–6529. doi:10.100710570-018-2014-x

Mohanraj, J., Durgalakshmi, D., Rakkesh, R. A., Balakumar, S., Rajendran, S., & Karimi-Maleh, H. (2020). Facile synthesis of paper based graphene electrodes for point of care devices: A double stranded DNA (dsDNA) biosensor. *Journal of Colloid and Interface Science*, *566*, 463–472. doi:10.1016/j.jcis.2020.01.089 PMID:32032811

Mojumdar, S. C., & Raki, L. (2005). Preparation and properties of calcium silicate hydrate- poly(vinyl alcohol) nanocomposite materials. *Journal of Thermal Analysis and Calorimetry*, *82*(1), 89–95. doi:10.1007/s10973-005-0846-8

Mo, L., Pang, H., Tan, Y., Zhang, S., & Li, J. (2019). 3D multi-wall perforated nanocellulose-based polyethylenimine aerogels for ultrahigh efficient and reversible removal of Cu (II) ions from water. *Chemical Engineering Journal*, *378*, 122157.

Monda, S., Roy, N., Laskar, R. A., Sk, I., Basu, S., Mandal, D., & Begum, N. A. (2011). Biogenic synthesis of Ag, Au and bimetallic Au/Ag alloy nanoparticles using aqueous extract of mahogany (Swietenia mahogani JACQ.) leaves. *Colloid Surf. B*, *82*(2), 497–504. doi:10.1016/j.colsurfb.2010.10.007 PMID:21030220

Monthioux, M., & Kuznetsov, V. L. (2006). Who should be given the credit for the discovery of carbon nanotubes? *Carbon*, *44*(9), 1621–1623. doi:10.1016/j.carbon.2006.03.019

Moon, R. J., Martini, A., Nairn, J., Simonsen, J., & Youngblood, J. (2011). Cellulose nanomaterials review: Structure, properties and nanocomposites. *Chemical Society Reviews*, *40*(7), 3941–3994.

Moon, S. I., Paek, K. K., Lee, Y. H., Park, H. K., Kim, J. K., Kim, S. W., & Ju, B. K. (2008). Biasheating recovery of MWCNT gas sensor. *Materials Letters*, *62*(16), 2422–2425. https://doi.org/10.1016/j.matlet.2007.12.027

Mor, G. K., Shankar, K., Paulose, M., Varghese, O. K., & Grimes, C. A. (2006). Use of highly-ordered TiO2 nanotube arrays in dye-sensitized solar cells. *Nano Letters*, *6*(2), 215–218. doi:10.1021/nl052099j PMID:16464037

Morones, J. R., Elechiguerra, J. L., Camacho, A., Holt, K., Kouri, J. B., Ramírez, J. T., & Yacaman, M. J. (2005). The bactericidal effect of silver nanoparticles. *Nanotechnology*, *16*(10), 2346–2353. https://doi.org/10.1088/0957-4484/16/10/059

Morozan, A., Stamatin, L., Nastase, F., Dumitru, A., Vulpe, S., Nastase, C., Stamatin, I., & Scott, K. (2007). The biocompatibility microorganisms-carbon nanostructures for applications in microbial fuel cells. *Physica Status Solidi*, *204*(6), 1797–1803. https://doi.org/10.1002/pssa.200675344

Mostafa, A. M., Mwafy, E. A., & Toghan, A. (2021). ZnO nanoparticles decorated carbon nanotubes via pulsed laser ablation method for degradation of methylene blue dyes. *Colloids and Surfaces. A, Physicochemical and Engineering Aspects*, *627*, 127204. doi:10.1016/j.colsurfa.2021.127204

Mostafavi, S. T., Mehrnia, M. R., & Rashidi, A. M. (2009). Preparation of nanofilter from carbon nanotubes for application in virus removal from water. *Desalination*, *238*(1–3), 271–280. https://doi.org/10.1016/j.desal.2008.02.018

Mousaa, S. A., Shalan, A. E., Hassan, H. H., Ebnawale, A. A., & Khairya S. A. (2022). Enhanced the photocatalytic degradation of titanium dioxide nanoparticles synthesized by different plant extracts for wastewater treatment. *Journal of Molecular Structure, 1250*(3).

Moztahida, M., Nawaz, M., Kim, J., Shahzad, A., Kim, S., Jang, J., & Lee, D. S. (2019). Reduced graphene oxide-loaded-magnetite: A Fenton-like heterogeneous catalyst for photocatalytic degradation of 2-methylisoborneol. *Chemical Engineering Journal*, *370*, 855–865. doi:10.1016/j.cej.2019.03.214

Mubarak, N. M., Sazila, N., & Nizamuddin, S. (2017). Adsorptive removal of phenol from aqueous solution by using carbon nanotubes and magnetic biochar. *NanoWorld J.* doi:10.17756/nwj.2017-043

Mubarak, N., Daniel, S., Khalid, M., & Tan, J. (2012). Comparative study of functionalize and non-functionalized carbon nanotube for Removal of Copper from polluted water. *Int. J. Chem. Environ. Eng., 3*(5).

Mubarak, N. M., Sahu, J. N., Abdullah, E. C., & Jayakumar, N. S. (2014). Removal of heavy metals from wastewater using carbon nanotubes. *Separation and Purification Reviews*, *43*(4), 311–338. https://doi.org/10.1080/15422119.2013.821996

Mubarak, N. M., Sahu, J. N., Abdullah, E. C., Jayakumar, N. S., & Ganesan, P. (2016). Microwave-assisted synthesis of multi-walled carbon nanotubes for enhanced removal of Zn(II) from wastewater. *Research on Chemical Intermediates*, *42*(4), 3257–3281. https://doi.org/10.1007/s11164-015-2209-9

Mubarak, N. M., Sahu, J. N., Wong, J. R., Jayakumar, N. S., Ganesan, P., & Abdullah, E. C. (2015). *Overview on the functionalization of carbon nanotubes. Chemical functionalization of carbon nanomaterials*. CRC Press.

Mubarak, N. M., Yusof, F., & Alkhatib, M. F. (2011). The production of carbon nanotubes using two-stage chemical vapor deposition and their potential use in protein purification. *Chemical Engineering Journal*, *168*(1), 461–469. https://doi.org/10.1016/j.cej.2011.01.045

Mubarak, N. M., Yusof, F., Alkhatib, M. F., Ameen, E., Khalid, M., Mohammed, A. S., Muataz, A., Qudsieh, I. Y., & Rashmi, W. (2010). Optimization of CNTs production using full factorial design and its advanced application in protein purification. *International Journal of Nanoscience*, *09*(3), 181–192. https://doi.org/10.1142/S0219581X10006648

Murty, B. S., Shankar, P., & Baldev Raj, R. B. (2013). *James Murdas. Text book of nano science and nano technology*. Academic Press.

Muyibi, S. A., Ambali, A. R., & Eissa, G. S. (2008). Development-induced water pollution in Malaysia: Policy implication from an econometric analysis. *Water Policy*, *10*(2), 193–206. doi:10.2166/wp.2008.039

Nabid, M. R., Golbabaee, M., Moghaddam, A. B., & Dinarvand, R., & Sedghi, R. (2008). P./TiO2 Nanocomposite: Enzymatic Synthesis and Electrochemical Properties. *International Journal of Electrochemical Science*, *3*, 1117–1126.

Nabi, G., Majid, A., Riaz, A., Alharbi, T., Arshad Kamran, M., & Al-Habardi, M. (2021). Green synthesis of spherical TiO_2 nanoparticles using Citrus Limetta extract: Excellent photocatalytic water decontamination agent for RhB dye. *Inorganic Chemistry Communications*, *129*, 108618. doi:10.1016/j.inoche.2021.108618

Nagarnaik, P. B., Bhole, A. G., & Natarajan, G. S. (2003). Adsorption of arsenic on flyash. *Indian Journal of Environmental Health*, *45*(1), 1–4. PMID:14723275

Nag, C. M., Chen, P. C., & Kumar, S. (2012). Hydrothermal crystallization of titania on silver nucleation sites for the synthesis of visible light nano-photocatalysts—Enhanced photoactivity using rhodamine 6G. *Applied Catalysis A*, *433–434*, 75–80.

Naguib, M., Adams, R. A., Zhao, Y., Zemlyanov, D., Varma, A., Nanda, J., & Pol, V. G. (2017). Electrochemical performance of MXenes as K-ion battery anodes. *Chemical Communications*, *53*(51), 6883–6886. doi:10.1039/C7CC02026K PMID:28607970

Naguib, M., Kurtoglu, M., Presser, V., Lu, J., Niu, J., Heon, M., Hultman, L., Gogotsi, Y., & Barsoum, M. W. (2011). Two-dimensional nanocrystals produced by exfoliation of Ti3 AlC2. *Advanced Materials*, *23*(37), 4248–4253. doi:10.1002/adma.201102306 PMID:21861270

Naguib, M., Mashtalir, O., Carle, J., Presser, V., Lu, J., Hultman, L., Gogotsi, Y., & Barsoum, M. W. (2012). Two-Dimensional Transition Metal Carbides. *ACS Nano*, *6*(2), 1322–1331. doi:10.1021/nn204153h PMID:22279971

Namasivayam, C., & Ranganathan, K. (1995). Removal of Cd (II) from wastewater by adsorption on "waste" Fe (III) Cr (III) hydroxide. *Water Research*, *29*(7), 1737–1744. doi:10.1016/0043-1354(94)00320-7

Naseem, T., & Durrani, T. (2020). The role of some important metal oxide nanoparticles for wastewater and antibacterial applications: A review. *Environmental Chemistry and Ecotoxicology*, *3*, 59–75. doi:10.1016/j.enceco.2020.12.001

Nasirian, S., & Milani Moghaddam, H. M. (2014). Hydrogen gas sensing based on polyaniline/anatase titania nanocomposite. *International Journal of Hydrogen Energy*, *39*(1), 630–642. https://doi.org/10.1016/j.ijhydene.2013.09.152

Nasrollahzadeh, M., Sajjadi, M., Iravani, S., & Varma, R. S. (2021). Starch, cellulose, pectin, gum, alginate, chitin and chitosan derived (nano) materials for sustainable water treatment: A review. *Carbohydrate Polymers*, *251*, 116986. doi:10.1016/j.carbpol.2020.116986 PMID:33142558

Nawaz, M., Shahzad, A., Tahir, K., Kim, J., Moztahida, M., & Lee, D. S. (2020). Photo-Fenton reaction for the degradation of sulfamethoxazole using a multi-walled carbon nanotubes- $NiFe_2O_4$ composite. *Chemical Engineering Journal*, *382*, 123053. doi:10.1016/j.cej.2019.123053

Nelson, G. (2002). Application of microencapsulation in textiles. *International Journal of Pharmaceutics*, *242*(1-2), 55–62.

Nepal, D., Balasubramanian, S., Simonian, A. L., & Davis, V. A. (2008). Strong antimicrobial coatings: Single-walled carbon nanotubes armored with biopolymers. *Nano Letters*, *8*(7), 1896–1901. https://doi.org/10.1021/nl080522t

Neves, T., Frantz, T., Schenque, E., Gelesky, M., & Mortola, V. (2017). An investigation into an alternative photocatalyst based on CeO_2/Al_2O_3 in dye degradation. *Environmental Technology & Innovation*, *8*, 349–359. Advance online publication. doi:10.1016/j.eti.2017.08.003

Ngah, W. S. W., Teong, L. C., & Hanafiah, M. A. K. M. (2011). Adsorption of dyes and heavy metal ions by chitosan composites: A review. *Carbohydrate Polymers*, *83*(4), 1446–1456.

Ngomsik, A. F., Bee, A., Talbot, D., & Cote, G. (2012). Magnetic solid-liquid extraction of Eu (III), La (III), Ni (II) and Co(II) with maghemite nanoparticles. *Separation and Purification Technology*, *86*, 1–8. doi:10.1016/j.seppur.2011.10.013

Nguyen, H. Q., & Huh, J. S. (2006). Behavior of singlewalled carbon nanotube-based gas sensors at various temperatures of treatment and operation. *Sensors and Actuators. Part B*, *117*(2), 426–430. doi:10.1016/j.snb.2005.11.056

Nguyen, L. H., Phi, T. V., Phan, P. Q., Vu, H. N., Nguyen-Duc, C., & Fossard, F. (2007). Synthesis of multi-walled carbon nanotubes for NH3 gas detection. *Physica E, Low-Dimensional Systems and Nanostructures*, *37*(1–2), 54–57. https://doi.org/10.1016/j.physe.2006.12.006

Nidheesh, P. V., Zhou, M., & Oturan, M. A. (2018). An overview on the removal of synthetic dyes from water by electrochemical advanced oxidation processes. *Chemosphere*, *197*, 210–227. doi:10.1016/j.chemosphere.2017.12.195 PMID:29366952

Nimbalkar, D. B., Lo, H.-H., Ramacharyulu, P. V. R. K., & Ke, S.-C. (2016). Improved photocatalytic activity of RGO/MoS2 nanosheets decorated on TiO2 nanoparticles. *RSC Advances*, *6*(38), 31661–31667. doi:10.1039/C6RA01591C

Niu, H., & Volesky, B. (2003). Characteristics of anionic metal species biosorption with waste crab shells. *Hydrometallurgy*, *71*(1-2), 209–215. doi:10.1016/S0304-386X(03)00158-0

Noohpisheh, Z., Amiri, H., Farhadi, S., & Mohammadi-gholami, A. (2020). Green synthesis of AgZnO nanocomposites using Trigonella foenum-graecum leaf extract and their antibacterial, antifungal, antioxidant and photocatalytic properties. *Spectrochimica Acta. Part A: Molecular and Biomolecular Spectroscopy*, *240*, 118595.

Noreen, S., Mustafa, G., Ibrahim, S. M., Naz, S., Iqbal, M., Yaseen, M., & Nisar, J. (2020). Iron oxide (Fe_2O_3) prepared via green route and adsorption efficiency evaluation for an anionic dye: Kinetics, isotherms and thermodynamics studies. *Journal of Materials Research and Technology*, *9*(3), 4206–4217. doi:10.1016/j.jmrt.2020.02.047

Norouzi, A., & Nezamzadeh-Ejhieh, A. (2020). α-Fe_2O_3/Cu_2O heterostructure: Brief characterization and kinetic aspect of degradation of methylene blue. *Physica B, Condensed Matter*, *599*, 412422. doi:10.1016/j.physb.2020.412422

Noruzi, M., Zare, D., Khoshnevisan, K., & Davoodi, D. (2011). Rapid green synthesis of gold nanoparticles using Rosa hybrida petal extract at room temperature. Spectrachem. *Acta Part A*, *79*(5), 1461–1465. doi:10.1016/j.saa.2011.05.001 PMID:21616704

Noy, A., Park, H. G., Fornasiero, F., Holt, J. K., Grigoropoulos, C. P., & Bakajin, O. (2007). Nanofluidics in carbon nanotubes. *Nano Today*, *2*(6), 22–29. https://doi.org/10.1016/S1748-0132(07)70170-6

Nuengmatcha, P., Porrawatkul, P., Chanthai, S., Sricharoen, P., & Limchoowong, N. (2019). Enhanced photocatalytic degradation of methylene blue using Fe_2O_3/graphene/CuO nanocomposites under visible light. *Journal of Environmental Chemical Engineering*, *7*(6), 103438. doi:10.1016/j.jece.2019.103438

Nwanya, A. C., Razanamahandry, L. C., Bashir, A. K. H., Ikpo, C. O., Nwanya, S. C., Botha, S., Ntwampe, S. K. O., Ezema, F. I., Iwuoha, E. I., & Maaza, M. (2019). Industrial textile effluent treatment and antibacterial effectiveness of Zea mays L. Dry husk mediated bio-synthesized copper oxide nanoparticles. *Journal of Hazardous Materials*, *375*, 281–289. doi:10.1016/j.jhazmat.2019.05.004 PMID:31078988

Nyairo, W. N., Eker, Y. R., Kowenje, C., Akin, I., Bingol, H., Tor, A., & Ongeri, D. M. (2018). Efficient adsorption of lead(II) and copper(II) from aqueous solution phase using oxidized multiwalled carbon nanotubes/polypyrrole composite. *Separation Science and Technology*, *58*(1), 1–13.

O'Connell, D. W., Birkinshaw, C., & O'Dwyer, T. F. (2008). Heavy metal adsorbents prepared from the modification of cellulose: A review. *Bioresource Technology*, *99*, 6709–6724.

O'Mahony, T., Guibal, E., & Tobin, J. M. (2002). Reactive dye biosorption by Rhizopus arrhizus biomass. *Enzyme and Microbial Technology*, *31*(4), 456–463. https://doi.org/10.1016/S0141-0229(02)00110-2

Oberlin, A., Endo, M., & Koyama, T. (1976). Filamentous growth of carbon through benzene decomposition. *Journal of Crystal Growth*, *32*(3), 335–349. doi:10.1016/0022-0248(76)90115-9

Okano, M., Itoh, K., Fujishima, A., & Honda, K. (1987). Photoelectrochemical polymerization of pyrrole on TiO2 and its application to conducting pattern generation. *Journal of the Electrochemical Society*, *134*(4), 837–841. doi:10.1149/1.2100582

Oliveira, J. T. A., Silveira, S. B., Vasconcelos, I. M., Cavada, B. S., & Moreira, R. A. (1999). Compositional and nutritional attributes of seeds from the multiple purpose tree Moringa oleifera Lamarck. *Journal of the Science of Food and Agriculture, 79*(6), 815–820. doi:10.1002/(SICI)1097-0010(19990501)79:6<815::AID-JSFA290>3.0.CO;2-P

Olivera, S., Muralidhara, H. B., Venkatesh, K., Guna, V. K., Gopalakrishna, K., & Kumar, Y. (2016). Potential applications of cellulose and chitosan nanoparticles/composites in wastewater treatment: A review. *Carbohydrate Polymers, 153*, 600–618.

Ong, W.-J., Yeong, J.-J., Tan, L.-L., Goh, B. T., Yong, S.-T., & Chai, S.-P. (2014). Synergistic effect of graphene as a co-catalyst for enhanced daylight-induced photocatalytic activity of $Zn_{0.5}Cd_{0.5}S$ synthesized via an improved one-pot co-precipitation-hydrothermal strategy. *RSC Advances, 4*(103), 59676-59685. doi:10.1039/C4RA10467F

Ong, C. B., Ng, L. Y., & Mohammad, A. W. (2018). A review of ZnO nanoparticles as solar photocatalysts: Synthesis, mechanisms and applications. *Renewable & Sustainable Energy Reviews, 81*, 536–551. doi:10.1016/j.rser.2017.08.020

Osterloh, F. E., & Parkinson, B. A. (2011). Recent developments in solar water splitting photocatalysis. *MRS Bulletin, 36*(1), 17–22. doi:10.1557/mrs.2010.5

Oyaro, N., Ogendi, J., Murago, E. N., & Gitonga, E. (2007). *The contents of Pb, Cu, Zn and Cd in meat in Nairobi.* Academic Press.

Oyewo, O. A., Elemike, E. E., Onwudiwe, D. C., & Onyango, M. S. (2020). Metal oxide-cellulose nanocomposites for the removal of toxic metals and dyes from wastewater. *International Journal of Biological Macromolecules.* j.ijbiomac.2020.08.074 doi:10.1016/

Özacar, M., & Şengil, İ. A. (2002). engil ˙ IA. *Adsorption, 8*(4), 301–308. https://doi.org/10.1023/A:1021585413857

Pagnanelli, F., Mainelli, S., Vegliò, F., & Toro, L. (2003). Heavy metal removal by olive pomace: Biosorbent characterisation and equilibrium modelling. *Chemical Engineering Science, 58*(20), 4709–4717. doi:10.1016/j.ces.2003.08.001

Pagnanelli, F., Toro, L., & Sara, M. (2002). Removal and Recovery of Ni (II), Pb (II) and Cr (III) by Olive Mill Residue. *Environmental Science & Technology, 15*, 47–59.

Pai, S., Kini, M. S., & Selvaraj, R. (2021). A review on adsorptive removal of dyes from wastewater by hydroxyapatite nanocomposites. *Environmental Science and Pollution Research International, 28*, 11835–11849.

Paknikar, K. M., Pethkar, A. V., & Puranik, P. R. (2003). *Bioremediation of metalliferous wastes and products using inactivated microbial biomass.* Academic Press.

Pallela, P. N. V. K., Ruddaraju, L. K., Veerla, S. C., Matangi, R., Kollu, P., Ummey, S., & Pammi, S. V. N. (2020). Synergetic antibacterial potential, dye degrading capability and biocompatibility of Asperagus racemosus root assisted ZnO nanoparticles. *Materials Today. Communications, 25*, 101574. doi:10.1016/j.mtcomm.2020.101574

Palmore, G. T. R., & Whitesides, G. M. (1994). *Enzymatic conversion of biomass for fuels cells, 566.* American Chemical Society.

Panchal, P., Paul, D. R., Sharma, A., Choudhary, P., Meena, P., & Nehra, S. P. (2020). Biogenic mediated Ag/ZnO nanocomposites for photocatalytic and antibacterial activities towards disinfection of water. *Journal of Colloid and Interface Science, 563*, 370–380. doi:10.1016/j.jcis.2019.12.079 PMID:31887701

Pandey, N., Shukla, S. K., & Singh, N. B. (2017). Water purification by polymer nanocomposites: An overview. *Nanocomposites, 3*(2), 47–66. doi:10.1080/20550324.2017.1329983

Pare, B., Singh, P., & Jonnalgadda, S. B. (2009). Degradation and mineralization of Victoria Blue B dye in a slurry photo reactor using advanced oxidation process. *Journal of Scientific and Industrial Research, 68*, 724–729.

Park, C. M., Heo, J., Wang, D., Su, C., & Yoon, Y. (2018). Heterogeneous activation of persulfate by reduced graphene oxide–elemental silver/magnetite nanohybrids for the oxidative degradation of pharmaceuticals and endocrine disrupting compounds in water. *Applied Catalysis B: Environmental, 225*, 91–99. doi:10.1016/j.apcatb.2017.11.058 PMID:32704206

Park, H. G., & Holt, J. K. (2010). Recent advances in nano-electrode architecture for photochemical hydrogen production. *Energy & Environmental Science, 3*(8), 1028–1036. doi:10.1039/b922057g

Park, H. S., Kweon, K. E., Ye, H., Paek, E., Hwang, G. S., & Bard, A. J. (2011). Factors in the metal doping of $BiVO_4$ for improved photoelectrocatalytic activity as studied by scanning electrochemical microscopy and first-principles density-functional calculation. *The Journal of Physical Chemistry C, 115*(36), 17870–17879. doi:10.1021/jp204492r

Park, Y., Lee, S. H., Kang, S. O., & Choi, W. (2010). Organic dye-sensitized TiO_2 for the redox conversion of water pollutants under visible light. *Chemical Communications, 46*(14), 2477–2479. doi:10.1039/b924829c PMID:20309473

Park, Y., McDonald, K. J., & Choi, K. S. (2013). Progress in bismuth vanadate photoanodes for use in solar water oxidation. *Chemical Society Reviews, 42*(6), 2321–2337. doi:10.1039/C2CS35260E PMID:23092995

Parthibavarman, M., Karthik, M., & Prabhakaran, S. (2018). Facile and one step synthesis of WO3 nanorods and nanosheets as an efficient photocatalyst and humidity sensing material. *Vacuum, 155*, 224–232. doi:10.1016/j.vacuum.2018.06.021

Patil, P. T., Anwane, R. S., & Kondawar, S. B. (2015). Development of Electrospun Polyaniline/ZnO Composite Nanofibers for LPG Sensing. *Procedia Materials Science, 10*, 195–204. https://doi.org/10.1016/j.mspro.2015.06.041

Patnaik, S., Martha, S., Acharya, S., & Parida, K. M. (2016). An overview of the modification of g-C3N4 with high carbon containing materials for photocatalytic applications. *Inorganic Chemistry Frontiers, 3*(3), 336–347. doi:10.1039/C5QI00255A

Patra, J. K., & Baek, K. H. (2015). Novel green synthesis of gold nanoparticles using Citrullus lanatus rind and investigation of proteasome inhibitory activity, antibacterial, and antioxidant potential. *International Journal of Nanomedicine, 10*, 7253–7264. PMID:26664116

Pavithra, M., & Jessie Raj, M. B. (2021). Influence of ultrasonication time on solar light irradiated photocatalytic dye degradability and antibacterial activity of Pb doped ZnO nanocomposites. *Ceramics International, 47*(22), 32324–32331. doi:10.1016/j.ceramint.2021.08.128

Pavoski, G., Baldisserotto, D. L. S., Maraschin, T., Brum, L. F. W., dos Santos, C., dos Santos, J. H. Z., Brandelli, A., & Galland, G. B. (2019). Silver nanoparticles encapsulated in silica: Synthesis, characterization and application as antibacterial fillers in the ethylene polymerization. *European Polymer Journal, 117*, 38–54. doi:10.1016/j.eurpolymj.2019.04.055

Pawar, R. C., Kang, S., Park, J. H., Kim, J., Ahn, S., & Lee, C. S. (2017). Evaluation of a multi-dimensional hybrid photocatalyst for enrichment of H2 evolution and elimination of dye/non-dye pollutants. *Catalysis Science & Technology, 7*(12), 2579–2590. doi:10.1039/C7CY00466D

Pendolino, F., & Armata, N. (2017). *Graphene Oxide in Environmental Remediation Process*. doi:10.1007/978-3-319-60429-9

Peng, C., Yang, X., Li, Y., Yu, H., Wang, H., & Peng, F. (2016). Hybrids of Two-Dimensional Ti3C2 and TiO2 Exposing {001} Facets toward Enhanced Photocatalytic Activity. *ACS Applied Materials & Interfaces, 8*(9), 6051–6060. doi:10.1021/acsami.5b11973 PMID:26859317

Peng, W., Li, Y., Zhang, F., Zhang, G., & Fan, X. (2017). Roles of Two-Dimensional Transition Metal Dichalcogenides as Cocatalysts in Photocatalytic Hydrogen Evolution and Environmental Remediation. *Industrial & Engineering Chemistry Research*, *56*(16), 4611–4626. doi:10.1021/acs.iecr.7b00371

Peng, W., Wang, X., & Li, X.-y. (2014). The synergetic effect of MoS2 and graphene on Ag3PO4 for its ultra-enhanced photocatalytic activity in phenol degradation under visible light. *Nanoscale*, *6*(14), 8311–8317. doi:10.1039/c4nr01654h PMID:24933179

Peng, X. J., Luan, Z. K., Ding, J., Di, Z. H., Li, Y. H., & Tian, B. H. (2005). Ceria nanoparticles supported on carbon nanotubes for the removal of arsenate from water. *Materials Letters*, *59*(4), 399–403. doi:10.1016/j.matlet.2004.05.090

Peng, X., Li, Y., Luan, Z., Di, Z., Wang, H., Tian, B., & Jia, Z. (2003). Adsorption of 1, 2- dichlorobenzene from water to carbon nanotubes. *Chemical Physics Letters*, *376*(1-2), 154–158. doi:10.1016/S0009-2614(03)00960-6

Peng, X., Li, Y., Luan, Z., Di, Z., Wang, H., Tian, B., & Jia, Z. (2003). Adsorption of 1,2-dichlorobenzene from water to carbon nanotubes. *Chemical Physics Letters*, *376*(1–2), 154.

Peng, X., Luan, Z., Di, Z., Zhang, Z., & Zhu, C. (2005). Carbon nanotubes–iron oxides magnetic composites as adsorbent for removal of Pb(II) and Cu(II) from water. *Carbon*, *43*(4), 880–883. https://doi.org/10.1016/j.carbon.2004.11.009

Penza, M., Rossi, R., Alvisi, M., Cassano, G., & Serra, E. (2009a). Functional characterization of carbon nanotube networked films functionalized with tuned loading of Au nanoclusters for gas sensing applications. *Sensors and Actuators. Part B*, *140*(1), 176–184. doi:10.1016/j.snb.2009.04.008

Penza, M., Rossi, R., Alvisi, M., Signore, M. A., Cassano, G., Dimaio, D., Pentassuglia, R., Piscopiello, E., Serra, E., & Falconieri, M. (2009b). Characterization of metal-modified and vertically- aligned carbon nanotube films for functionally enhanced gas sensor applications. *Thin Solid Films*, *517*(22), 6211–6216. https://doi.org/10.1016/j.tsf.2009.04.009

Pérez, U. M. G., Guzmán, S. S., Cruz, A. M. l., & Peral, J. (2012). Selective Synthesis of Monoclinic Bismuth Vanadate Powders by Surfactant-Assisted Co-Precipitation Method: Study of Their Electrochemical and Photocatalytic Properties. *International Journal of Electrochemical Science*, *7*, 9622–9632.

Perreault, F., Fonseca de Faria, A., & Elimelech, M. (2015). Environmental applications of graphene-based nanomaterials. *Chemical Society Reviews*, *44*(16), 5861–5896. doi:10.1039/C5CS00021A PMID:25812036

Pete, S., Kattil, R. A., & Thomas, L. (2021). Polyaniline-multiwalled carbon nanotubes (PANI-MWCNTs) composite revisited: An efficient and reusable material for methyl orange removal. *Diamond and Related Materials*, *117*, 108455. doi:10.1016/j.diamond.2021.108455

Pham, T.-T., Nguyen-Huy, C., & Shin, E. W. (2016). Facile one-pot synthesis of nickel-incorporated titanium dioxide/graphene oxide composites: Enhancement of photodegradation under visible-irradiation. *Applied Surface Science*, *377*, 301–310. doi:10.1016/j.apsusc.2016.03.144

Phang, S. J., & Tan, L.-L. (2019). Recent advances in carbon quantum dot (CQD)-based two dimensional materials for photocatalytic applications. *Catalysis Science & Technology*, *9*(21), 5882-5905. doi:10.1039/C9CY01452G

Pilli, S. K., Furtak, T. E., Brown, L. D., Deutsch, T. G., Turner, J. A., & Herring, A. M. (2011). Cobalt-phosphate (Co-Pi) catalyst modified Mo-doped $BiVO_4$photoelectrodes for solar water oxidation. *Energy & Environmental Science*, *4*(12), 5028–5034. doi:10.1039/c1ee02444b

Pincus, L. N., Melnikov, F., Yamani, J. S., & Zimmerman, J. B. (2018). Multifunctional photoactive and selective adsorbent for arsenite and arsenate: Evaluation of nano titanium dioxide-enabled chitosan cross-linked with copper. *Journal of Hazardous Materials*, *358*, 145–154.

Pi, Y., Ma, L., Zhao, P., Cao, Y., Gao, H., Wang, C., Li, Q., Dong, S., & Sun, J. (2018). Facile green synthetic graphene-based Co-Fe Prussian blue analogues as an activator of peroxymonosulfate for the degradation of levofloxacin hydrochloride. *Journal of Colloid and Interface Science*, *526*, 18–27. doi:10.1016/j.jcis.2018.04.070 PMID:29709668

Planeix, J. M., Coustel, N., Coq, B., Brotons, V., Kumbhar, P. S., Dutartre, R., Geneste, P., Bernier, P., & Ajayan, P. M. (1994). Application of carbon nanotubes as supports in heterogeneous catalysis. *Journal of the American Chemical Society*, *116*(17), 7935–7936. https://doi.org/10.1021/ja00096a076

Pokropivny, V. V., & Skorokhod, V. V. (2007). Classification of nanostructures by dimensionality and concept of surface forms engineering in nanomaterial science. *Materials Science and Engineering C*, *27*(5-8), 990–993. doi:10.1016/j.msec.2006.09.023

Pookmanee, P., Kojinok, S., Punthaeod, R., Sangsricharan, S., & Phanichapat, S. (2013). Preparation and Characterization of $BiVO_4$ Powder by the Sol- gel Method. *Ferroelectrics*, *456*(1), 45–54. doi:10.1080/00150193.2013.846197

Pookmanee, P., Longchin, P., Kangwansupamonkon, W., Puntharod, R., & Phanichphant, S. (2013). Microwave-assisted synthesis Bismuth Vanadate ($BiVO_4$) Powder. *Ferroelectrics*, *455*(1), 35–42. doi:10.1080/00150193.2013.843414

Poon, C. P. C. (1986). Removal of cadmium from wastewaters. In Cadmium in the Environment (pp. 46-55). doi:10.1007/978-3-0348-7238-6_7

Poonam, K., Rose, N. M., & Singh, S. S. J. (2015). Microencapsulation of lime essential oil for fragrant textiles. *Annals of Agri Bio Research*, *20*(1), 152–157.

Prabhuraj, T., Prabhu, S., Dhandapani, E., Duraisamy, N., Ramesh, R., Kumar, K. A. R., & Maadeswaran, P. (2021). Bifunctional ZnO sphere/r-GO composites for supercapacitor and photocatalytic activity of organic dye degradation. *Diamond and Related Materials*, *120*, 108592. doi:10.1016/j.diamond.2021.108592

Pradhan, G. K., Padhi, D. K., & Parida, K. M. (2013). Fabrication of #-Fe2O3 nanorod/RGO composite: A novel hybrid photocatalyst for phenol degradation. *ACS Applied Materials & Interfaces*, *5*(18), 9101–9110. doi:10.1021/am402487h PMID:23962068

Pradhan, S., Shukla, S. S., & Dorris, K. L. (2005). Removal of nickel from aqueous solutions using crab shells. *Journal of Hazardous Materials*, *125*(1–3), 201–204. https://doi.org/10.1016/j.jhazmat.2005.05.029

Pragathiswaran, C., Smitha, C., Mahin Abbubakkar, B., Govindhan, P., & Anantha Krishnan, N. (2021). Synthesis and characterization of TiO_2/ZnO–Ag nanocomposite for photocatalytic degradation of dyes and anti-microbial activity. *Materials Today: Proceedings*, *45*, 3357–3364. doi:10.1016/j.matpr.2020.12.664

Prasad, G. K., Takei, T., Yonesaki, Y., Kumada, N., & Kinomura, N. (2006). Hybrid nanocomposite based on NbWO6 nanosheets and polyaniline. *Materials Letters*, *60*(29–30), 3727–3730. https://doi.org/10.1016/j.matlet.2006.03.097

Prasek, J., Drbohlavova, J., Chomoucka, J., Hubalek, J., Jasek, O., Adam, V., & Kizek, R. (2011). Methods for carbon nanotubes synthesis-Review. *Journal of Materials Chemistry*, *21*(40), 15872–15884. doi:10.1039/c1jm12254a

Primo, J. O., Bittencourt, C., Acosta, S., Sierra-Castillo, A., Colomer, J. F., Jaerger, S., Teixeira, V. C., & Anaissi, F. J. (2020). Synthesis of Zinc Oxide Nanoparticles by Ecofriendly Routes: Adsorbent for Copper Removal From Wastewater. *Frontiers in Chemistry*, *8*, 571790. doi:10.3389/fchem.2020.571790 PMID:33330360

Pugazhendhi, A., Prabhu, R., Muruganantham, K., Shanmuganathan, R., & Natarajan, S. (2019). Anticancer, antimicrobial and photocatalytic activities of green synthesized magnesium oxide nanoparticles (MgONPs) using aqueous extract of Sargassum wightii. *Journal of Photochemistry and Photobiology. B, Biology*, *190*, 86–97. doi:10.1016/j.jphotobiol.2018.11.014 PMID:30504053

Pumera, M. (2007). Carbon Nanotubes Contain Residual Metal Catalyst Nanoparticles even after Washing with Nitric Acid at Elevated Temperature because These Metal Nanoparticles Are Sheathed by Several Graphene Sheets. *Langmuir, 23*(11), 6453–6458. doi:10.1021/la070088v PMID:17455966

Purkait, M. K., DasGupta, S., & De, S. (2005). Adsorption of eosin dye on activated carbon and its surfactant based desorption. *Journal of Environmental Management, 76*, 135.

Pyrzynska, K. (2008). Carbon nanotubes as a new solid phase extraction material for removal and enrichment of organic pollutants in water. *Separation and Purification Reviews, 37*(4), 372–389. doi:10.1080/15422110802178843

Qasem, N. A., Mohammed, R. H., & Lawal, D. U. (2021). Removal of heavy metal ions from wastewater: A comprehensive and critical review. *Npj Clean Water, 4*(1), 1–15.

Qiao, Y., Li, C. M., Bao, S. J., & Bao, Q. L. (2007). Carbon nanotube/polyaniline composite as anode material for microbial fuel cells. *Journal of Power Sources, 170*(1), 79–84. https://doi.org/10.1016/j.jpowsour.2007.03.048

Qin, J., Hu, X., Li, X., Yin, Z., Liu, B., & Lam, K. (2019). 0D/2D AgInS2/MXene Z-scheme heterojunction nanosheets for improved ammonia photosynthesis of N2. *Nano Energy, 61*, 27–35. doi:10.1016/j.nanoen.2019.04.028

Qi, P., Vermesh, O., Grecu, M., Javey, A., Wang, Q., Dai, H., Peng, S., & Cho, K. J. (2003). Toward large arrays of multiplex functionalized carbon nanotube sensors for highly sensitive and selective molecular detection. *Nano Letters, 3*(3), 347–351. https://doi.org/10.1021/nl034010k

Quang, N. H., Van Trinh, M., Lee, B. H., & Huh, J. S. (2006). Effect of NH3 gas on the electrical properties of single-walled carbon nanotube bundles. *Sensors and Actuators. Part B, 113*(1), 341–346. doi:10.1016/j.snb.2005.03.089

Qu, L., Zhu, G., Ji, J., Yadav, T. P., Chen, Y., Yang, G., Xu, H., & Li, H. (2018). Recyclable Visible Light-Driven O-g-C3N4/Graphene Oxide/N-Carbon Nanotube Membrane for Efficient Removal of Organic Pollutants. *ACS Applied Materials & Interfaces, 10*(49), 42427–42435. doi:10.1021/acsami.8b15905 PMID:30444339

Qumar, U., Hassan, J. Z., Bhatti, R. A., Raza, A., Nazir, G., Nabgan, W., & Ikram, M. (2022). Photocatalysis vs adsorption by metal oxide nanoparticles. *Journal of Materials Science and Technology, 131*, 122–166. doi:10.1016/j.jmst.2022.05.020

Qu, R., Xu, B., Meng, L., Wang, L., & Wang, Z. (2015). Ozonation of indigo enhanced by carboxylated carbon nanotubes: Performance optimization, degradation products, reaction mechanism and toxicity evaluation. *Water Research, 68*, 316–327. doi:10.1016/j.watres.2014.10.017 PMID:25462739

Qu, X., Brame, J., Li, Q., & Alvarez, P. J. (2013). Nanotechnology for a safe and sustainable water supply: Enabling integrated water treatment and reuse. *Accounts of Chemical Research, 46*(3), 834–843. https://doi.org/10.1021/ar300029v

Radaei, E., Moghaddam, M. R. A., & Arami, M. (2017). Application of artifcial neural network on modeling of reactive blue 19 removal by modified pomegranate residual. *Environmental Engineering and Management Journal, 16*(9), 2113–2122. https://doi.org/10.30638/eemj.2017.218

Radushkevich, L. V., & Lukyanovich, V. M. (1952). Soviet. *Journal of Physical Chemistry, 26*, 88–95.

Rafatullah, M., Sulaiman, O., Hashim, R., & Ahmad, A. (2010). Adsorption of methylene blue on low-cost adsorbents: A review. *Journal of Hazardous Materials, 177*, 70–80.

Rafiq, A., Ikram, M., Ali, S., Niaz, F., Khan, M., Khan, Q., & Maqbool, M. (2021). Photocatalytic degradation of dyes using semiconductor photocatalysts to clean industrial water pollution. *Journal of Industrial and Engineering Chemistry, 97*, 111–128. doi:10.1016/j.jiec.2021.02.017

Rahmanian, E., Malekfar, R., & Pumera, M. (2018). Nanohybrids of Two-Dimensional Transition-Metal Dichalcogenides and Titanium Dioxide for Photocatalytic Applications. *Chemistry (Weinheim an der Bergstrasse, Germany)*, *24*(1), 18–31. doi:10.1002/chem.201703434 PMID:28872715

Rajasekhar, B., Nambi, I. M., & Govindarajan, S. K. (2018). Human health risk assessment of ground water contaminated with petroleum PAHs using Monte Carlo simulations: A case study of an Indian metropolitan city. *Journal of Environmental Management, 205*, 183–191. doi:10.1016/j.jenvman.2017.09.078 PMID:28985597

Rajasulochana, P., & Preethy, V. (2016). Comparison on efficiency of various techniques in treatment of waste and sewage water – A comprehensive review. *Resource-Efficient Technologies, 2*(4), 175–184. doi:10.1016/j.reffit.2016.09.004

Rajawat, D. S., Kardam, A., Srivastava, S., & Satsangee, S. P. (2013). Nanocellulosic fibermodified carbon paste electrode for ultra trace determination of Cd (II) and Pb (II) in aqueous solution. *Environmental Science and Pollution Research International*, *20*(5), 3068–3076.

Rajendran, R., Vignesh, S., Sasireka, A., Priya, P., Suganthi, S., Raj, V., & AlFaify, S. (2021). Investigation on novel Cu_2O modified g-C_3N_4/ZnO heterostructures for efficient photocatalytic dye degradation performance under visible-light exposure. *Colloid and Interface Science Communications*, *44*, 100480. doi:10.1016/j.colcom.2021.100480

Rajeswari, A., Amalraj, A., & Pius, A. (2016). Adsorption studies for the removal of nitrate using chitosan/PEG and chitosan/PVA polymer composites. *Journal of Water Process Engineering*, *9*, 123–134.

Raji, C., Manju, G. N., & Anirudhan, T. S. (1997). *Removal of heavy metal ions from water using sawdust-based activated carbon*. Academic Press.

Ramesh, S., Kim, H. S., & Kim, J. H. (2018). Cellulose–Polyvinyl Alcohol–Nano-TiO_2 Hybrid Nanocomposite: Thermal, Optical, and Antimicrobial Properties against Pathogenic Bacteria. *Polymer-Plastics Technology and Engineering*, *57*, 669–681.

Ran, J., Gao, G., Li, F.-T., Ma, T.-Y., Du, A., & Qiao, S.-Z. (2017). Ti3C2 MXene co-catalyst on metal sulfide photoabsorbers for enhanced visible-light photocatalytic hydrogen production. *Nature Communications*, *8*(1), 13907. doi:10.1038/ncomms13907 PMID:28045015

Ran, R., McEvoy, J. G., & Zhang, Z. (2015). Synthesis and Optimization of Visible Light Active $BiVO_4$Photocatalysts for the Degradation of Rh B. *International Journal of Photoenergy*, 612857.

Rao, M., Parwate, A. V., & Bhole, A. G. (2002). Removal of Cr6+ and Ni2+ from aqueous solution using bagasse and fly ash. *Waste Management, 22*(7), 821–830. doi:10.1016/s0956-053x(02)00011-9

Rao, G. P., Lu, C., & Su, F. (2007). Sorption of divalent metal ions from aqueous solution by carbon nanotubes: A review. *Separation and Purification Technology*, *58*(1), 224–231. https://doi.org/10.1016/j.seppur.2006.12.006

Rashid Al-Mamun, M., Shofikul Islam, M., Rasel Hossain, M., Kader, S., Shahinoor Islam, M., & Zaved Hossain Khan, M. (2021). A novel and highly efficient Ag and GO co-synthesized ZnO nano photocatalyst for methylene blue dye degradation under UV irradiation. *Environmental Nanotechnology, Monitoring & Management*, *16*, 100495. doi:10.1016/j.enmm.2021.100495

Rautela, A., & Rani, J., & Debnath (Das), M. (2019). Green synthesis of silver nanoparticles from Tectona grandis seeds extract: Characterization and mechanism of antimicrobial action on different microorganisms. *Journal of Analytical Science and Technology*, *10*, 5. doi:10.118640543-018-0163-z

Ray, B., Jones, N., Manna, A. C., & Ranjit, K. T. (2008). Antibacterial activity of ZnO nanoparticle suspensions on a broad spectrum of microorganisms. *FEMS Microbiology Letters*, *279*(1), 71–76. doi:10.1111/j.1574-6968.2007.01012.x PMID:18081843

Ray, P. Z., & Shipley, H. J. (2015). Inorganic nano-adsorbents for the removal of heavy metals and arsenic: A review. *RSC Advances*, *5*(38), 29885–29907.

Ray, S. S., Vaudreuil, S., Maazouz, A., & Bousmina, M. (2006). Dispersion of multi-walled carbon nanotubes in biodegradable poly(butylene succinate) matrix. *Journal of Nanoscience and Nanotechnology*, *6*(7), 2191–2195. https://doi.org/10.1166/jnn.2006.368

Ray, S., & Chatterjee, B. P. (1995). Saracin: A lectin from Saraca indica seed integument recognizes complex carbohydrates. *Phytochemistry*, *40*(3), 643–649. doi:10.1016/0031-9422(95)98168-G PMID:7576454

Raza, A., Zhang, X., Ali, S., Cao, C., Rafi, A. A., & Li, G. (2022). Photoelectrochemical Energy Conversion over 2D Materials. *Photochem*, *2*(2).

Raza, A., Altaf, S., Ali, S., Ikram, M., & Li, G. (2022). Recent advances in carbonaceous sustainable nanomaterials for wastewater treatments. *Sustainable Materials and Technologies*, *32*, e00406. doi:10.1016/j.susmat.2022.e00406

Raza, A., Qin, Z., Ahmad, S. O. A., Ikram, M., & Li, G. (2021). Recent advances in structural tailoring of BiOX-based 2D composites for solar energy harvesting. *Journal of Environmental Chemical Engineering*, *9*(6), 106569. doi:10.1016/j.jece.2021.106569

Raza, A., Qumar, U., Haider, A., Naz, S., Haider, J., Ul-Hamid, A., Ikram, M., Ali, S., Goumri-Said, S., & Kanoun, M. B. (2021). Liquid-phase exfoliated MoS2 nanosheets doped with p-type transition metals: A comparative analysis of photocatalytic and antimicrobial potential combined with density functional theory. *Dalton Transactions (Cambridge, England)*, *50*(19), 6598–6619. doi:10.1039/D1DT00236H PMID:33899890

Raza, A., Qumar, U., Hassan, J., Ikram, M., Ul-Hamid, A., Haider, J., Imran, M., & Ali, S. (2020). A comparative study of dirac 2D materials, TMDCs and 2D insulators with regard to their structures and photocatalytic/sonophotocatalytic behavior. *Applied Nanoscience*, *10*(10), 3875–3899. doi:10.100713204-020-01475-y

Raza, A., Qumar, U., Rafi, A. A., & Ikram, M. (2022). MXene-based nanocomposites for solar energy harvesting. *Sustainable Materials and Technologies*, *33*, e00462. doi:10.1016/j.susmat.2022.e00462

Raza, A., Rafiq, A., Qumar, U., & Hassan, J. Z. (2022). 2D hybrid photocatalysts for solar energy harvesting. *Sustainable Materials and Technologies*, *33*, e00469. doi:10.1016/j.susmat.2022.e00469

Reber, J. F., & Rusek, M. (1986). Photochemical hydrogen production with platinized suspensions of cadmium sulphide and cadmium zinc sulphide modified by silver sulphide. *Journal of Physical Chemistry*, *90*(5), 824–834. doi:10.1021/j100277a024

Reddy, K. M., Manorama, S. V., & Reddy, A. R. (2002). Bandgap studies on anatase titanium dioxide nanoparticles. *Materials Chemistry and Physics*, *78*(1), 239–245. doi:10.1016/S0254-0584(02)00343-7

Regmi, C., Kim, T. H., Ray, S. K., Yamaguchi, T., & Lee, S. W. (2017). Cobalt-doped $BiVO_4$ (Co-$BiVO_4$) as a visible-light-driven photocatalyst for the degradation of malachite green and inactivation of harmful microorganisms in wastewater. *Research on Chemical Intermediates*, *43*(9), 5203–5216. doi:10.100711164-017-3036-y

Regmi, C., Kshetri, Y. K., Kim, T. H., Pandey, R. P., & Lee, S. W. (2017). Visible-light-induced Fe-doped $BiVO_4$photocatalyst for contaminated water treatment. *Mol. Catal.*, *432*, 220–231. doi:10.1016/j.mcat.2017.02.004

Regmi, C., Kshetri, Y. K., Kim, T. H., Pandey, R. P., Ray, S. K., & Lee, S. W. (2017). Fabrication of Ni-doped $BiVO_4$ semiconductors with enhanced visible-light photocatalytic performances for wastewater treatment. *Applied Surface Science, 413*, 253–265. doi:10.1016/j.apsusc.2017.04.056

Rendón-Villalobos, R., Ortíz-Sánchez, A., Tovar-Sánchez, E., & Flores-Huicochea, E. (2016). The role of biopolymers in obtaining environmentally friendly materials. *Composites from renewable and sustainable materials*, 151-159.

Ren, G., Hu, D., Cheng, E. W. C., Vargas-Reus, M. A., Reip, P., & Allaker, R. P. (2009). Characterization of copper oxide nanoparticles for antimicrobial applications. *International Journal of Antimicrobial Agents, 33*(6), 587–590. doi:10.1016/j.ijantimicag.2008.12.004 PMID:19195845

Ren, X. M., Shao, D. D., Zhao, G. X., Sheng, G. D., Hu, J., Yang, S. T., & Wang, X. K. (2011). Plasma induced multiwalled carbon nanotube grafted with 2-vinylpyridine for preconcentration of Pb(II) from aqueous solutions. *Plasma Processes and Polymers, 8*(7), 589–598. doi:10.1002/ppap.201000192

Ritter, K. S., Sibley, P., Hall, K., Keen, P., Mattu, G., & Linton, B. L. (2002). Sources, pathways, and relative risks of contaminants in surface water and groundwater: A perspective prepared for the Walkerton inquiry. *Journal of Toxicology and Environmental Health. Part A., 65*(1), 1–142. doi:10.1080/152873902753338572 PMID:11809004

Rivas, B. L., Urbano, B. F., & Sánchez, J. (2018). Water-soluble and insoluble polymers, nanoparticles, nanocomposites and hybrids with ability to remove Hazardous inorganic pollutants in water. *Frontiers in Chemistry, 6*, 320. doi:10.3389/fchem.2018.00320 PMID:30109224

Rodney, J. D., Deepapriya, S., Annie Vinosha, P., Krishnan, S., Janet Priscilla, S., Daniel, R., & Jerome Das, S. (2018). Photo-Fenton degradation of nano-structured La doped CuO nanoparticles synthesized by combustion technique. *Optik (Stuttgart), 161*, 204–216. doi:10.1016/j.ijleo.2018.01.125

Rodriguez-Proteau, R., & Grant, R. L. (2006). Toxicity evaluation and human health risk assessment of surface and ground water contaminated by recycled hazardous waste materials. Handb. Environ. Chem. Vol. 5 Water Pollut., 5 F, 133–189.

Rojas, J., Bedoya, M., & Ciro, Y. (2015). Current Trends in the Production of Cellulose Nanoparticles and Nanocomposites for Biomedical Applications. In M. Poletto, & H. L. Ornaghi (Eds.), *Cellulose - Fundamental Aspects and Current Trends*. IntechOpen. doi:10.5772/61334

Romero-Gonzalez, J., Peralta-Videa, J. R., Rodriguez, E., Delgado, M., & Gardea-Torresdey, J. L. (2006). Potential of Agave lechuguilla biomass for Cr (III) removal from aqueous solutions: Thermodynamic studies. *Bioresource Technology, 97*(1), 178–182. doi:10.1016/j.biortech.2005.01.037 PMID:16154514

Roy, J. C., Ferri, A., Salaün, F., Giraud, S., Chen, G., & Jinping, G. (2017, October). Chitosan-carboxymethylcellulose based microcapsules formulation for controlled release of active ingredients from cosmeto textile. *IOP Conference Series. Materials Science and Engineering, 254*(7), 072020. doi:10.1088/1757-899X/254/7/072020

Ruan, W., Hu, J., Qi, J., Hou, Y., Zhou, C., & Wei, X. (2019). Wenqian Ruan Removal of Dyes from Wastewater by. Nanomaterials: A Review. Jiwei Hu Jimei Qi Yu Hou Chao Zhou Xionghui Wei*Advanced. Materials Letters, 10*, 9–20.

Ruíz-Baltazar, Á. J. (2020). Green synthesis assisted by sonochemical activation of Fe_3O_4-Ag nano-alloys: Structural characterization and studies of sorption of cationic dyes. *Inorganic Chemistry Communications, 120*, 108148. doi:10.1016/j.inoche.2020.108148

Russell, D. L. (2006). *Practical wastewater treatment*. John Wiley & Sons, Inc.

Saad, A. M., Abukhadra, M. R., Abdel-Kader Ahmed, S., Elzanaty, A. M., Mady, A. H., Betiha, M. A., Shim, J.-J., & Rabie, A. M. (2020). Photocatalytic degradation of malachite green dye using chitosan supported ZnO and Ce–ZnO nano-flowers under visible light. *Journal of Environmental Management*, *258*, 110043. doi:10.1016/j.jenvman.2019.110043 PMID:31929075

Sabarinathan, M., Harish, S., Archana, J., Navaneethan, M., Ikeda, H., & Hayakawa, Y. (2017). Highly efficient visible-light photocatalytic activity of MoS2–TiO2 mixtures hybrid photocatalyst and functional properties. *RSC Advances*, *7*(40), 24754–24763. doi:10.1039/C7RA03633G

Saga, S., Matsumoto, H., Saito, K., Minagawa, M., & Tanioka, A. (2008). Polyelectrolyte membranes based on hydrocarbon polymer containing fullerene. *Journal of Power Sources*, *176*(1), 16–22. https://doi.org/10.1016/j.jpowsour.2007.10.017

Sakoda, A., Kawazoe, K., & Suzuki, M. (1987). Adsorption of tri- and tetra-chloroethylene from aqueous solutions on activated carbon fibers. *Water Research*, *21*(6), 717–722. https://doi.org/10.1016/0043-1354(87)90084-4

Salah Azab, M., & Peterson, P. J. (1989). The removal of cadmium from water by the use of biological sorbents. *Water Science and Technology*, *21*(12), 1705–1706. doi:10.2166/wst.1989.0149

Salaün, F., Devaux, E., Bourbigot, S., & Rumeau, P. (2008). Preparation of multinuclear microparticles using a polymerization in emulsion process. *Journal of Applied Polymer Science*, *107*(4), 2444–2452.

Salehi, E., Daraei, P., & Shamsabadi, A. A. (2016). A review on chitosan-based adsorptive membranes. *Carbohydrate Polymers*, *152*, 419–432.

Saleh, R., & Djaja, N. F. (2014). Transition-metal-doped ZnO nanoparticles: Synthesis, characterization and photocatalytic activity under UV light. *Spectrochimica Acta. Part A: Molecular and Biomolecular Spectroscopy*, *130*, 581–590. doi:10.1016/j.saa.2014.03.089 PMID:24813289

Saleh, R., & Taufik, A. (2019). Ultraviolet-light-assisted heterogeneous Fenton reaction of Ag-Fe3O4/graphene composites for the degradation of organic dyes. *Journal of Environmental Chemical Engineering*, *7*(1), 102895. doi:10.1016/j.jece.2019.102895

Salem, S. S., & Fouda, A. (2021). Green Synthesis of Metallic Nanoparticles and Their Prospective Biotechnological Applications: An Overview. *Biological Trace Element Research*, *199*(1), 344–370. doi:10.100712011-020-02138-3 PMID:32377944

Sanou, Y., Nguyen, T. T. P., Pare, S., & Nguyen, V. P. (2019). The removal As (V) from aqueous solutions using Ferromagnetic Activated Carbon: Equilibrium and kinetic studies. *Revue des Sciences de l'Eau*, *32*(2), 179–192. doi:10.7202/1065206ar

Sanou, Y., Pare, S., Baba, G., Segbeaya, N. K., & Bonzi-Coulibaly, L. Y. (2016). Removal of COD in Wastewaters by Activated Charcoal from Rice Husk. *Revue des Sciences de l'Eau*, *29*(3), 265–277. doi:10.7202/1038927ar

Santhosh, C., Velmurugan, V., Jacob, G., Jeong, S. K., Grace, A. N., & Bhatnagar, A. (2016). Role of nanomaterials in water treatment applications: A review. *Chemical Engineering Journal*, *306*, 1116–1137. doi:10.1016/j.cej.2016.08.053

Saquib, M., & Muneer, M. (2003). TiO_2-mediated photocatalytic degradation of a triphenylmethane dye (gentian violet), in aqueous suspensions. *Dyes and Pigments*, *56*(1), 37–49. doi:10.1016/S0143-7208(02)00101-8

Sarkar, C., Basu, J. K., & Samanta, A. N. (2021). Synthesis of novel ZnO/Geopolymer nanocomposite photocatalyst for degradation of congo red dye under visible light. *Environmental Nanotechnology, Monitoring & Management*, *16*, 100521. doi:10.1016/j.enmm.2021.100521

Sarkar, C., Bora, C., & Dolui, S. K. (2014). Selective Dye Adsorption by pH Modulation on Amine-Functionalized Reduced Graphene Oxide–Carbon Nanotube Hybrid. *Industrial & Engineering Chemistry Research*, *53*(42), 16148–16155. doi:10.1021/ie502653t

Sarkar, S., Ponce, N. T., Banerjee, A., Bandopadhyay, R., Rajendran, S., & Lichtfouse, E. (2020). Green polymeric nanomaterials for the photocatalytic degradation of dyes: A review. *Environmental Chemistry Letters*, *18*, 1569–1580.

Sarmah, S., & Kumar, A. (2011). Photocatalytic activity of polyaniline-TiO2 nanocomposites. *Indian Journal of Physics*, *85*(5), 713–726. doi:10.100712648-011-0071-1

Sathishkumar, M., Sneha, K., Won, S. W., Cho, C. W., Kim, S., & Yun, Y. S. (2009). Cinnamon zeylanicum bark extract and powder mediated green synthesis of nano-crystalline silver particles and its bactericidal activity. *Colloid Surf. B*, *73*(2), 332–338. doi:10.1016/j.colsurfb.2009.06.005 PMID:19576733

Sathiyavimal, S., Vasantharaj, S., Kaliannan, T., & Pugazhendhi, A. (2020). Eco-biocompatibility of chitosan coated biosynthesized copper oxide nanocomposite for enhanced industrial (Azo) dye removal from aqueous solution and antibacterial properties. *Carbohydrate Polymers*, *241*, 116243. Advance online publication. doi:10.1016/j.carbpol.2020.116243 PMID:32507166

Sato, J., Saito, N., Yamada, Y., Maeda, K., Takata, T., Kondo, J. N., Hara, M., Kobayashi, H., Domen, K., & Inoue, Y. (2005). RuO_2-loaded β-Ge_3N_4 as a non-oxide photocatalyst for overall water splitting. *Journal of the American Chemical Society*, *127*(12), 4150–4151. doi:10.1021/ja042973v PMID:15783179

Sato, S. (1986). Photocatalytic activity of NO_x-doped TiO_2 in the visible light region. *Chemical Physics Letters*, *123*(1-2), 126–128. doi:10.1016/0009-2614(86)87026-9

Savage, N., & Diallo, M. S. (2005). Nanomaterials and water purification: Opportunities and challenges. *Journal of Nanoparticle Research*, *7*(4–5), 331–342. https://doi.org/10.1007/s11051-005-7523-5

Savage, T., Bhattachanya, S., Sanadan, B., Gaillard, J., Tritt, T., Sun, Y.-P., & Rao, A. M. (2003). Photo-induced oxidation of nanotubes. *Journal of Physics Condensed Matter*, *15*(35), 5915–5921. doi:10.1088/0953-8984/15/35/301

Sawalha, M. F., Gardea-Torresdey, J. L., Parsons, J. G., Saupe, G., & Peralta-Videa, J. R. (2005). Determination of adsorption and speciation of chromium species by saltbush (Atriplex canescens) biomass using a combination of XAS and ICP–OES. *Microchemical Journal*, *81*(1), 122–132. doi:10.1016/j.microc.2005.01.008

Saxena, G., Purchase, D., Mulla, S. I., Saratale, G. D., & Bharagava, R. N. (2019). Phytoremediation of heavy metal-contaminated sites: Eco-environmental concerns, field studies, sustainability issues, and future prospects. *Reviews of Environmental Contamination and Toxicology*, *249*, 71–131. PMID:30806802

Saxena, M., Sharma, N., & Saxena, R. (2020). Highly efficient and rapid removal of toxic dye: Adsorption kinetics, isotherm, and mechanism studies of functionalized multiwalled carbon nanotubes. *Surfaces and Interfaces*, *21*, 100639. doi:10.1016/j.surfin.2020.100639

Sayadi, M. H., Amadpour, N., & Homaeigohar, S. (2021). Photocatalytic and Antibacterial Properties of Ag-$CuFe_2O_4$@WO_3 Magnetic Nanocomposite. *Nanomaterials (Basel, Switzerland)*, *11*(2), 298. doi:10.3390/nano11020298 PMID:33498950

Sayama, K., & Arakawa, H. (1994). Effect of Na_2CO_3 addition on photocatalytic decomposition of liquid water over various semiconductor catalysis. *Journal of Photochemistry and Photobiology A Chemistry*, *77*(2-3), 243–247. doi:10.1016/1010-6030(94)80049-9

Sayes, C. M., Liang, F., Hudson, J. L., Mendez, J., Guo, W., Beach, J. M., Moore, V. C., Doyle, C. D., West, J. L., Billups, W. E., Ausman, K. D., & Colvin, V. L. (2006). Functionalization density dependence of single-walled carbon nanotubes cytotoxicity in vitro. *Toxicology Letters, 161*(2), 135–142. https://doi.org/10.1016/j.toxlet.2005.08.011

Schmidt, N. A., & Brown, J. M. (2017). *Evidence-based practice for nurses: Appraisal and application of research* (4th ed.). Jones & Bartlett Learning, LLC.

Sedlačík, M., Mrlík, M., Pavlínek, V., Sáha, P., & Quadrat, O. (2012). Electrorheological properties of suspensions of hollow globular titanium oxide/polypyrrole particles. *Colloid & Polymer Science, 290*(1), 41–48. doi:10.100700396-011-2521-x

Seki, K., Saito, N., & Aoyama, M. (1997). Removal of heavy metal ions from solutions by coniferous barks. *Wood Science and Technology, 31*(6), 441–447. doi:10.1007/BF00702566

Sengupta, A., & Sarkar, C. K. (Eds.). (2015). *Introduction to nano: Basics to nanoscience and nanotechnology*. Springer.

Sethy, N. K., Arif, Z., Mishra, P. K., & Kumar, P. (2020). Green synthesis of TiO2 nanoparticles from Syzygium cumini extract for photo-catalytic removal of lead (Pb) in explosive industrial wastewater. *Green Process Synth., 9*(1), 171–181. doi:10.1515/gps-2020-0018

Shaban, M., Abukhadra, M. R., Hamd, A., Amin, R. R., & Abdel Khalek, A. (2017). Photocatalytic removal of Congo red dye using MCM-48/Ni_2O_3 composite synthesized based on silica gel extracted from rice husk ash; fabrication and application. *Journal of Environmental Management, 204*, 189–199.

Shah, A., Haq, S., Rehman, W., Waseem, M., Shoukat, S., & Rehman, M. (2019). Photocatalytic and antibacterial activities of paeonia emodi mediated silver oxide nanoparticles. *Mater. Res. Express, 6*(4). . doi:10.1088/2053-1591/aafd42

Shahzad, F., Alhabeb, M., Hatter Christine, B., Anasori, B., Man Hong, S., Koo Chong, M., & Gogotsi, Y. (2016). Electromagnetic interference shielding with 2D transition metal carbides (MXenes). *Science, 353*(6304), 1137–1140. doi:10.1126cience.aag2421 PMID:27609888

Shameli, K., Ahmad, M., Al-Mulla, E. A. J., Ibrahim, N. A., Shabanzadeh, P., Rustaiyan, A., & Abdollahi, Y. (2012). Green biosynthesis of silver nanoparticles using Callicarpa maingayi stem bark extraction. *Molecules (Basel, Switzerland), 17*(7), 8506–8517. doi:10.3390/molecules17078506 PMID:22801364

Shanmugam, V., & Jeyaperumal, K. S. (2018). Investigations of visible light driven Sn and Cu doped ZnO hybrid nanoparticles for photocatalytic performance and antibacterial activity. *Applied Surface Science, 449*, 617–630. doi:10.1016/j.apsusc.2017.11.167

Shao, B., Wang, J., Liu, Z., Zeng, G., Tang, L., Liang, Q., He, Q., Wu, T., Liu, Y., & Yuan, X. (2020). Ti3C2Tx MXene decorated black phosphorus nanosheets with improved visible-light photocatalytic activity: Experimental and theoretical studies. *Journal of Materials Chemistry. A, Materials for Energy and Sustainability, 8*(10), 5171–5185. doi:10.1039/C9TA13610J

Shao, L., & Chen, G. Q. (2013). Water footprint assessment for wastewater treatment: Method, indicator, and application. *Environmental Science & Technology, 47*(14), 7787–7794. doi:10.1021/es402013t PMID:23777208

Sharifi-Rad, M., Pawel, P., Francesco, E., & Álvarez-Suarez, J. M. (2020). Green Synthesis of Silver Nanoparticles Using Astragalus tribuloides Delile. Root Extract: Characterization, Antioxidant, Antibacterial, and Anti-Inflammatory Activities. *Nanomaterials, 10*(12). . doi:10.3390/nano10122383

Sharkawy, A., Fernandes, I. P., Barreiro, M. F., Rodrigues, A. E., & Shoeib, T. (2017). Aroma-loaded microcapsules with antibacterial activity for eco friendly textile application: Synthesis, characterization, release, and green grafting. *Industrial & Engineering Chemistry Research, 56*(19), 5516–5526. doi:10.1021/acs.iecr.7b00741

Sharma, V. K., Jinadatha, C., & Lichtfouse, E. (2020). Environmental chemistry is most relevant to study coronavirus pandemics. *Environmental Chemistry Letters.* doi:10.1007/s10311-020-01017-6

Sharma, M., Poddar, M., Gupta, Y., Nigam, S., Avasthi, D. K., Adelung, R., Abolhassani, R., Fiutowski, J., Joshi, M., & Mishra, Y. K. (2020). Solar light assisted degradation of dyes and adsorption of heavy metal ions from water by CuO–ZnO tetrapodal hybrid nanocomposite. *Materials Today. Chemistry, 17*, 100336. doi:10.1016/j.mtchem.2020.100336

Sharma, R., & Goel, A. (2018). Development of insect repellent finish by a simple coacervation microencapsulation technique. *International Journal of Clothing Science and Technology, 30*(2), 152–158. doi:10.1108/IJCST-02-2017-0022

Sharma, S., Kumar, K., Thakur, N., Chauhan, S., & Chauhan, M. S. (2019). The effect of shape and size of ZnO nanoparticles on their antimicrobial and photocatalytic activities: A green approach. *Bulletin of Materials Science, 43*(1), 20. doi:10.100712034-019-1986-y

Sharma, T., Mohanareddy, A. L. M., Chandra, T. S., & Ramaprabhu, S. (2008). Development of carbon nanotubes and nanofluids based microbial fuel cell. *International Journal of Hydrogen Energy, 33*(22), 6749–6754. https://doi.org/10.1016/j.ijhydene.2008.05.112

Shatkin, J. A., Wegner, T. H., & Neih, W. (2013). Incorporating life-cycle thinking into risk assessment for nanoscale materials: Case study of nanocellulose. *2013 TAPPI International Conference on Nanotechnology for Renewable Materials.*

Shawky, H. A., Chae, S.-R., Lin, S., & Wiesner, M. R. (2011). Synthesis and characterization of a carbon nanotube/polymer nanocomposite membrane for water treatment. *Desalination, 272*(1–3), 46–50. https://doi.org/10.1016/j.desal.2010.12.051

Sheldon, R. A. (2007). Enzyme immobilization: The quest for optimum performance. *Advanced Synthesis & Catalysis, 349*(8–9), 1289–1307. https://doi.org/10.1002/adsc.200700082

Shende, S., Ingle, A. P., Gade, A., & Rai, M. (2015). Green synthesis of copper nanoparticles by Citrus medica Linn. (Idilimbu) juice and its antimicrobial activity. *World Journal of Microbiology & Biotechnology, 31*(6), 865–873. doi:10.100711274-015-1840-3 PMID:25761857

Shen, J., Shen, J., Zhang, W., Yu, X., Tang, H., Zhang, M., Zulfiqar, & Liu, Q. (2019). Built-in electric field induced CeO2/Ti3C2-MXene Schottky-junction for coupled photocatalytic tetracycline degradation and CO2 reduction. *Ceramics International, 45*(18), 24146–24153. doi:10.1016/j.ceramint.2019.08.123

Shen, Y., & Chen, B. (2015). Sulfonated Graphene Nanosheets as a Superb Adsorbent for Various Environmental Pollutants in Water. *Environmental Science & Technology, 49*(12), 7364–7372. doi:10.1021/acs.est.5b01057 PMID:26008607

Shikuku, V. O., & Mishra, T. (2021). Adsorption isotherm modeling for methylene blue removal onto magnetic kaolinite clay: A comparison of two-parameter isotherms. *Applied Water Science, 11*(6), 103. doi:10.100713201-021-01440-2

Shi, L., Zhuo, S., Abulikemu, M., Mettela, G., Palaniselvam, T., Rasul, S., Tang, B., Yan, B., Saleh, N. B., & Wang, P. (2018). Annealing temperature effects on photoelectrochemical performance of bismuth vanadate thin film photoelectrodes. *RSC Advances, 8*(51), 29179–29188. doi:10.1039/C8RA04887H PMID:35548013

Shirmardi, M., Mahvi, A. H., Hashemzadeh, B., Naeimabadi, A., Hassani, G., & Niri, M. V. (2013). The adsorption of malachite green (MG) as a cationic dye onto functionalized multi walled carbon nanotubes. *Korean Journal of Chemical Engineering, 30*(8), 1603–1608. doi:10.100711814-013-0080-1

Shirmardi, M., Mesdaghinia, A., Mahvi, A. H., Nasseri, S., & Nabizadeh, R. (2012). Kinetics and equilibrium studies on adsorption of Acid Red 18 (azo-dye) using multiwall carbon nanotubes (MWCNTs) from aqueous solution. *E-Journal of Chemistry*, *9*(4), 9. doi:10.1155/2012/541909

Shi, W., Yan, Y., & Yan, X. (2013). Microwave- assisted synthesis of nano-scale $BiVO_4$ photocatalysts and their excellent visible-light-driven photocatalytic activity for the degradation of ciprofloxacin. *Chemical Engineering Journal*, *215*, 740–746. doi:10.1016/j.cej.2012.10.071

Shi, Y., Huang, X.-K., Wang, Y., Zhou, Y., Yang, D.-R., Wang, F.-B., Gao, W., & Xia, X.-H. (2019). Electronic Metal–Support Interaction To Modulate MoS2-Supported Pd Nanoparticles for the Degradation of Organic Dyes. *ACS Applied Nano Materials*, *2*(6), 3385–3393. doi:10.1021/acsanm.9b00297

Shrimali, K., & Dedhia, E. M. (2015). Microencapsulating for Textile Finishing. *IOSR Journal of Polymer and Textile Engineering*, *2*(2). https://www.iosrjournals.org/iosr-jpte/papers/Vol2-issue2/A0220104.pdf

Shukla, S. R., & Pai, R. S. (2005). Adsorption of Cu (II), Ni (II) and Zn (II) on modified jute fibres. *Bioresource Technology*, *96*(13), 1430–1438. doi:10.1016/j.biortech.2004.12.010 PMID:15939269

Simamora, A. J., Hsiung, T. L., Chang, F. C., Yang, T. C., Liao, C. Y., & Wang, H. P. (2012). Photocatalytic splitting of seawater and degradation of methylene blue on CuO/nano TiO_2. *International Journal of Hydrogen Energy*, *37*(18), 13855–13858. doi:10.1016/j.ijhydene.2012.04.091

Singh, A., Ahmed, A., Sharma, A., Sharma, C., Paul, S., Khosla, A., & Arya, S. (2021). Promising photocatalytic degradation of methyl orange dye via sol-gel synthesized Ag–CdS@Pr-TiO_2 core/shell nanoparticles. *Physica B, Condensed Matter*, *616*, 413121. doi:10.1016/j.physb.2021.413121

Singha, K., Mondal, A., Ghosh, S. C., & Panda, A. B. (2018). Visible-light-driven Efficient Photocatalytic Reduction of Organic Azides to Amines over CdS Sheet–rGO Nanocomposite. *Chemistry, an Asian Journal*, *13*(3), 255–260. doi:10.1002/asia.201701614 PMID:29265682

Singhal, S., Dixit, S., & Shukla, A. K. (2018). Self-assembly of the Ag deposited ZnO/carbon nanospheres: A resourceful photocatalyst for efficient photocatalytic degradation of methylene blue dye in water. *Advanced Powder Technology*, *29*(12), 3483–3492. doi:10.1016/j.apt.2018.09.031

Singh, I., & Mishra, P. K. (2020). Nano-membrane filtration a novel application of nanotechnology for waste water treatment. *Materials Today: Proceedings*, *29*, 327–332. doi:10.1016/j.matpr.2020.07.284

Singh, J., Dutta, T., Kim, K. H., Rawat, M., Samddar, P., & Kumar, P. (2018). Green' synthesis of metals and their oxide nanoparticles: Applications for environmental remediation. *Journal of Nanobiotechnology*, *16*(1), 84. doi:10.118612951-018-0408-4 PMID:30373622

Singh, R. L., Singh, P. K., & Singh, R. P. (2015). Enzymatic decolorization and degradation of azo dyes—A review. *International Biodeterioration & Biodegradation*, *104*, 21–31. doi:10.1016/j.ibiod.2015.04.027

Sinha, A., Dhanjai, Zhao, H., Huang, Y., Lu, X., Chen, J., & Jain, R. (2018). MXene: An emerging material for sensing and biosensing. *Trends in Analytical Chemistry*, *105*, 424–435. doi:10.1016/j.trac.2018.05.021

Sirajudheen, P., Karthikeyan, P., Vigneshwaran, S., & Meenakshi, S. (2020). Synthesis and characterization of La (III) supported carboxymethylcellulose-clay composite for toxic dyes removal: Evaluation of adsorption kinetics, isotherms, and thermodynamics. *International Journal of Biological Macromolecules*, *161*, 1117–1126. doi:10.1016/j.ijbiomac.2020.06.103 PMID:32553962

Sitaaraman, S. R., Shanmugapriyan, M. I., Varunkumar, K., Grace, A. N., & Sellappan, R. (2021). Synthesis of heterojunction tungsten oxide (WO_3) and Bismuth vanadate ($BiVO_4$) photoanodes by spin coating method for solar water splitting applications. *Materials Today: Proceedings*, *45*, 3920–3926. doi:10.1016/j.matpr.2020.07.601

Sitharaman, B., Shi, X., Walboomers, X. F., Liao, H., Cuijpers, V., Wilson, L. J., Mikos, A. G., & Jansen, J. A. (2008). In vivo biocompatibility of ultrashort singlewalled carbon nanotube/biodegradable polymer nanocomposites for bone tissue engineering. *Bone*, *43*(2), 362–370. https://doi.org/10.1016/j.bone.2008.04.013

Sivakami, M., Gomathi, T., Venkatesan, J., Jeong, H.-S., Kim, S.-K., & Sudha, P. (2013). Preparation and characterization of nano chitosan for treatment wastewaters. *International Journal of Biological Macromolecules*, *57*, 204–212.

Son, B. C., Park, K., Song, S. H., & Yoo, Y. J. (2004). Selective biosorption of mixed heavy metal ions using polysaccharides. *Korean Journal of Chemical Engineering*, *21*(6), 1168–1172. doi:10.1007/BF02719489

Song, J. Y., & Kim, B. S. (2009). Biological synthesis of bimetallic Au/Ag nanoparticles using Persimmon (Diospyros kaki) leaf extract. *Korean Journal of Chemical Engineering*, *25*(4), 808–811. doi:10.100711814-008-0133-z

Song, L., & Qiu, Z. (2009). Crystallization behavior and thermal property of biodegradable poly(butylene succinate)/ functional multi-walled carbon nanotubes nanocomposite. *Polymer Degradation & Stability*, *94*(4), 632–637. https://doi.org/10.1016/j.polymdegradstab.2009.01.009

Souza, J. S., Hirata, F. T. H., & Corio, P. (2019). Microwave-assisted synthesis of bismuth vanadate nano-flowers decorated with gold nanoparticles with enhanced photocatalytic activity. *Journal of Nanoparticle Research*, *35*, 1–9.

Sreeja, V. G., Vinitha, G., Reshmi, R., Jayaraj, M. K., & Anila, E. I. (2019). Structural, Spectral, Electrical and Nonlinear Optical Characterizations of rGO-PANI Composites. *Materials Today: Proceedings*, *10*, 456–465. doi:10.1016/j.matpr.2019.03.010

Srinivasan, A., & Viraraghavan, T. (2010). Decolorization of Dye Wastewaters by Biosorbents, A Review. *Journal of Environmental Management*, *91*, 1915–1929.

Srivastava, A., Srivastava, O. N., Talapatra, S., Vajtai, R., & Ajayan, P. M. (2004). Carbon nanotube filters. *Nature Materials*, *3*(9), 610–614. https://doi.org/10.1038/nmat1192

Srivastava, A., Srivastava, O. N., Talapatra, S., Vajtai, R., & Ajayan, P. M. (2004). Carbon nanotubes filters. *Nature Materials*, *3*(9), 610–614. https://doi.org/10.1038/nmat1192

Srivastava, V., & Choubey, A. K. (2021). Investigation of adsorption of organic dyes present in wastewater using chitosan beads immobilized with biofabricated CuO nanoparticles. *Journal of Molecular Structure*, *1242*, 130749. doi:10.1016/j.molstruc.2021.130749

Stanley, R., Jebasingh, J. A., Stanley, P. K., Ponmani, P., Shekinah, M. E., & Vasanthi, J. (2021). Excellent Photocatalytic degradation of Methylene Blue, Rhodamine B and Methyl Orange dyes by Ag-ZnO nanocomposite under natural sunlight irradiation. *Optik (Stuttgart)*, *231*, 166518. doi:10.1016/j.ijleo.2021.166518

Sternberg, S. P., & Dorn, R. W. (2002). Cadmium removal using Cladophora in batch, semi-batch and flow reactors. *Bioresource Technology*, *81*(3), 249–255. doi:10.1016/S0960-8524(01)00131-6 PMID:11806406

Stewart, M. H., Wolfe, R. L., & Means, E. G. (1990). Assessment of the bacteriological activity associated with granular activated carbon treatment of drinking water. *Applied and Environmental Microbiology*, *56*(12), 3822–3829. https://doi.org/10.1128/aem.56.12.3822-3829.1990

Stoller, M., Sacco, O., Vilardi, G., Pulido, J. M. O., & Di Palma, L. (2018). Technical–economic evaluation of chromium recovery from tannery wastewater streams by means of membrane processes, Desalin. *Water Treat*, *127*, 57–63. doi:10.5004/dwt.2018.22533

Su, C. (2017). Environmental implications and applications of engineered nanoscale magnetite and its hybrid nanocomposites: A review of recent literature. *Journal of Hazardous Materials, 322*(A), 48–84. https://doi.org/10.1016/j.jhazmat.2016.06.060

Subha, V., Divya, K., Gayathri, S., Jagan Mohan, E., Keerthanaa, N., Vinitha, M., Kirubanandan, S., & Renganathan, S. (2018). Applications of iron oxide nano composite in waste water treatment–dye decolourisation and anti–microbial activity. *MOJ Drug Design Development & Therapy*, *2*, 178–184.

Subodh, N. K., Mogha, N. K., Chaudhary, K., Kumar, G., & Masram, D. T. (2018). Fur-Imine-Functionalized Graphene Oxide-Immobilized Copper Oxide Nanoparticle Catalyst for the Synthesis of Xanthene Derivatives. *ACS Omega, 3*(11), 16377–16385. doi:10.1021/acsomega.8b01781

Suehiro, J., Zhou, G., & Hara, M. (2005). Detection of partial discharge in SF6 gas using a carbon nanotube-based gas sensor. *Sensors and Actuators. Part B, 105*(2), 164–169. doi:10.1016/S0925-4005(04)00415-0

Su, F. C., Zhou, H. J., Zhang, Y. X., & Wang, G. Z. (2016). Three-dimensional honeycomb-like structured zero-valent iron/chitosan composite foams for effective removal of inorganic arsenic in water. *Journal of Colloid and Interface Science*, *478*, 421–429.

Su, F., & Lu, C. (2007). Adsorption kinetics, thermodynamics and desorption of natural dissolved organic matter by multiwalled carbon nanotubes. *Journal of Environmental Science and Health. Part A, Toxic/Hazardous Substances & Environmental Engineering*, *42*(11), 1543–1552. https://doi.org/10.1080/10934520701513381

Su, L., & Gan, Y. X. (2012). Experimental study on synthesizing TiO2 nanotube/polyaniline (PANI) nanocomposites and their thermoelectric and photosensitive property characterization. *Composites. Part B, Engineering*, *43*(2), 170–182. https://doi.org/10.1016/j.compositesb.2011.07.015

Sultana, S., Rafiuddin, Z., Khan, M., & Umar, K. (2012). Synthesis and characterization of copper ferrite nanoparticles doped polyaniline. *Journal of Alloys and Compounds, 535*, 44–49. https://doi.org/10.1016/j.jallcom.2012.04.081

Sun, G., & Shi, W. (1998). Sunflower stalks as adsorbents for the removal of metal ions from wastewater. *Industrial & Engineering Chemistry Research*, *37*(4), 1324–1328. doi:10.1021/ie970468j

Sun, L., Shi, Y., Li, B., Li, X., & Wang, Y. (2013). Preparation and characterization of polypyrrole/TiO_2 nanocomposites by reverse microemulsion polymerization and its photocatalytic activity for the degradation of methyl orange under natural light. *Polymer Composites*, *34*(7), 1076–1080. doi:10.1002/pc.22515

Sun, M., & Li, J. (2018). Graphene oxide membranes: Functional structures, preparation and environmental applications. *Nano Today*, *20*, 121–137. doi:10.1016/j.nantod.2018.04.007

Sun, X., Yang, L., Li, Q., Zhao, J., Li, X., Wang, X., & Liu, H. (2014). Amino-functionalized magnetic cellulose nanocomposite as adsorbent for removal of Cr (VI): Synthesis and adsorption studies. *Chemical Engineering Journal*, *241*, 175–183.

Sun, Y., Gao, S., Lei, F., & Xie, Y. (2015). Atomically-thin two-dimensional sheets for understanding active sites in catalysis. *Chemical Society Reviews*, *44*(3), 623–636. doi:10.1039/C4CS00236A PMID:25382246

Sun, Y., Lin, H., Wang, C., Wu, Q., Wang, X., & Yang, M. (2018). Morphology-controlled synthesis of TiO2/MoS2 nanocomposites with enhanced visible-light photocatalytic activity. *Inorganic Chemistry Frontiers*, *5*(1), 145–152. doi:10.1039/C7QI00491E

Sun, Y., Sun, Y., Meng, X., Gao, Y., Dall'Agnese, Y., Chen, G., Dall'Agnese, C., & Wang, X.-F. (2019). Eosin Y-sensitized partially oxidized Ti3C2 MXene for photocatalytic hydrogen evolution. *Catalysis Science & Technology*, *9*(2), 310–315. doi:10.1039/C8CY02240B

Su, Q., Pang, S., Alijani, V., Li, C., Feng, X., & Müllen, K. (2009). Composites of graphene with large aromatic molecules. *Advanced Materials*, *21*(31), 3191–3195. https://doi.org/10.1002/adma.200803808

Sure, B., Easterling, L., Dowell, J., Crudup, M., & Scrimshaw, N. S. (1953). Protein Efficiency, Improvement in Whole Yellow Corn with Lysine, Tryptophan, and Threonine. *Journal of Agricultural and Food Chemistry*, *1*(9), 626–629. doi:10.1021/jf60009a007

Su, T., Hood, Z. D., Naguib, M., Bai, L., Luo, S., Rouleau, C. M., Ivanov, I. N., Ji, H., Qin, Z., & Wu, Z. (2019). Monolayer Ti3C2Tx as an Effective Co-catalyst for Enhanced Photocatalytic Hydrogen Production over TiO2. *ACS Applied Energy Materials*, *2*(7), 4640–4651. doi:10.1021/acsaem.8b02268

Su, T., Peng, R., Hood, Z. D., Naguib, M., Ivanov, I. N., Keum, J. K., Qin, Z., Guo, Z., & Wu, Z. (2018). One-Step Synthesis of Nb2O5/C/Nb2C (MXene) Composites and Their Use as Photocatalysts for Hydrogen Evolution. *ChemSusChem*, *11*(4), 688–699. doi:10.1002/cssc.201702317 PMID:29281767

Su, T., Shao, Q., Qin, Z., Guo, Z., & Wu, Z. (2018). Role of Interfaces in Two-Dimensional Photocatalyst for Water Splitting. *ACS Catalysis*, *8*(3), 2253–2276. doi:10.1021/acscatal.7b03437

Suyana, P., Ganguly, P., Nair, B. N., Mohamed, A. P., Warrier, K. G. K., & Hareesh, U. S. (2017). Co3O4–C3N4 p–n nano-heterojunctions for the simultaneous degradation of a mixture of pollutants under solar irradiation. *Environmental Science. Nano*, *4*(1), 212–221. doi:10.1039/C6EN00410E

Szpyrkowicz, L., Juzzolino, C., & Kaul, S. N. (2001). A Comparative study on oxidation of disperse dyes by electrochemical process, ozone, hypochlorite and fenton reagent. *Water Research*, *35*(9), 2129–2136. doi:10.1016/S0043-1354(00)00487-5 PMID:11358291

Taghizadeh, M. T., Siyahi, V., Ashassi-Sorkhabi, H., & Zarrini, G. (2020). ZnO, AgCl and AgCl/ZnO nanocomposites incorporated chitosan in the form of hydrogel beads for photocatalytic degradation of MB, E. coli and S. aureus. *International Journal of Biological Macromolecules*, *147*, 1018–1028. doi:10.1016/j.ijbiomac.2019.10.070 PMID:31739064

Tahaikt, M., El Habbani, R., Ait Haddou, A., Achary, I., Amor, Z., Taky, M., Alami, A., Boughriba, A., Hafsi, M., & Elmidaoui, A. (2007). Fluoride removal from groundwater by nanofiltration. *Desalination*, *212*(1–3), 46–53. https://doi.org/10.1016/j.desal.2006.10.003

Tai, H., Jiang, Y., Xie, G. Y., Yu, J., & Zhao, M. (2007). Self-assembly of TiO2/polypyrrole nanocomposite ultrathin films and application for an NH3 gas sensor. *International Journal of Environmental Analytical Chemistry*, *87*(8), 539–551. doi:10.1080/03067310701272954

Takeuchi, Y., Hino, M., Yoshimura, Y., Otowa, T., Izuhara, H., & Nojima, T. (1999). Removal of single component chlorinated hydrocarbon vapor by activated carbon of very high surface area. *Separation and Purification Technology*, *15*(1), 79–90. https://doi.org/10.1016/S1383-5866(98)00082-3

Talapatra, S., & Migone, A. D. (2002). Adsorption of methane on bundles of closed-ended single-wall carbon nanotubes. *Physical Review. Part B*, *65*(4), 045416-045421. doi:10.1103/PhysRevB.65.045416

Tan, B., Ye, X., Li, Y., Ma, X., Wang, Y., & Ye, J. (2018). Defective Anatase TiO2−x Mesocrystal Growth In Situ on g-C3N4 Nanosheets: Construction of 3D/2D Z-Scheme Heterostructures for Highly Efficient Visible-Light Photocatalysis. *Chemistry (Weinheim an der Bergstrasse, Germany)*, *24*(50), 13311–13321. doi:10.1002/chem.201802366 PMID:29957872

Tan, C., Cao, X., Wu, X.-J., He, Q., Yang, J., Zhang, X., Chen, J., Zhao, W., Han, S., Nam, G.-H., Sindoro, M., & Zhang, H. (2017). Recent Advances in Ultrathin Two-Dimensional Nanomaterials. *Chemical Reviews*, *117*(9), 6225–6331. doi:10.1021/acs.chemrev.6b00558 PMID:28306244

Tan, C., Lai, Z., & Zhang, H. (2017). Ultrathin Two-Dimensional Multinary Layered Metal Chalcogenide Nanomaterials. *Advanced Materials*, *29*(37), 1701392. doi:10.1002/adma.201701392 PMID:28752578

Tan, C., & Zhang, H. (2015). Wet-chemical synthesis and applications of non-layer structured two-dimensional nanomaterials. *Nature Communications*, *6*(1), 7873. doi:10.1038/ncomms8873 PMID:26303763

Tang, F., Cheng, W., Su, H., Zhao, X., & Liu, Q. (2018). Smoothing surface trapping states in 3D coral-like CoOOH-wrapped-$BiVO_4$ for efficient photoelectrochemical water oxidation. *ACS Applied Materials & Interfaces*, *10*(7), 6228–6234. doi:10.1021/acsami.7b15674 PMID:29384358

Tantawy, H. R., Nada, A. A., Baraka, A., & Elsayed, M. A. (2021). Novel synthesis of bimetallic Ag-Cu nanocatalysts for rapid oxidative and reductive degradation of anionic and cationic dyes. *Applied Surface Science Advances*, *3*, 100056. doi:10.1016/j.apsadv.2021.100056

Tayyebi, A., Soltani, T., & Lee, B. K. (2019). Effect of p^H on photocatalytic and photoelectrochemical (PEC) properties of monoclinic bismuth vanadate. *Journal of Colloid and Interface Science*, *534*, 37–46. doi:10.1016/j.jcis.2018.08.095 PMID:30205253

Tegbe, T. S. B., Adeyinka, I. A., Baye, K. D., & Alawa, J. P. (2006). Evaluation of Feeding Graded Levels of Dried and Milled FicusthonningiiLeaves on Growth Performance, Carcass Characteristics and Organs of Weaner Rabbits. *Pakistan Journal of Nutrition*, *5*(6), 548–550. doi:10.3923/pjn.2006.548.550

Terrones, M. (2003). Science and technology of the twenty-first century: Synthesis, properties and applications of carbon nanotubes. *Annual Review of Materials Research*, *33*(1), 419–501. https://doi.org/10.1146/annurev.matsci.33.012802.100255

Teunou, E., & Poncelet, D. (2002). Batch and continuous fluid bed coating–review and state of the art. *Journal of Food Engineering*, *53*(4), 325–340.

Thandapani, K., Kathiravan, M., Namasivayam, E., Padiksan, I. A., Natesan, G., Tiwari, M., Giovanni, B., & Perumal, V. (2018). Enhanced larvicidal, antibacterial, and photocatalytic efficacy of TiO2 nanohybrids green synthesized using the aqueous leaf extract of Parthenium hysterophorus. *Environmental Science and Pollution Research International*, *25*(11), 10328–10339. doi:10.100711356-017-9177-0 PMID:28537028

Thennarasu, G., & Sivasamy, A. (2016). Enhanced visible photocatalytic activity of cotton ball like nano structured Cu doped ZnO for the degradation of organic pollutant. *Ecotoxicology and Environmental Safety*, *134*, 412–420. doi:10.1016/j.ecoenv.2015.10.030 PMID:26560433

Thines, R. K., Mubarak, N. M., Nizamuddin, S., Sahu, J. N., Abdullah, E. C., & Ganesan, P. (2017). Application potential of carbon nanomaterials in water and wastewater treatment: A review. *Journal of the Taiwan Institute of Chemical Engineers*, *72*, 116–133. https://doi.org/10.1016/j.jtice.2017.01.018

Thirumalai, K., Balachandran, S., & Swaminathan, M. (2016). Superior photocatalytic, electrocatalytic, and self-cleaning applications of Fly ash supported ZnO nanorods. *Materials Chemistry and Physics*, *183*, 191–200. doi:10.1016/j.matchemphys.2016.08.018

Thoke, S. B. (2012 April 11). *A Seminar on Microencapsulation Techniques and Application* [PowerPoint slides]. Department of Pharmaceuticals, S.N.D College of Pharmacy. https://www.slideshare.net/thokesagar/sagar-thoke

Thompson, T. L., & Yates, J. T. (2006). Surface science studies of the photoactivation of TiO2 new photochemical processes. *Chemical Reviews*, *106*(10), 4428–4453. doi:10.1021/cr050172k PMID:17031993

Tian, H., Liu, M., & Zheng, W. (2018). Constructing 2D graphitic carbon nitride nanosheets/layered MoS2/graphene ternary nanojunction with enhanced photocatalytic activity. *Applied Catalysis B: Environmental*, *225*, 468–476. doi:10.1016/j.apcatb.2017.12.019

Tian, Q., Wu, W., Liu, J., Wu, Z., Yao, W., Ding, J., & Jiang, C. (2017). Dimensional heterostructures of 1D CdS/2D ZnIn2S4 composited with 2D graphene: Designed synthesis and superior photocatalytic performance. *Dalton Transactions (Cambridge, England)*, *46*(9), 2770–2777. doi:10.1039/C7DT00018A PMID:28168251

Tiemann, K. J., Gamez, G., Dokken, K., Parsons, J. G., & Gardea-Torresdey, J. L. (2002). Chemical modification and X-ray absorption studies for lead (II) binding by Medicago sativa (alfalfa) biomass. *Microchemical Journal*, *71*(2-3), 287–293. doi:10.1016/S0026-265X(02)00021-8

Timoszyk, A. (2018). A review of the biological synthesis of gold nanoparticles using fruit extracts: Scientific potential and application. *Bulletin of Materials Science*, *41*(6), 154. doi:10.100712034-018-1673-4

Tiraferri, A., Vecitis, C. D., & Elimelech, M. (2011). Covalent binding of single-walled carbon nanotubes to polyamide membranes for antimicrobial surface properties. *ACS Applied Materials & Interfaces*, *3*(8), 2869–2877. https://doi.org/10.1021/am200536p

Tiwari, D., Mishra, S. P., Mishra, M., & Dubey, R. S. (1999). Biosorptive behaviour of Mango (Mangifera indica) and Neem (Azadirachta indica) bark for Hg2+, Cr3+ and Cd2+ toxic ions from aqueous solutions: A radiotracer study. *Applied Radiation and Isotopes*, *50*(4), 631–642. doi:10.1016/S0969-8043(98)00104-3

Tjong, S. C. (2006). Structural and mechanical properties of polymer nanocomposites. *Materials Science and Engineering R Reports*, *53*(3–4), 73–197. https://doi.org/10.1016/j.mser.2006.06.001

Tome, S., Dzoujo, H., Shikuku, V., & Otieno, S. (2021). Synthesis, characterization and application of acid and alkaline activated volcanic ash-based geopolymers for adsorptive remotion of cationic and anionic dyes from water. *Ceramics International*, *47*(15), 20965–20973. Advance online publication. doi:10.1016/j.ceramint.2021.04.097

Tong, Z., Yang, D., Shi, J., Nan, Y., Sun, Y., & Jiang, Z. (2015). Three-Dimensional Porous Aerogel Constructed by g-C3N4 and Graphene Oxide Nanosheets with Excellent Visible-Light Photocatalytic Performance. *ACS Applied Materials & Interfaces*, *7*(46), 25693–25701. doi:10.1021/acsami.5b09503 PMID:26545166

Trache, D., Hussin, M. H., Haafiz, M. M., & Thakur, V. K. (2017). Recent progress in cellulose nanocrystals: Sources and production. *Nanoscale*, *9*(5), 1763–1786.

Tryba, B., & Morawski, A. W., & Kalenczuk. (2003). Exfoliated graphite as a new sorbent for removal of engine oils from wastewater. *Spill Science & Technology Bulletin*, *8*(5–6), 569–571.

Tsai, H. Y., Wu, C. C., Lee, C. Y., & Shih, E. P. (2009). Microbial fuel cell performance of multiwall carbon nanotubes on carbon cloth as electrodes. *Journal of Power Sources*, *194*(1), 199–205. https://doi.org/10.1016/j.jpowsour.2009.05.018

Tso, C., Zhung, C., Shih, Y., Tseng, Y.-M., Wu, S., & Doong, R. (2010). Stability of metal oxide nanoparticles in aqueous solutions. *Water Science and Technology*, *61*(1), 127–133. doi:10.2166/wst.2010.787 PMID:20057098

Türkoğlu, G. C., Sariişik, A. M., Erkan, G., Kayalar, H., Kontart, O., & Öztuna, S. (2017). *Determination of antioxidant capacity of capsule loaded textiles*. Academic Press.

Tuzen, M., & Soylak, M. (2007). Multiwalled carbon nanotubes for speciation of chromium in environmental samples. *Journal of Hazardous Materials*, *147*(1–2), 219–225. https://doi.org/10.1016/j.jhazmat.2006.12.069

Udayabhanu, L. R., Lakshmana Reddy, N., Shankar, M. V., Sharma, S. C., & Nagaraju, G. (2020). One-pot synthesis of Cu–TiO_2/CuO nanocomposite: Application to photocatalysis for enhanced H2 production, dye degradation & detoxification of Cr (VI). *International Journal of Hydrogen Energy*, *45*(13), 7813–7828. doi:10.1016/j.ijhydene.2019.10.081

Ueda, T., Bhuiyan, M. M. H., Norimatsu, H., Katsuki, S., Ikegami, T., & Mitsugi, F. (2008a). Development of carbon nanotube-based gas sensors for NOx gas detection working at low temperature. *Physica E, Low-Dimensional Systems and Nanostructures*, *40*(7), 2272–2277. https://doi.org/10.1016/j.physe.2007.12.006

Ueda, T., Katsuki, S., Takahashi, K., Narges, H. A., Ikegami, T., & Mitsugi, F. (2008b). Fabrication and characterization of carbon nanotube based high sensitive gas sensors operable at room temperature. *Diamond and Related Materials*, *17*(7–10), 1586–1589. https://doi.org/10.1016/j.diamond.2008.03.009

Ullah, M. W., Park, J. K., Ul-Islam, M., Khan, S., & Khattak, W. A. (2013). Synthesis of regenerated bacterial cellulose-zinc oxide nanocomposite films for biomedical applications. *Cellulose (London, England)*, *21*, 433–447.

Ullah, R., & Dutta, J. (2008). Photocatalytic degradation of organic dyes with manganese-doped ZnO nanoparticles. *Journal of Hazardous Materials*, *156*(1), 194–200. doi:10.1016/j.jhazmat.2007.12.033 PMID:18221834

Umer, H., Nigam, H., Tamboli, A. M., & Nainar, M. S. M. (2011). Microencapsulation: Process, techniques and applications. *International Journal of Research in Pharmaceutical and Biomedical Sciences*, *2*(2), 474–481.

Upadhyayula, V. K. K., Deng, S., Mitchell, M. C., & Smith, G. B. (2009). Application of carbon nanotube technology for removal of contaminants in drinking water: A review. *The Science of the Total Environment*, *408*(1), 1–13. doi:10.1016/j.scitotenv.2009.09.027 PMID:19819525

Urbankowski, P., Anasori, B., Hantanasirisakul, K., Yang, L., Zhang, L., Haines, B., May, S. J., Billinge, S. J. L., & Gogotsi, Y. (2017). 2D molybdenum and vanadium nitrides synthesized by ammoniation of 2D transition metal carbides (MXenes). *Nanoscale*, *9*(45), 17722–17730. doi:10.1039/C7NR06721F PMID:29134998

V, P. P., R, P., Sathe, V., & Mahalingam, U. (2019). Graphene boosted silver nanoparticles as surface enhanced Raman spectroscopic sensors and photocatalysts for removal of standard and industrial dye contaminants. *Sensors and Actuators B: Chemical, 281*, 679-688.

Valentini, L., Cantalini, C., Armentano, I., Kenny, J. M., Lozzi, L., & Santucci, S. (2004). Highly sensitive and selective sensors based on carbon nanotubes thin films for molecular detection. *Diamond and Related Materials*, *13*(4–8), 1301–1305. https://doi.org/10.1016/j.diamond.2003.11.011

Van Duy, N., Van Hieu, N., Huy, P. T., Chien, N. D., Thamilselvan, M., & Yi, J. (2008). Mixed SnO2/TiO2 included with carbon nanotubes for gas-sensing application. *Physica E, Low-Dimensional Systems and Nanostructures*, *41*(2), 258–263. https://doi.org/10.1016/j.physe.2008.07.007

Van Hieu, N., Thuy, L. T. B., & Chien, N. D. (2008). Highly sensitive thin film NH_3 gas sensor operating at room temperature based on SnO_2/MWCNTs composite. *Sensors and Actuators. Part B*, *129*(2), 888–895. doi:10.1016/j.snb.2007.09.088

Vanaja, M., Rajeshkumar, S., Paulkumar, K., Gnanajobitha, G., Malarkodi, C., & Annadurai, G. (2013). Phytosynthesis and characterization of silver nanoparticles using stem extract of Coleus aromaticus. *Int. J. Mater. Biomat. Appl.*, *3*, 1–4.

Vankar, P. S., & Bajpai, D. (2010). Preparation of gold nanoparticles from Mirabilis jalapa flowers. *Indian Journal of Biochemistry & Biophysics*, *47*, 157–160. PMID:20653286

Vankar, P. S., Tiwari, V., & Srivastava, J. (2007). Antioxidants from supercritical carbon dioxide fluid extracts (SCFE) of bark-peel of Eucalyptus globulus. *EJEAF Chem.*, *6*(11), 2550–2556.

Varghese, O. K., Kichambre, P. D., Gong, D., Ong, K. G., Dickey, E. C., & Grimes, C. A. (2001). Gas sensing characteristics of multi-wall carbon nanotubes. *Sensors and Actuators. Part B*, *81*(1), 32–41. doi:10.1016/S0925-4005(01)00923-6

Varghese, A. G., Paul, S. A., & Latha, M. S. (2019). Remediation of heavy metals and dyes from wastewater using cellulose-based adsorbents. *Environmental Chemistry Letters*, *17*(2), 867–877.

Varjani, S. J., Gnansounou, E., & Pandey, A. (2017). Comprehensive review on toxicity of persistent organic pollutants from petroleum refinery waste and their degradation by microorganisms. *Chemosphere*, *188*, 280–291. doi:10.1016/j.chemosphere.2017.09.005 PMID:28888116

Vaudreuil, S., Labzour, A., Sinha-Ray, S., El Mabrouk, K. E., & Bousmina, M. (2007). Dispersion characteristics and properties of poly(methylmethacrylate)/multiwalled carbon nanotubes nanocomposites. *Journal of Nanoscience and Nanotechnology*, *7*(7), 2349–2355. https://doi.org/10.1166/jnn.2007.419

Vautier, M., Guillard, C., & Herrmann, J. M. (2001). Photocatalytic degradation of dyes in water: Case study of indigo and of indigo Carmine. *Journal of Catalysis*, *201*(1), 46–59. doi:10.1006/jcat.2001.3232

Velumani, A., Sengodan, P., Arumugam, P., Rajendran, R., Santhanam, S., & Palanisamy, M. (2020). Carbon quantum dots supported ZnO sphere based photocatalyst for dye degradation application. *Current Applied Physics*, *20*(10), 1176–1184. doi:10.1016/j.cap.2020.07.016

Velusamy, S., Roy, A., Sundaram, S., & Kumar Mallick, T. (2021). A Review on Heavy Metal Ions and Containing Dyes Removal Through Graphene Oxide-Based Adsorption Strategies for Textile Wastewater Treatment. *Chemical Record (New York, N.Y.)*, *21*, 1570–1610.

Vert, M., Doi, Y., Hellwich, K. H., Hess, M., Hodge, P., Kubisa, P., Rinaudo, M., & Schué, F. (2012). Terminology for biorelated polymers and applications (IUPAC Recommendations 2012). *Pure and Applied Chemistry*, *84*(2), 377–410. doi:10.1351/PAC-REC-10-12-04

Vidya, C., Hiremath, S., & Chandraprabha, M. N. (2013). Green synthesis of ZnO nanoparticles by Calotropis gigantea. *Int J Curr Eng Technol.*, *1*, 118–120.

Vidya, C., Manjunatha, C., Chandraprabha, M. N., & Rajshekar, M., & Raj, M. (2017). Hazard free green synthesis of ZnO nano-photo-catalyst using Artocarpus Heterophyllus leaf extract for the degradation of Congo red dye in water treatment applications. *Journal of Environmental Chemical Engineering*, *5*(4), 3172–3180. doi:10.1016/j.jece.2017.05.058

Vieira, R. H., & Volesky, B. (2000). Biosorption: A solution to pollution? *International Microbiology*, *3*(1), 17–24. PMID:10963329

Vijayarangamuthu, K., Ahn, S., Seo, H., Yoon, S.-H., Park, C.M., & Jeon, K. J. (2016). Temporospatial control of graphene wettability. *Advanced Materials*, *28*(4), 661–667. doi:10.1002/adma.201503444

Vilardi, G., Ochando-Pulido, J. M., Stoller, M., Verdone, N., & Di Palma, L. (2018). Fenton oxidation and chromium recovery from tannery wastewater by means of iron-based coated biomass as heterogeneous catalyst in fixed-bed columns. *Chemical Engineering Journal*, *351*, 1–11. doi:10.1016/j.cej.2018.06.095

Vilardi, G., Stoller, M., Verdone, N., & Di Palma, L. (2017). Production of nano zero-valent iron particles by means of a spinning disk reactor. *Chemical Engineering Transactions*, *57*, 751–756. doi:10.3303/CET1757126

Villaescusa, I., Fiol, N., Martínez, M., Miralles, N., Poch, J., & Serarols, J. (2004). Removal of copper and nickel ions from aqueous solutions by grape stalks wastes. *Water Research, 38*(4), 992–1002. doi:10.1016/j.watres.2003.10.040 PMID:14769419

Vinay, S. P., & Chandrasekhar, N. (2019). Facile Green Chemistry Synthesis of Ag Nanoparticles Using Areca Catechu Extracts for the Antimicrobial Activity and Photocatalytic Degradation of Methylene Blue Dye. *Materials Today: Proceedings, 9*, 499–505. doi:10.1016/j.matpr.2018.10.368

Vinodgopal, K., Wynkoop, D. E., & Kamat, P. V. (1996). Environmental photochemistry on semiconductor surfaces: Photosensitized degradation of a textile azo dye, acid orange 7, on TiO2 particles using visible light. *Environmental Science & Technology, 30*(5), 1660–1666. doi:10.1021/es950655d

Vinuth, M., Naik, H. S. B., Vinoda, B. M., Pradeepa, S. M., Arun, K. G., & … . (2016). Rapid removal of hazardous rose bengal dye using Fe(III)–montmorillonite as an effective adsorbent in aqueous solution. *Journal of Environmental & Analytical Toxicology, 6*(02), 355. doi:10.4172/2161-0525.1000355

Volesky, B. (2001). Detoxification of metal-bearing effluents: Biosorption for the next century. *Hydrometallurgy, 59*(2-3), 203–216. doi:10.1016/S0304-386X(00)00160-2

Vu, A.-T., Xuan, T. N., & Lee, C.-H. (2019). Preparation of mesoporous $Fe_2O_3 \cdot SiO_2$ composite from rice husk as an efficient heterogeneous Fenton-like catalyst for degradation of organic dyes. *Journal of Water Process Engineering, 28*, 169–180. doi:10.1016/j.jwpe.2019.01.019

Vuković, G. D., Marinković, A. D., Čolić, M., Ristić, M. Đ., Aleksić, R., Perić-Grujić, A. A. P., & Uskoković, P. S. (2010). Removal of cadmium from aqueous solutions by oxidized and ethylenediamine-functionalized multi-walled carbon nanotubes. *Chemical Engineering Journal, 157*(1), 238–248. https://doi.org/10.1016/j.cej.2009.11.026

Vukovic, G. D., Marinkovic, A. D., Skapin, S. D., Ristic, M. D., Aleksic, R., Peric-Grujic, A. A., & Uskokovic, P. S. (2011). Removal of lead from water by amino modified multiwalled carbon nanotubes. *Chemical Engineering Journal, 173*(3), 855–865. doi:10.1016/j.cej.2011.08.036

Vulliet, E., Chovelon, J. M., Guillard, C., & Herrmann, J. M. (2003). Factors influencing the photocatalytic degradation of sulfonylurea herbicides by TiO_2 aqueous suspension. *Journal of Photochemistry and Photobiology A Chemistry, 159*(1), 71–79. doi:10.1016/S1010-6030(03)00108-4

Walker, D. J., Clemente, R., & Bernal, M. P. (2004). Contrasting effects of manure and compost on soil pH, heavy metal availability and growth of *Chenopodiumalbum L.* in a soil contaminated by pyritic mine waste. *Chemosphere, 57*(3), 215–224. doi:10.1016/j.chemosphere.2004.05.020 PMID:15312738

Walter, M. G., Warren, E. L., McKone, J. R., Boettcher, S. W., Mi, Q., Santori, E. A., & Lewis, N. S. (2010). Solar water splitting cells. *Chemical Reviews, 110*(11), 6446–6473. doi:10.1021/cr1002326 PMID:21062097

Wang, B., Jiang, Y. S., Li, F. Y., & Yang, D. Y. (2017b). Preparation of biochar by simultaneous carbonization, magnetization and activation for norfloxacin removal in water. *Bioresource Technology, 233*, 159–165. https://doi.org/10.1016/j.biortech.2017.02.103

Wang, B., Li, C., Pang, J., Qing, X., Zhai, J., & Li, Q. (2012). Novel polypyrrole-sensitized hollow TiO2/fly ash cenospheres: Synthesis, characterization, and photocatalytic ability under visible light. *Applied Surface Science, 258*(24), 9989–9996. doi:10.1016/j.apsusc.2012.06.061

Wang, C., Shen, J., Chen, R., Cao, F., & Jin, B. (2020). Self-assembled BiOCl/Ti3C2Tx composites with efficient photo-induced charge separation activity for photocatalytic degradation of p-nitrophenol. *Applied Surface Science, 519*, 146175. doi:10.1016/j.apsusc.2020.146175

Wang, D., Wang, Y., Li, X., Luo, Q., An, J., & Yue, J. (2008). Sunlight photocatalytic activity of polypyrrole–TiO_2 nanocomposites prepared by "in situ" method. *Catalysis Communications*, *9*(6), 1162–1166. doi:10.1016/j.catcom.2007.10.027

Wang, G. L., Shan, L. W., Wu, Z., & Dong, L. M. (2017). Enhanced photocatalytic properties of molybdenum-doped $BiVO_4$ prepared by sol-gel method. *J. Alloys Comp.*, *36*(2), 129–133. doi:10.100712598-015-0669-0

Wang, H., Lin, J., & Shen, Z. X. (2016). Journal of Science. *Advanced Materials and Devices*, *1*, 225–255.

Wang, H., Wu, Y., Xiao, T., Yuan, X., Zeng, G., Tu, W., Wu, S., Lee, H. Y., Tan, Y. Z., & Chew, J. W. (2018). Formation of quasi-core-shell In2S3/anatase TiO2@metallic Ti3C2Tx hybrids with favorable charge transfer channels for excellent visible-light-photocatalytic performance. *Applied Catalysis B: Environmental*, *233*, 213–225. doi:10.1016/j.apcatb.2018.04.012

Wang, H., Xu, J.-Z., Zhu, J.-J., & Chen, H.-Y. (2002). Preparation of CuO nanoparticles by microwave irradiation. *Journal of Crystal Growth*, *244*(1), 88–94. doi:10.1016/S0022-0248(02)01571-3

Wang, H., Xu, P., Zhong, W., Shen, L., & Du, Q. (2005). Transparent poly(methylmethacrylate)/silica/zirconia nanocomposites with excellent thermal stabilities. *Polymer Degradation & Stability*, *87*(2), 319–327. https://doi.org/10.1016/j.polymdegradstab.2004.08.015

Wang, K., Li, J., & Zhang, G. (2019). Ag-Bridged Z-Scheme 2D/2D Bi5FeTi3O15/g-C3N4 Heterojunction for Enhanced Photocatalysis: Mediator-Induced Interfacial Charge Transfer and Mechanism Insights. *ACS Applied Materials & Interfaces*, *11*(31), 27686–27696. doi:10.1021/acsami.9b05074 PMID:31282639

Wang, K., Xu, X., Lu, L., Li, A., Han, X., Wu, Y., & Jiang, Y. (2019). Magnetically recoverable Ag/Bi2Fe4O9 nanoparticles as a visible-light-driven photocatalyst. *Chemical Physics Letters*, *715*, 129–133. doi:10.1016/j.cplett.2018.11.021

Wang, M., Guo, P., Chai, T., Xie, Y., Han, J., You, M., Wang, Y., & Zhu, T. (2017). Effects of Cu dopants on the structures and photocatalytic performance of cocoon-like Cu-$BiVO_4$ prepared via ethylene glycol solvothermal method. *Journal of Alloys and Compounds*, *691*, 8–14. doi:10.1016/j.jallcom.2016.08.198

Wang, M., Niu, C., Liu, J., Wang, Q., Yang, C., & Zheng, H. (2015). Effective visible light- active nitrogen and samarium co-doped $BiVO_4$ for the degradation of organic pollutants. *Journal of Alloys and Compounds*, *648*, 1109–1115. doi:10.1016/j.jallcom.2015.05.115

Wang, P., Qi, C., Hao, L., Wen, P., & Xu, X. (2019). Sepiolite/Cu_2O/Cu photocatalyst: Preparation and high performance for degradation of organic dye. *Journal of Materials Science and Technology*, *35*(3), 285–291. doi:10.1016/j.jmst.2018.03.023

Wang, Q. H., Setlur, A. A., Lauerhaas, J. M., Dai, J. Y., Seelig, E. W., & Chang, R. P. H. (1998). A nanotube-based field-emission flat panel display. *Applied Physics Letters*, *72*(22), 2912–2913. https://doi.org/10.1063/1.121493

Wang, T., Pan, X., Ben, W., Wang, J., Hou, P., & Qiang, Z. (2017a). Adsorptive removal of antibiotics from water using magnetic ion exchange resin. *Journal of Environmental Sciences (China)*, *52*, 111–117. https://doi.org/10.1016/j.jes.2016.03.017

Wang, W.-K., Zhu, W., Mao, L., Zhang, J., Zhou, Z., & Zhao, G. (2019). Two-dimensional TiO2-g-C3N4 with both TiN and CO bridges with excellent conductivity for synergistic photoelectrocatalytic degradation of bisphenol A. *Journal of Colloid and Interface Science*, *557*, 227–235. doi:10.1016/j.jcis.2019.08.088 PMID:31521972

Wang, X., Chen, C., Hu, W., Ding, A., Xu, D., & Zhou, X. (2005). Sorption of 234Am(III) to multiwall carbon nanotubes. *Environmental Science & Technology*, *39*(8), 2856–2860. https://doi.org/10.1021/es048287d

Wang, X., Lu, J., & Xing, B. (2008). Sorption of organic contaminants by carbon nanotubes: Influence of adsorbed organic matter. *Environmental Science & Technology*, *42*(9), 3207–3212. https://doi.org/10.1021/es702971g

Wang, X., Wang, H., Yu, K., & Hu, X. (2018). Immobilization of 2D/2D structured g-C3N4 nanosheet/reduced graphene oxide hybrids on 3D nickel foam and its photocatalytic performance. *Materials Research Bulletin*, *97*, 306–313. doi:10.1016/j.materresbull.2017.09.024

Wang, Y., Liu, T., & Liu, J. (2020). Synergistically Boosted Degradation of Organic Dyes by CeO_2 Nanoparticles with Fluoride at Low pH. *ACS Applied Nano Materials*, *3*(1), 842–849. doi:10.1021/acsanm.9b02356

Wang, Y., Wei, F., Luo, G., Yu, H., & Gu, G. (2002). The large-scale production of carbon nanotubes in a nano-agglomerate fluidized-bed reactor. *Chemical Physics Letters*, *364*(5–6), 568–572. https://doi.org/10.1016/S0009-2614(02)01384-2

Wang, Z., & Mi, B. (2017). Environmental Applications of 2D Molybdenum Disulfide (MoS2) Nanosheets. *Environmental Science & Technology*, *51*(15), 8229–8244. doi:10.1021/acs.est.7b01466 PMID:28661657

Wan, H., Wang, F., Chen, Y., Zhao, Z., Zhang, G., Dou, M., & Xue, B. (2021). Enhanced Reactive Red 2 anaerobic degradation through improving electron transfer efficiency by nano-Fe_3O_4 modified granular activated carbon. *Renewable Energy*, *179*, 696–704. doi:10.1016/j.renene.2021.07.046

Wan, X., Yang, J., Huang, X., Tie, S., & Lan, S. (2019). A high-performance room temperature thermocatalyst Cu_2O/Ag_0@Ag-NPs for dye degradation under dark condition. *Journal of Alloys and Compounds*, *785*, 398–409. doi:10.1016/j.jallcom.2019.01.215

Watanabe, K. (2008). Recent developments in microbial fuel cell technologies for sustainable bioenergy. *Journal of Bioscience and Bioengineering*, *106*(6), 528–536. https://doi.org/10.1263/jbb.106.528

Wawrzkiewicz, M., Wiśniewska, M., Wołowicz, A., Gun'ko, V. M., & Zarko, V. I. (2017). Mixed silica-alumina oxide as sorbent for dyes and metal ions removal from aqueous solutions and wastewaters. *Microporous and Mesoporous Materials*, *250*, 128–147. doi:10.1016/j.micromeso.2017.05.016

Wei, B. Y., Hsu, M. C., Su, P. G., Lin, H. M., Wu, R. J., & Lai, H. J. (2004). A novel SnO2 gas sensor doped with carbon nanotubes operating at room temperature. *Sensors and Actuators. Part B*, *101*(1–2), 81–89. doi:10.1016/j.snb.2004.02.028

Wei, C., Dai, L., Roy, A., & Tolle, T. B. (2006). Multifunctional chemical vapor sensors of aligned carbon nanotube and polymer composites. *Journal of the American Chemical Society*, *128*(5), 1412–1413. https://doi.org/10.1021/ja0570335

Wei, S., Mavinakuli, P., Wang, Q., Chen, D., Asapu, R., Mao, Y., Haldolaarachchige, N., Young, D. P., & Guo, Z. (2011). Polypyrrole-Titania nanocomposites derived from different oxidants. *Journal of the Electrochemical Society*, *158*(11), K205–K212. doi:10.1149/2.048111jes

Wei, X., Akbar, M. U., Raza, A., & Li, G. (2021). A review on bismuth oxyhalide based materials for photocatalysis. *Nanoscale Advances*, *3*(12), 3353–3372. doi:10.1039/D1NA00223F

Weller, M., Weller, M. T., Overton, T., Rourke, J., & Armstrong, F. (2014). *Inorganic chemistry*. Oxford University Press.

Wepasnick, K. A., Smith, B. A., Schrote, K. E., Wilso, H. K., Diegemann, S. R., & Fairbrothe, D. H. (2011). Surface and structural characterization of multi-walled carbon nanotubes following different oxidative treatments. *Carbon*, *49*(1), 24–36. doi:10.1016/j.carbon.2010.08.034

Wick, P., Manser, P., Limbach, L. K., Dettlaff-Weglikowska, U., Krumeich, F., Roth, S., Stark, W. J., & Bruinink, A. (2007). The degree and kind of agglomeration affect carbon nanotube cytotoxicity. *Toxicology Letters*, *168*(2), 121–131. https://doi.org/10.1016/j.toxlet.2006.08.019

Wijesirigunawardana, P. B., & Perera, B. G. K. (2018). Development of a cotton smart textile with medicinal properties using lime oil microcapsules. *Acta Chimica Slovenica*, *65*(1), 150–159. doi:10.17344/acsi.2017.3727 PMID:29562116

Wongkaew, A., Kongsi, W., & Limsuwan, P. (2013). Physical Properties and Selective CO Oxidation of Coprecipitated CuO/CeO_2 Catalysts Depending on the CuO in the Samples. *Advances in Materials Science and Engineering, 2013*, 1–8. Advance online publication. doi:10.1155/2013/374080

Wu, C. H. (2007). Adsorption of reactive dye onto carbon nanotubes: Equilibrium, kinetics and thermodynamics. *Journal of Hazardous Materials, 144*(1–2), 93–100. https://doi.org/10.1016/j.jhazmat.2006.09.083

Wu, F. C., Tseng, R. L., & Juang, R. S. (2010). A review and experimental verification of using chitosan and its derivatives as adsorbents for selected heavy metals. *Journal of Environmental Management, 91*, 798–806.

Wu, J., Wang, T., Wang, J., Zhang, Y., & Pan, W. P. (2021). A novel modified method for the efficient removal of Pb and Cd from wastewater by biochar: Enhanced the ion exchange and precipitation capacity. *The Science of the Total Environment, 754*, 142150. doi:10.1016/j.scitotenv.2020.142150 PMID:32920404

Wu, L., Luo, Y., Zhou, S., Wu, Z., & Chen, X. (2021). Fabrication of Ag-TiO_2 functionalized activated carbon for dyes degradation based on tea residues. *Colloids and Surfaces. A, Physicochemical and Engineering Aspects, 627*, 127130. doi:10.1016/j.colsurfa.2021.127130

Wu, M., Li, L., Liu, N., Wang, D., Xue, Y., & Tang, L. (2018). Molybdenum disulfide (MoS2) as a co-catalyst for photocatalytic degradation of organic contaminants: A review. *Process Safety and Environmental Protection, 118*, 40–58. doi:10.1016/j.psep.2018.06.025

Wu, S., Su, Y., Zhu, Y., Zhang, Y., & Zhu, M. (2020). In-situ growing Bi/BiOCl microspheres on Ti3C2 nanosheets for upgrading visible-light-driven photocatalytic activity. *Applied Surface Science, 520*, 146339. doi:10.1016/j.apsusc.2020.146339

Wutich, A., Rosinger, A. Y., Stoler, J., Jepson, W., & Brewis, A. (2019). Measuring human water needs. *American Journal of Human Biology, 3*, e23350. https://doi.org/10.1002/ajhb.23350

Wu, Y. T., Lin, L. Y., Tao, S. M., Chen, Y. S., Ma, J. S., Lee, P. Y., & Sung, Y. S. (2020). Novel in situ synthesis of $BiVO_4$photocatalyst/$Co_3(PO_4)_2$ co-catalyst powder via the one-step solid-state process for photoelectrochemicalcatalyzing water oxidation. *ACS Sustainable Chemistry & Engineering, 8*(7), 2948–2956. doi:10.1021/acssuschemeng.9b07517

Wu, Z., Liang, Y., Yuan, X., Zou, D., Fang, J., Jiang, L., zhang, J., Yang, H., & Xiao, Z. (2020). MXene Ti3C2 derived Z–scheme photocatalyst of graphene layers anchored TiO2/g–C3N4 for visible light photocatalytic degradation of refractory organic pollutants. *Chemical Engineering Journal, 394*, 124921. doi:10.1016/j.cej.2020.124921

Xiao, B., & Thomas, K. M. (2004). Competitive adsorption of aqueous metal ions on an oxidized nanoporous activated carbon. *Langmuir, 20*(11), 4566–4578. https://doi.org/10.1021/la049712j

Xiao, B., & Thomas, K. M. (2005). Adsorption of aqueous metal ions on oxygen and nitrogen functionalized nanoporous activated carbons. *Langmuir, 21*(9), 3892–3902. https://doi.org/10.1021/la047135t

Xiao, Q. K., Wang, H. Y., Li, F., & Gao, Y. (2011). 3D object retrieval based on a graph model descriptor. *Neurocomputing, 74*(17), 3486–3493. doi:10.1016/j.neucom.2011.06.002

Xiao, Z., Yang, Z., Li, Z., Li, P., & Wang, R. (2019). Synchronous Gains of Areal and Volumetric Capacities in Lithium-Sulfur Batteries Promised by Flower-like Porous Ti3C2T x Matrix. *ACS Nano, 13*(3), 3404–3412. doi:10.1021/acsnano.8b09296 PMID:30790514

Xin, S., Yin, Y.-X., Guo, Y.-G., & Wan, L.-J. (2014). A High-Energy Room-Temperature Sodium-Sulfur Battery. *Advanced Materials, 26*(8), 1261–1265. doi:10.1002/adma.201304126 PMID:24338949

Xu, C., Wang, L., Liu, Z., Chen, L., Guo, J., Kang, N., Ma, X.-L., Cheng, H.-M., & Ren, W. (2015). Large-area high-quality 2D ultrathin Mo2C superconducting crystals. *Nature Materials*, *14*(11), 1135–1141. doi:10.1038/nmat4374 PMID:26280223

Xue, S., He, H., Wu, Z., Yu, C., Fan, Q., Peng, G., & Yang, K. (2017). An interesting Eu, F-codoped $BiVO_4$ microsphere with enhanced photocatalytic performance. *Journal of Alloys and Compounds*, *694*, 989–997. doi:10.1016/j.jallcom.2016.10.146

Xu, J., Cao, Z., Zhang, Y., Yuan, Z., Lou, Z., Xu, X., & Wang, X. (2018). A review of functionalized carbon nanotubes and graphene for heavy metal adsorption from water: Preparation, application, and mechanism. *Chemosphere*, *195*, 351–364. doi:10.1016/j.chemosphere.2017.12.061 PMID:29272803

Xu, L., Yang, L., Johansson, E. M. J., Wang, Y., & Jin, P. (2018). Photocatalytic activity and mechanism of bisphenol a removal over TiO2−x/rGO nanocomposite driven by visible light. *Chemical Engineering Journal*, *350*, 1043–1055. doi:10.1016/j.cej.2018.06.046

Xu, S., Zhu, Y., Jiang, L., & Dan, Y. (2010). Visible light induced photocatalytic degradation of methyl orange by polythiophene/TiO_2 composite particles. *Water, Air, and Soil Pollution*, *213*(1–4), 151–159. doi:10.100711270-010-0374-4

Xu, X., Du, M., Chen, T., Xiong, S., Wu, T., Zhao, D., & Fan, Z. (2016). New insights into Ag-doped $BiVO_4$ microspheres as visible light photocatalysts. *RSC Advances*, *6*(101), 98788–98796. doi:10.1039/C6RA20850A

Xu, X., Jia, K., Chen, S., Lang, D., Yang, C., Wang, L., & Wang, J. (2021). Ultra-fast degradation of phenolics and dyes by Cu_2O/Cu catalysts: Synthesis and degradation kinetics. *Journal of Environmental Chemical Engineering*, *9*(4), 105505. doi:10.1016/j.jece.2021.105505

Xu, X., Qin, J., Wei, Y., Ye, S., Shen, J., Yao, Y., Ding, B., Shu, Y., He, G., & Chen, H. (2019). Heterogeneous activation of persulfate by NiFe2−xCoxO4-RGO for oxidative degradation of bisphenol A in water. *Chemical Engineering Journal*, *365*, 259–269. doi:10.1016/j.cej.2019.02.019

Xu, Y., Wu, S., Li, X., Meng, H., Zhang, X., Wang, Z., & Han, Y. (2017). Ag nanoparticle-functionalized ZnO microflowers for enhanced photodegradation of herbicide derivatives. *Chemical Physics Letters*, *679*, 119–126. doi:10.1016/j.cplett.2017.04.091

Yadav, M., Gupta, R., Arora, G., Yadav, P., Srivastava, A., & Sharma, R. K. (2020). Current status of heavy metal contaminants and their removal/recovery techniques. In Contaminants in Our Water: Identification and Remediation Methods (pp. 41-64). American Chemical Society.

Yamamoto, Y., Inoue, S., & Matsumura, Y. (2017). Thermal decomposition products of various carbon sources in chemical vapor deposition synthesis of carbon nanotube. *Diamond and Related Materials*, *75*, 1–5. doi:10.1016/j.diamond.2016.11.017

Yamjala, K., Nainar, M. S., & Ramisetti, N. R. (2016). Methods for the analysis of azo dyes employed in food industry: A review. *Food Chemistry*, *192*, 813–824.

Yan, D., Li, Y., Huo, J., Chen, R., Dai, L., & Wang, S. (2017). Defect Chemistry of Nonprecious-Metal Electrocatalysts for Oxygen Reactions. *Advanced Materials*, *29*(48), 1606459. doi:10.1002/adma.201606459 PMID:28508469

Yang, K., Zhu, L., & Xing, B. (2006b). Adsorption of polycyclic aromatic hydrocarbons by carbon nanomaterials. *Environmental Science & Technology*, *40*(6), 1855–1861. https://doi.org/10.1021/es052208w

Yang, L., Xu, L., Bai, X., & Jin, P. (2019). Enhanced visible-light activation of persulfate by Ti3+ self-doped TiO2/ graphene nanocomposite for the rapid and efficient degradation of micropollutants in water. *Journal of Hazardous Materials*, *365*, 107–117. doi:10.1016/j.jhazmat.2018.10.090 PMID:30412807

Yang, M., Guo, Z., Qiu, K., Long, J., Yin, G., Guan, D., & Zhou, S. (2010). Synthesis and characterization of Mn-doped ZnO column arrays. *Applied Surface Science*, *256*(13), 4201–4205. doi:10.1016/j.apsusc.2010.01.125

Yang, N., Liu, Y., Wen, H., Tang, Z., Zhao, H., Li, Y., & Wang, D. (2013). Photocatalytic Properties of Graphdiyne and Graphene Modified TiO2: From Theory to Experiment. *ACS Nano*, *7*(2), 1504–1512. doi:10.1021/nn305288z PMID:23350627

Yang, S., Li, J., Shao, D., Hu, J., & Wang, X. (2009). Adsorption of Ni(II) on oxidized multi- -walled carbon nanotubes: Effect of contact time, pH, foreign ions and PAA. *Journal of Hazardous Materials*, *166*(1), 109–116. https://doi.org/10.1016/j.jhazmat.2008.11.003

Yang, S., Yang, X., Shao, X., Niu, R., & Wang, L. (2011). Activated carbon catalyzed persulfate oxidation of azo dye acid orange 7 at ambient temperature. *Journal of Hazardous Materials*, *186*(1), 659–666. doi:10.1016/j.jhazmat.2010.11.057 PMID:21145652

Yang, X. H., Yang, H. G., & Li, C. (2011). Controllable Nanocarving of Anatase TiO2 Single Crystals with Reactive {001} Facets. *Chemistry (Weinheim an der Bergstrasse, Germany)*, *17*(24), 6615–6619. doi:10.1002/chem.201100134 PMID:21538611

Yang, X., Cai, H., Bao, M., Yu, J., Lu, J., & Li, Y. (2018). Insight into the highly efficient degradation of PAHs in water over graphene oxide/Ag3PO4 composites under visible light irradiation. *Chemical Engineering Journal*, *334*, 355–376. doi:10.1016/j.cej.2017.09.104

Yang, X.-Y., & Al-Duri, B. (2001). Application of branched pore diffusion model in the adsorption of reactive dyes on activated carbon. *Chemical Engineering Journal*, *83*(1), 15–23. https://doi.org/10.1016/S1385-8947(00)00233-3

Yang, Y., Sun, M., Zhou, J., Ma, J., & Komarneni, S. (2020). Degradation of orange II by Fe@Fe_2O_3 core shell nanomaterials assisted by $NaHSO_3$. *Chemosphere*, *244*, 125588. doi:10.1016/j.chemosphere.2019.125588 PMID:32050354

Yang, Y., Wen, J., Wei, J., Xiong, R., Shi, J., & Pan, C. (2013). Polypyrrole-decorated Ag-TiO_2 nanofibers exhibiting enhanced photocatalytic activity under visible-light illumination. *ACS Applied Materials & Interfaces*, *5*(13), 6201–6207. doi:10.1021/am401167y PMID:23767991

Yang, Y., Zeng, Z., Zeng, G., Huang, D., Xiao, R., Zhang, C., Zhou, C., Xiong, W., Wang, W., Cheng, M., Xue, W., Guo, H., Tang, X., & He, D. (2019). Ti3C2 Mxene/porous g-C3N4 interfacial Schottky junction for boosting spatial charge separation in photocatalytic H2O2 production. *Applied Catalysis B: Environmental*, *258*, 117956. doi:10.1016/j.apcatb.2019.117956

Yan, L., Liu, B., Li, W., Zhao, T., Wang, Y., & Zhao, Q. (2020). Multiscale cellulose based self-assembly of hierarchical structure for photocatalytic degradation of organic pollutant. *Cellulose (London, England)*, *27*(9), 5241–5253.

Yan, M., Yan, Y., Wu, Y., Shi, W., & Hua, Y. (2015). Microwave-assisted synthesis of monoclinic-tetragonal $BiVO_4$ hetero-junctions with enhanced visible-light-driven photocatalytic degradation of tetracycline. *RSC Advances*, *5*(110), 90255–90264. doi:10.1039/C5RA13684A

Yao, W., Iwai, H., & Ye, J. (2008). Effects of molybdenum substitution on the photocatalyticbehaviour of $BiVO_4$. *Dalton Transactions (Cambridge, England)*, (11), 1426–1430. doi:10.1039/b713338c

Yao, Y., Lu, F., Qin, J., Wei, F., Xu, C., & Wang, S. (2014). Magnetic ZnFe2O4–C3N4 Hybrid for Photocatalytic Degradation of Aqueous Organic Pollutants by Visible Light. *Industrial & Engineering Chemistry Research*, *53*(44), 17294–17302. doi:10.1021/ie503437z

Yasir, M., Masar, M., Sopik, T., Ali, H., Urbanek, M., Antos, J., Machovsky, M., & Kuritka, I. (2022). ZnO nanowires and nanorods based ZnO/WO3/Pt heterojunction for efficient photocatalytic degradation of estriol (E3) hormone. *Materials Letters*, *319*, 132291. doi:10.1016/j.matlet.2022.132291

Yeamin, M. B., Islam, M. M., Chowdhury, A. N., & Awual, M. R. (2021). Efficient encapsulation of toxic dyes from wastewater using several biodegradable natural polymers and their composites. *Journal of Cleaner Production*, *291*, 125920. doi:10.1016/j.jclepro.2021.125920

Ye, K. H., Yu, X., Qiu, Z., Zhu, Y., Lu, X., & Zhang, Y. (2015). Facile synthesis of bismuth oxide/bismuth vanadate hetero-structures for efficient photoelectrochemical cells. *RSC Advances*, *5*(43), 34152–34156. doi:10.1039/C5RA03500G

Ye, S., Zeng, G., Wu, H., Liang, J., Zhang, C., Dai, J., Xiong, W., Song, B., Wu, S., & Yu, J. (2019). The effects of activated biochar addition on remediation efficiency of co-composting with contaminated wetland soil. *Resources, Conservation and Recycling*, *140*, 278–285. doi:10.1016/j.resconrec.2018.10.004

Yi, L., Zuo, L., Wei, C., Fu, H., Qu, X., Zheng, S., Xu, S., Guo, Y., Li, H., & Zhu, D. (2020). Enhanced adsorption of bisphenol A, tylosin, and tetracycline from aqueous solution to nitrogen- doped multiwall carbon nanotubes via cation-π and π-π electron-donor-acceptor (EDA) interaction. *The Science of the Total Environment*, *719*, 137389. doi:10.1016/j.scitotenv.2020.137389 PMID:32120097

Yim, W. L., & Liu, Z. F. (2004). A reexamination of the chemisorptions and desorption of ozone on the exterior of a (5,5) single-walled carbon nanotube. *Chemical Physics Letters*, *398*(4–6), 297–303. https://doi.org/10.1016/j.cplett.2004.09.082

Yin, C., Zhu, S., Chen, Z., Zhang, W., Gu, J., & Zhang, D. (2013). One step fabrication of C-doped $BiVO_4$ with hierarchical structures for a high-performance photocatalyst under visible-light irradiation. *Journal of Materials Chemistry. A, Materials for Energy and Sustainability*, *1*(29), 8367–8378. doi:10.1039/c3ta11833a

Youssef, A. M., & Malhat, F. M. (2014). Selective removal of heavy metals from drinking water using titanium dioxide nanowire. *Macromolecular Symposia*, *337*(1), 96–101. doi:10.1002/masy.201450311

Yuan, W., Cheng, L., An, Y., Lv, S., Wu, H., Fan, X., Zhang, Y., Guo, X., & Tang, J. (2018). Laminated Hybrid Junction of Sulfur-Doped TiO2 and a Carbon Substrate Derived from Ti3C2 MXenes: Toward Highly Visible Light-Driven Photocatalytic Hydrogen Evolution. *Advancement of Science*, *5*(6), 1700870. doi:10.1002/advs.201700870 PMID:29938169

Yuan, W., Cheng, L., An, Y., Wu, H., Yao, N., Fan, X., & Guo, X. (2018). MXene Nanofibers as Highly Active Catalysts for Hydrogen Evolution Reaction. *ACS Sustainable Chemistry & Engineering*, *6*(7), 8976–8982. doi:10.1021/acssuschemeng.8b01348

Yuan, W., Cheng, L., Zhang, Y., Wu, H., Lv, S., Chai, L., Guo, X., & Zheng, L. (2017). 2D-Layered Carbon/TiO2 Hybrids Derived from Ti3C2MXenes for Photocatalytic Hydrogen Evolution under Visible Light Irradiation. *Advanced Materials Interfaces*, *4*(20), 1700577. doi:10.1002/admi.201700577

Yu, F., Chen, J., Chen, L., Huai, J., Gong, W., Yuan, Z., & Ma, J. (2012). Magnetic carbon nanotubes synthesis by Fenton's reagent method and their potential application for removal of azo dye from aqueous solution. *Journal of Colloid and Interface Science*, *378*(1), 175–183. doi:10.1016/j.jcis.2012.04.024 PMID:22564767

Yuge, R., Toyama, K., Ichihashi, T., Ohkara, T., Aoki, Y., & Manako, T. (2012). Characterization and field mission properties of multiwalled carbon nanotubes with fine crystallinity prepared by CO_2 laser ablation. *Applied Surface Science*, *258*(18), 6958–6962. doi:10.1016/j.apsusc.2012.03.143

Yu, J., & Kudo, A. (2005). Hydrothermal Synthesis of Nano-fibrous Bismuth Vanadate. *Chemistry Letters*, *34*(6), 850–851. doi:10.1246/cl.2005.850

Yuwei, C., & Jianlong, W. (2011). Preparation and characterization of magnetic chitosan nanoparticles and its application for Cu (II) removal. *Chemical Engineering Journal*, *168*(1), 286–292.

Yu, X., Tong, S., Ge, M., Zuo, J., Cao, C., & Song, W. (2013). One-step synthesis of magnetic composites of cellulose@ iron oxide nanoparticles for arsenic removal. *Journal of Materials Chemistry. A, Materials for Energy and Sustainability*, *1*, 959–965.

Zang, Z., & Tang, X. (2015). Enhanced fluorescence imaging performance of hydrophobic colloidal ZnO nanoparticles by a facile method. *Journal of Alloys and Compounds*, *619*, 98–101. doi:10.1016/j.jallcom.2014.09.072

Zare, K., Sadegh, H., Shahryari-ghoshekandi, R., Maazinejad, B., Ali, V., Tyagi, I., Agarwal, S., & Gupta, V. K. (2015). Enhanced removal of toxic Congo red dye using multi walled carbon nanotubes: Kinetic, equilibrium studies and its comparison with other adsorbents. *Journal of Molecular Liquids*, *212*, 266–271. doi:10.1016/j.molliq.2015.09.027

Zhang, A., & Zhang, J. (2010a). Characterization and photocatalytic properties of Au/$BiVO_4$ composites. *Journal of Alloys and Compounds*, *491*(1-2), 631–635. doi:10.1016/j.jallcom.2009.11.027

Zhang, A., & Zhang, J. (2010b). Effects of europium doping on the photocatalyticbehaviour of $BiVO_4$. *Journal of Hazardous Materials*, *173*(1-3), 265–272. doi:10.1016/j.jhazmat.2009.08.079 PMID:19729243

Zhang, C., & Nicolosi, V. (2019). Graphene and MXene-based transparent conductive electrodes and supercapacitors. *Energy Storage Materials*, *16*, 102–125. doi:10.1016/j.ensm.2018.05.003

Zhang, F., Huang, L., Ding, P., Wang, C., Wang, Q., Wang, H., Li, Y., Xu, H., & Li, H. (2019). One-step oxygen vacancy engineering of WO3-x/2D g-C3N4 heterostructure: Triple effects for sustaining photoactivity. *Journal of Alloys and Compounds*, *795*, 426–435. doi:10.1016/j.jallcom.2019.04.297

Zhang, G., Wu, S., Li, Y., & Zhang, Q. (2020). Significant improvement in activity, durability, and light-to-fuel efficiency of Ni nanoparticles by La2O3 cluster modification for photothermocatalytic CO2 reduction. *Applied Catalysis B: Environmental*, *264*, 118544. doi:10.1016/j.apcatb.2019.118544

Zhang, H., Li, M., Zhu, C., Tang, Q., Kang, P., & Cao, J. (2020). Preparation of magnetic α-Fe2O3/ZnFe2O4@Ti3C2 MXene with excellent photocatalytic performance. *Ceramics International*, *46*(1), 81–88. doi:10.1016/j.ceramint.2019.08.236

Zhang, J., Li, L., Li, Y., & Yang, C. (2017). Microwave-assisted synthesis of hierarchical mesoporous nano-TiO_2/cellulose composites for rapid adsorption of Pb2+. *Chemical Engineering Journal*, *313*, 1132–1141. doi:10.1016/j.cej.2016.11.007

Zhang, L., Liu, P., & Su, Z. (2006). Preparation of PANI–TiO2 nanocomposites and their solid-phase photocatalytic degradation. *Polymer Degradation & Stability*, *91*(9), 2213–2219. https://doi.org/10.1016/j.polymdegradstab.2006.01.002

Zhang, L., Lv, F., Zhang, W., Li, R., Zhong, H., Zhao, Y., Zhang, Y., & Wang, X. (2009). Photo degradation of methyl orange by attapulgite-SnO_2-TiO_2 nanocomposites. *Journal of Hazardous Materials*, *171*(1–3), 294–300. doi:10.1016/j.jhazmat.2009.05.140 PMID:19577837

Zhang, L., Zeng, Y. X., & Cheng, Z. J. (2016). Removal of heavy metal ions using chitosan and modified chitosan: A review. *Journal of Molecular Liquids*, *214*, 175–191.

Zhang, M., Qin, J., Rajendran, S., Zhang, X., & Liu, R. (2018). Heterostructured d-Ti3C2/TiO2/g-C3N4 Nanocomposites with Enhanced Visible-Light Photocatalytic Hydrogen Production Activity. *ChemSusChem*, *11*(24), 4226–4236. doi:10.1002/cssc.201802284 PMID:30334348

Zhang, S. J., Shao, T., Kose, H. S., & Karanfil, T. (2012). Adsorption kinetics of aromatic compounds on carbon nanotubes and activated carbons. *Environmental Toxicology and Chemistry*, *31*(1), 79–85. doi:10.1002/etc.724 PMID:22021047

Zhang, S., Gao, H., Huang, Y., Wang, X., Hayat, T., Li, J., Xu, X., & Wang, X. (2018). Ultrathin g-C3N4 nanosheets coupled with amorphous Cu-doped FeOOH nanoclusters as 2D/0D heterogeneous catalysts for water remediation. *Environmental Science. Nano*, *5*(5), 1179–1190. doi:10.1039/C8EN00124C

Zhang, S., Li, J., Wang, X., Huang, Y., Zeng, M., & Xu, J. (2015). Rationally designed 1D Ag@AgVO3 nanowire/graphene/protonated g-C3N4 nanosheet heterojunctions for enhanced photocatalysis via electrostatic self-assembly and photochemical reduction methods. *Journal of Materials Chemistry. A, Materials for Energy and Sustainability*, *3*(18), 10119–10126. doi:10.1039/C5TA00635J

Zhang, S., Quan, X., Zheng, J. F., & Wang, D. (2017). Probing the interphase "HO• zone" originated by carbon nanotubes during catalytic ozonation. *Water Research*, *122*, 86–95. doi:10.1016/j.watres.2017.05.063 PMID:28595124

Zhang, W. D., & Zhang, W. H. (2009). Carbon nanotubes as active components for gas sensors. *Journal of Sensors*, *2009*, 1–16. https://doi.org/10.1155/2009/160698

Zhang, W., Xiao, X., Li, Y., Zeng, X., Zheng, L., & Wan, C. (2016). Liquid exfoliation of layered metal sulphide for enhanced photocatalytic activity of TiO2 nanoclusters and DFT study. *RSC Advances*, *6*(40), 33705–33712. doi:10.1039/C6RA03534E

Zhang, X., Pan, J., Zhu, C., Sheng, Y., Yan, Z., Wang, Y., & Feng, B. (2015). The visible light catalytic properties of carbon quantum dots/ZnO nanoflowers composites. *Journal of Materials Science Materials in Electronics*, *26*(5), 2861–2866. doi:10.100710854-015-2769-x

Zhang, Y. J., Xue, J. Q., Li, F., & Dai, J. Z. (2019). Preparation of polypyrrole/chitosan/carbon nanotube composite nano-electrode and application to capacitive deionization process for removing Cu2+. *Chemical Engineering and Processing-Process Intensification*, *139*, 121–129.

Zhang, Y., Wang, L., Zhang, N., & Zhou, Z. (2018). Adsorptive environmental applications of MXene nanomaterials: A review. *RSC Advances*, *8*(36), 19895–19905. doi:10.1039/C8RA03077D PMID:35541640

Zhang, Y., Wu, B., Xu, H., Liu, H., Wang, M., He, Y., & Pan, B. (2016). Nanomaterials-enabled water and wastewater treatment. *NanoImpact*, *3*, 22–39. doi:10.1016/j.impact.2016.09.004

Zhang, Y., Zhang, N., Tang, Z.-R., & Xu, Y.-J. (2012). Graphene Transforms Wide Band Gap ZnS to a Visible Light Photocatalyst. The New Role of Graphene as a Macromolecular Photosensitizer. *ACS Nano*, *6*(11), 9777–9789. doi:10.1021/nn304154s PMID:23106763

Zhao, B., Li, H., Qin, X., Li, Z., Zhang, S., Wang, A., & Zhu, Z. (2021). Performance enhancement and catalytic mechanism identification of Cu-based composite for degradation of organic contaminants. *Powder Technology*, *389*, 11–20. doi:10.1016/j.powtec.2021.04.092

Zhao, G., Li, J., Ren, X., Chen, C., & Wang, X. (2011). Few-layered graphene oxide nanosheets as superior sorbents for heavy metal ion pollution management. *Environmental Science & Technology*, *45*(24), 10454–10462. https://doi.org/10.1021/es203439v

Zhao, H. Y., Zhou, H. M., Zhang, J. X., Zheng, W., & Zheng, Y. F. (2009). Carbon nanotube-hydroxyapatite nanocomposite: A novel platform for glucose/O2 biofuel cell. *Biosensors & Bioelectronics*, *25*(2), 463–468. https://doi.org/10.1016/j.bios.2009.08.005

Zhao, M., & Liu, P. (2009). Adsorption of methylene blue from aqueous solutions by modified expanded graphite powder. *Desalination, 249*(1), 331–336. https://doi.org/10.1016/j.desal.2009.01.037

Zhao, Y., Chen, G., Bian, T., Zhou, C., Waterhouse, G. I. N., Wu, L.-Z., Tung, C.-H., Smith, L. J., O'Hare, D., & Zhang, T. (2015). Defect-Rich Ultrathin ZnAl-Layered Double Hydroxide Nanosheets for Efficient Photoreduction of CO2 to CO with Water. *Advanced Materials, 27*(47), 7824–7831. doi:10.1002/adma.201503730 PMID:26509528

Zhao, Y., Zhang, X., Wang, C., Zhao, Y., Zhou, H., Li, J., & Jin, H. (2017). The synthesis of hierarchical nanostructured MoS2/Graphene composites with enhanced visible-light photo-degradation property. *Applied Surface Science, 412*, 207–213. doi:10.1016/j.apsusc.2017.03.181

Zhong, D. K., Choi, S., & Gamelin, D. R. (2011). Near-complete suppression of surface recombination in solar photoelectrolysis by "Co-Pi" catalyst-modified W:$BiVO_{4.}$ *Journal of the American Chemical Society, 133*(45), 18370–18377. doi:10.1021/ja207348x PMID:21942320

Zhong, Q., Li, Y., & Zhang, G. (2021). Two-dimensional MXene-based and MXene-derived photocatalysts: Recent developments and perspectives. *Chemical Engineering Journal, 409*, 128099. doi:10.1016/j.cej.2020.128099

Zhong, R., Zhong, Q., Huo, M., Yang, B., & Li, H. (2020). Preparation of biocompatible nano-ZnO/chitosan microspheres with multi-functions of antibacterial, UV-shielding and dye photodegradation. *International Journal of Biological Macromolecules, 146*, 939–945. doi:10.1016/j.ijbiomac.2019.09.217 PMID:31726126

Zhou, W., Yin, Z., Du, Y., Huang, X., Zeng, Z., Fan, Z., Liu, H., Wang, J., & Zhang, H. (2013). Synthesis of Few-Layer MoS2 Nanosheet-Coated TiO2 Nanobelt Heterostructures for Enhanced Photocatalytic Activities. *Small, 9*(1), 140–147. doi:10.1002mll.201201161 PMID:23034984

Zhou, X., Wang, J., Sheng, N., Cui, R., Deng, Y., & Dai, J. (2019). Corrigendum to "Subchronic reproductive effects of 6:2 chlorinated polyfluorinated ether sulfonate (6:2 Cl-PFAES), an alternative to PFOS, on adult male mice". *Journal of Hazardous Materials, 365*, 972. doi:10.1016/j.jhazmat.2018.10.085 PMID:30448035

Zhu, W., Sun, F., Goei, R., & Zhou, Y. (2017). Construction of WO3–g-C3N4 composites as efficient photocatalysts for pharmaceutical degradation under visible light. *Catalysis Science & Technology, 7*(12), 2591–2600. doi:10.1039/C7CY00529F PMID:29129990

Zia, Z., Hartland, A., & Mucalo, M. R. (2020). Use of low-cost biopolymers and biopolymeric composite systems for heavy metal removal from water. *International Journal of Environmental Science and Technology, 17*(10), 4389–4406.

Zięba, A., Drelinkiewicz, A., Konyushenko, E. N., & Stejskal, J. (2010). Activity and stability of polyaniline-sulfate-based solid acid catalysts for the transesterification of triglycerides and esterification of fatty acids with methanol. *Applied Catalysis A, 383*(1–2), 169–181. https://doi.org/10.1016/j.apcata.2010.05.042

Zimniewska, M., Pawlaczyk, M., Krucinska, I., Frydrych, I., Mikolajczak, P., Schmidt-Przewozna, K., Komisarczyk, A., Herczynska, L., & Romanowska, B. (2019). The influence of natural functional clothing on some biophysical parameters of the skin. *Textile Research Journal, 89*(8), 1381–1393. doi:10.1177/0040517518770680

Zou, Y., Shi, J.-W., Sun, L., Ma, D., Mao, S., Lv, Y., & Cheng, Y. (2019). Energy-band-controlled ZnxCd1−xIn2S4 solid solution coupled with g-C3N4 nanosheets as 2D/2D heterostructure toward efficient photocatalytic H2 evolution. *Chemical Engineering Journal, 378*, 122192. doi:10.1016/j.cej.2019.122192

Zyoud, A., Dwikat, M., Al-Shakhshir, S., Ateeq, S., Shteiwi, J., Zu'bi, A., & Hilal, H. S. (2016). Natural dye-sensitized ZnO nano-particles as photo-catalysts in complete degradation of E. coli bacteria and their organic content. *Journal of Photochemistry and Photobiology A Chemistry, 328*, 207–216. doi:10.1016/j.jphotochem.2016.05.020

About the Contributors

Azad Kumar is currently serving as an assistant professor in the Department of Chemistry, Siddharth University, Kapilvastu, Siddharth Nagar. Dr. Azad has obtained B.Sc. Hons (2007), M.Sc. (2009), M.Phil. (2010) in Chemistry from Dayalbagh Educational Institute, Dayalbagh Agra, and Ph.D. (2018) in Chemistry from Babasaheb Bhimrao Ambedkar University, Lucknow. Dr. Azad is also a member of several scientific societies. His research and teaching interests include the theory and application of nanomaterial, hybrid composite, and polymer. He has published 20 research papers in peer-reviewed & refereed journals of repute. He has served as an editorial member of some journals. He has teaching experience of 10 years in various institutes.

* * *

Anil Kumar Bahe was born in Balaghat, (M.P.) India in 1988. He has done his graduation in 2011 from J. S. T. College Balaghat, (M.P.) and completed a post-graduation in chemistry from Dr. Harisingh Gour Central University Sagar MP in 2014. He has completed his Ph.D. from Dr. Harisingh Gour Central University Sagar M.P. (India) in 2022. Currently, he is teaching as a guest faculty in the department of chemistry (Dr. Harisingh Gour Central University, Sagar, MP). He has 3 years of teaching experience and also qualify CSIR NET & GATE 2016. He passed 10 years and have excelled in the field of organic chemistry. He has published 07 publications and 2 book chapters.

Megha Das is working as Assistant Professor in Department of Education, Indira Gandhi National Tribal University [IGNTU], Amarkantak, Madhya Pradesh. She has completed post- graduation [M. Sc] in Biotechnology from Guru Ghasidas University Bilaspur (C.G) and [M.A] in English Literature from Barkatullah University Bhopal [M.P], M. Ed. from Dr. Harisingh Gour University Sagar (M.P). She has qualified UGC-NET and awarded Ph.D. in Education from Dr. HarisinghGour University Sagar (M.P). She has completed her UGC Women's Postdoctoral Research with fellowship from UGC, New Delhi. She has published eight research papers andarticles in National and International journals and published six chapters in different books related to the area of Education. Her doctoral thesis has been published by Lambert Academic Publishing, Germany. She has participated and presented fourteen papers at National and International Seminars/Conferences. She has been awarded as Bharat Shiksha Ratan by GAF India in 2012. She is life member of AIAER Association Bhubaneswar. Her area of specialization includes Environmental Education, Nutrition Education and Teacher Education.

Ratnesh Das is a Professor in the Department of Chemistry, Dr. Harisingh Gour Central University, Sagar, India. He commands a rich experience in teaching, and research of about 18 years during which he has supervised many sponsored research projects. His active research areas include Heterocyclic synthesis, medicinal chemistry, electro-organic chemistry, synthesis of nano-catalysts and green chemistry. He has authored about 70 research papers in peer- reviewed national and international journals and refereed conferences organized by professional societies around the world. He is an active member of several professional bodies and societies, both in India and abroad. He is a vibrant speaker and delivered many lectures in conferences, workshops, and seminars organized both in India and abroad.

Manish K. Dwivedi, Young Scientist (ICMR-DHR), has completed PhD in the Department of Biotechnology, Indira Gandhi National Tribal University, Amarkantak, Madhya Pradesh. He has completed his M.Phil. and M.Sc. in Biochemistry from Awadesh Pratap Singh University, Rewa, MP. He has been awarded gold medals during M.Sc. and M. Phil. His areas of interest are infectious disease biology and drug development. He has published several national and international papers.

Muhammad Ikram obtained Master's degree (M. Phil Physics) from BZU Multan, Pakistan in 2010. He obtained his PhD degree in Physics from Department of Physics, Government College University (GCU) Lahore through Pak-US joint project between Department of Physics, GC U Lahore, Pakistan and University of Delaware, USA in 2015. In 2017, Ikram joined Department of Physics, GC University Lahore as Assistant Professor Physics in 2017. Ikram published over 100 manuscripts in international well reputed journal, 10 book chapters and one international book. Ikram received Seal of Excellence Marie Skłodowska-Curie Individual Fellowship in 2017 and 2020. His research work involves the synthesis and characterization of inorganic semiconductor nanomaterials, sensor, 2D materials for water treatment and optoelectronic applications.

Sushil K. Kashaw is an Associate Professor in Department of Pharmaceutical Sciences, Sagar, M.P. He has completed his Raman Postdoctoral Research at the Department of Pharmaceutical Sciences, Wayne State University, Detroit, USA. He has been conferred with many prestigious awards including Young Scientist Award by APP and presented papers in APPS (San Diego, USA), RPM (Chicago, USA) and FIP (Glasgow,UK). He has 110 research papers, 6 books, 6 book chapters, completed five government funded research projects and one granted and one filed patents. He has 35 invited lectures, chaired 27 scientific sessions, 45 papers presented in the conferences and developed 10 e-contents for CEC-MHRD. His total citations are 3798 and h-index of 27. He has supervised four PhD students and 53 M. Pharm students. His area of research is medicinal and computational chemistry.

Gyanendra Kumar, Department of Chemistry (Faculty of Science), University of Delhi. Gyanendra does research in Metal organic frameworks and Graphene oxide in drug delivery, and Catalytic applications.

Indu Kumari is currently working as an Assistant Professor at Department of Biotechnology, Chandigarh Group of Colleges, Mohali, Punjab, India. She received her Ph.D. in Computational and Theoretical Chemsitry from Department of Chemistry, Panjab University, Chandigarh, India in 2019. The nature of her Ph.D. research involved the density functional theory studies on structural, electronic, and oxidative properties of transition metal oxide clusters. Her research interests focuses on DFT study, metal oxide clusters, QM/MM study, quantum mechanics and reaction mechanistic study of chemical reactions.

Gao Li received his BSc (2004) from Hunan Normal University, and PhD (2016) from Shanghai Jiaotong University. After his postdoctoral research in Carnegie Mellon University (United States, 2011-2014), he joined State Key Laboratory of Catalysis, Dalian Institute of Chemical Physics, CAS as a professor in 2014. His current research interests focused on the preparation and application of small-sized nanoparticle catalysts, e.g., noble metal clusters and metal oxides. His bibliography includes over 100 publications.

Arunesh Kumar Mishra was born in Allahabad, India in 1991. He received his M.Sc. degree from NGBU, Allahabad in 2017. Currently, he is pursuing Ph.D. under the supervision of Prof. Ratnesh Das in the Department of Chemistry, Dr. Harisingh Gour Central University, Sagar (M.P.), India, where his research focused on the synthesis, characterization, and applications of novel heterocyclic compound using nanocatalysts. He has co-authored book chapter on the adsorption of toxic metals from water by the adsorbent. He has co-authored 04 publications and 03 book chapter.

Pratibha Mishra was born in the Indian city of Prayagraj in 1989. In 2012, she obtained his M.Sc. from Kanpur University. She is currently pursuing her Ph.D. at the Department of Chemistry, Dr. Harisingh Gour Central University, Sagar (MP), India, under the supervision of Prof. Ratnesh Das, where her research focuses on the green synthesis and biological characterization of novel heterocyclic compounds using nanocatalysts. She has contributed to four publications and three book chapter.

Usman Qumar obtained his M.Sc. degree in physics from the University of Gujrat (Gujrat, Pakistan) in 2017 and his M.Phil Physics degree from Riphah International University (Islamabad, Pakistan) in 2020. Qumar completed his M.Phil research work in the Solar Cell Applications Lab at GCU Lahore (Punjab, Pakistan) and later worked as Research Associate at department of physics Riphah international university Lahore Campus. Currently, he is a lecturer (visiting) at the Department of Physics, University of Gujarat, Pakistan. His current research area is the fabrication of 2D materials and their applications in water treatment and antimicrobial analysis, and he is a member of Materials Innovation (an interdisciplinary journal) and a member of the American Physical Society (APS).

Asma Rafiq obtained her Ph.D. degree in Physics from the Department of Physics, Government College University (GCU) Lahore, Pakistan. She served as visiting lecturer at Department of Physics, GCU Lahore (2016 - 2018). Asma joined Nanomagnetism and Spintronics research group at Durham University UK in 2018 as a visiting researcher. Later on, she joined Department of Physics, Riphah International University Lahore as Research Associate. Currently, she is serving as Assistant Professor at Centre of Excellence in Solid State Physics, University of the Punjab, Lahore, Pakistan. Her research interest involves the synthesis and characterization of semiconductor nanomaterials, evaluating their catalytic and photocatalytic properties.

Ali Raza obtained his B.S. (Physics) degree from the University of Gujrat (Punjab, Pakistan) in 2016 and his Master's (M. Phil) degree in Material Physics from Riphah International University, Islamabad Pakistan in 2019. Raza completed his M. Phil research work in the Solar Cell Applications Lab at GCU Lahore (Punjab, Pakistan) and later worked as Research Associate in the same lab on the field of catalytic and energy harvesting applications of pure and engineered 2D materials (TMDCs, Graphene, h-BN, and MXenes). Currently, he is serving as Lecturer (Research) at the Department of Physics, University of

Sialkot, Punjab, and has enrolled as a Ph.D. scholar at CAS Key Laboratory of Design and Assembly of Functional Nanostructures, Fujian Institute of Research on the Structure of Matter, CAS. His current research directions are the fabrication of functional 2D materials for catalytic and other energy-harvesting applications.

Victor Shikuku holds a Ph.D. degree in Chemistry and is a Lecturer of Physical Chemistry at the Department of Physical Sciences, Kaimosi Friends University College, Kenya. Victor has published more than 20 peer-reviewed papers, 11 book chapters, and one edited University-level textbook. He has post-doctoral research experience at Trier University in Germany supported by the Alexander Humboldt Foundation. He has experience in water and wastewater treatment approaches and materials chemistry for water purification.

Nirupama Singh is a qualified, highly-skilled and well organized academician and educator having several years of research and academic experience. Presently she is working as Assistant Professor at RCCV Girls College, Ghaziabad. She is a passionate teacher and she has a vision for developing students into a better human-beings Dr. Nirupama, has earned graduation, post graduation and M.Phil in chemistry from Dayalbagh Educational Institute, Dayalbagh, Agra. She has worked at Bhabha Atomic Research Centre, Mumbai and Dayalbagh Educational Institite to earn her PhD degree. Her research involved preparation, characterization of Zn nano-materials and utilizing them to study their efficiency for Hydrogen generation via photo-splitting of water. Dr. Nirupama has authored & co-authored 13 international journal research papers. She has 4 book chapters in her credit. She has presented her work in more than 20 International and National conferences. Dr. Nirupama has been awarded with young systems scientist award in the year 2008, by systems society of India.

Index

I

M

N

P

R

S

www.ingramcontent.com/pod-product-compliance
Lightning Source LLC
LaVergne TN
LVHW080535300926
849162LV00007B/1609
* 9 7 8 1 6 6 8 4 4 5 5 3 2 *